Lecture Notes in Physics

Volume 852

For further volumes:
http://www.springer.com/series/5304

The Lecture Notes in Physics

The series Lecture Notes in Physics (LNP), founded in 1969, reports new developments in physics research and teaching—quickly and informally, but with a high quality and the explicit aim to summarize and communicate current knowledge in an accessible way. Books published in this series are conceived as bridging material between advanced graduate textbooks and the forefront of research and to serve three purposes:

- to be a compact and modern up-to-date source of reference on a well-defined topic
- to serve as an accessible introduction to the field to postgraduate students and nonspecialist researchers from related areas
- to be a source of advanced teaching material for specialized seminars, courses and schools

Both monographs and multi-author volumes will be considered for publication. Edited volumes should, however, consist of a very limited number of contributions only. Proceedings will not be considered for LNP.

Volumes published in LNP are disseminated both in print and in electronic formats, the electronic archive being available at springerlink.com. The series content is indexed, abstracted and referenced by many abstracting and information services, bibliographic networks, subscription agencies, library networks, and consortia.

Proposals should be sent to a member of the Editorial Board, or directly to the managing editor at Springer:

Christian Caron
Springer Heidelberg
Physics Editorial Department I
Tiergartenstrasse 17
69121 Heidelberg/Germany
christian.caron@springer.com

Janos Polonyi · Achim Schwenk
Editors

Renormalization Group and Effective Field Theory Approaches to Many-Body Systems

Editors
Janos Polonyi
Laboratoire de Physique Therique
Université Louis Pasteur
rue de l'Université 3
67084 Strasbourg
France

Achim Schwenk
Institut für Kernphysik
Technische Universität Darmstadt
64289 Darmstadt
Germany

and

ExtreMe Matter Institute EMMI
GSI Helmholtzzentrum für
Schwerionenforschung GmbH
64291 Darmstadt
Germany

ISSN 0075-8450 ISSN 1616-6361 (electronic)
ISBN 978-3-642-27319-3 ISBN 978-3-642-27320-9 (eBook)
DOI 10.1007/978-3-642-27320-9
Springer Heidelberg New York Dordrecht London

Library of Congress Control Number: 2012935017

Printed on acid-free paper

Springer is part of Springer Science+Business Media (www.springer.com)

Preface

Over the past years, there have been many important developments based on effective field theory and the renormalization group in atomic, condensed matter, nuclear and high-energy physics. These powerful and versatile methods offer novel approaches to study complex and strongly interacting many-body systems in a controlled manner.

These Springer Lecture Notes in Physics combine selected introductory and interdisciplinary presentations focused on recent applications of effective field theory and the renormalization group to many-body problems in

- atomic physics,
  Jean-Paul Blaizot: Nonperturbative Renormalization Group and Bose-Einstein Condensation.
- condensed matter physics,
  Bertrand Delamotte: An Introduction to the Nonperturbative Renormalization Group.
- nuclear physics,
  Richard Furnstahl: Effective Field Theory for Density Functional Theory,
  Thomas Schaefer: Effective Theories of Dense and Very Dense Matter,
  Bengt Friman, Kai Hebeler and Achim Schwenk: Renormalization Group and Fermi Liquid Theory for Many-Nucleon Systems.
- and high-energy physics,
  Holger Gies: Introduction to the Functional Renormalization Group and Applications to Gauge Theories.

The discussions of these Lecture Notes are aimed at graduate students and junior researchers, and hopefully offer an opportunity to explore physics across subfield boundaries at an early stage in their career.

We would like to thank Jean-Paul Blaizot and Wolfram Weise for their encouragement with this volume, and Christian Caron for his kind help with putting the volume together.

Darmstadt and Strasbourg, September 2011

Janos Polonyi
Achim Schwenk

Contents

Chapter 1
Nonperturbative Renormalization Group and Bose-Einstein Condensation

Jean-Paul Blaizot

1.1 Introduction

These lectures are centered around a specific problem, the effect of weak repulsive interactions on the transition temperature T_c of a Bose gas. This problem provides indeed a beautiful illustration of many of the techniques which have been discussed at this school on effective theories and renormalization group. Effective theories are used first in order to obtain a simple hamiltonian describing the atomic interactions: because the typical atomic interaction potentials are short range, and the systems that we consider are dilute, these potentials can be replaced by a contact interaction whose strength is determined by the s-wave scattering length. Effective theories are used next in order to obtain a simple formula for the shift in T_c: this comes from the fact that near T_c the physics is dominated by low momentum modes whose dynamics is most economically described in terms of classical fields. The ingredients needed to calculate the shift of T_c can be obtained from this classical field theory. Finally the renormalization group is used both to obtain a qualitative understanding, and also as a non perturbative tool to evaluate quantitatively the shift in T_c.

In the first lecture, I recall some known aspects of Bose-Einstein condensation of the ideal gas. Then I turn to an elementary discussion of the interaction effects, and introduce an effective theory with a contact interaction tuned to reproduce the scattering length of the atom-atom interaction. I show that at the mean field

J.-P. Blaizot (✉)
Institute of Theoretical Physics (IPhT),
CEA Saclay, 91191 Gif-sur-Yvette Cedex,
France
e-mail: Jean-Paul.Blaizot@cea.fr

J. Polonyi and A. Schwenk (eds.), *Renormalization Group and Effective Field Theory Approaches to Many-Body Systems*, Lecture Notes in Physics 852,
DOI: 10.1007/978-3-642-27320-9_1,

level, weak repulsive interactions produce no shift in T_c. Finally, I briefly explain why approaching the transition from the low temperature phase is delicate and may lead to erroneous conclusions.

In the second lecture I establish the general formula for the shift of T_c:

$$\frac{\Delta T_c}{T_c^0} = \frac{T_c - T_c^0}{T_c^0} = c(an^{1/3}), \tag{1.1}$$

where a is the s-wave scattering length for the atom-atom interaction, and T_c^0 the transition temperature of the ideal gas at density n. This formula holds in leading order in the parameter $an^{1/3}$ which measures the diluteness of the system. Getting formula (1.1) involves a number of steps. First I explain why perturbation theory cannot be used to calculate ΔT_c, however small a is. Then I show that the problem with the perturbative expansion is localized in a particular subset of Feynman diagrams that are conveniently resummed by an effective theory of a classical 3-dimensional field. The outcome of the analysis is the formula (1.1) where c is given by the following integral

$$c \propto \int \frac{d^3p}{(2\pi)^3} \left(\frac{1}{p^2 + \Sigma(p)} - \frac{1}{p^2} \right), \tag{1.2}$$

where the proportionality coefficient, not written here, is a known numerical factor, and Σ is the self-energy of the classical field, whose calculation requires non perturbative techniques.

In the last lecture, I use the non perturbative renormalization group (NPRG) in order to estimate c. This requires the knowledge of the 2-point function of the effective 3-dimensional field theory for all momenta, and in particular in the cross-over between the critical region of low momenta and perturbative region of high momenta. This cross-over region is where the dominant contribution to the integral (1.2) comes from. In order to obtain an accurate determination of $\Sigma(p)$, it has been necessary to develop new techniques to solve the NPRG equations. Describing those techniques in detail would take us too far. I shall only present in this lecture the material that can help the student not familiar with the NPRG to understand how it works, and how it can be used. I shall do so by discussing several simple cases that at the same time provide indications on the approximation scheme that we have developed in order to calculate $\Sigma(p)$. I shall end by reporting and discussing the results obtained with the NPRG, and compare them to those obtained using other non perturbative techniques.

Recent discussion of Bose-Einstein condensation can be found in [1–5]. Equation (1.1) for the shift of T_c is derived and discussed in the series of papers [6–10]. Much of the material of the last lecture is borrowed from the papers [11–15].

1.2 LECTURE 1: Bose-Einstein Condensation

1.2.1 *Bose-Einstein Condensation for the Non Interacting Gas*

- The discussion of Bose-Einstein condensation of the ideal Bose gas in the grand canonical ensemble is standard. We consider a homogeneous system of identical spinless bosons of mass m, at temperature T. The occupation factor of a single particle state with momentum $\mathbf{p}$ is

$$n_{\mathbf{p}} = \frac{1}{e^{(\varepsilon^0_{\mathbf{p}}-\mu)/T} - 1}, \qquad \varepsilon^0_{\mathbf{p}} = \frac{p^2}{2m}, \tag{1.3}$$

where μ is the chemical potential. The number density of non condensed particles is given by

$$n = \int \frac{d^3\mathbf{p}}{(2\pi)^3} n_{\mathbf{p}} \equiv n(\mu, T). \tag{1.4}$$

For small density, the chemical potential is negative and large in absolute value, $e^{-\mu/T} \gg 1$. The gas is then described by Boltzmann statistics:

$$n_{\mathbf{p}} \approx e^{-(\varepsilon^0_{\mathbf{p}}-\mu)/T}, \qquad n \approx e^{\mu/T}\lambda^{-3}, \tag{1.5}$$

where λ is the thermal wavelength:

$$\lambda = \sqrt{\frac{2\pi}{mT}}. \tag{1.6}$$

Unless specified otherwise, we use units such that $\hbar = 1, k_B = 1$. Boltzmann's statistics applies as long as $n\lambda^3 \ll 1$, that is, as long as the thermal wavelength is small compared to the interparticle distance. As one lowers the temperature, keeping the density fixed, the chemical potential increases, and so does the thermal wavelength. Eventually, as $\mu \to 0_-$, the density of non condensed particles reaches a maximum

$$n = \int \frac{d^3p}{(2\pi)^3} \frac{1}{e^{\varepsilon^0_{\mathbf{p}}/T} - 1} = n(\mu = 0, T) = \frac{\zeta(3/2)}{\lambda^3}, \tag{1.7}$$

where $\zeta(z)$ is the Riemann zeta-function and λ the thermal wavelength. As one keeps lowering the temperature, particles start to accumulate in the lowest energy single particle state. This is the onset of Bose-Eisntein condensation, which takes place then when

$$n\lambda^3 = \zeta(3/2) \approx 2.612. \tag{1.8}$$

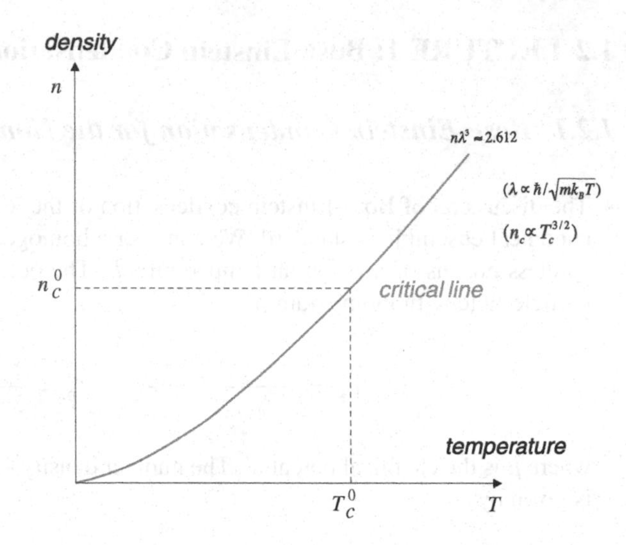

Fig. 1.1 Critical line in the (μ, T) plane for the ideal Bose gas

At this point the thermal wavelength has become comparable to the interparticle spacing. Equation (1.8) defines the critical line in the n, T plane (see Fig. 1.1). In particular, the critical temperature is given by

$$T_c^0 = \frac{2\pi}{m}\left(\frac{n}{\zeta(3/2)}\right)^{2/3}. \tag{1.9}$$

It is a function only of the mass of the atom, and of the density. For $T < T_c^0$, the chemical potential stays equal to zero and the single particle state $\mathbf{p} = 0$ is macroscopically occupied, with density $n_0 \propto n$ given by

$$n_0(T) = n\left(1 - \left(\frac{T}{T_c^0}\right)^{3/2}\right). \tag{1.10}$$

- The occurrence of condensation relies on the existence of a maximum of the integral (1.4) fixing the number of non condensed atoms. This depends on the number of spatial dimensions. In 2 dimensions, the integral diverges logarithmically as $\mu \to 0_-$:

$$n\lambda^2 = -\ln\left(1 - e^{\mu/T}\right). \tag{1.11}$$

In this case, there is no limit to the number of thermal particles, hence no condensation in the lowest energy mode.

- The Bose-Einstein condensation of the non-interacting gas has some pathological features. In particular the compressibility

$$\frac{1}{n^2}\frac{\partial n}{\partial \mu} = \frac{1}{n^2}\frac{1}{T}\int \frac{d^3p}{(2\pi)^3} n_{\mathbf{p}}(1+n_{\mathbf{p}}), \tag{1.12}$$

diverges when $\mu \to 0_-$. Also the fluctutation of the number of particles in the condensate,

$$\sqrt{\Delta\langle N^2\rangle} = \sqrt{N_0(N_0+1)} \simeq N_0, \tag{1.13}$$

is of the same order of magnitude as the average value N_0. Such features can be related to the large degeneracy of states in Fock space that appears as $\mu \to 0$.

Consider indeed the ground state in Fock space of the hamiltonian of the non interacting Bose gas,

$$H_0 - \mu N = \sum_{\mathbf{p}} (\varepsilon^0_{\mathbf{p}} - \mu) a^\dagger_{\mathbf{p}} a_{\mathbf{p}}. \tag{1.14}$$

If $\mu < 0$, then the ground state is the vacuum, with no particle present in the system: adding a particle costs a positive "free-energy" $-\mu$. When $\mu = 0$, there appears a huge degeneracy in Fock space: all the states with an arbitrary number of particles in the single particle state $\mathbf{p} = 0$ are degenerate. If $\mu > 0$, then there is no minimum, the more particle one adds in the state $\mathbf{p} = 0$, the more free energy one gains ($-\mu$ for each added particle). Note that here the chemical potential does not fully plays its role of controlling the density: either it is negative, and the density is zero, or it is positive and the density is infinite; if it vanishes the density is arbitrary.

To relate the large degeneracy to the large fluctuations (1.13), one may use a simple maximum entropy technique to determine the most likely state with N_0 particles in the degenerate space. This calculation parallels the corresponding calculation done in the grand canonical ensemble at finite temperature. Maximizing

$$\hat{S} = -\sum_n p_n \ln p_n + \mu \sum_n n p_n - \lambda \sum_n p_n, \tag{1.15}$$

with the constraints

$$\sum_n p_n = 1, \qquad \sum_n n p_n = N_0, \tag{1.16}$$

one finds

$$p_n = \frac{1}{Z} \mathrm{e}^{\mu n}, \qquad Z = \frac{1}{1-\mathrm{e}^{\mu}}, \qquad \mu = -\ln\left(1 + \frac{1}{N_0}\right). \tag{1.17}$$

For large N_0,

$$p_n \approx \frac{e^{-n/N_0}}{N_0}, \tag{1.18}$$

which leads indeed to the large fluctuations (1.13).

Of course, some of the pathologies of the ideal Bose gas in the grand canonical ensemble could be eliminated by working in the canonical ensemble, where the particle number is fixed (see e.g. [16]). However, we shall keep the discussion within the grand canonical ensemble, as it is closer to familiar field theoretical techniques. This allows us in particular to treat Bose-Eisntein condensation (of the interacting gas) as a symmetry breaking phenomenon.

The large degeneracy of states in Fock space implies that infinitesimal interactions could have a large effect, and indeed they do: as soon as weak repulsive interactions are present the large fluctuations are damped, and the compressibility becomes finite.

1.2.2 Interactions in the Dilute Gas

- As we shall discuss later in this lecture, the dominant effect of the interactions in the dilute gas can be accounted for by an effective contact interaction whose strength g is proportional to the s-wave scattering length a:

$$V(\mathbf{r_1} - \mathbf{r_2}) = g\delta(\mathbf{r_1} - \mathbf{r_2}), \qquad g = \frac{4\pi a}{m}. \tag{1.19}$$

- In the mean field (Hartree-Fock) approximation, which is also the leading order in a, the effect of the interaction is a simple shift of the single particle energies:

$$\varepsilon^0_{\mathbf{p}} \longrightarrow \varepsilon^0_{\mathbf{p}} + 2gn, \tag{1.20}$$

where the factor 2 comes from the exchange term.
It is easy to see that this produces no shift in T_c: because the shift of the single particle energy is constant, independent of the momentum, Eq. (1.4) which gives the number density of non condensed particles, can be written

$$n = n(\mu - \Delta\mu, T) \tag{1.21}$$

with

$$\Delta\mu = 2gn. \tag{1.22}$$

Bose-Einstein condensation now occurs when $\mu - \Delta\mu = 0$, instead of $\mu = 0$ in the non interacting case, but clearly the critical line is identical to that given by Eq. (1.8).

Of course, because the Hartree-Fock self-energy in Eq. (1.20) depends on the density, the relation between the chemical potential at the transition and the critical density is more complicated than in the non-interacting case. The transition now takes place at a finite value of the chemical potential, and some of the pathologies of the ideal gas disappear. In particular the compressibility remains finite at T_c (fluctuations remain important though):

$$\frac{\partial n}{\partial \mu} = \frac{\frac{1}{T}\int_p n_p(1+n_p)}{1+\frac{2g}{T}\int_p n_p(1+n_p)} \simeq \frac{1}{2g}. \tag{1.23}$$

- The presence of the interactions allows us to treat the phenomenon of Bose-Einstein condensation as a symmetry breaking phenomenon. Let us return first to the ground state, and calculate the free energy

$$\frac{\Omega(n_0)}{\mathcal{V}} = -\mu n_0 + \frac{g}{2}n_0^2, \tag{1.24}$$

where $\mathcal{V}$ is the volume. $\Omega(n_0)$ is the expectation value of the hamiltonian (1.14) to which is added the interaction (1.19), $V \sim (g/2)a_0^\dagger a_0^\dagger a_0 a_0$, in a coherent state containing an average number of particles $N_0 = n_0\mathcal{V}$ in the state $\mathbf{p} = \mathbf{0}$, $|n_0\rangle \sim \exp\sqrt{N_0}a_0^\dagger|0\rangle$. In contrast to what happened for the ideal gas, it is now possible to minimize $\Omega(n_0)$ w.r.t. n_0 in order to find the optimum ground state for a given μ. One gets

$$n_0 = \mu/g. \tag{1.25}$$

Thus the ground state has now a finite number of particles, and the fluctuations are normal, proportional to the square root of the mean value. The density in the ground state is $n = \partial\Omega/\partial\mu = n_0$, and the pressure is $P = -\Omega(n_0 = \mu/g)/\mathcal{V} = gn^2/2$, so that also at zero temperature the compressibility is finite, $\chi^{-1} = n dP/dn = gn^2$. Note however that the coherent state is a state with a non definite number of particles: the symmetry related to particle number conservation is spontaneously broken. We shall return later to the similar picture at finite temperature, and come back to this issue of symmetry breaking.

1.2.3 Atoms in a Trap

Although our main discussion concerns homogeneous systems, it is instructive to contrast the situation in homogeneous systems to what happens in a trap. We shall consider here a spherical harmonic trap, corresponding to the following external potential

$$V(r) = \frac{1}{2} m\omega_0^2 r^2. \tag{1.26}$$

- Consider first the non interacting gas. We assume the validity of a semiclassical approximation allowing us to express the particle density as the following phase space integral:

$$n(r) = \int \frac{d^3p}{(2\pi)^3} \frac{1}{e^{(\varepsilon(r,p)-\mu)/T} - 1}, \tag{1.27}$$

where $\varepsilon(r, p) = p^2/2m + (1/2)m\omega_0^2 r^2$. This requires that the temperature is large compared to the level spacing, $k_B T \gg \hbar\omega_0$, a condition well satisfied near the transition if N, the number of particles in the trap, is large enough. The number density in the trap can be written as

$$N = \int d^3\mathbf{r}\, n(\mu - \tfrac{1}{2} m\omega_0^2 r^2, T), \tag{1.28}$$

where $n(\mu, T)$ is the function (1.4). Thus in a wide trap for which the semiclassical approximation is valid, the particles experience the same conditions as in a uniform system with a local effective chemical potential $\mu - \frac{1}{2} m\omega_0^2 r^2$. In particular, Eq. (1.27) shows that the density at the center of the trap is related to the chemical potential μ by the same relation as in the homogeneous gas. It follows that T_c^0 is determined by the same condition as for the homogeneous gas, that is, $n(0)\lambda^3 = 2.612$ where $n(0)$ is the density at the center of the trap. An explicit calculation gives

$$\frac{k_B T_c^0}{\hbar\omega_0} = \left(\frac{N}{\zeta(3)}\right)^{1/3}, \tag{1.29}$$

with $\zeta(3) = 1.202$. Note that while the condensation condition is "universal" when expressed in terms of the density at the center of the trap, the dependence of T_c^0 on N depends on the form of the confining potential. In the present case, the N dependence can be obtained from the following heuristic argument. For a temperature $T \gtrsim T_c^0$ the particle density is approximately given by the classical formula

$$n \sim e^{-m\omega_0^2 r^2/2T}, \tag{1.30}$$

so that the thermal particles occupy a cloud of radius $R_{th} \sim \sqrt{T/m\omega_0^2} \sim a_{ho}^2/\lambda$, where $a_{ho} = 1/\sqrt{m\omega_0}$ is the characteristic radius of the harmonic trap and $\lambda \sim 1/\sqrt{mT}$ is the thermal wavelength. The average density in the thermal cloud is $\bar{n} \sim N R_{th}^{-3}$. At the transition, the interparticle distance is of the order of the thermal wavelength, so that, $\bar{n}^{1/3} \sim N^{1/3}\lambda/a_{ho}^2 \sim \lambda^{-1}$, or $N^{1/3}\lambda^2 \sim a_{ho}^2$, from which the relation $T_c^0 \sim \omega N^{1/3}$ follows.

Note that the confining potential makes condensation easier than in the uniform case. This is related to the fact that the density of single particle states in a trap decreases more rapidly with decreasing energy than in a uniform system: it goes as ε^{d-1} in a trap and as $\varepsilon^{d/2-1}$ in a uniform system, where d is the number of spatial dimensions. It follows in particular that, in a trap, condensation can occur in $d = 2$, in contrast to the homogeneous case; the heuristic argument presented above yields then $T_c^0 \sim \omega N^{1/2}$. Note however that the effects of the interactions in a 2-dimensional trap are subtle (for a recent discussion see [17]).

- The leading effect of repulsive interactions in a trap is to push the particles away from the center of the trap, thereby decreasing the central density. This effect is analogous to that produced by an increase of the temperature. One expects therefore the interactions to lead to a decrease of the transition temperature. To estimate this, we note that, in the presence of interactions, the density is still given by Eq. (1.30) after substituting $(1/2)m\omega_0^2 r^2 \to (1/2)m\omega_0^2 r^2 + 2gn(r)$. A simple calculation gives then the density $n(r)$ (for r not too large, and to leading order in a) in terms of the density $n(0)$ at the center of the trap [10]:

$$
\begin{aligned}
n(\mathbf{r}) &\approx n(0)\mathrm{e}^{-\beta(m\omega_0^2 r^2/2)(1-2gn(0)/T)} \\
&\approx n(0)\left[1 - \frac{1}{2T}\frac{m\omega_0^2 r^2}{1+4a\lambda^2 n(0)}\right].
\end{aligned}
\tag{1.31}
$$

This result suggests that the effect of the interaction can also be viewed as a modification of the oscillator frequency, $\omega_0^2 \to \omega_0^2/(1+4a\lambda^2 n(0))$. This is enough to estimate the shift in T_c:

$$
\frac{\Delta T_c}{T_c^0} = \frac{\Delta\omega}{\omega_0} \sim -a\lambda^2 n(0) \sim -\frac{a}{a_{ho}} N^{1/6},
\tag{1.32}
$$

where, in the last step, we have used the fact that $n(0) \sim \lambda^{-3}$ at the transition. A more explicit calculation yields the proportionality coefficient -1.32 [18]. It is important to keep in mind that this effect of mean field interactions on T_c is very different from the one that leads to Eq. (1.1) (indeed the sign of the effect is different). If we were comparing systems at fixed central density rather than at fixed particle number, there would be no shift of T_c (see the discussion in [10]).

1.2.4 The Two-Body Problem

We now come back to the construction of the effective interaction that can be used in many-body calculations of the dilute gas. We shall in particular recall how effective field theory can be used to relate the effective interaction to the low energy scattering data. More on the use of effective theories can be found in the lecture by Schäfer in this volume [19]. A pedagogical introduction is given in Ref. [20].

Consider two atoms of mass m interacting with the central two-body potential $V(r)$, with $\mathbf{r}$ the relative coordinate and $r = |\mathbf{r}|$. The relative wave function satisfies the Schrödinger equation

$$\left[-\frac{\hbar^2\,\nabla^2}{m} + V(r)\right]\psi(\mathbf{r}) = E\psi(\mathbf{r}). \tag{1.33}$$

The scattering wave function is given by ($E = \hbar^2 k^2/m$)

$$\psi_{\mathbf{k}}(\mathbf{r}) = \mathrm{e}^{i\mathbf{k}\cdot\mathbf{r}} + f(\mathbf{k}', \mathbf{k})\frac{\mathrm{e}^{ikr}}{r}, \tag{1.34}$$

where $\mathbf{k}$ is the initial relative momentum (with $k = |\mathbf{k}|$), while $\mathbf{k}'$ is the final relative momentum. The scattering is elastic and because the potential is rotationally invariant, the scattering amplitude $f(\mathbf{k}', \mathbf{k})$ is a function only of the scattering angle θ in the center of mass frame, and of the energy $E = \hbar^2 k^2/m$. For short-range interactions, the interaction takes place predominantly in the s-wave, and the scattering amplitude becomes a function of the energy only. It has then the following low momentum expansion:

$$f(E) = \frac{1}{-\frac{1}{a} + \frac{r_0}{2}k^2 - ik} \qquad \hbar k = \sqrt{mE}, \tag{1.35}$$

where a is the scattering length, r_0 the effective range, and the neglected terms in the denominator involve higher powers of k. In the very low momentum limit, when $kr_0\, ka \ll 1$, one can ignore the effective range. Then the scattering amplitude depends on a single parameter, a.

The scattering amplitude can be expressed as the matrix element between plane wave states of the T-matrix:

$$\langle \mathbf{k}'|T|\mathbf{k}\rangle = -\frac{4\pi\hbar^2}{m} f(\mathbf{k}', \mathbf{k}), \tag{1.36}$$

and the T-matrix itself can be calculated in terms of the Green function G_0

$$T = V - VG_0T, \tag{1.37}$$

where

$$G_0(E) = \frac{1}{H_0 - E}, \tag{1.38}$$

and $E \to E + i\eta$ with E real for the retarded Green's function G_0^R. The following formal relations are useful

$$H = H_0 + V, \qquad G^{-1} = H - E = G_0^{-1} + V, \tag{1.39}$$

and

$$T = V - VGV = V - VG_0T \qquad T^{-1} = V^{-1} + G_0. \tag{1.40}$$

Note that both $G(E)$ and $T(E)$ are analytic functions of E, with poles on the negative real axis corresponding to the energies of the bound states, and a cut on the real positive axis.

Let us now turn to the many-body problem. Assuming that the atoms interact only through the two body potential $V(r)$, we can write the interaction hamiltonian as

$$V = \frac{1}{2}\int d^3\mathbf{r}_1 d^3\mathbf{r}_2\, \psi^\dagger(\mathbf{r}_1)\psi^\dagger(\mathbf{r}_2)V(\mathbf{r}_1 - \mathbf{r}_2)\psi(\mathbf{r}_2)\psi(\mathbf{r}_1), \tag{1.41}$$

We demand that the local effective theory

$$V_{\text{eff}} = \frac{g_0}{2}\int d^3\mathbf{r}\, \psi^\dagger(\mathbf{r})\psi^\dagger(\mathbf{r})\psi(\mathbf{r})\psi(\mathbf{r}), \tag{1.42}$$

reproduce the same scattering data in the two particle channel at low momentum as the original potential $V(\mathbf{r}_1 - \mathbf{r}_2)$. We know that in the long wavelength limit the scattering amplitude depends only on the scattering length a, so we expect g_0 to be related to a.

To establish this relation we calculate the scattering amplitude for the effective theory. For a contact interaction $T(E)$ is given by

$$\frac{1}{T(E)} = \frac{1}{g_0} + G_0^R(E), \tag{1.43}$$

where

$$G_0^R(E) = \int \frac{d^3p}{(2\pi)^3}\frac{1}{p^2/m - E - i\eta} \tag{1.44}$$

To calculate $G_0^R(E)$, which is divergent, we introduce a cut-off Λ on the momentum integral

$$\begin{aligned} G^R(E) &= m\int \frac{d^3p}{(2\pi)^3}\frac{1}{p^2 - Em} = \frac{m}{2\pi^2}\int_0^\Lambda dp \frac{p^2}{p^2 - mE - i\eta} \\ &= \frac{m\Lambda}{2\pi^2} + \frac{m^2E}{4\pi^2}\int_{-\infty}^{+\infty} dp \frac{1}{p^2 - mE - i\eta}. \end{aligned} \tag{1.45}$$

It is then convenient to define a "renormalized" strength g_R by

$$\frac{1}{g_R}=\frac{1}{g_0}+\frac{m\Lambda}{2\pi^2},\qquad g_R=\frac{4\pi a}{m},\tag{1.46}$$

where in the last relation a is the scattering length. This relation between g_R and a is obtained by comparing $T(E)$ obtained from Eq. (1.44) ($E>0$)

$$\frac{1}{T(E)}=\frac{1}{g_R}+i\frac{m}{4\pi}\sqrt{mE},\tag{1.47}$$

with Eq. (1.35), and using the relation (1.36) to relate $T(E)$ and $f(E)$.

Remark One may improve the description by including the effective range correction. This is done by adding to the hamiltonian a term of the form [21]

$$\frac{g_0'}{2}\int d^3r\,\nabla(\psi^\dagger\psi)\cdot\nabla(\psi^\dagger\psi),\tag{1.48}$$

and adjusting g_0' so as to reproduce the scattering amplitude of the original two body problem, at the precision of the effective range. At tree level in the effective theory, the calculation of the scattering amplitude yields

$$\begin{aligned}\langle\mathbf{k}_3\mathbf{k}_4|T|\mathbf{k}_1\,\mathbf{k}_2\rangle&=-2g_0'\,[(\mathbf{k}_1-\mathbf{k}_3)\cdot(\mathbf{k}_2-\mathbf{k}_4)+(\mathbf{k}_1-\mathbf{k}_4)\cdot(\mathbf{k}_2-\mathbf{k}_3)]\\&=8g_0'k^2,\end{aligned}\tag{1.49}$$

where in the second line k is the magnitude of the relative momentum. Note that in order to make the identification with the two-body problem discussed above, we have to pay attention that the two-body problem traditionally treats the two particles as distinguishable, whereas the present calculation involves matrix element of the T-matrix between symmetric two particle states. Thus the tree level calculation of the T-matrix for the hamiltonian (1.41) yields $\langle\mathbf{k}_3\mathbf{k}_4|T|\mathbf{k}_1\,\mathbf{k}_2\rangle=2g_0$ which differs by a factor 2 with the conventional definition of the scattering length. Staying with the usual convention, we therefore write

$$T\approx\frac{4\pi}{m}\frac{a}{1-r_0ak^2/2+ika}\approx\frac{a}{1+ika}\left(1+\frac{ar_0}{2}k^2\right),\tag{1.50}$$

from which the identification of g_0' follows

$$g_0'=g_0\frac{ar_0}{8}.\tag{1.51}$$

For recent discussions on the application of effective field theory techniques to the Bose gas, see [21–23].

1.2.5 One-Loop Calculation

Having at our disposal an effective many-body hamiltonian, we may now perform detailed calculations of the effect of the interactions. The one loop calculation that we present here gives us the opportunity to come back to the issue of symmetry breaking, illustrates the use of the delta potential and points to the difficulty of approaching the phase transition from below.

The grand canonical partition function can be written as a path integral:

$$Z = \mathrm{Tr} e^{-\beta\hat{H}-\mu N} = \int_{\psi(\beta)=\psi(0)} \mathcal{D}(\psi,\psi^*)\, \mathrm{e}^{-\mathcal{S}}\,, \tag{1.52}$$

with

$$\mathcal{S} = \int_0^\beta d\tau \int d^3\mathbf{r} \left\{ \psi^*(x)\left(\frac{\partial}{\partial\tau} - \frac{\nabla^2}{2m} - \mu\right)\psi(x) + \frac{g_0}{2}\psi^*(x)\psi^*(x)\psi(x)\psi(x)\right\}, \tag{1.53}$$

and $\beta = 1/T$ is the inverse temperature. The complex field $\psi(x) = \psi(\tau,\mathbf{r})$ to be integrated over is a periodic function of the imaginary time τ, with period β. The action $\mathcal{S}$ is invariant under a $U(1)$ symmetry:

$$\psi(x) \to \mathrm{e}^{i\alpha}\psi(x), \qquad \psi(x)^* \to \mathrm{e}^{-i\alpha}\psi^*(x). \tag{1.54}$$

It is convenient to add an external source $j(x)$ linearly coupled to the bosonic field (thereby breaking the $U(1)$ symmetry):

$$Z[j] = \int_{\psi(\beta)=\psi(0)} \mathcal{D}(\psi^*,\psi)\, \mathrm{e}^{-\mathcal{S}_j}, \tag{1.55}$$

where

$$\mathcal{S} \to \mathcal{S} = \mathcal{S} - \int \mathrm{d}^3x \left[j(x)\psi^*(x) + j^*(x)\psi(x)\right]. \tag{1.56}$$

The loop expansion is an expansion around the field configurations which make the action stationary. These field configurations, which we denote by $\phi_0(x)$ are determined by the equation

$$\left[\frac{\partial}{\partial\tau} - \frac{\hbar^2}{2m}\nabla^2 - \mu\right]\phi_0(x) + g_0|\phi_0|^2\phi_0(x) = j(x) \tag{1.57}$$

and its complex conjugate. For uniform, time-independent, external sources j, Eq. (1.57) admits constant field solutions (compare Eq. (1.24)):

$$[-\mu + g_0|\phi_0|^2]\phi_0 = j. \tag{1.58}$$

There are two solutions. For $j = 0$, $\phi_0 = 0$ or $|\phi_0|^2 = \mu/g_0$. The solution $\phi_0 = 0$ corresponds to the vacuum state and is the stable solution when $\mu < 0$. For $\mu > 0$ (and $g_0 > 0$) the second solution corresponds to broken $U(1)$ symmetry and BE condensation; in that case the solution $\phi_0 = 0$ is a maximum of the free energy.

We now expand the field around the solution ϕ_0 which corresponds to a minimum of the (classical) free energy:

$$\psi(x) \longrightarrow \phi_0 + \psi(x) \qquad \psi(x)^* \longrightarrow \phi_0^* + \psi(x)^*, \tag{1.59}$$

with $\psi(x)$ having only $k \neq 0$ Fourier components, and keep in $\mathcal{S}$ terms which are at most quadratic in the fluctuations. That is, we write $\mathcal{S}_j = \mathcal{S}_0 + \mathcal{S}_1$, with $\mathcal{S}_0$ the classical action

$$\mathcal{S}_0 = \beta V\Big[-\mu|\phi_0|^2 + \frac{g_0}{2}\big(|\phi_0|^2\big)^2 + j\phi_0^* + j^*\phi_0\Big], \tag{1.60}$$

and S_1 the one-loop correction

$$\begin{aligned}\mathcal{S}_1 = \int_0^\beta \mathrm{d}\tau \int \mathrm{d}^3x \Bigg[\psi^*(x)\left(\frac{\partial}{\partial\tau} - \frac{\hbar^2}{2m}\nabla^2 - \mu + 2g_0|\phi_0|^2\right)\psi(x) \\ + \frac{g_0}{2}\left((\phi_0^*)^2\,\psi(x)\psi(x) + (\phi_0)^2\,\psi^\dagger(x)\psi^\dagger(x)\right)\Bigg].\end{aligned} \tag{1.61}$$

The gaussian integral is standard and, after a Legendre transform to eliminate j, it yields the following expression of the thermodynamic potential:

$$\frac{\Omega}{\mathcal{V}} = \frac{g_R n_0^2}{2} - \mu n_0 + \frac{1}{2\mathcal{V}}\sum_{k\neq 0}(E_k - \varepsilon_k) - \frac{g_R^2 n_0^2}{4\mathcal{V}}\sum_k \frac{1}{\varepsilon_k^0} + \frac{1}{\beta\mathcal{V}}\sum_k \ln\left(1 - \mathrm{e}^{-\beta E_k}\right) \tag{1.62}$$

where E_k is the Bogoliubov quasiparticle energy:

$$E_k^2 = (\varepsilon_k^0 + 2gn_0 - \mu)^2 - (gn_0)^2. \tag{1.63}$$

Note that in this calculation we have used the relation (1.46) in order to replace the bare coupling constant by the renormalized one. This is the origin of the fourth term in the r.h.s. of Eq. (1.62) which eliminates the divergence in the sum over the zero point energies. Note that this replacement assumes that g_R and g_0 are perturbatively related, which can only be possible if the cut-off Λ in (1.46) is not too large, i.e., if $m\Lambda g_R/(2\pi^2) \ll 1$, or $\Lambda \ll \pi/(2a)$.

This calculation was first made in the context of Bose-Einstein condensation by Toyoda [24], and led him to the erroneous conclusion that $\Delta T_c \sim -\sqrt{a n^{1/3}}$.

It is essentially a mean field calculation, the mean field being entirely due to the condensed particles. This approximation, equivalent to the lowest order Bogoliubov theory (see e.g. [25]), describes correctly the ground state at zero temperature and its elementary excitations. However, its extension near the critical temperature meets several difficulties; in particular it predicts a first order phase transition [26], a point apparently overlooked in Ref. [24]. In fact, above T_c, $\phi_0 = 0$, and the present one-loop calculation yields the free energy of an ideal gas, with no shift in the critical temperature.

1.3 LECTURE 2: The Formula for ΔT_c

Let us start by considering again the phase diagram in Fig. 1.1. We want to determine the change of the critical line due to weak repulsive interactions. For small and positive values of the s-wave scattering length a, we expect the change illustrated in Fig. 1.2, with the two critical lines close to each other. One can then relate, in leading order, the shift of the critical temperature at fixed density, ΔT_c, to that of the critical density at fixed temperature, Δn_c. Since, for $a = 0$, $n_c^0 \propto (T_c^0)^{3/2}$, we have

$$\frac{\Delta T_c}{T_c^0} = -\frac{2}{3}\frac{\Delta n_c}{n_c^0}. \tag{1.64}$$

It turns out that it is easier to calculate at fixed temperature than at fixed density, and in the following we shall set up the calculation of Δn_c.

In the previous lecture, we have seen that, under appropriate conditions, the effective hamiltonian is of the form

$$H = \int d^3\mathbf{r} \left\{ \psi^\dagger(\mathbf{r}) \left(\frac{\nabla^2}{2m} - \mu \right) \psi(\mathbf{r}) + \frac{g_0}{2} \psi^\dagger(\mathbf{r})\psi^\dagger(\mathbf{r})\psi(\mathbf{r})\psi(\mathbf{r}) \right\}, \tag{1.65}$$

where g_0 is related to the scattering length by (see Eq. (1.46)):

$$\frac{1}{g_R} = \frac{1}{g_0} + \frac{m\Lambda}{2\pi^2}, \qquad g_R = \frac{4\pi a}{m}. \tag{1.66}$$

This effective hamiltonian provides an accurate description of phenomena where the dominant degrees of freedom have long wavelength, $kr_0 \ll 1$, and the system is dilute, $an^{1/3} \ll 1$. Recall also that we shall be working in the vicinity of the transition where $n^{1/3}\lambda \simeq 1$, with λ the thermal wavelength.

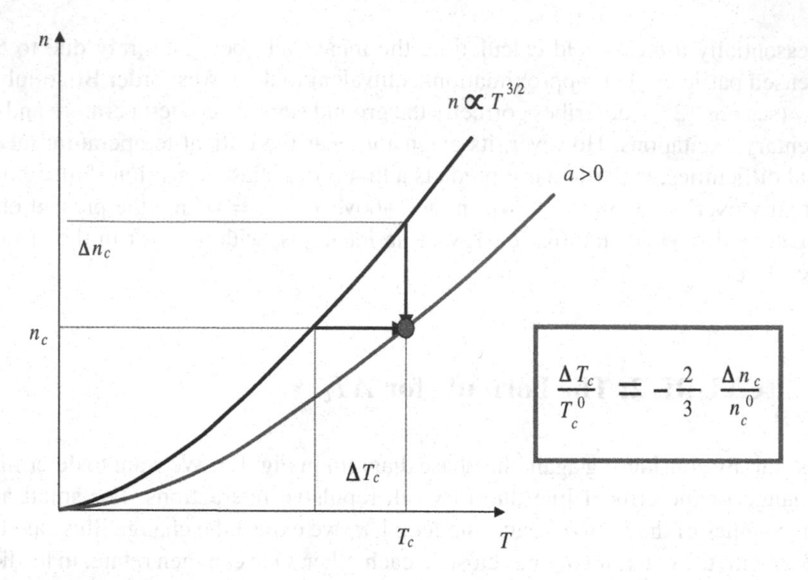

Fig. 1.2 Expected change in the critical line caused by weak repulsive interactions characterized by a small scattering length a

1.3.1 Condensation Condition and Critical Density

In order to exploit standard techniques of quantum field theory and many-body physics (see e.g. [25, 27–29]), we shall first relate the particle density to the single particle propagator. The particle density can be written as

$$\begin{aligned} n &= \langle \psi^\dagger(\mathbf{r})\psi(\mathbf{r})\rangle, \\ &= \lim_{\tau\to 0_-} \langle \mathrm{T}\psi(\tau,\mathbf{r})\psi^\dagger(0,\mathbf{r})\rangle = \lim_{\tau\to 0_-} G(\tau,0), \end{aligned} \tag{1.67}$$

where

$$\psi^\dagger(\tau,\mathbf{r}) = \mathrm{e}^{\tau H}\psi^\dagger(\mathbf{r})\mathrm{e}^{-\tau H}, \qquad \psi(\tau,\mathbf{r}) = \mathrm{e}^{\tau H}\psi(\mathbf{r})\,\mathrm{e}^{-\tau H}, \tag{1.68}$$

are respectively the creation and annihilation operators in the Heisenberg representation and T denotes (imaginary) time ordering (see e.g. [29]). The single particle propagator is

$$\begin{aligned} G(\tau_1-\tau_2,\mathbf{r_1}-\mathbf{r_2}) &= \langle \mathrm{T}\psi(\tau_1,\mathbf{r_1})\psi^\dagger(\tau_2,\mathbf{r_2})\rangle \\ &= \mathrm{Tr}\left(\frac{\mathrm{e}^{-\beta H}}{Z}\mathrm{T}\psi(\tau_1,\mathbf{r_1})\psi^\dagger(\tau_2,\mathbf{r_2})\right). \end{aligned} \tag{1.69}$$

It is a periodic function of $\tau = \tau_1 - \tau_2$: for $0 < \tau < \beta$, $G(\tau - \beta, \mathbf{r}) = G(\tau, \mathbf{r})$, where $\beta = 1/T$ is the inverse temperature. Because of its periodicity, it can be represented by a Fourier series

$$G(\tau, \mathbf{p}) = T \sum_n e^{-i\omega_n \tau} G_\alpha(i\omega_n, \mathbf{p}), \tag{1.70}$$

where the ω_n's are called the Matsubara frequencies:

$$\omega_n = 2n\pi T, \tag{1.71}$$

and we have also taken a Fourier transform with respect to the spatial coordinates. We are making here an abuse of notation: we denote by the same symbol the function and its Fourier transform, with the implicit understanding that the arguments, whether space-time or energy-momentum variables, are enough to specify which function one is considering. The inverse transform is given by

$$G(i\omega_n, \mathbf{p}) = \int_0^\beta \mathrm{d}\tau \, e^{i\omega_n \tau} G(\tau, \mathbf{p}). \tag{1.72}$$

In the absence of interactions, the hamiltonian is of the form

$$\begin{aligned} H_0 - \mu N &= \int d^3\mathbf{r}\, \psi^\dagger(\mathbf{r}) \left(-\frac{\hbar^2 \nabla^2}{2m} - \mu \right) \psi(\mathbf{r}) \\ &= \sum_{\mathbf{p}} (\varepsilon^0_{\mathbf{p}} - \mu)\, a^\dagger_{\mathbf{p}} a_{\mathbf{p}} = \sum_p \varepsilon_{\mathbf{p}}\, a^\dagger_{\mathbf{p}} a_{\mathbf{p}}, \end{aligned} \tag{1.73}$$

where

$$\psi(\mathbf{r}) = \sum_{\mathbf{p}} \frac{\mathrm{e}^{i\mathbf{p}\cdot\mathbf{r}}}{\sqrt{\mathcal{V}}} a_{\mathbf{p}}, \qquad \psi^\dagger(\mathbf{r}) = \sum_{\mathbf{p}} \frac{\mathrm{e}^{-i\mathbf{p}\cdot\mathbf{r}}}{\sqrt{\mathcal{V}}} a^\dagger_{\mathbf{p}}, \qquad \varepsilon^0_{\mathbf{p}} = \frac{\mathbf{p}^2}{2m}. \tag{1.74}$$

In these formulae, $\mathcal{V}$ is the volume of the system, and the creation and annihilation operators satisfy $[a_{\mathbf{p}}, a^\dagger_{\mathbf{p}'}] = \delta_{\mathbf{p},\mathbf{p}'}$. The free single particle propagator can be obtained by a direct calculation. It reads

$$G_0^{-1}(i\omega_n, \mathbf{p}) = \varepsilon_{\mathbf{p}} - i\omega_n, \tag{1.75}$$

or, in imaginary time,

$$\begin{aligned} G_0(\tau_1 - \tau_2, \mathbf{p}) &= \langle \mathrm{T} a_{\mathbf{p}}(\tau_1) a^\dagger_{\mathbf{p}}(\tau_2) \rangle_0 \\ &= e^{-\varepsilon_{\mathbf{p}}(\tau_1 - \tau_2)} \left[(1 + n_{\mathbf{p}})\theta(\tau_1 - \tau_2) + n_{\mathbf{p}}\theta(\tau_2 - \tau_1) \right], \end{aligned} \tag{1.76}$$

where:

$$n_{\mathbf{p}} \equiv \langle a_{\mathbf{p}}^{\dagger} a_{\mathbf{p}} \rangle_0 = \frac{\mathrm{Tr} e^{-\beta(H_0-\mu N)} a_{\mathbf{p}}^{\dagger} a_{\mathbf{p}}}{\mathrm{Tr} e^{-\beta(H_0-\mu N)}} = \frac{1}{e^{\beta \varepsilon_{\mathbf{p}}} - 1}, \qquad \varepsilon_{\mathbf{p}} = \varepsilon_{\mathbf{p}}^0 - \mu. \tag{1.77}$$

Thus, for non interacting particles,

$$\lim_{\tau \to 0_-} G_0(\tau, \mathbf{p}) = n_{\mathbf{p}}, \tag{1.78}$$

so that the formula (1.67) yields the familiar formula (1.4) of the density.

The full propagator is related to the bare propagator by Dyson's equation

$$G^{-1}(i\omega_n, \mathbf{p}) = G_0^{-1}(i\omega_n, \mathbf{p}) + \Sigma(i\omega_n, \mathbf{p}), \tag{1.79}$$

where $\Sigma(i\omega, \mathbf{p})$ is the self-energy. In this case Eq. (1.67) yields the following expression for the density

$$n = \lim_{\tau \to 0_-} T \sum_n \int \frac{d^3\mathbf{p}}{(2\pi)^3} \frac{e^{-i\omega_n \tau}}{\varepsilon_{\mathbf{p}} - i\omega_n + \Sigma(i\omega_n, \mathbf{p})}, \tag{1.80}$$

or equivalently the occupation factor in the interacting system

$$n_{\mathbf{p}} = \lim_{\tau \to 0_-} T \sum_n e^{-i\omega_n \tau} G(i\omega_n, \mathbf{p}). \tag{1.81}$$

We shall approach the condensation from the high temperature phase. Then the system remains in the normal state all the way down to T_c. The Bose-Einstein condensation occurs when the chemical potential reaches a value such that (see e.g. [30]):

$$G^{-1}(\omega = 0, \mathbf{p} = 0) = 0 \quad \text{or} \quad \Sigma(\omega = 0, \mathbf{p} = 0) = \mu\,. \tag{1.82}$$

At that point,

$$G^{-1}(i\omega_n, \mathbf{p}) = i\omega_n - \varepsilon_{\mathbf{p}}^0 - \left[\Sigma(i\omega_n, \mathbf{p}) - \Sigma(0,0)\right]. \tag{1.83}$$

By using the general relation (1.80) between the Green function and the density, one can then write the following formula for the shift Δn_c in the critical density caused by the interaction:

$$\Delta n_c = \lim_{\tau \to 0_-} T \sum_n e^{-i\omega_n \tau} \int \frac{d^3\mathbf{p}}{(2\pi)^3} \times \left\{ \frac{1}{\varepsilon_{\mathbf{p}}^0 + \Sigma(i\omega_n, \mathbf{p}) - \Sigma(0,0) - i\omega_n} - \frac{1}{\varepsilon_{\mathbf{p}}^0 - i\omega_n} \right\}. \tag{1.84}$$

The first term in this expression is the critical density of the interacting system at temperature T, the second term is the critical density of the non interacting system at the same temperature. This formula makes it obvious that Δn_c vanishes if the self-energy is independent of energy and momentum. This is the case in particular when the interactions are treated at the mean field level, as we already observed. In this case, $\Sigma = 2gn$, and the formula above yields

$$\Delta n_c = \int \frac{d^3\mathbf{p}}{(2\pi)^3} \left\{ \frac{1}{e^{\beta(\varepsilon^0_{\mathbf{p}}+2gn-\mu_c)} - 1} - \frac{1}{e^{\beta\varepsilon^0_{\mathbf{p}}} - 1} \right\} = 0, \tag{1.85}$$

which vanishes since $\mu_c = 2gn$.

1.3.2 Breakdown of Perturbation Theory

Because the interactions are weak, one may imagine calculating Δn_c by perturbation theory. However the perturbative expansion for a critical theory does not exist for any fixed dimension $d < 4$; infrared divergences prevent a complete calculation, as we shall recall. If one introduces an infrared cutoff k_c to regulate the momentum integrals, one finds that perturbation theory breaks down when $k_c \sim a/\lambda^2$, all terms being then of the same order of magnitude.

The leading order on a is given by the diagram in Fig. 1.3. As we just saw, the contribution of this diagram to Σ is just the mean field value $2gn$, and the net effect on Δn_c is zero. We shall then examine the second order contribution, given by the diagram displayed in Fig. 1.4. We shall see that this diagram is infrared divergent. Next, we shall show, using simple power counting, that such infrared divergences occur in higher orders and signal a breakdown of perturbation theory as one approaches the critical point.

Second Order Perturbation Theory

The second order self-energy diagram is the lowest order diagram that is momentum dependent and can therefore yield corrections to the critical density. It is displayed in Fig. 1.4. Its contribution is given by

$$\Sigma(i\omega_n, \mathbf{p}) = -2g^2T^2 \sum_{n'n''} \int \frac{d^3\mathbf{k}}{(2\pi)^3} \frac{d^3\mathbf{q}}{(2\pi)^3} \times \frac{1}{\varepsilon_{\mathbf{k}-\mathbf{q}} - i(\omega_{n'} - \omega_{n''})} \frac{1}{\varepsilon_{\mathbf{k}} - i\omega_{n'}} \frac{1}{\varepsilon_{\mathbf{p}+\mathbf{q}} - i(\omega_n + \omega_{n''})}. \tag{1.86}$$

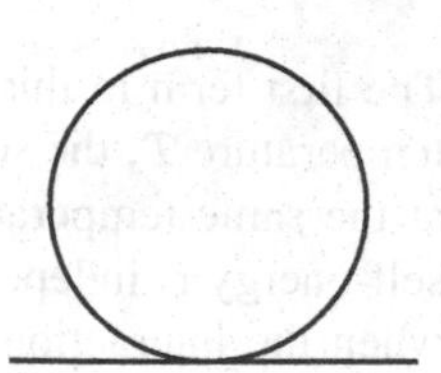

Fig. 1.3 The contribution to the self-energy Σ that is of leading order in a

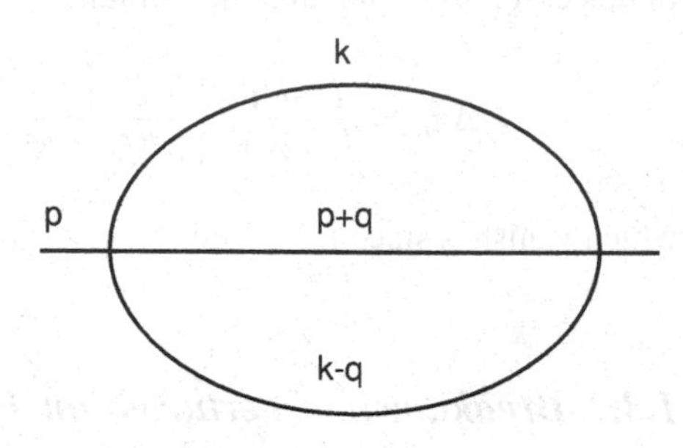

Fig. 1.4 The contribution to the self-energy Σ that is, a priori, of second order in a

Anticipating that infrared divergence can occur, we focus on the contribution where $\omega_{n'} = \omega_{n''} = 0$, and calculate the difference $\Sigma(i\omega_n = 0, \mathbf{p}) - \Sigma(i\omega_n = 0, \mathbf{p} = 0)$, which we write from now on simply as $\Sigma_{cl}(\mathbf{p}) - \Sigma_{cl}(0)$:

$$\Sigma_{cl}(\mathbf{p}) - \Sigma_{cl}(0) = -2g^2T^2 \int \frac{d^3\mathbf{k}}{(2\pi)^3} \frac{d^3\mathbf{q}}{(2\pi)^3} \times \frac{1}{(\varepsilon^0_{\mathbf{k-q}} - \mu')(\varepsilon^0_{\mathbf{k}} - \mu')} \left[\frac{1}{\varepsilon^0_{\mathbf{p+q}} - \mu'} - \frac{1}{\varepsilon^0_{\mathbf{q}} - \mu'} \right]. \tag{1.87}$$

In this calculation we have used Hartree-Fock propagators, and set

$$\mu' = \mu - \Sigma_{HF} \equiv -\frac{k_c^2}{2m}, \tag{1.88}$$

where $\Sigma_{HF} = 2gn$. The quantity k_c, whose significance will appear more clearly as we progress in the discussion, plays the role of an infrared cutoff in the integrals. Note that $k_c \to 0$ ($\mu' \to 0$) when $T \to T_c^0$. With this new notation

$$\varepsilon_k^0 - \mu' = (k^2 + k_c^2)/2m. \tag{1.89}$$

We shall also set

$$U(p) \equiv 2m(\Sigma_{cl}(p) - \Sigma_{cl}(0)). \tag{1.90}$$

The integral in Eq. (1.87) can be calculated analytically (see [8]) and yields

$$U(p) = 128\pi^2 \left(\frac{a}{\lambda^2}\right)^2 \left\{ \frac{3k_c}{p} \arctan\frac{p}{3k_c} + \frac{1}{2} \ln\left[1 + \left(\frac{p}{3k_c}\right)^2\right] - 1 \right\}. \quad (1.91)$$

This equation shows that $U(p)$ is a monotonically increasing function of p, $\sim p^2$ at small p, and growing logarithmically at large p. This logarithmic behavior, obtained in perturbation theory, remains in general the dominant behavior of $U(p)$ at large p, i.e., for $k_c \ll p \lesssim 1/\lambda$. This equation (1.91) also reveals the anticipated infrared divergence: $U(p)$ diverges logarithmically as $k_c \to 0$.

The condensation condition (1.82), $\mu' = \Sigma_{cl}(0)$, reads

$$k_c^2 \approx 128\pi^2 \left(\frac{a}{\lambda^2}\right)^2 \left[\ln\left(\frac{\Lambda}{3k_c}\right) - \gamma\right], \quad (1.92)$$

where the right hand side is $-2m\Sigma_{cl}(0)$, $\gamma = 0.577\ldots$ is Euler's constant, and the approximate equality is valid when $\Lambda \gg k_c$, which we assume to be the case. Here Λ is an ultraviolet cut-off that we have introduced to handle the divergence of $\Sigma_{cl}(0)$, and whose origin is the neglect of the non vanishing Matsubara frequencies. Typically, $\Lambda \sim 1/\lambda$. Equation (1.92) shows that $k_c \sim a/\lambda^2$ when a is small [8].

This simple calculation illustrates the limits of a pertubative approach. The infrared cutoff $k_c \sim a/\lambda^2$ introduces spurious a dependence. The condensation condition which relates the infrared cutoff to the microscopic length λ, induces a spurious logarithmic correction which does not vanish as $a \to 0$. In fact such logarithms do appear as higher order corrections, as we shall see later, but are absent in the leading order result.

Higher Orders

The infrared divergences that we have identified in the second order calculation persist, and worsen, in higher order contributions. This may be seen by using a simple power counting argument. Let us first consider diagrams in which all the internal lines carry vanishing Matsubara frequencies. We use again HF propagators so that all the functions that are integrated in the diagrams are products of fractions of the form

$$\left[K^2 + k_c^2\right]^{-1}, \quad (1.93)$$

where K denotes a generic combination of momenta; it is then natural to use the dimensionless products K/k_c as new integration variables. Consider then a diagram of order a^n. The lowest order $n = 2$ has just been calculated, and, for large p/k_c, it is proportional to $(a/\lambda)^2 \ln(p/k_c)$, where p is the external momentum. For $n > 2$, every additional order brings in one factor a from the vertex, one integration over three-momenta, a factor T, and two internal propagators. The contribution of the diagram can thus be written as:

$$T\left(\frac{a}{\lambda}\right)^2\left(\frac{a}{k_c\lambda^2}\right)^{n-2} F(p/k_c), \tag{1.94}$$

where F is a dimensionless function, which we do not explicitly need here. The main point is that when one approaches the critical temperature, the coherence length becomes large so that the summation of terms (1.94) diverges. In the critical region, $k_c \sim a/\lambda^2$, so that all the terms in the perturbative expansion are of the same order of magnitude. Therefore, at the critical point, perturbation theory is not valid.

Let us now assume that in a given diagram some propagators carry non-zero Matsubara frequencies so that one momentum integration (k) will be altered. For that integration, the presence of a non vanishing Matsubara frequency in the denominators of the propagators ensures that no singularity at $k = 0$ can take place. Essentially, in the corresponding propagators, k_c is replaced by a term proportional to $1/\lambda$, so that one factor $a/\lambda^2 k_c$ in (1.94) is now replaced by a/λ. Compared to the diagram with only vanishing Matsubara frequencies, this diagram is down by a factor a/λ, and thus negligible in a leading order calculation of Σ. This reasoning generalizes trivially to diagrams containing more non vanishing Matsubara frequencies.

Formula for Δn_c

It follows from the previous discussion that in order to obtain the leading order shift in the critical density, one may retain in Eq. (1.80) the contribution of the zero Matsubara frequency only. That is,

$$\begin{aligned}
\Delta n_c &\simeq T\int\frac{d^3p}{(2\pi)^3}\left\{\frac{1}{\varepsilon_p+\Sigma_{cl}(p)-\Sigma_{cl}(0)}-\frac{1}{\varepsilon_p}\right\}\\
&\simeq -\frac{2}{\pi\lambda^2}\int_0^\infty dp\frac{U(p)}{p^2+U(p)},
\end{aligned} \tag{1.95}$$

with $U(p)$ given by Eq. (1.90). Note that this integral is finite: $U(p) \sim \ln p$ at large p, and $p^2 + U(p) \sim p^{2-\eta}$ at small p. Note also that since $U(p) > 0$ (in fact the general qualitative behavior of $U(p)$ is correctly given by Eq. (1.91)), the correction Δn_c is negative, implying a positive shift of T_c.

1.3.3 Classical Field Approximation

Once restricted to their zero Matsubara frequency components, the fields ψ and $\psi^\dagger$ can be considered as classical fields, and the entire calculation can be cast in terms of a classical field theory. To see that, let us expand the field variables in the path

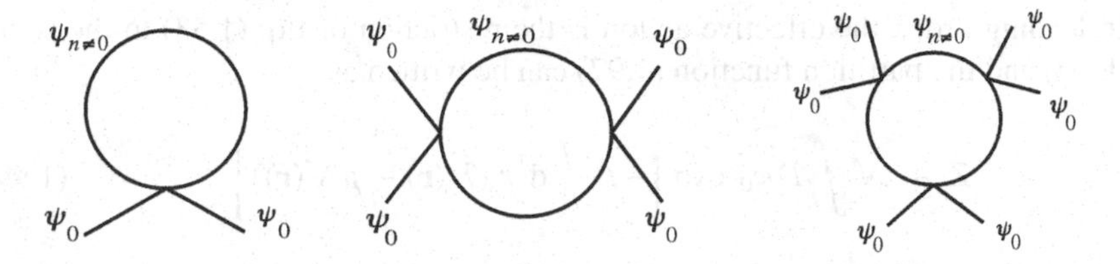

Fig. 1.5 Some diagrams contributing to the parameters of the effective theory. The external legs represent the field ψ_0 carrying vanishing Matsubara frequency. The internal lines carry non vanishing Matsubara frequencies ω_n. The *first* diagram from the *left* is a correction to the chemical potential; it is of order a, but does not contribute to the shift in T_c. The *second* diagram is a correction to the ψ_0^4 coupling constant. The *last* diagram is the contribution to the ψ^6 vertex. The *last* two contributions are of order a^2 and a^3 respectively, and can be ignored in a leading order analysis

integral (1.52) in terms of their Fourier components:

$$\psi(\tau, \mathbf{r}) = \psi_0(\mathbf{r}) + T \sum_{n \neq 0} \mathrm{e}^{-i\omega_n \tau} \psi_n(\mathbf{r}), \tag{1.96}$$

where the ω_n's are the Matsubara frequencies, and we have called $\psi_0(\mathbf{r})$ the τ-independent part of the field.

The partition function (1.52) can then be written as:

$$Z = \mathcal{N}_1 \int \mathcal{D}\psi_0 \exp\{-\mathcal{S}_{\text{eff}}[\psi_0]\}, \tag{1.97}$$

where ψ_0 depends only on spatial coordinates, and

$$\exp\{-\mathcal{S}_{\text{eff}}[\psi_0]\} = \mathcal{N}_2 \int \mathcal{D}\psi_{n\neq 0}\, \mathrm{e}^{-\mathcal{S}} \tag{1.98}$$

where $\mathcal{S}$ is the action (1.53). $\mathcal{N}_1$ and $\mathcal{N}_2$ are (infinite) normalization constants. The quantity $\mathcal{S}_{\text{eff}}[\psi_0]$ is the effective action for ψ_0. Aside from the direct classical field contribution to which we shall return shortly, this effective action receives also contributions which, diagrammatically, correspond to connected diagrams whose external lines are associated to ψ_0, and the internal lines are the propagators of the non-static modes ψ_n. A few examples are displayed in Fig. 1.5. Thus, a priori, $\mathcal{S}_{\text{eff}}[\psi_0]$ contains operators of arbitrarily high order in ψ_0. In the present case, however, it is easy to verify that the contributions beyond those kept in the classical action, are of higher order in a, and can therefore be ignored in a leading order calculation.

The strategy which consists of integrating out the non-static modes in perturbation theory in order to obtain an effective three-dimensional theory for the soft static modes, is referred to, in another context, as "dimensional reduction" (see e.g. [31–33]).

In leading order, the effective action is the restriction of Eq. (1.53) to the static mode ψ_0 and the partition function (1.97) can be written as:

$$Z \approx \mathcal{N} \int \mathcal{D}\psi_0 \exp\left\{-\beta \int \mathrm{d}^3 r \left(\mathcal{H}(\mathbf{r}) - \mu \mathcal{N}(\mathbf{r})\right)\right\}, \tag{1.99}$$

where $\psi_0(\mathbf{r})$ is a three-dimensional field, and

$$\mathcal{H} - \mu\mathcal{N} = \frac{|\nabla\psi_0|^2}{2m} - \mu' |\psi_0|^2 + \frac{g_0}{2}\left(|\psi_0|^2\right)^2. \tag{1.100}$$

We shall refer to this limit as the *classical field approximation*. The zero Matsubara component of the density is given by $\langle|\psi_0(r)|^2\rangle$, which diverges in the effective theory. However, recall that by assumption, the wavenumbers of the classical field are limited to k less than an ultraviolet cutoff $\Lambda \sim \lambda^{-1}$. In fact we shall not need to use a cut-off since the variation of the critical density is a finite quantity (see Eq. (1.110)).

Remark Ignoring the time dependence of the fields is equivalent to retaining only the zero Matsubara frequency in their Fourier decomposition. Then the Fourier transform of the free propagator is simply:

$$G_0(\mathbf{p}) = \frac{T}{\varepsilon_{\mathbf{p}}}. \tag{1.101}$$

This may be obtained directly from (1.70) and (1.75) keeping only the term with $\omega_n = 0$, or from Eq. (1.99). The classical field approximation corresponds also to the following approximation of the statistical factor (see Eq. (1.81))

$$n_{\mathbf{p}} = \frac{1}{\mathrm{e}^{\beta\varepsilon_{\mathbf{p}}} - 1} \approx \frac{T}{\varepsilon_{\mathbf{p}}}. \tag{1.102}$$

Both approximations make sense only for $\varepsilon_{\mathbf{p}} \ll T$, implying $n_{\mathbf{p}} \gg 1$. In this limit, the energy per mode is $\propto \varepsilon_{\mathbf{p}} n_{\mathbf{p}} \approx T$, as expected from the classical equipartition theorem. This long wavelength limit can also be viewed as a high temperature limit (the time dependence of the field becomes indeed unimportant as $\beta \to 0$). One should not confuse this classical field approximation with the classical limit reached when the thermal wavelength of the particles becomes small compared to their average separation distances. In this limit, the occupation of the single particle states becomes small, and the statistical factors can be approximated by their Boltzmann form:

$$\frac{1}{\mathrm{e}^{\beta(\varepsilon_p - \mu)} \pm 1} \approx \mathrm{e}^{\beta(\varepsilon_p - \mu)} \ll 1, \tag{1.103}$$

where we have used the fact that $\mathrm{e}^{-\beta\mu}$ is large in the classical limit.

At this point it is easy to understand the origin of the breakdown of perturbations theory. The critical region is characterized by the fact that all the terms in the integrand of (1.100) become of the same order of magnitude. This occurs for $k \lesssim k_c$, with k_c such that:

$$\frac{k_c^2}{2m} \sim \mu' \sim \frac{a}{m}\frac{T}{\mu'}k_c^3, \tag{1.104}$$

where $(T/\mu')k_c^3$ is the contribution to the density of the modes with $k \sim k_c$. From Eq. (1.104) we see that $k_c \sim a/\lambda^2$. For $k \simeq k_c$ perturbation theory in a makes no sense, and in fact all terms in the perturbative expansion are infrared divergent. For $k_c \ll k \ll \lambda^{-1}$, perturbation theory is applicable.

Remark In a trap the effect of critical fluctuations is subleading. To see that let us estimate the size of the critical region in a trap [34]. Recall that at the transition the thermal wavelength is $\lambda \sim a_{ho}N^{-1/6}$, where $a_{ho} = 1/\sqrt{m\omega_0}$ is the characteristic size of the harmonic oscillator trap (see the discussion after Eq. (1.30)), and the size R_{cl} of the thermal cloud where most of the particles sit is $R_{cl} \sim a_{ho}N^{1/6} \sim \lambda N^{1/3}$. According to Eq. (1.104), the critical region is reached when the chemical potential deviates from its critical value by an amount $\lesssim k_c^2/2m \sim a^2/(m\lambda^4)$. In a trap the effective local potential is of the form $\mu - \frac{1}{2}m\omega_0^2 r^2$. Taking μ at its critical value, one finds that the particles will be in the critical region as long as $r \lesssim R_{cr}$ with $m\omega_0^2 R_{cr}^2 \sim a^2/(m\lambda^4)$, or $R_{cr} \sim a(a_{ho}/\lambda)^2 \sim aN^{1/3}$. Thus the relative size of the critical region is $R_{cr}/R_{cl} \sim a/\lambda$, and the ratio of the particles in the critical region to the total number of particles is $\sim (a/\lambda)^3$. Under such conditions, it can be shown that one can use perturbation theory to estimate the corrections due to the interactions to the relation (1.4) in a trap; the resulting contributions to the shift of T_c are then subleading in a, as compared to the mean field effect discussed in the first lecture [34].

In view of the forthcoming discussions, it is convenient to rescale the field ψ_0 and to parametrize it in terms of two real fields φ_1, φ_2: $\psi_0 = \sqrt{mT}(\varphi_1 + i\varphi_2)$. The partition function then reads

$$\mathcal{Z} = \int \mathcal{D}\varphi\, \mathrm{e}^{-\mathcal{S}}, \tag{1.105}$$

where $\mathcal{S}(\varphi) = (H - \mu N)/T$ is given by:

$$\mathcal{S}(\varphi) = \int \left\{ \frac{1}{2}[\partial_\mu \varphi(x)]^2 + \frac{1}{2} r\phi^2(x) + \frac{u}{4!}\left[\varphi^2(x)\right]^2 \right\} \mathrm{d}^d x\,, \tag{1.106}$$

where $\varphi^2 = \varphi_1^2 + \varphi_2^2$ and:

$$r = -2mT\mu, \qquad u = 96\pi^2 \frac{a}{\lambda^2}. \tag{1.107}$$

In Eq. (1.106) we have kept the dimension d of the spatial integration arbitrary for the convenience of forthcoming discussions. The single particle Green's function $G(p)$ is related to the inverse two-point function $\Gamma^{(2)}(p)$ of the classical field theory by

$$2mTG^{-1}(p) = \Gamma^{(2)}(p), \qquad p^2 + U(p) = \Gamma^{(2)}(p) - \Gamma^{(2)}(0). \tag{1.108}$$

As it stands this field theory suffers from UV divergences. These are absent in the original theory, the higher frequency modes providing a large momentum cutoff $\sim 1/\lambda$. This cutoff may be restored when needed, but, as we show later, since the shift of the critical temperature is dominated by long distance properties it is independent of the precise cutoff procedure.

We shall also find useful to consider the $O(N)$ symmetric generalization of the Euclidean action (1.106). The field $\varphi(x)$ then has N real components, and, e.g.,

$$\varphi^2 = \sum_{i=1}^{N} \varphi_i^2, \tag{1.109}$$

and the shift in the critical density is given by

$$\begin{aligned}\Delta n_c &= 2mT \sum_{i=1}^{N} \left[\langle \phi_i^2 \rangle_{a \neq 0} - \langle \phi_i^2 \rangle_{a=0} \right] \\ &= 2mT\, N \int \frac{\mathrm{d}^d p}{(2\pi)^d} \left(\frac{1}{\Gamma^{(2)}(p)} - \frac{1}{p^2} \right),\end{aligned} \tag{1.110}$$

with $\delta_{ij}/\Gamma^{(2)}(k)$ the connected two-point correlation function.

The advantage of this generalization is that it provides us with a tool, the large N expansion, which allows us to calculate at the critical point (for a recent review see e.g. [35]).

Linear Dependence of the Density Correction

It is now easy to see the origin of the linear relation between Δn_c and a. Note first that the action (1.106) contains a single dimensionfull parameter, u, r being adjusted for any given u to be at criticality. In fact the effective three dimensional theory is ultraviolet divergent, so there is a priori another parameter, the ultraviolet cut-off $\Lambda \sim 1/\lambda$. It follows then from dimensional analysis that $U(p)$ defined in Eq. (1.90) can be written as

$$U(p = xu) = u^2 \sigma(x, u/\Lambda). \tag{1.111}$$

Now, the diagrams involved in the calculation of U are ultraviolet convergent, so that U is in fact independent of the cut-off Λ, and the infinite cut-off limit can be taken.

Note however that the validity of the classical field approximation requires that all momenta involved in the various integrations are small in comparison with $\Lambda \sim \lambda^{-1}$ or, in other words, that the integrands are negligibly small for momenta $k \sim \lambda^{-1}$. Only then can we ignore the effects of non vanishing Matsubara frequencies and use for instance the approximate form of the statistical factors (1.102). In other words, the infinite cut-off limit is meaningful only if letting the cut-off becoming bigger than λ^{-1} does not affect the results. This implies that $u\lambda \sim a/\lambda$ is sufficiently small.

In the region of validity of the classical field approximation, that is, for small enough u, $\sigma(x, u/\Lambda)$ becomes a universal function $\sigma(x)$, independent of u, and Δn_c in Eq. (1.95) takes the form

$$\Delta n_c = -\frac{2u}{\pi\lambda^2} \int dx \frac{\sigma(x)}{x^2 + \sigma(x)}, \tag{1.112}$$

showing that the change in the critical density is indeed linear in a.

Renormalization Group Argument

The linearity of the relation between the shift in T_c and the scattering length can also be understood from a simple renormalization group analysis. Let us introduce a large momentum cutoff $\Lambda \sim 1/\lambda$, and a dimensionless coupling constant g

$$g = \Lambda^{d-4} u \propto \left(\frac{a}{\lambda}\right)^{d-2}. \tag{1.113}$$

At T_c the two-point function in momentum space satisfies the renormalization group equation [36]

$$\left(\Lambda \frac{\partial}{\partial \Lambda} + \beta(g) \frac{\partial}{\partial g} - \eta(g)\right) \Gamma^{(2)}(p, \Lambda, g) = 0. \tag{1.114}$$

This equation, together with dimensional analysis, implies that the two-point function has the general form

$$\Gamma^{(2)}(p, \Lambda, g) = p^2 Z(g) F(p/k_c), \tag{1.115}$$

where $k_c = k_c(g)$ is a function of g which, on dimensional grounds, is proportional to Λ, so that $\Lambda \partial_\Lambda k_c = k_c$. The ansatz (1.115) for $\Gamma^{(2)}(p, \Lambda, g)$ provides then a solution of Eq. (1.114) if $Z(g)$ and $k_c(g)$ obey the equations

$$\frac{\partial \ln Z(g)}{\partial g} = \frac{\eta(g)}{\beta(g)}, \tag{1.116}$$

$$\frac{\partial \ln k_c(g)}{\partial g} = -\frac{1}{\beta(g)}. \tag{1.117}$$

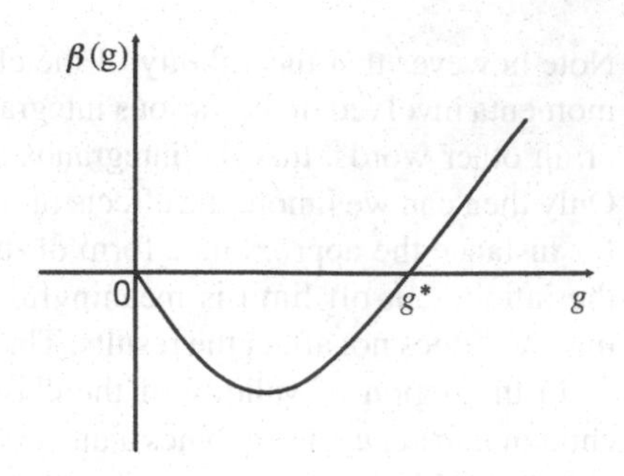

Fig. 1.6 The β-function with the two fixed points at $g = 0$ and $g = g^*$

Since $\beta(g) = -(4-d)g + (N+8)g^2/48\pi^2 + \mathcal{O}(g^3)$, $\beta(g)$ is of order g for small g in $d < 4$; similarly $\eta(g) = (N+2)g^2/(72(8\pi^2)^2) + \mathcal{O}(g^3)$. Therefore

$$Z(g) = \exp \int_0^g \frac{\eta(g')}{\beta(g')} dg' = 1 + \mathcal{O}(g^2); \tag{1.118}$$

to leading order $Z(g) = 1$. The function $k_c(g)$ is then obtained by integrating Eq. (1.117),

$$k_c(g) = g^{1/(4-d)} \Lambda \exp\left[- \int_0^g dg' \left(\frac{1}{\beta(g')} + \frac{1}{(4-d)g'} \right) \right]. \tag{1.119}$$

In $d = 3$,

$$k_c(g) = \Lambda \frac{gg^*}{g^* - g}, \tag{1.120}$$

where g^* is the infrared fixed point (see Fig. 1.6). The scale $k_c(g)$ plays a specific role in the analysis as the crossover separating a universal long-distance regime, where

$$\Gamma^{(2)}(p) \propto p^{2-\eta} \qquad p \ll k_c(g), \tag{1.121}$$

governed by the non-trivial zero, g^*, of the β-function, from a universal short distance regime governed by the Gaussian fixed point, $g = 0$, where

$$\Gamma^{(2)}(p) \propto p^2 \qquad k_c(g) \ll p \ll \Lambda\,. \tag{1.122}$$

However such a regime exists only if $k_c(g) \ll \Lambda$, i.e., if there is an intermediate scale between the infrared and the microscopic scales; otherwise only the infrared behavior can be observed. In a generic situation g is of order unity, and thus $k_c(g)$ is of order Λ, and the universal large momentum region is absent. Instead $k_c(g) \ll \Lambda$ implies that g be small. Since $g \sim a/\lambda \ll 1$, see Eq. (1.113), this condition is satisfied in the present situation.

It is then easy to show, repeating essentially the same analysis as before, that with this condition, $\Delta T_c \propto k_c(g)$. We set $p = xk_c(g)$, and find for the integral in Eq. (1.110) the general form

$$\int \frac{d^d p}{(2\pi)^d}\left(\frac{1}{\Gamma^{(2)}(k)} - \frac{1}{p^2}\right) = k_c(g)\int \frac{d^3 x}{(2\pi)^3}\frac{1}{x^2}\left(\frac{1}{F(x)} - 1\right); \qquad (1.123)$$

the g dependence is entirely contained in $k_c(g)$, and for small g, $k_c(g) \simeq u$.

Recall that both the perturbative large momentum region and the non-perturbative infrared region contribute to the integrand in Eq. (1.123), or equivalently in Eq. (1.112), and that the functions $F(x)$ or $\sigma(x)$ cannot be calculated using perturbation theory.

The $1/N$ expansion allows an explicit calculation, and yields, in leading order for the coefficient c in Eq. (1.1) the value $c = 2.3$ [7]. However, the best numerical estimates for c are those which have been obtained using the lattice technique by two groups, with the results: $c = 1.32 \pm 0.02$ [37] and $c = 1.29 \pm 0.05$ [38]. The availability of these results has turned the calculation of c into a testing ground for other non perturbative methods: expansion in $1/N$ [7, 39], optimized perturbation theory [40, 41], resummed perturbative calculations to high loop orders [42]. Note that while the latter methods yield critical exponents with several significant digits, they predict c with only a 10% accuracy. This illustrates the difficulty of getting an accurate determination of c using (semi) analytical techniques.

Remark The linear dependence in a of the shift of T_c holds only if a is small enough, as we have already indicated. When a is not small enough various corrections need to be taken into account that alter the simple linear law. In particular, corrections come from the non vanishing Matsubara frequencies, and their impact on the effective theory for ψ_0. Such corrections have been analyzed in detail in [43] (see also [9]). The net result is the following expression for ΔT_c:

$$\frac{\Delta T_c}{T_c^0} = c(an^{1/3}) + \left[c_2' \ln(an^{1/3}) + c_2''\right](an^{1/3})^2 + \mathcal{O}(a^3 n). \qquad (1.124)$$

Aside from such corrections, the evaluation of ΔT_c in higher order requires that one improves the accuracy of the effective hamiltonian and include for instance effective range corrections.

1.4 LECTURE 3: The Non Perturbative Renormalization Group and the Calculation of *c*

The analysis of the previous lectures has shown that the coefficient c in the formula (1.1) can be written as:

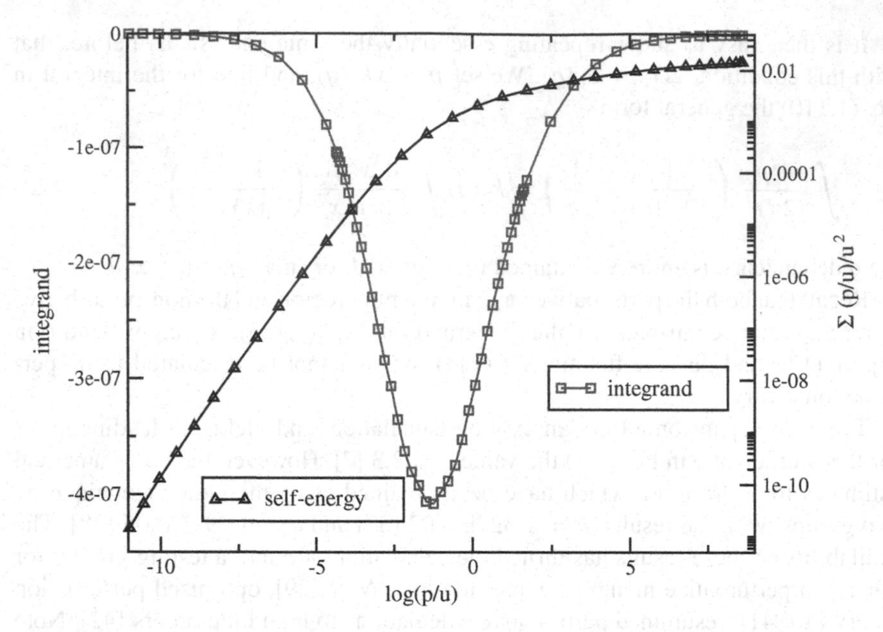

Fig. 1.7 The function $\sigma(x = p/u)$ at criticality, and the integrand of Eq. (1.126) as a function of $\ln(p/u)$, for $u \lesssim 10^{-4}$ for which $\sigma(x)$ is independent of u. From Ref. [11]

$$c \equiv -\frac{256\pi^3}{(\zeta(3/2))^{4/3}} \frac{\Delta\langle\varphi_i^2\rangle}{Nu}, \tag{1.125}$$

where $\Delta\langle\varphi_i^2\rangle$ represents the change, due to interactions, of the fluctuations of the scalar field of the $O(N)$ symmetric model, and is given by:

$$\frac{\Delta\langle\varphi_i^2\rangle}{Nu} = -\int \frac{dx}{2\pi^2} \frac{\sigma(x)}{x^2 + \sigma(x)}, \tag{1.126}$$

where $\sigma(x) = u^{-2}U(p = xu)$ and $U(p)$ is essentially the self-energy of the field at criticality (see Eqs. (1.110) and (1.108)). In order to get the numerical factor in Eq. (1.125), we have combined Eqs. (1.1), (1.64), (1.8) and (1.107). Equation (1.125) has been written for arbitray N in order to be able to compare with all available results. Bose-Einstein condensation corresponds to $N = 2$.

The integrand of Eq. (1.126), calculated in the approximate scheme described in [11], is shown in Fig. 1.7. The minimum occurs at the typical scale k_c which separates the scaling region from the high momentum region where perturbation theory applies. As we have already emphasized, the difficulty in getting a precise evaluation of the integral (1.126) is that it requires an accurate determination of $\sigma(x)$ in a large region of momenta including in particular the crossover region between two different physical regimes; as we have seen in the previous lecture, this cannot be done using perturbation theory, eventhough the coupling constant, $\sim a$, is small.

In this last lecture, I shall show how the non perturbative renormalization group (NPRG) can be used to calculate $\sigma(x)$. The NPRG [44–48] (sometimes called "exact" or "functional", depending on which aspect of the formalism one wishes to emphasize) stands out as a promising tool, suggesting new approximation schemes which are not easily formulated in other, more conventional, approaches in field theory or many body physics. It has been applied successfully to a variety of problems, in condensed matter, particle or nuclear physics (for reviews, see e.g. [49–51]). In most of these problems however, the focus is on long wavelength modes and the solution of the NPRG equations involves generally a derivative expansion which only allows for the determination of the n-point functions and their derivatives essentially only at vanishing external momenta. This is not enough in the present situation where, as we have seen, a full knowledge of the momentum dependence of the 2 point function is required. We have therefore developed new methods to solve the NPRG equations. The following sections will briefly present some of the ingredients involved, without going into the technical details of their implementation. These can be found in Refs. [11–15], from which much of the material presented here is borrowed. For definiteness, we focus on the $O(N)$ symmetric scalar field theory with action (1.106). See also the lecture by H. Gies in this volume [52] for further introduction to these techniques, and their application to other theories, in particular non-abelian gauge theories. Other applications of the renormalization group to Bose-Einstein condensation can be found in [53, 54].

1.4.1 The NPRG Equations

The NPRG allows the construction of a set of effective actions $\Gamma_\kappa[\phi]$ which interpolate between the classical action S and the full effective action $\Gamma[\phi]$: In $\Gamma_\kappa[\phi]$ the magnitude of long wavelength fluctuations of the field is controlled by an infrared regulator depending on a continuous parameter κ which has the dimension of a momentum. The full effective action is obtained for the value $\kappa = 0$, the situation with no infrared cut-off. In the other limit, corresponding to a value of κ of the order of the microscopic scale Λ at which fluctuations are suppressed, $\Gamma_{\kappa=\Lambda}[\phi]$ reduces to the classical action.[1]

In practice the fluctuations are controlled by adding to the classical action (1.106) the $O(N)$ symmetric regulator

$$\Delta S_\kappa[\varphi] = \frac{1}{2}\int \frac{\mathrm{d}^d q}{(2\pi)^d}\, \varphi_i(q)\, R_\kappa(q)\, \varphi_i(-q), \tag{1.127}$$

[1] Note that depending on the choice of the regulator, not all fluctuations may be suppressed when $\kappa = \Lambda$. However, for renormalisable theories, and if Λ is large enough, the effects of these remnant fluctuations can be absorbed into a redefinition of the parameters of the classical action.

where R_κ denotes a family of "cut-off functions" depending on κ. As we just said, the role of ΔS_κ is to suppress the fluctuations with momenta $q \lesssim \kappa$, while leaving unaffected those with $q \gtrsim \kappa$. Thus, typically, $R_\kappa(q) \to \kappa^2$ when $q \ll \kappa$, and $R_\kappa(q) \to 0$ when $q \gtrsim \kappa$. There is a large freedom in the choice of $R_\kappa(q)$, abundantly discussed in the literature [55–58]. The choice of the cut-off function matters when approximations are done, as is the case in all situations of practical interest. Most of the results that will be presented here have been obtained with the cut-off function proposed in [57]:

$$R_\kappa(q) \propto (\kappa^2 - q^2)\theta(\kappa^2 - q^2). \tag{1.128}$$

This regulator has the advantage of allowing some calculations to be done analytically.

For each value of κ, one defines the generating functional of connected Green's functions

$$W_\kappa[J] = \ln \int D\varphi \, \exp\left\{-S[\varphi] - \Delta S_\kappa[\varphi] + \int d^d x \varphi(x) J(x)\right\}. \tag{1.129}$$

The Feynman diagrams contributing to W_κ are those of ordinary perturbation theory, except that the propagators contain the infrared regulator. We also define the effective action, through a modified Legendre transform that includes the explicit subtraction of ΔS_κ:

$$\Gamma_\kappa[\phi] = -W_\kappa[J_\phi] + \int d^d x \, \phi(x) J_\phi(x) - \Delta S_\kappa[\phi], \tag{1.130}$$

where J_ϕ is obtained by inverting the relation

$$\phi_{\kappa,J}(x) \equiv \langle\varphi(x)\rangle_{\kappa,J} = \frac{\delta W_\kappa}{\delta J(x)}, \tag{1.131}$$

for $\phi_{\kappa,J}(x) = \phi$. Note that, in this inversion, ϕ is considered as a given variable, so that J_ϕ becomes implicitly dependent on κ. Taking this in to account, it is easy to derive the following exact flow equation for $\Gamma_\kappa[\phi]$:

$$\partial_\kappa \Gamma_\kappa[\phi] = \frac{1}{2} \mathrm{tr} \int \frac{d^d q}{(2\pi)^d} \, \partial_\kappa R_\kappa(q) \, G(\kappa, q), \tag{1.132}$$

where the trace tr runs over the O(N) indices, and

$$G_{ij}^{-1}(q; \kappa; \phi) = \Gamma_{ij}^{(2)}(\kappa; q; \phi) + \delta_{ij} R_\kappa(q), \tag{1.133}$$

with $\Gamma^{(2)}$ the second functional derivative of Γ_κ with respect to ϕ (see Eq. (1.134)). Equation (1.132) is the master equation of the NPRG. Its solution yields the effective

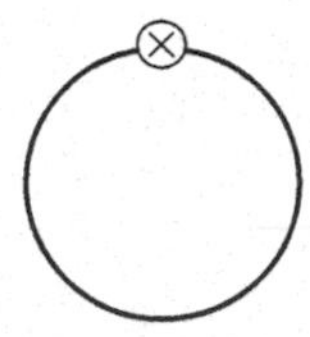

Fig. 1.8 Diagrammatic illustration of the r.h.s. of the flow equation of the effective action, Eq. (1.132). The *crossed circle* represents an insertion of $\partial_\kappa R_\kappa$, and the *thick line* a full propagator in an arbitrary background field

action $\Gamma[\phi] = \Gamma_{\kappa=0}[\phi]$ starting with the initial condition $\Gamma_{\kappa=\Lambda}[\phi] = S[\phi]$. Its right hand side has the structure of a one loop integral, with one insertion of $\partial_\kappa R_\kappa(q^2)$ (see Fig. 1.8). This simple structure should not hide the fact that Eq. (1.132) is an exact equation (G in the r.h.s. is the exact propagator), and as such it is of limited use unless some approximation is made. Before we turn to such approximation, let us further analyze the content of Eq. (1.132) in terms of the n-point functions.

As well known (see e.g. [36]), the effective action $\Gamma[\phi]$ is the generating functional of the one-particle irreducible n-point functions. This property extends trivially to $\Gamma_\kappa[\phi]$:

$$\Gamma_\kappa[\phi] = \sum_n \frac{1}{n!} \int d^d x_1 \ldots \int d^d x_n \, \Gamma_\kappa^{(n)}[\phi; x_1, \cdots, x_n] \, \phi(x_1) \ldots \phi(x_n). \tag{1.134}$$

By differentiating Eq. (1.132) with respect to ϕ, letting the field be constant, and taking a Fourier transform, one gets the flow equations for all n-point functions in a constant background field ϕ. For instance, the flow of the 2-point function in a constant external field reads:

$$\begin{aligned}\partial_\kappa \Gamma_{ab}^{(2)}(p,-p;\kappa;\phi) = \int \frac{d^d q}{(2\pi)^d} \partial_\kappa R_\kappa(q) \Big\{ & G_{ij}(q;\kappa;\phi)\Gamma_{ajk}^{(3)}(p,q,-p-q;\kappa;\phi) \\ & \times G_{kl}(q+p;\kappa;\phi)\Gamma_{blm}^{(3)}(-p,p+q,-q;\kappa;\phi)G_{mi}(q;\kappa;\phi) \\ & -\frac{1}{2} G_{ij}(q;\kappa;\phi)\Gamma_{abjk}^{(4)}(p,-p,q,-q;\kappa;\phi)G_{ki}(q;\kappa;\phi) \Big\}.\end{aligned} \tag{1.135}$$

The corresponding diagrams contributing to the flow are shown in Fig. 1.9.

When the external field vanishes, this equation simplifies greatly since then $\Gamma^{(3)}$ vanishes, and $\Gamma^{(2)}$ is diagonal:

$$\Gamma_{ab}^{(2)}(\kappa;q) = \delta_{ab}(q^2 + \Sigma(\kappa;q)), \tag{1.136}$$

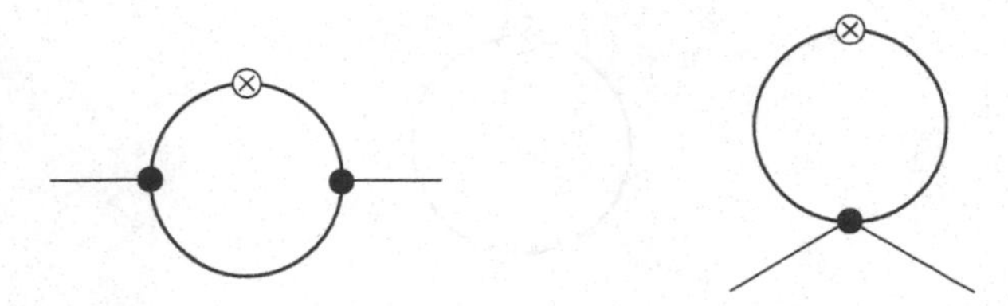

Fig. 1.9 The two diagrams contributing to the flow of the 2-point function in a constant background field. The *lines* represent dressed propagators, G_κ. The *cross* represents an insertion of $\partial_\kappa R_k$. The vertices denoted by *black dots* are $\Gamma_\kappa^{(3)}$ and $\Gamma_\kappa^{(4)}$. When the background field vanishes, so does the diagram on the *left*

which defines the self-energy Σ (this Σ differs from Σ_{cl} introduced in Sect. 1.3.2 by a factor $2m$; see e.g. Eqs. (1.90) and (1.108)). We get then

$$\partial_\kappa \Gamma_{ab}^{(2)}(\kappa; p) = -\frac{1}{2}\int \frac{d^d q}{(2\pi)^d}\, \partial_\kappa R_\kappa(q)\, G^2(\kappa; q)\, \Gamma_{abll}^{(4)}(\kappa; p, -p, q, -q), \quad (1.137)$$

where

$$G^{-1}(\kappa, q) = q^2 + R_\kappa(q) + \Sigma(\kappa; q). \quad (1.138)$$

Similarly, the flow of the the 4-point function in vanishing field reads:

$$\begin{aligned}\partial_\kappa \Gamma_{abcd}^{(4)}(\kappa; p_1, p_2, p_3, p_4) &= \int \frac{d^d q}{(2\pi)^d}\, \partial_\kappa R_k(q^2)\, G^2(\kappa; q)\\ &\times \Big\{ G(\kappa; q + p_1 + p_2)\Gamma_{abij}^{(4)}(\kappa; p_1, p_2, q, \cdot)\Gamma_{cdij}^{(4)}(\kappa; p_3, p_4, -q, \cdot)\\ &+ G(\kappa; q + p_1 + p_3)\Gamma_{acij}^{(4)}(\kappa; p_1, p_3, q, \cdot)\Gamma_{bdij}^{(4)}(\kappa; p_2, p_4, -q, \cdot)\\ &+ G(\kappa; q + p_1 + p_4)\Gamma_{adij}^{(4)}(\kappa; p_1, p_4, q, \cdot)\Gamma_{cbij}^{(4)}(\kappa; p_3, p_2, -q, \cdot)\Big\}\\ &- \frac{1}{2}\int \frac{d^d q}{(2\pi)^d}\partial_\kappa R_\kappa(q) G^2(\kappa; q)\Gamma_{abcdii}^{(6)}(\kappa; p_1, p_2, p_3, p_4, q, -q),\end{aligned} \quad (1.139)$$

where we have used the abbreviated notation $\Gamma_{abij}^{(4)}(\kappa; p_1, p_2, q, -p_1 - p_2 - q) \to \Gamma_{abij}^{(4)}(\kappa; p_1, p_2, q, \cdot)$. The four contributions in the r.h.s. of Eq. (1.139) are represented in the diagrams shown in Figs. 1.10 and 1.11.

Equations (1.137) and (1.139) for the 2- and 4-point functions constitute the beginning of an infinite hierarchy of exact equations for the n-point functions, reminiscent of the Schwinger-Dyson hierarchy, with the flow equation for the n-point function involving all the m-point functions up to $m = n + 2$. Clearly, solving this hierarchy requires approximations. A most natural approximation would rely on a truncation, in order to close the infinite hierarchy, as commonly done for instance in solving the

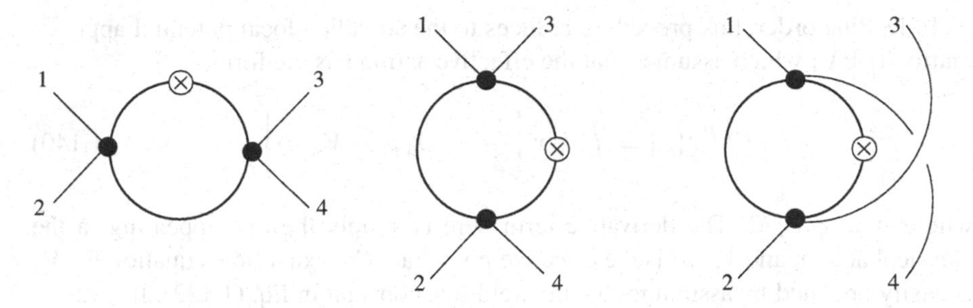

Fig. 1.10 Diagrammatic illustration of the r.h.s. of the flow equation for the 4-point function, Eq. (1.139): contribution of the 4-point functions (represented by *black disks*) in the three possible channels, from *left* to *right*. The *crossed circle* represents an insertion of $\partial_\kappa R_\kappa$, and the *thick line* a full propagator

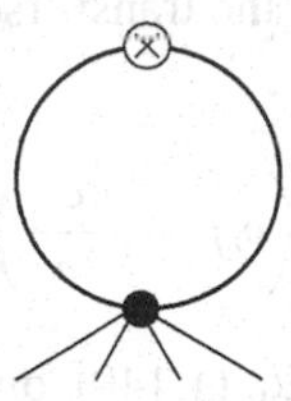

Fig. 1.11 Diagrammatic illustration of the r.h.s. of the flow equation for the 4-point function, Eq. (1.139): contribution of the 6-point function $\Gamma^{(6)}$ (represented by a *black disk*). The *crossed circle* represents an insertion of $\partial_\kappa R_\kappa$, and the *thick line* a full propagator

Schwinger Dyson equations. For instance, one could ignore in Eq. (1.139) the effect of the 6-point function on the flow of the 4-point function, arguing for instance that the 6-point function is perturbatively of order u^3. However such truncations prove to be not accurate enough for the present problem. Besides, the NPRG offers the possibility of exploring new approximation schemes that involve the entire functional $\Gamma[\phi]$ rather than the individual n-point functions. An example of such an approximation is provided by the derivative expansion described in the next section.

1.4.2 The Local Potential Approximation

The derivative expansion exploits the fact that the regulator in the flow equations (e.g. Eqs. (1.137) or (1.139)) forces the loop momentum q to be smaller than κ, i.e., only momenta $q \lesssim \kappa$ contribute to the flow. Besides, in general, the regulator insures that all vertices are smooth functions of momenta. Then, in the calculation of the n-point functions at vanishing external momenta p_i, it is possible to expand the n-point functions in the r.h.s. of the flow equations in terms of q^2/κ^2, or equivalently in terms of the derivatives of the field.

In leading order, this procedure reduces to the so-called local potential approximation (LPA), which assumes that the effective action has the form:

$$\Gamma_\kappa^{LPA}[\phi] = \int d^d x \left\{ \frac{1}{2} \partial_\mu \phi_i \partial_\mu \phi_i + V_\kappa(\rho) \right\}, \tag{1.140}$$

where $\rho \equiv \phi_i \phi_i / 2$. The derivative term here is simply the one appearing in the classical action, and $V_\kappa(\rho)$ is the effective potential. The exact flow equation for V_κ is easily obtained by assuming that the field ϕ is constant in Eq. (1.132). It reads:

$$\partial_\kappa V_\kappa(\rho) = \frac{1}{2} \int \frac{d^d q}{(2\pi)^d} \partial_\kappa R_\kappa(q) \{(N-1) G_T(\kappa; q) + G_L(\kappa; q)\}, \tag{1.141}$$

where G_T and G_L are, respectively, the transverse and longitudinal components of the propagator:

$$G_{ij}(\kappa; q) = G_T(\kappa; q) \left(\delta_{ij} - \frac{\phi_i \phi_j}{2\rho} \right) + G_L(\kappa; q) \frac{\phi_i \phi_j}{2\rho}. \tag{1.142}$$

By using the LPA effective action, Eq. (1.140), one gets

$$\begin{aligned} G_T^{-1}(\kappa; q) &= q^2 + V'(\rho) + R_k(q), \\ G_L^{-1}(\kappa; q) &= q^2 + V'(\rho) + 2\rho V''(\rho) + R_\kappa(q), \end{aligned} \tag{1.143}$$

with $V'(\rho) = dV/d\rho$ and $V''(\rho) = d^2V/d\rho^2$.

The solution of the LPA is well documented in the literature (see e.g. [50, 58]). In this section, we shall just, for illustrative purposes, solve approximately these equations, by keeping only a few terms in the expansion of the effective potential in powers of ρ (thereby effectively implementing a truncation which ignores the effect of higher n-point functions on the flow). The derivatives of $V_\kappa(\rho)$ with respect to ρ give the n-point functions at zero external momenta as a function of κ. We shall introduce a special notation for these n-point functions in vanishing external field:

$$m_\kappa^2 \equiv \left. \frac{dV_\kappa}{d\rho} \right|_{\rho=0}, \qquad g_\kappa \equiv \left. \frac{d^2 V_\kappa}{d\rho^2} \right|_{\rho=0}, \qquad h_\kappa \equiv \left. \frac{d^3 V_\kappa}{d\rho^3} \right|_{\rho=0}. \tag{1.144}$$

The equations for these n-point functions are obtained by differentiating once and twice Eq. (1.141) with respect to ρ, then setting $\rho = 0$, and using the definitions in Eq. (1.144). One gets, respectively:

$$\kappa \partial_\kappa m_\kappa^2 = -\frac{(N+2)}{2} g_\kappa I_d^{(2)}, \tag{1.145}$$

and

$$\kappa\partial_\kappa g_\kappa = (N+8)g_\kappa^2 I_d^{(3)}(\kappa) - \frac{1}{2}(N+4)\, h_\kappa I_d^{(2)}(\kappa), \tag{1.146}$$

where we have defined

$$\begin{aligned} I_d^{(n)}(\kappa) &\equiv \int \frac{d^d q}{(2\pi)^d} \kappa\partial_\kappa R_\kappa(q^2) G^n(\kappa; q) \\ &= 2K_d \frac{\kappa^{d+2}}{(\kappa^2+m_\kappa^2)^n}, \end{aligned} \tag{1.147}$$

the explicit form in the second line being obtained for the regulator (1.128) and

$$K_d^{-1} \equiv 2^{d-1}\, \pi^{d/2}\, d\, \Gamma(d/2). \tag{1.148}$$

We have used the fact that for vanishing external field, the propagator is diagonal and $\Sigma(\kappa, q) = m_\kappa^2$ (see Eqs. (1.136) and (1.138)). Equations (1.145) and (1.146) are solved starting from the initial condition at $\kappa = \Lambda$:

$$m_\Lambda^2 = r \qquad g_\Lambda = \frac{u}{3}. \tag{1.149}$$

In order to factor out the large variations which arise when κ varies from the microscopic scale Λ to the physical scale $\kappa = 0$, and also to exhibit the fixed point structure, it is convenient to isolate the explicit scale factors and to define dimensionless quantities:

$$m_\kappa^2 \equiv \kappa^2\, \hat{m}_\kappa^2, \qquad g_\kappa \equiv K_d^{-1} \kappa^{4-d}\, \hat{g}_\kappa, \qquad h_\kappa \equiv K_d^{-2} \kappa^{6-2d}\, \hat{h}_\kappa. \tag{1.150}$$

If one assumes that $\hat{m}_\kappa \ll 1$, and ignore the contribution of h_κ, then the equation for $\hat{g}_\kappa$ becomes:

$$\kappa \frac{d\hat{g}_\kappa}{d\kappa} = (d-4)\, \hat{g}_\kappa + 2(N+8)\, \hat{g}_\kappa^2, \tag{1.151}$$

and can be solved explicitly:

$$\hat{g}_\kappa = \frac{\hat{g}^*}{1 + \left(\frac{\kappa}{\kappa_c}\right)^{4-d}}, \tag{1.152}$$

where $\hat{g}^*$ is the value of $\hat{g}$ at the IR fixed point, $\hat{g}^* = (4-d)/(2(N+8))$, and κ_c the value of κ for which $\hat{g}_\kappa = \hat{g}^*/2$. We have:

$$\left(\frac{\kappa_c}{\Lambda}\right)^{d-4} = \frac{\hat{g}^* - \hat{g}_\Lambda}{\hat{g}_\Lambda} \approx \frac{\hat{g}^*}{\hat{g}_\Lambda}, \tag{1.153}$$

where the last approximate equality holds if $\hat{g}^* \gg \hat{g}_\Lambda$. In this regime, one recovers the qualitative feature already discussed in the previous lecture: there exists a well defined scale $\kappa_c \ll \Lambda$, $\kappa_c^{4-d} = uK_d/(3\hat{g}^*)$, that separates the scaling region, dominated by the IR fixed point, where $\hat{g} = \hat{g}^*$, from the perturbative region, dominated by the UV fixed point $\hat{g} = 0$ (when $\kappa \gg \kappa_c$, one can expand $\hat{g}_\kappa$ in powers of κ_c/κ; in leading order $g_\kappa = (u/3)(1 - (\kappa_c/\kappa)^{4-d})$).

The local potential approximation, or a refined version of it, enters in an essential way in the approximation scheme that we have developed in order to calculate $\Sigma(p)$. Further insight can be gained by studying the n-point functions in the large N limit, to which we now turn.

1.4.3 Correlation Functions in the Large N Limit

The LPA, as well as the higher orders of the derivative expansion, give accurate results for the correlation functions and their derivatives only at zero external momenta. In order to get insight into the effect of non vanishing external momenta we consider now the correlation functions in the large N limit (at fixed uN) [14, 59–61].

For vanishing field, the inverse propagator has the same form as in the LPA, Eq. (1.138), where the running mass m_κ is here given by a gap equation

$$m_\kappa^2 = r + \frac{Nu}{6}\int \frac{d^dq}{(2\pi)^d}\,(G(\kappa;q) - G(\Lambda;q)). \tag{1.154}$$

The 4-point function has the following structure:

$$\begin{aligned} &\Gamma^{(4)}_{1234}(\kappa; p_1, p_2, p_3, p_4) \\ &\quad = \delta_{12}\delta_{34} g_\kappa(p_1+p_2) + \delta_{13}\delta_{24} g_\kappa(p_1+p_3) + \delta_{14}\delta_{23} g_\kappa(p_1+p_4), \end{aligned} \tag{1.155}$$

where $g_\kappa(p)$ is given by

$$g_\kappa(p) = \frac{u}{3}\frac{1}{1 + \frac{Nu}{6}B_d(\kappa;p)}, \tag{1.156}$$

with

$$B_d(\kappa;p) \equiv \int \frac{d^dq}{(2\pi)^d}\,G(\kappa;q)G(\kappa;p+q). \tag{1.157}$$

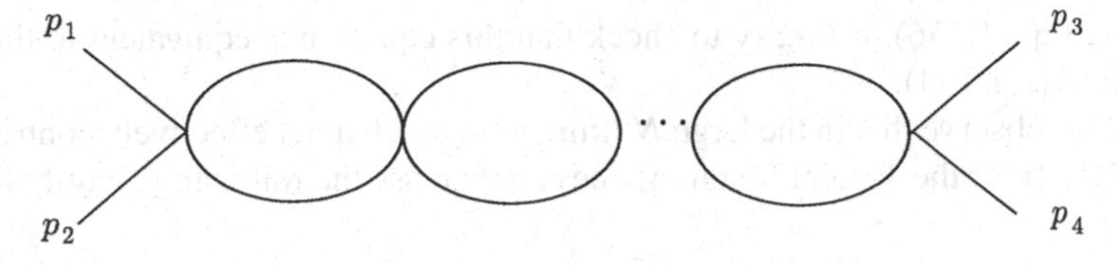

Fig. 1.12 Diagrams representing the contribution to $\Gamma^{(4)}_{1234}(\kappa; p_1, p_2, p_3, p_4)$ in the *first* channel of Eq. (1.155). This diagram represents also a typical contribution to the function $g_\kappa(p)$ of Eq. (1.156). External momenta are counted as incoming momenta, so that $p_1 + p_2$ flows into the bubble chain, and $p_3 + p_4 = -(p_1 + p_2)$

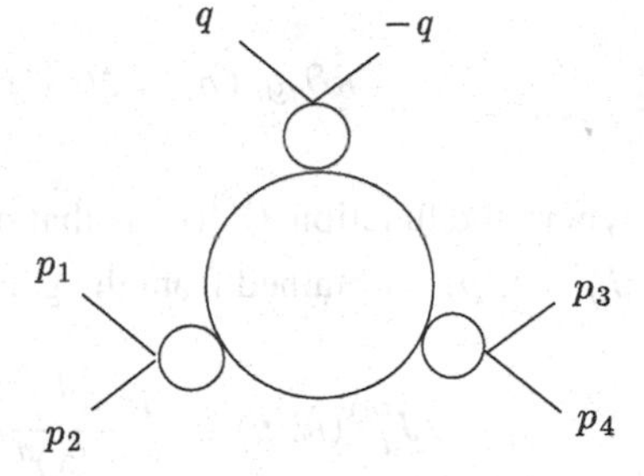

Fig. 1.13 Diagrams representing the contribution to $\Gamma^{(6)}_{1234mm}(\kappa; p_1, p_2, p_3, p_4, q, -q)$ in Eq. (1.158). The *small circles* represent the function $g_\kappa(p)$ with p the momentum flowing through the vertex. The convention for momenta is as in Fig. 1.12

Finally the 6-point function $\Gamma^{(6)}_{1234mm}(\kappa; p_1, p_2, p_3, p_4, q, -q)$ (summation over repeated indices is understood) is of the form

$$\frac{1}{N}\Gamma^{(6)}_{1234mm}(\kappa; p_1, p_2, p_3, p_4, q, -q)$$
$$= h_\kappa(p_1 + p_2)\delta_{12}\delta_{34} + h_\kappa(p_1 + p_3)\delta_{13}\delta_{24} + h_\kappa(p_1 + p_4)\delta_{14}\delta_{23}, \quad (1.158)$$

with

$$h_\kappa(p) = N g_\kappa(0) g^2_\kappa(p) \int \frac{d^d q'}{(2\pi)^d} G^2(\kappa; q') G(\kappa; q' + p). \quad (1.159)$$

Note that a single function, $g_\kappa(p)$, suffices to calculate all the n-point functions in the large N limit. These can be obtained in a straightforward fashion by calculating the corresponding Feynman diagrams with a regulator (see Figs. 1.12, 1.13). It is however easy to verify that the various n-point functions that we have just written are indeed solutions of the flow equations in the large N limit.

To this aim, one notes first that Eq. (1.137) reduces to an equation for the running mass:

$$\partial_\kappa m^2_\kappa = -\frac{1}{2} N g_\kappa(0) \int \frac{d^d q}{(2\pi)^d} \partial_\kappa R_\kappa(q) G^2(\kappa; q), \quad (1.160)$$

and using Eq. (1.156), it is easy to check that this equation is equivalent to the gap equation, Eq. (1.154).

Next, we observe that in the large N limit, a single channel effectively contributes in Eq. (1.139) for the 4-point function; one can then use the following identity in this limit:

$$\begin{aligned}\Gamma^{(4)}_{12ij}(\kappa; p_1, p_2, q, -q-p_1-p_2)\Gamma^{(4)}_{34ij}(\kappa; p_3, p_4, -q, q-p_3-p_4)\\ = N g_\kappa^2(p_1+p_2)\delta_{12}\delta_{34},\end{aligned} \tag{1.161}$$

together with Eq. (1.158) for $\Gamma^{(6)}$, and one obtains:

$$\kappa\partial_\kappa g_\kappa(p) = N g_\kappa^2(p) J_d^{(3)}(\kappa; p) - \frac{N}{2} h_\kappa(p) I_d^{(2)}(\kappa), \tag{1.162}$$

where the function $I_d^{(2)}(\kappa)$ is that defined in Eq. (1.147), with $n = 2$. The function $J_d^{(3)}(\kappa; p)$ is obtained from the general definition

$$J_d^{(n)}(\kappa; p) \equiv \int \frac{d^d q}{(2\pi)^d} \kappa\partial_\kappa R_\kappa(q) G^{n-1}(\kappa; q) G(\kappa; p+q). \tag{1.163}$$

Note that $J_d^{(n)}(\kappa; p=0) = I_d^{(n)}(\kappa)$.

At this point we remark that the flow equation for $g_\kappa(p)$ can also be obtained directly from the explicit expression (1.156), in the form:

$$\partial_\kappa g_\kappa(p) = -\frac{N}{2} g_\kappa^2(p)\partial_\kappa \int \frac{d^d q}{(2\pi)^d} G(\kappa; q) G(\kappa; q+p). \tag{1.164}$$

It is then straightforward to verify, using Eqs. (1.160) and (1.159) that Eqs. (1.162) and (1.164) are indeed equivalent. The first term in Eq. (1.162) comes from the derivative of the cut-off function in the propagators in Eq. (1.164), while the second term, which involves the 6-point vertex, comes from the derivative of the running mass in the propagators.

Note that Eqs. (1.160) for m_κ and (1.162) for $g_\kappa(p=0)$ become identical respectively to Eqs. (1.145) and (1.146) of the LPA in the large N limit, a well know property [61].

Let us now analyze characteristic features of the function $g_\kappa(p)$. For simplicity we specialize for the rest of this subsection to $d = 3$. Furthermore, for the purpose of the present, qualitative, discussion, one may assume $m_\kappa = 0$. This allows us to obtain easily $g_\kappa(p)$ from Eq. (1.156) in the two limiting cases $p = 0$ and $\kappa = 0$. In the first case, we have

$$g_\kappa(0) = \frac{u}{3} \frac{1}{1 + \frac{uN}{9\pi^2}\frac{1}{\kappa}}. \tag{1.165}$$

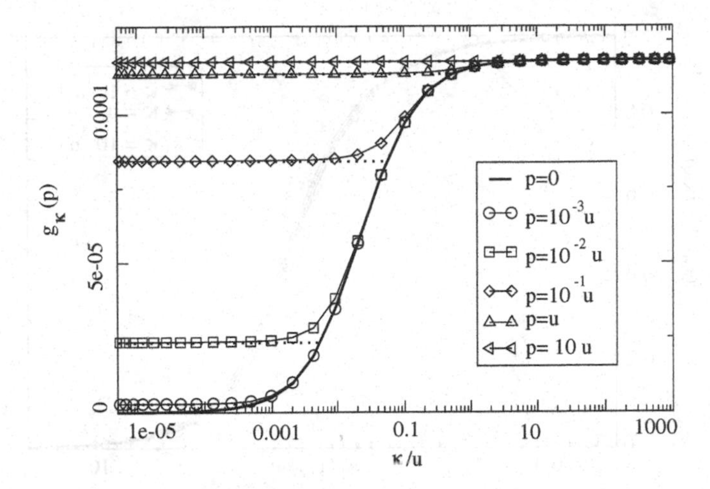

Fig. 1.14 The function $g_\kappa(p)$ (in units of Λ) obtained from a complete numerical solution of Eqs. (1.160) and (1.162), as a function of κ/u (in a logarithmic scale) for five values of p: from *bottom* to *top*, $p/u = 0.001, 0.01, 0.1, 1$ and 10. The envelope corresponds to $p = 0$. This figure illustrates the decoupling of modes: for each value of p, the flow stops when $\kappa \lesssim \alpha p$. The various horizontal asymptotes (*dotted llines*) correspond to the single value $\alpha = 0.54$. From Ref. [12]

This is identical to Eq. (1.152), with here $\hat{g}^* = 1/(2N)$ and $\kappa_c = Nu/9\pi^2$. In the other case, we have

$$g_{\kappa=0}(p) = \frac{u}{3}\frac{1}{1+\frac{uN}{48}\frac{1}{p}} = \frac{u}{3}\frac{p}{p+p_c}, \tag{1.166}$$

with $p_c \equiv uN/48$.

One sees on Eqs. (1.165) and (1.166) that the dependence on p of $g_{\kappa=0}(p)$ is quite similar to the dependence on κ of $g_\kappa(p=0)$. In particular both quantities vanish linearly as $\kappa \to 0$ or $p \to 0$, respectively. The result of the complete (numerical) calculation, including the effect of the running mass (i.e., solving the gap equation (1.154) and calculating $g_\kappa(p)$ from Eq. (1.156)), can in fact be quite well represented (to within a few percents) for arbitrary p and κ by the following approximate formula

$$g_\kappa(p) \approx \frac{u}{3}\frac{X}{1+X} \qquad X \equiv \frac{\kappa}{\kappa_c} + \frac{p}{p_c}. \tag{1.167}$$

This simple expression shows that p, when it is non vanishing, plays the same role as κ as an infrared regulator. In particular, at fixed p, the flow of $g_\kappa(p)$ stops when X becomes independent of κ, i.e., when $\kappa \lesssim p(\kappa_c/p_c)$, with $\kappa_c/p_c = 16/3\pi^2 \approx 0.54$. This important property of decoupling of the short wavelength modes is illustrated in Fig. 1.14. As shown by this figure, and also by the expression (1.167), the momentum dependence of the 4-point function can be obtained from its cut-off dependence at zero momentum. In fact Fig. 1.14 suggests that, to a very good approximation, there

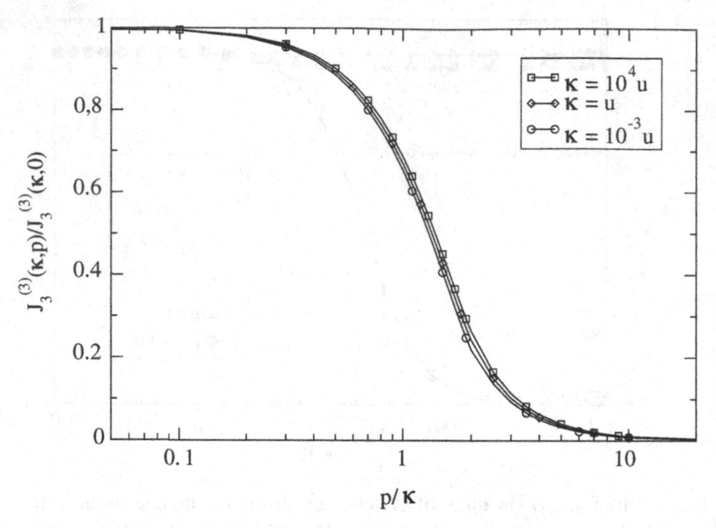

Fig. 1.15 The function $J_3^{(3)}(\kappa; p)/I_3^{(3)}(\kappa)$ as a function of p/κ (in a logarithmic scale), for different values of κ: $\kappa = 10^{-3}u$ *(circles)*, $\kappa = u$ *(diamonds)* and $\kappa = 10^4 u$ *(squares)*. From Ref. [12]

exists a parameter α such that $g(\kappa; p) \approx g(\kappa; 0)$ when $\kappa > \alpha p$, and $g(\kappa; p) \approx g(\kappa = \alpha p; 0)$ when $\kappa < \alpha p$. From the discussion above, one expects $\alpha \approx \kappa_c/p_c = 16/3\pi^2 \approx 0.54$, which is indeed in agreement with the analysis in Fig. 1.14.

The decoupling of modes can also be visualized in Fig. 1.15. The p-dependence of $J_d^{(n)}(\kappa; p)$ (Eq. (1.163)) shown in this figure is relatively simple: when $p \ll \kappa$, $J_d^{(n)}(\kappa; p) \simeq I_d^{(n)}(\kappa)$; when $p \gg \kappa$, $J_d^{(n)}(\kappa; p)$ vanishes as $1/p^2$. On a logarithmic scale the transition between these two regimes occurs rapidly at momentum $p \sim \kappa$, as illustrated on Fig. 1.15. A similar analysis can be made for the contribution of the 6-point function in Eq. (1.162), and one can show that the ratio of the p-dependence of the second term in the right hand side (the one proportional to h_κ) is, whenever it is significant, proportional to that of the first term, the proportionality coefficient being a function of κ only.

All this suggests that one can rewrite Eq. (1.162) for $g_\kappa(p)$ as follows:

$$\partial_\kappa g_\kappa(p) \approx N g_\kappa^2(p) \Theta\left(1 - \frac{\alpha^2 p^2}{\kappa^2}\right) I_d^{(3)}(\kappa)(1 - F_\kappa), \tag{1.168}$$

where α is a parameter of order unity. The Θ-function ensures that the flow exists only when $\kappa > \alpha p$, and stops for smaller values of κ.

These approximations, together with another one that we shall discuss more thoroughly in the next section have been used in order to construct approximate flow equations in Refs. [12, 13].

1.4.4 Beyond the Derivative Expansion

The arguments on which the derivative expansion is based can be generalized in order to set up a much more powerful approximation scheme [14] that we now briefly present. This scheme allows us to obtain the momentum dependence of the n-point function with a single approximation.

We observe that: (i) the momentum q circulating in the loop integral of a flow equation is limited by κ; (ii) the smoothness of the n-point functions allows us to make an expansion in powers of q^2/κ^2, independently of the value of the external momenta p. Now, a typical n-point function entering a flow equation is of the from $\Gamma_\kappa^{(n)}(p_1, p_2, ..., p_{n-1}+q, p_n-q; \phi)$, where q is the loop momentum. The proposed approximation scheme, *in its leading order*, consists in neglecting the q-dependence of such vertex functions:

$$\Gamma_\kappa^{(n)}(p_1, p_2, ..., p_{n-1}+q, p_n-q; \phi) \sim \Gamma_\kappa^{(n)}(p_1, p_2, ..., p_{n-1}, p_n; \phi). \quad (1.169)$$

Note that this approximation is a priori well justified. Indeed, when all the external momenta vanish $p_i = 0$, Eq. (1.169) is the basis of the LPA which, as stated above, is a good approximation. When the external momenta p_i begin to grow, the approximation in Eq. (1.169) becomes better and better, and it is trivial when all momenta are much larger than κ. With this approximation, Eq. (1.137) becomes (for simplicity we set here $N = 1$):

$$\begin{aligned}\partial_\kappa \Gamma_\kappa^{(2)}(p, -p; \phi) = \int \frac{d^d q}{(2\pi)^d} \partial_\kappa R_k(q^2) \Big\{ & G_\kappa(q^2; \phi)\Gamma_\kappa^{(3)}(p, 0, -p; \phi) \\ & \times G_\kappa((q+p)^2; \phi)\Gamma_\kappa^{(3)}(-p, p, 0; \phi)G_\kappa(q^2; \phi) \\ & -\frac{1}{2} G_\kappa(q^2; \phi)\Gamma_\kappa^{(4)}(p, -p, 0, 0; \phi)G_\kappa(q^2; \phi) \Big\}. \end{aligned} \quad (1.170)$$

Now comes the second ingredient of the approximation scheme, which exploits the advantage of working with a non vanishing background field: vertices evaluated at zero external momenta can be obtained as derivatives of vertex functions with a smaller number of legs:

$$\Gamma_\kappa^{(n+1)}(p_1, p_2, ..., p_n, 0; \phi) = \frac{\partial \Gamma_\kappa^{(n)}(p_1, p_2, ... p_n; \phi)}{\partial \phi}. \quad (1.171)$$

By exploiting Eq. (1.171), one easily transforms Eq. (1.170) into a *closed equation* (recall that G_κ and $\Gamma_\kappa^{(2)}$ are related by Eq. (1.133)):

$$\partial_\kappa \Gamma_\kappa^{(2)}(p^2;\phi) = \int \frac{d^d q}{(2\pi)^d}\, \partial_\kappa R_\kappa(q^2)\, G_\kappa^2(q^2;\phi)$$
$$\times \left\{ \left(\frac{\partial \Gamma_\kappa^{(2)}(p,-p;\phi)}{\partial \phi} \right)^2 G_\kappa((p+q)^2;\phi) - \frac{1}{2}\frac{\partial^2 \Gamma_\kappa^{(2)}(p,-p;\phi)}{\partial \phi^2} \right\}. \quad (1.172)$$

Note the similarity of this equation with Eq. (1.141): both are closed equations, the vertices appearing in the r.h.s. being expressed as derivatives of the function in the l.h.s..

The approximation scheme presented here is similar to that used in [12, 13]. There also the momentum dependence of the vertices was neglected in the leading order. However further approximations were needed in order to close the hierarchy. The progress realized here is to bypass these extra approximations by working in a constant background field.

The resulting equations can be solved with a numerical effort comparable to that involved in solving the equations of the derivative expansion. The preliminary results obtained so far are encouraging [15].

1.4.5 Calculation of $\Delta\langle\varphi^2\rangle$

We come now to the conclusion of these lectures, where results concerning the numerical value of c obtained with the NPRG will be presented.

The approximation developed in [11–13] is based on an iteration scheme, where one starts by building approximate equations for the n-point functions, that are then solved exactly. To give a flavor of this method, consider more specifically the equation (1.139) for the 4-point function. An approximate flow equation is obtained with the following three approximations:

Our first approximation is that discussed in Sect. 1.4.4: we ignore the q dependence of the vertices in the flow equation, i.e., we set $q = 0$ in the vertices $\Gamma^{(4)}$ and $\Gamma^{(6)}$ and factor them out of the integral in the r.h.s. of Eq. (1.139). The second approximation concerns the propagators in the flow equation, for which we make the replacements:

$$G(p+q) \longrightarrow G_{LPA'}(q)\, \Theta\left(1 - \frac{\alpha^2 p^2}{\kappa^2}\right) \quad (1.173)$$

where α is an adjustable parameter. A motivation for this approximation may be obtained from the analysis of the n-point function presented in Sect. 1.4.3 (see Eq. (1.168) and Fig. 1.15). This approximation introduces a dependence of the results on the value of α. The third approximation concerns the function $\Gamma^{(6)}$ for which we use an ansatz inspired by the expressions of the various n-point functions in the large N limit (see again the discussion at the end of Sect. 1.4.3): one assumes that the contribution of $\Gamma^{(6)}$ to the r.h.s. of Eq. (1.139) is proportional to that of the

Table 1.1 Summary of available results for the coefficient c

c	$N=1$	$N=2$	$N=3$	$N=4$	$N=10$	$N=40$	$N=\infty$
Lattice [37]		1.32 ± 0.02					
Lattice [38]		1.29 ± 0.05					
Lattice [62]	1.09 ± 0.09			1.60 ± 0.10			
7-loops [42]	1.07 ± 0.10	1.27 ± 0.10	1.43 ± 0.11	1.54 ± 0.11			
Large N [7]							$c = 2.33$
NPRG	1.11	1.30	1.45	1.57	1.91	2.12	

The last line contains the results obtained in [13]

other terms, the proportionality coefficient F_κ being only a function of κ. The same proportionality also holds in the LPA regime, which allows us to use the LPA to determine F_κ.

These three approximations result then in the following approximate equation for $\Gamma^{(4)}$:

$$\begin{aligned}\partial_\kappa \Gamma^{(4)}_{ijkl}(\kappa; p_1, p_2, p_3, p_4) &= I^{(3)}_\kappa(0)\,(1 - F_\kappa) \\ &\times \Big\{ \Theta\big(\kappa^2 - \alpha^2(p_1 + p_2)^2\big) \Gamma^{(4)}_{ijmn}(\kappa; p_1, p_2, 0, -p_1 - p_2) \\ &\quad \times \Gamma^{(4)}_{klnm}(\kappa; p_3, p_4, -p_3 - p_4, 0) + \text{permutations} \Big\}. \qquad (1.174)\end{aligned}$$

This equation can be solved analytically in terms of the solution of the LPA (actually a refined version of it). This is done by steps, starting from the momentum domain $\alpha^2(p_1 + p_2)^2, \alpha^2(p_1 + p_3)^2, \alpha^2(p_1 + p_4)^2 \leq \kappa^2$, where it can be verified that the solution is that of the LPA itself. The solution of this equation is then inserted in Eq. (1.137) for the 2-point function and leads to the leading order determination of $\Sigma(p)$. The scheme is improved through an iteration procedure. At next-to-leading order, one improves $\Gamma^{(4)}$ by establishing an approximate equation for $\Gamma^{(6)}$. The solution of this equation is then used in Eq. (1.139) for $\Gamma^{(4)}$, and the resulting $\Gamma^{(4)}$ is used in turn in Eq. (1.137) to get Σ at next-to-leading order.

The results obtained with this method for the value of c are reported in Table 1.1 and Fig. 1.16 for various values of N for which results have been obtained with other techniques, either the lattice technique [37, 38, 62], or variationally improved 7-loops perturtbative calculations [42] (other optimized perturbative calculations have also been recently performed, and are in agreement with those quoted here; see [40, 41]). The results reported in Table 1.1 have been obtained with an improved next-to-leading order calculation where, among other things, the parameter α of Eq. (1.173) has been fixed by a principle of minimum sensitivity (see [13] for details). A recent (approximate) calculation along the lines described in Sect. 1.4.4 yields, for $N = 1$, the value $c = 1.2$ [15]. Let me also mention the result $c = 1.23$ obtained, for $N = 2$ in Ref. [63]; however, as discussed in [13] it is difficult to gauge the quality of the approximation made in [63].

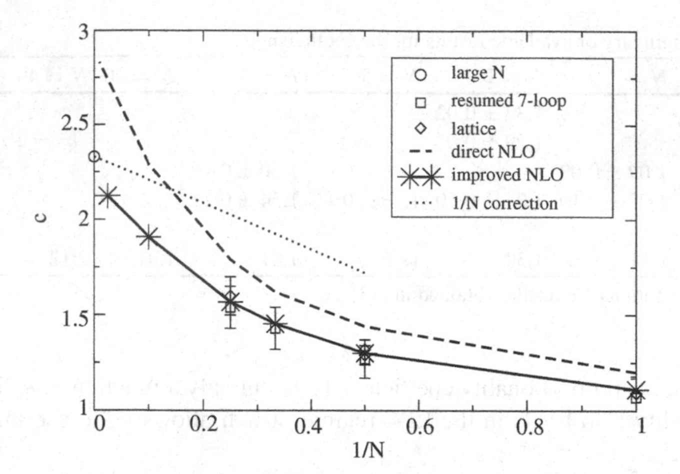

Fig. 1.16 The coefficient c (obtained with the fastest apparent convergence procedure) as a function of $1/N$. Our NLO results, are compared with results obtained with other methods: lattice [37, 38, 62] (*diamonds*) and 7-loops perturbation theory [42] (*squares*), all of them with their corresponding error *bars*, together with the $N \to \infty$ result [7] (*circle*), and the extrapolation following the $1/N$ correction calculated in Ref. [39]. From Ref. [13]

For $N \leq 4$, where we can compare with other results, the values of c obtained with the present improved NLO calculation are in excellent agreement with those obtained from lattice and "7-loops" calculations. What happens at large values of N deserves a special discussion. As seen in Fig. 1.16 the curve showing the improved leading order results extrapolates when $N \to \infty$ to a value that is about 4% below the known exact result [7]. A direct calculation at very large values of N is difficult in the present approach for numerical reasons: since the coefficient c represents in effect an order $1/N$ correction (see [7]), it is necessary to insure the cancellation of the large, order N, contributions to the self-energy, in order to extract the value of c. This is numerically demanding when $N \gtrsim 100$. Figure 1.16 also reveals an intriguing feature: there seems to be no natural way to reconcile the present results, and for this matter the results from lattice calculations or 7-loop calculations, with the calculation of the $1/N$ correction presented in Ref. [39]: the dependence in $1/N$ of our results, be they obtained from the direct NLO or the improved NLO, appear to be incompatible with the slope predicted in Ref. [39] for the $1/N$ expansion.

Acknowledgments Most of the material of these lectures is drawn from work done in a most enjoyable collaboration with several people during these last years: G. Baym, F. Laloë, M. Holzmann, R. Mendez-Galain, D. Vautherin, N. Wschebor and J. Zinn-Justin. I would also like to express my gratitude to Achim Schwenk for insisting on having these lecture notes....and putting such a high cut-off on his patience as he waited for them.

References

1. Pitaevskii, L., Stringari, S.: Bose-Einstein Condensation. Oxford University Press, Oxford (2003)
2. Pethick, C.J., Smith, H.: Bose-Einstein Condensation in Dilute Gases. Cambridge University Press, Cambridge (2003)
3. Dalfovo, F., Giorgini, S., Stringari, S., Pitaevskii, L.: Rev. Mod. Phys. **71**, 463 (1999)
4. Legget, A.J.: Bose-Einstein condensation in the alkali gases: some fundamental concepts. Rev. Mod. Phys. **73**, 307 (2001)
5. Andersen, J.O.: Rev. Mod. Phys. **76**, 599 (2004)
6. Baym, G., Blaizot, J.-P., Holzmann, M., Laloë, F., Vautherin, D.: Phys. Rev. Lett. **83**, 1703 (1999)
7. Baym, G., Blaizot, J.-P., Zinn-Justin, J.: Europhys. Lett. **49**, 150 (2000)
8. Baym, G., Blaizot, J.-P., Holzmann, M., Laloë, F., Vautherin, D.: Eur. Phys. J. B **24**, 107 (2001)
9. Holzmann, M., Baym, G., Blaizot, J.-P., Laloë, F.: Phys. Rev. Lett. **87**, 120403 (2001)
10. Holzmann, M., Fuchs, J.N., Baym, G., Blaizot, J.-P., Laloë, F.: Comptes Rend. Phys. **5**, 21 (2004)
11. Blaizot, J.P., Mendez Galain, R., Wschebor, N.: Europhys. Lett. **72**(5), 705–711 (2005)
12. Blaizot, J.P., Mendez-Galain, R., Wschebor, N.: Phys. Rev. E **74**, 051116 (2006)
13. Blaizot, J.P., Mendez-Galain, R., Wschebor, N.: Phys. Rev. E **74**, 051117 (2006)
14. Blaizot, J.P., Mendez Galain, R., Wschebor, N.: Phys. Lett. B **632**, 571 (2006)
15. Blaizot, J.P., Mendez-Galain, R., Wschebor, N.: Eur. Phys. J. B **58**, 297–309 (2007). doi:10.1140/epjb/e2007-00223-3
16. Ziff, R.M., Uhlenbeck, G.E., Kac, M.: Phys. Rep. **32**, 169 (1977)
17. Holzmann, M., Baym, G., Blaizot, J.-P., Laloë, F.: Proc. Natl. Acad. Sci. **104**, 1476–1481 (2007)
18. Giorgini, S., Pitaevskii, L., Stringari, S.: Phys. Rev. A **54**, 4633 (1996)
19. Schafer, T.: arXiv:nucl-th/0609075
20. Kaplan, D.B.: Lectures Delivered at the 17th National Nuclear Physics Summer School 2005, Berkeley, CA, 6–17 June 2005
21. Braaten, E., Hammer, H.W., Hermans, S.: Phys. Rev. A **63**, 063609 (2001)
22. Braaten, E., Nieto, A.: Phys. Rev. B **55**, 8090–8093 (1997). doi:10.1103/PhysRevB.55.8090
23. Braaten, E., Nieto, A.: Phys. Rev. B **56**, 14745–14765 (1997). doi:10.1103/PhysRevB.56.14745
24. Toyoda, T.: Ann. Phys. N.Y. **141**, 154 (1982)
25. Fetter, A.L., Walecka, J.D.: Quantum Theory of Many-Particle Systems, vol. 28. McGraw-Hill, New York (1971)
26. Baym, G., Grinstein, G.: Phys. Rev. D **15**, 2897 (1977)
27. Kadanoff, L.P., Baym, G.: Quantum Statistical Mechanics. Benjamin, New York (1962)
28. Abrikosov, A.A., Gorkov, L.P., Dzyaloshinski, I.E.: Methods of Quantum Field Theory in Statistical Physics. Prentice Hall, Englewood Cliffs (1963)
29. Blaizot, J.P., Ripka, G.: Quantum Theory of Finite Systems. MIT Press, Cambridge (1986)
30. Patashinskii, A.Z., Pokrovskii, V.L.: Fluctuation Theory of Phase Transitions. Pergamon Press, Oxford (1979)
31. Ginsparg, P.: Nucl. Phys. B **170**, 388 (1980)
32. Appelquist, T., Pisarski, R.D.: Phys. Rev. D **23**, 2305 (1981)
33. Nadkarni, S.: Phys. Rev. D **27**, 917 (1983); Phys. Rev. D **38**, 3287 (1988)
34. Arnold, P., Tomasik, B.: Phys. Rev. A **64**, 053609 (2001)
35. Moshe, M., Zinn-Justin, J.: Phys. Rept. **385**, 69 (2003)
36. Zinn-Justin, J.: Quantum field theory and critical phenomena. Int. Ser. Monogr. Phys. **113**, 1 (2002)
37. Arnold, P., Moore, G.: Phys. Rev. Lett. **87**, 120401 (2001); Phys. Rev. E **64**, 066113 (2001)
38. Kashurnikov, V.A., Prokof'ev, N.V., Svistunov, B.V.: Phys. Rev. Lett. **87**, 160601 (2001); Phys. Rev. Lett. **87**, 120402 (2001)

39. Arnold, P., Tomasik, B.: Phys. Rev. A **62**, 063604 (2000)
40. Kneur, J.-L., Neveu, A., Pinto, M.B.: Phys. Rev. A **69**, 053624 (2004)
41. de Souza Cruz, F., Pinto, M.B., Ramos, R.O.: Phys. Rev. B **64**, 014515 (2001); Phys. Rev. A **65**, 053613 (2002)
42. Kastening, B.: Phys. Rev. A **68**, 061601 (R) (2003); Phys. Rev. A **69**, 043613 (2004)
43. Arnold, P., Moore, G.D., Tomasik, B.: Phys. Rev. A **65**, 013606 (2002)
44. Wetterich, C.: Phys. Lett. B **301**, 90 (1993)
45. Ellwanger, U.: Z. Phys. C **58**, 619 (1993)
46. Tetradis, N., Wetterich, C.: Nucl. Phys. B **422**, 541 (1994)
47. Morris, T.R.: Int. J. Mod. Phys. A **9**, 2411 (1994)
48. Morris, T.R.: Phys. Lett. B **329**, 241 (1994)
49. Bagnuls, C., Bervillier, C.: Phys. Rept. **348**, 91 (2001)
50. Berges, J., Tetradis, N., Wetterich, C.: Phys. Rept. **363**, 223–386 (2002)
51. Delamotte, B., Mouhanna, D., Tissier, M.: Phys. Rev. B **69**, 134413 (2004); Canet, L., Delamotte, B.: Condensed Matter Phys. **8**, 163–179 (2005)
52. Gies, H.: arXiv:hep-ph/0611146
53. Bijlsma, M., Stoof, H.T.C.: Phys. Rev. A **54**, 5085 (1996)
54. Andersen, J.O., Strickland, M.: Phys. Rev. A **60**, 1442 (1999)
55. Ball, R.D., Haagensen, P.E., Latorre, J.I., Moreno, E.: Phys. Lett. B **347**, 80 (1995)
56. Comellas, J.: Nucl. Phys. B **509**, 662 (1998)
57. Litim, D.: Phys. Lett. B **486**, 92 (2000); Phys. Rev. D **64**, 105007 (2001); Nucl. Phys. B **631**, 128 (2002); Int. J. Mod. Phys. A **16**, 2081 (2001)
58. Canet, L., Delamotte, B., Mouhanna, D., Vidal, J.: Phys. Rev. D **67**, 065004 (2003)
59. Ellwanger, U., Wetterich, C.: Nucl. Phys. B **423**, 137 (1994)
60. Tetradis, N., Litim, D.F.: Nucl. Phys. B **464**, 492–511 (1996)
61. D'Attanasio, M., Morris, T.R.: Phys. Lett. B **409**, 363–370 (1997)
62. Sun, X.: Phys. Rev. E **67**, 066702 (2003)
63. Ledowski, S., Hasselmann, N., Kopietz, P.: Phys. Rev. A **69**, 061601(R) (2004); Phys. Rev. A **70**, 063621 (2004)

Chapter 2
An Introduction to the Nonperturbative Renormalization Group

Bertrand Delamotte

2.1 Wilson's Renormalization Group

"What can I do, what can I write,
Against the fall of night?"
A.E. Housman

2.1.1 Introduction

We give in these notes a short presentation of both the main ideas underlying Wilson's renormalization group (RG) and their concrete implementation under the form of what is now called the non-perturbative renormalization group (NPRG) or sometimes the functional renormalization group (which can be perturbative). Prior knowledge of perturbative field theory is not required for the understanding of the heart of the article. However, some basic knowledge about phase transitions in Ising and $O(N)$ models is supposed [1–3]. We shall mainly work in the framework of statistical field theory but when it will turn out to be illuminating we shall also use the language of particle physics.

The beginning of this article will be rather elementary and known to most physicists working in the field of critical phenomena. Nevertheless, both for completeness and to set up a language (actually a way of thinking at renormalization group) it has appeared necessary to include it. The first subsection of this article deals with a comparison between perturbative and Wilson's RG. Then, in the next subsection is presented the implementation of Kadanoff-Wilson's RG on the very peculiar case

B. Delamotte (✉)
Laboratoire de Physique Théorique de la Matière Condensée,
CNRS-UMR 7600, Université Pierre et Marie Curie,
4 Place Jussieu, 75252 Paris Cedex 05, France
e-mail: delamotte@lptl.jussieu.fr

J. Polonyi and A. Schwenk (eds.), *Renormalization Group and Effective Field Theory Approaches to Many-Body Systems*, Lecture Notes in Physics 852,
DOI: 10.1007/978-3-642-27320-9_2,

of the two-dimensional Ising model on the triangular lattice. This allows us to introduce the ideas of decimation and block spins and also those of RG flow of coupling constants and fixed point. The third subsection, which is also the heart of this article, deals with the "modern" implementation of Wilson's ideas. The general framework as well as detailed calculations for the $O(N)$ models will be given.

2.1.2 The Perturbative Method in Field Theory

The idea behind perturbation theory is to consider an exactly soluble model, either the gaussian or the mean field "model", and to add, in a perturbation expansion, the term(s) present in the model under study and which are not taken into account in the exactly soluble model taken as a reference [1, 4, 5]. For instance, in the "ϕ^4" model (which belongs to the same universality class as the Ising model) the partition function writes:

$$Z[B] = \int \mathcal{D}\phi\, e^{-H(\phi)+\int B\phi} \tag{2.1}$$

with

$$H(\phi) = \int d^d x \left(\frac{1}{2}(\nabla\phi)^2 + \frac{1}{2}r_0\phi^2 + \frac{1}{4!}u_0\phi^4\right) \tag{2.2}$$

and B an external field and it is possible to take as a reference model the gaussian model:

$$Z_0 = \int \mathcal{D}\phi\, e^{-H_0(\phi)+\int B\phi} \tag{2.3}$$

where

$$H_0(\phi) = \int d^d x \left(\frac{1}{2}(\nabla\phi)^2 + \frac{1}{2}r_0\phi^2\right). \tag{2.4}$$

Z is then developed as a series in u_0 around Z_0:

$$Z = \int \mathcal{D}\phi \left(1 - \frac{u_0}{4!}\int_{x_1}\phi^4(x_1) + \frac{1}{2}\left(\frac{u_0}{4!}\right)^2 \int_{x_1,x_2}\phi^4(x_1)\phi^4(x_2) + \dots\right) e^{-H_0(\phi)+\int B\phi}. \tag{2.5}$$

This expansion leads to the series of Feynman diagrams for the Green functions.

The problem with this approach is that the "fluctuations" induced by the ϕ^4 term around the gaussian model are large. In the perturbation expansion they lead to integrals—corresponding to the loops in the Feynman diagrams—of the form:

$$\int^{\Lambda} d^d q_1 \dots d^d q_L \prod_i (\text{propagator}(q_i)) \tag{2.6}$$

where

$$\text{propagator}(q_i) \sim \frac{1}{(q_i + Q)^2 + r_0}. \tag{2.7}$$

These integrals are supposed to be cut-off at the upper bound by Λ which is an ultra-violet regulator. In the following, it will be convenient to think at Λ as the (analog of the) inverse of a lattice spacing, lattice that would be used to regularize the field theory. In statistical mechanics, it is actually the other way around: the microscopic model is very often a lattice model whereas the field theory is an effective model only useful to describe the long-distance physics.

If Λ was sent to infinity, the integral in Eq. (2.6) would be generically divergent for d sufficiently large. This means that for Λ finite but large, the integrals are large and depend crucially on the value of Λ. This is very unpleasant for at least two reasons:

(i) This invalidates the perturbation expansion even if u_0 is small. For instance, in the ϕ^4 model and for the four-point connected correlation function $G_c^{(4)}(x_1, \dots, x_4) = \langle \phi(x_1) \dots \phi(x_4) \rangle_c$, the one-loop approximation writes in Fourier space at zero momentum:

$$G_c^{(4)} \sim u_0 + (\text{constant}).u_0^2. \int^{\Lambda} \frac{d^d q}{(2\pi)^d} \frac{1}{(q^2 + r_0)^2} + \dots \tag{2.8}$$

This integral is divergent for $d \geq 4$ in the limit $\Lambda \to \infty$.

(ii) The universal quantities (critical exponents, etc.) are expected to be independent of the underlying lattice and thus, at least for these quantities, it is paradoxical that the lattice spacing ($\sim \Lambda^{-1}$) plays such a crucial role.

Perturbative renormalization is the method that allows to reparametrize the perturbation expansion in such a way that the sensitive dependence on Λ has been eliminated.[1] Then, the renormalization group allows to partially resum the perturbation expansions [6] and thus to compute universal behaviors [1–3, 5].

Let us now make a list of questions that are not very often addressed in the literature. Some answers are explicitly given in this text. Some others are worth thinking over. They are mainly there to nourish the reader's imagination...

Q1: The occurrence of ultra-violet divergences in field theory is often considered as a fundamental property of the theory. Thus, why do they play no role in the few known exact solutions of field theories or statistical models? For instance, in Onsager's solution of the two dimensional Ising model no divergence occurs. This is also the case when Wilson's RG is implemented in field theoretical models.

[1] Let us emphasize that apart from the field renormalization, the whole renormalization process is nothing but a reparametrization [6].

Q2: Ultra-violet divergences are often said to be related to the infinite number of degrees of freedom of a field theory (the value of the field at each point). But then, why does a classical field theory that also involves an infinite number of degrees of freedom show no divergence?

Q3: The answer to the last question often relies in the literature on the fact that a statistical (or quantum) field theory involves fluctuations contrary to a classical field theory. Fluctuations are thus supposed to be responsible for the divergences. The computation of the contributions of the fluctuations—the Feynman diagrams—is thus often considered to be the reason why field theoretical techniques are relevant in statistical mechanics. But there always exist thermal fluctuations in a statistical system whereas field theoretical techniques are most of the time useless in statistical mechanics. Then, which types of fluctuations require field theory and which ones do not?

Q4: Ultra-violet divergences are also often said to be related to the fact that we multiply fields at the same point (in a lagrangian) while fields are distributions the product of which is ill-defined. But what are the distributions in the case of the Ising model? And since the interaction takes place between spins that are not on the same site but on two neighboring sites why should we take care about this difficulty?

Q5: In the ϕ^4 theory for instance, the renormalization group flow of the coupling constant—given by the β-function—is determined (in $d = 4$) by the UV divergences. But then why is the (IR stable) zero of the β-function, that is the non-gaussian fixed point, useful to describe the infrared behavior of a field theory and in particular the critical behavior?

Q6: Why should we bother about the continuum limit in a statistical system—its ultra-violet behavior—for which on one hand there always exist a natural ultra-violet cut-off (such as a lattice spacing or a typical range of interaction) and for which on the other hand we are interested only in its long-distance physics?

2.1.3 Coarse-Graining and Effective Theories

There are two crucial remarks behind Wilson's method [7–9]:

(i) in general, we cannot compute exactly the contributions of the fluctuations (otherwise we could solve exactly the model): approximations are necessary;

(ii) the way fluctuations are summed over in perturbation theory is not appropriate since fluctuations of *all wavelengths are treated on the same footing* in Feynman diagrams. This is what produces integrals: at a given order of the perturbation expansion *all* fluctuations are summed over.[2]

[2] In quantum field theory, Feynman diagrams represent the summation over probability amplitudes corresponding to all possible exchanges of virtual particles compatible with a given process at a given order. Note that these integrals are cut-off in the ultraviolet by Λ and in the infrared by the "mass" r_0 (see Eq. (2.6)). In statistical mechanics, the mass is related to the correlation length ξ by $r_0 \sim \xi^{-2}$ (at the mean-field approximation).

Wilson's idea is to organize the summation over fluctuations in a better way. Note that because of remark (i) above, "better way" means "with respect to an approximation scheme".[3] What is the idea behind Wilson's method of summation over the fluctuations? Before answering, let us notice that

- in strongly correlated systems (e.g. close to a second order phase transition) the two relevant scales are (i) the microscopic scale $a \sim \Lambda^{-1}$—a lattice spacing, an intermolecular distance, the Planck length, etc.—and (ii) the correlation length ξ (the mass(es) in the language of particle physics). These two scales are very different for $T \simeq T_c$ and fluctuations exist on *all* wavelengths between a and ξ. In particle physics, ξ corresponds to the Compton wavelength of the particle $(mc/\hbar)^{-1}$ and a to the typical (inverse) energy scale of the underlying "fundamental theory": 10^{16} GeV for a Grand Unified Theory or 10^{19} GeV for quantum gravity;
- for the long distance physics and for universal quantities (magnetization, susceptibility, etc.) the short distance "details" of the model have been completely washed out. This means that these "details" (existence and shape of the lattice for instance) do matter for the short distance physics but that they are averaged out at large distances: there must exist *average processes* that eliminate the microscopic details as the scale at which we "observe" the system is enlarged.[4]

Wilson's idea is therefore to build an *effective theory for the long-distance degrees of freedom* we are interested in [8–12]. This is achieved by integrating out the short distance ones. Since, at least for universal quantities, these short distance quantities do not matter crucially, it should be possible to devise approximations that preserve the physics at long distance.

Actually, Wilson's idea is more general: it consists in saying that the "best" (approximate) way to study a subset of degrees of freedom of a system is to build an effective theory for them by integrating out the others. For instance, in molecular physics, one should build an effective hamiltonian for the valence electrons obtained by "integrating out" the core electrons (corresponding to high energy degrees of freedom).

For the Ising model, this consists in integrating out in the partition function the "high energy modes" of the field $\phi(p)$—those for which $p \in [\Lambda - d\Lambda, \Lambda]$—and in computing the effective hamiltonian for the remaining modes. By iterating this procedure down to a scale k, one should obtain an *effective hamiltonian* for the "low energy modes", those corresponding to $p < k$. The long distance physics, obtained for $p \to 0$, should then be readable on the effective hamiltonian corresponding to $k \to 0$ since no fluctuation would remain in this limit.

Once again, let us emphasize that if we could perform exactly the integration on these "rapid" modes, we could iterate this integration and obtain the exact solution of the model. In most cases, this is impossible and the interest of this method, beyond its

---

[3] It is extremely rare that renormalization group enables to solve exactly a model that was not already solved by another and simpler method.

[4] Note that this is true only for universal quantities. The critical temperatures for instance, which are non-universal, depend on microscopic details such as the shape of the lattice. We shall come back on this notion in the following.

conceptual aspect, lies in the possibility to implement new approximation schemes better than the usual perturbation expansion.[5]

Schematically, to implement Wilson's method, we divide $\phi(p)$ into two pieces: $\phi_>(p)$ that involves the rapid modes $p \in [\Lambda/s, \Lambda]$ of $\phi(p)$ and $\phi_<(p)$ that involves the slow modes $p \in [0, \Lambda/s]$:

$$Z = \int \mathcal{D}\phi\, e^{-H[\phi,\boldsymbol{K},\Lambda]} = \int \mathcal{D}\phi_< \mathcal{D}\phi_> e^{-H[\phi_<,\phi_>,\boldsymbol{K},\Lambda]} \tag{2.9}$$

where $\boldsymbol{K} = (K_1, K_2, \dots)$ represents *all possible coupling constants* compatible with the symmetries of the system. Here, we have supposed that H involves all these couplings although the initial hamiltonian (that is, at scale Λ) can involve only a finite number of them. For instance, in the ϕ^4 model all (initial) couplings K_i are vanishing, but those corresponding to the terms $(\nabla\phi)^2$, ϕ^2 and ϕ^4. The integration of the rapid modes consists in integrating out the $\phi_>$ field. In this integration, *all the couplings K_i that were initially vanishing start to grow* (this is why we have considered them from the beginning) but since we have considered the most general hamiltonian H (compatible with the symmetries of the problem), its functional form remains unchanged. Let us call $\boldsymbol{K}'$ the new coupling constants obtained after integrating out $\phi_>$. By definition of $\boldsymbol{K}'$:

$$Z = \int \mathcal{D}\phi_<\, e^{-H[\phi_<,\boldsymbol{K}',\Lambda/s]} \tag{2.10}$$

with

$$e^{-H[\phi_<,\boldsymbol{K}',\Lambda/s]} = \int \mathcal{D}\phi_>\, e^{-H[\phi_<,\phi_>,\boldsymbol{K},\Lambda]}. \tag{2.11}$$

We thus build a series of coupling constants, each associated with a given scale:

$$\Lambda \to \boldsymbol{K}, \quad \frac{\Lambda}{s} \to \boldsymbol{K}', \quad \frac{\Lambda}{s^2} \to \boldsymbol{K}'', \qquad \text{etc.} \tag{2.12}$$

This method has several advantages compared with the usual, *à la* Feynman, approach:

- There is no longer any summation over all length scales since the integration is performed on a momentum shell, $|q| \in [\Lambda/s, \Lambda]$. Thus, there can be no divergence and there is no need for any renormalization in the usual sense (subtraction of divergences).
- The coupling constants K_i are naturally associated with a scale whereas this comes out in the perturbative scheme as a complicated by-product of regularization and renormalization. Wilson's method by-passes completely renormalization to directly deals with renormalization group.

[5] Let us already mention that if Wilson's RG equations are truncated in a perturbation expansion, all the usual perturbative results are recovered as expected.

- The method is not linked with a particular expansion and there is therefore a hope to go beyond perturbation expansion.
- The "flow" of coupling constants $\boldsymbol{K} \to \boldsymbol{K}' \to \boldsymbol{K}'' \to \ldots$ is sufficient to obtain much information on the physics of the system under study. In particular, the notion of "fixed point" of this flow will play a particularly important role in statistical mechanics.

2.1.4 Renormalization Group Transformations

Blocks of Spins

As a pedagogical introduction, let us start by a simple and illuminating example of Wilson's method implemented in x-space instead of momentum space and without having recourse to field theory [1, 2] (Fig. 2.1).

We consider a triangular lattice with Ising spins to examplify block-spin transformations:

$$H = -J \sum_{\langle ij \rangle} S_i S_j - B \sum_i S_i \tag{2.13}$$

where S_i are Ising spins: $S_i = \pm 1$, the summation $< \ldots >$ runs only on nearest neighbors and B is a uniform magnetic field. The lattice is partitioned into triangular plaquettes labelled by capital letters $I, J, \ldots$. We call S_i^I, $i = 1, 2, 3$ the spin number i of the I-th plaquette. As a first step, we separate the 8 configurations of the three spins S_i^I of plaquette I into 2×4 configurations of (i) the block-spin $\mathcal{S}_I = \pm 1$—which is chosen here to be an Ising spin—and (ii) the four configurations, called $\sigma_I^{\alpha\pm}$, corresponding to a given value of $\mathcal{S}_I$ either $+1$ or -1. We choose (and this will be modified in the following) to define $\mathcal{S}_I$ by a majority rule:

$$\mathcal{S}_I = \text{sign}\,(S_1^I + S_2^I + S_3^I) = \pm 1 \tag{2.14}$$

Thus, for the four configurations of the S_i^I compatible with $\mathcal{S}_I = +1$, we define the four variables $\sigma_I^{\alpha+}$ by:

$$\begin{aligned}
\sigma_I^{1+} &\text{ corresponds to } \uparrow\uparrow\uparrow \\
\sigma_I^{2+} &\text{ corresponds to } \uparrow\uparrow\downarrow \\
\sigma_I^{3+} &\text{ corresponds to } \uparrow\downarrow\uparrow \\
\sigma_I^{4+} &\text{ corresponds to } \downarrow\uparrow\uparrow\,.
\end{aligned} \tag{2.15}$$

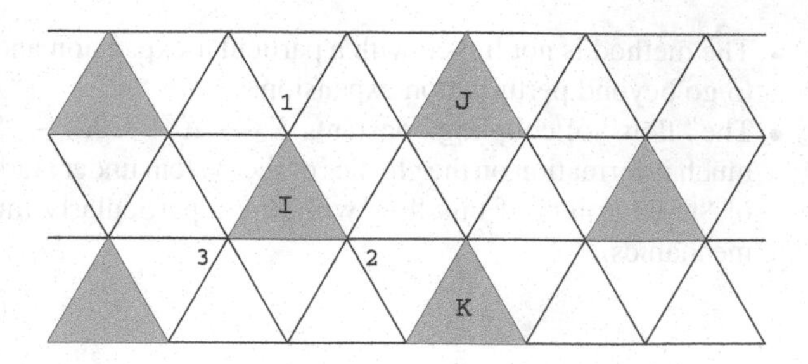

Fig. 2.1 Partition of the triangular lattice into plaquettes. The plaquettes are labelled by capital letters $I, J, K, \ldots$ and the spins inside the plaquettes are denoted, in obvious notations, S_i^I, $i = 1, 2, 3$. The lattice of plaquettes is again triangular with a lattice spacing $a\sqrt{3}$

For $\mathcal{S}_I = -1$, the $\sigma_I^{\alpha-}$ correspond to the opposite configurations of the spins. Note that it will not be necessary in the following to compute the S_i^I in terms of the $\sigma_I^{\alpha\pm}$. The sum over all spin configurations in the partition function can be written as:

$$\sum_{\{S_i\}} = \sum_{\{S_i^I\}} = \sum_{\{\mathcal{S}_I\}} \sum_{\{\sigma_I^{\alpha\pm}\}} \tag{2.16}$$

The partition function can thus be rewritten as

$$Z[B, T, n] = \sum_{\{\mathcal{S}_I\}} \sum_{\{\sigma_I^{\alpha\pm}\}} e^{-H[\mathcal{S}_I, \sigma_I^{\alpha\pm}, B, T, n, a]}. \tag{2.17}$$

where n is the number of lattice sites ($n \to \infty$ corresponds to the thermodynamical limit) and a is the lattice spacing. We have also chosen to redefine the coupling constant J and the magnetic field B so that the prefactor $1/k_B T$ of H in the Boltzmann weight is absorbed into the normalisation of these quantities. In this example, the lattice spacing of the lattice of plaquettes is $a\sqrt{3}$. The summation over the short distance degrees of freedom $\sigma_I^{\alpha\pm}$ can be formally performed

$$Z[B, T, n] = \sum_{\{\mathcal{S}_I\}} e^{-H'[\mathcal{S}_I, B, T, n/3, a\sqrt{3}]} \tag{2.18}$$

with, by definition of H':

$$e^{-H'[\mathcal{S}_I, B, T, n/3, a\sqrt{3}]} = \sum_{\{\sigma_I^{\alpha\pm}\}} e^{-H[\mathcal{S}_I, \sigma_I^{\alpha\pm}, B, T, n, a]}. \tag{2.19}$$

Let us make some remarks here.

- We have chosen the majority rule to define the block spin $\mathcal{S}_I$ so that it is again an Ising spin. The price to pay is that relation (2.14) between $\mathcal{S}_I$ and the S_I^i's is non linear. This is a source of many difficulties in more complicated systems than the

Ising model on the triangular lattice. Moreover, it is difficult to generalize this rule to continuous, N-component, spins.

- The explicit computation of H', Eq. (2.18), shows that it involves infinitely many interaction terms, even if H involves only a nearest neighbor interaction. Thus, as it stands, the form of H is not stable under block-spin transformations.

There is a solution to both problems.

- We define S_I through a linear transformation instead of the majority rule:

$$S_I \propto \sum_{i \in I} S_i^I \tag{2.20}$$

  The Ising character is lost but the relation (2.20) is simpler than (2.14) and can be generalized to other models.
- We take for H a hamiltonian involving all possible couplings among the S_i compatible with the $\mathbb{Z}_2$ symmetry ($S_i \to -S_i$ for all i) of the Ising model (they will all be generated):

$$H = -K_1 \sum_{\langle ij \rangle} S_i S_j + K_2 \sum_{\langle\langle ij \rangle\rangle} S_i S_j + K_3 \sum_{\langle ijkl \rangle} S_i S_j S_k S_l + \dots \tag{2.21}$$

  where $\langle\langle ij \rangle\rangle$ means summation over the next nearest neighbors.

Now $H = H(\boldsymbol{K}, S_i, n)$ where $\boldsymbol{K} = (K_1, K_2, \dots)$ represents the set of all $\mathbb{Z}_2$-symmetric coupling constants as well as the magnetic field if necessary. For the initial hamiltonian: $K_{i \neq 1} = 0$. Of course, this seems extremely complicated but it is the only possibility to have a form-invariant hamiltonian under block-spin transformations. The fact that all couplings are generated when fluctuations are integrated out simply means that even if an hamiltonian involves a finite number of couplings, all correlation functions, involving an arbitrary number of spins, are non-trivial. We shall come back on this point later.

Let us finally remark that for $n = \infty, n/3 = \infty, n/3^2 = \infty$, etc., and the "number of spins" remains identical. However, the lattice spacing varies: $a \to \sqrt{3}a \to 3a \to \dots$. Since, we shall look for *fixed point hamiltonians*[6] in the following, it will be necessary to rescale the lattice spacing by a factor $1/\sqrt{3}$ after each summation over the rapid modes in such a way that we obtain, after summation over the $\sigma_I^{\alpha\pm}$'s, the same hamiltonian for the same system. We shall come back on this point later.

---

[6] Fixed point hamiltonian is meant in the usual sense. If $H(\boldsymbol{K}^*, S_i)$ is a fixed point hamiltonian this means that $\boldsymbol{K}^* \to \boldsymbol{K}^*$ by summation over the $\sigma_I^{\alpha\pm}$'s, Eq. (2.19). The rescaling of the lattice spacing, which is equivalent to measuring all dimensionful quantities in terms of the *running* lattice spacing and not in terms of the (fixed) initial lattice spacing, is a necessary step to obtain fixed point hamiltonians.

We can now rewrite Eq. (2.19) for $n = \infty$ as

$$e^{-H[K',S_I,a\sqrt{3}]} = \sum_{\{\sigma_I^{\alpha\pm}\}} e^{-H[K,S_I,\sigma_I^{\alpha\pm},a]}. \tag{2.22}$$

This transformation, together with the rescaling of the lattice spacing, is called a renormalization group transformation.[7] Such a RG transformation

- preserves the partition function Z and thus its singularities and thus the critical behavior; more generally, all thermodynamical quantities are preserved[8];
- maps a hamiltonian onto another hamiltonian (a system onto another system) in such a way that they have the *same long distance physics*;
- consists in integrating out (averaging over) short distance degrees of freedom to obtain an *effective hamiltonian* for the long distance degrees of freedom;
- can be summarized in a change of (infinitely many) coupling constants: $\boldsymbol{K} \to \boldsymbol{K}'$.

And now, two questions:

Question 1: "Why is it interesting to integrate out the $\sigma_I^{\alpha\pm}$'s? Isn't it as complicated to integrate them out as would be the full integration over all degrees of freedom?"

It is true that integrating out *exactly* the $\sigma_I^{\alpha\pm}$'s is of the same order of difficulty as calculating Z completely. However

- the full calculation of Z contains much more informations than what we want to obtain to get a satisfactory description of the critical physics. Moreover, for universal quantities, we guess that we shall be able to make rather drastic approximations as for the microscopic details of the model, that is for the integration of the short distance degrees of freedom, since they probably play a minor role; this opens the possibility of new *approximation schemes*;
- the qualitative (or semi-quantitative) behavior of the RG flow of coupling constants $\boldsymbol{K} \to \boldsymbol{K}' \to \boldsymbol{K}'' \to \dots$ is enough to predict many non-trivial behaviors occurring around a second order phase transition.

Question 2: "Why should we make a series of small block-spins (coarse-graining) instead of directly a large one?"

This question, which is not independent of the first one, is a little subtle and requires some developments. Once again, if we were able to perform exactly the integration over the $\sigma_I^{\alpha\pm}$'s, small or large blocks would make no difference. Thus, the problem comes from the approximations and is therefore not fully under control before precise calculations are performed. However, the general idea is not difficult to grasp.

Let us call $\boldsymbol{T}(\,.\,, p)$ the function that maps $\boldsymbol{K} = \boldsymbol{K}^{(0)}$ onto $\boldsymbol{K}^{(p)}$ after p iterations of the RG transformations:

---

[7] We shall see in the following that this rescaling induces a rescaling of all coupling constants as well as of the magnitude of the spin, the so-called field renormalization.

[8] If we wanted to compute correlation functions of the original spins, we would have first to couple the system to an arbitrary magnetic field (in order to be able to compute derivatives of Z with respect to the magnetic field B_i). This is a complicated task.

$$\boldsymbol{K}^{(p)} = \boldsymbol{T}\left(\boldsymbol{K}^{(0)}, p\right) \tag{2.23}$$

We, of course, have the property

$$\boldsymbol{K}^{(p)} = \boldsymbol{T}\left(\boldsymbol{K}^{(r)}, p-r\right) = \boldsymbol{T}\left(\boldsymbol{T}\left(\boldsymbol{K}^{(0)}, r\right), p-r\right) \tag{2.24}$$

and thus

$$\boldsymbol{T}\,(\,.\,, p) = \boldsymbol{T}\Big(\boldsymbol{T}\,(\,.\,, r)\,, p-r\Big). \tag{2.25}$$

This is called a *self-similarity* [6, 13] property.[9] If $\boldsymbol{T}$ was known exactly, this property would be trivially satisfied. However, once approximations are performed, it is generically violated as is the case for instance in most perturbative expansions.

Let us illustrate the concept of self-similarity on the simple example of differential equations [6]. We consider the trivial differential equation:

$$\dot{y} = -\epsilon y \tag{2.26}$$

with $y(t_0) = y_0$. The solution is

$$y = f(t - t_0, y_0) = y_0 e^{-\epsilon(t-t_0)}. \tag{2.27}$$

Of course, f satisfies a self-similarity property which means that we can either (i) first integrate (2.26) between t_0 and τ to obtain $y(\tau) = y_\tau$ and then integrate again (2.26) between τ and t with y_τ as new initial condition or (ii) directly integrate (2.26) between t_0 and t:

$$y(t) = f(t - t_0, y_0) = f\big(t - \tau, f(\tau - t_0, y_0)\big). \tag{2.28}$$

This is trivially satisfied by the exact solution (2.27) since

$$f(\alpha, \beta) = \beta e^{-\epsilon\alpha} = \beta e^{-\epsilon a} e^{-\epsilon(\alpha - a)} = f\big(\alpha - a, f(a, \beta)\big) \tag{2.29}$$

However, this property is violated *at any finite order* of the perturbation expansion in ϵ of $y(t)$. Let us show this at first order in ϵ for which we, of course, obtain:

$$y(t) = y_0(1 - \epsilon(t - t_0)) + O(\epsilon^2). \tag{2.30}$$

---

9 Something is said to be self-similar if it looks everywhere the same. In our case, the self-similar character comes from the fact that the functional form of the RG flow does not depend on the initial couplings $\boldsymbol{K}^{(0)}$ since the same function $\boldsymbol{T}$ is used to transform $\boldsymbol{K}^{(0)}$ into $\boldsymbol{K}^{(p)}$ or $\boldsymbol{K}^{(r)}$ into $\boldsymbol{K}^{(p)}$. This results in the fact that the right hand side of Eq. (2.25) is independent of r since the left hand side is. This independence is completely similar to the independence of the bare theory on the renormalization scale in perturbative renormalization (or of the renormalized theory on the bare scale). This is what allows to derive the Callan-Symanzik RG equations in the perturbative context.

This defines the approximation of order one of f:

$$f^{(1)}(t-t_0, y_0) = y_0(1-\epsilon(t-t_0)). \qquad (2.31)$$

We obtain at this order:

$$f^{(1)}\big(\alpha-a, f^{(1)}(a,\beta)\big) = \beta(1-\epsilon a)(1-\epsilon(\alpha-a)) = f^{(1)}(\alpha,\beta)+\epsilon^2\beta a(\alpha-a). \qquad (2.32)$$

By comparing this result with Eq. (2.29), we find that self-similarity is satisfied at order ϵ, as expected, but is violated at order ϵ^2. The problem is that this violation can be arbitrarily large if α (which represents $t-t_0$) is large. Thus, even if ϵ is small, the self-similarity property is violated for large time intervals. This is true at *any finite order* of perturbation theory. This large violation comes ultimately from the fact that the perturbation expansion is *not an expansion in* ϵ *but in* $\epsilon(t-t_0)$. This is completely reminiscent of the perturbation expansion in field theory where the expansion is not performed in terms of u_0 but in terms of $u_0 \log \Lambda$ where u_0 is the bare coupling constant and Λ the cut-off (see Eq. (2.8) for the ϕ^4 theory in $d=4$). Reciprocally, it is clear that if $\alpha = t-t_0$ is small, so is the violation in Eq. (2.32) since, in this case, both a and $\alpha-a$ are small. Thus, using perturbation expansions on small or, even better, on infinitesimal time intervals preserves self-similarity at each step. In geometrical terms, this means that we can safely use perturbation theory to compute the envelope of the curve $f(\alpha,\beta)$—the field of tangent vectors given by the so-called β-function—but not the curve itself [6]. The curve $f(\alpha,\beta)$ can only be obtained in a second step by integration of its envelope.

The analogue for the RG is that small blocks will be under control. Coarse graining in this case respects self-similarity *even when approximations are used* while large ones lead inevitably to large errors as soon as approximations are used.

Before studying the structure of the RG flow, let us make two remarks about the RG transformations and their physical meaning.

Two Remarks Concerning RG Transformations

The first remark is that it is still widely believed that the correlation length $\xi(T)$ is a measure of the typical size of clusters of spins having the same orientation, that is of ordered domains (in the Ising case). As a consequence, it is believed that the divergence of ξ at T_c is a consequence of the divergence of the size of these (so-called) naive clusters. The traditional metaphor is that at T_c there would exist oceans of up spins with continents of down spins that would contain themselves lakes of up spins with islands of down spins, etc., with some kind of fractal geometry. This is wrong. It has been shown long ago that the distribution of cluster boundaries *does not scale at criticality*. Rather, at a temperature T_p well below T_c the clusters of spins having the opposite sign of the spontaneous magnetization merge into a large percolating cluster. An important point is that, strictly speaking, no phase transition occurs at T_p since no local order parameter of the Ising model can be built out the spins in order to describe this transition: there is no singularity of the partition function at T_p. In fact, it is possible to construct clusters of spins that are critical at T_c. These are the famous Fortuin and Kasteleyn clusters [14]. They are used in the Swendsen-Wang algorithm of Monte Carlo simulations of the Ising model since they partially defeat critical slowing down [15].

The second remark is that the "microscope analogy" is often used to give an intuition of the physical meaning of the RG transformations. In this analogy, the coarse-graining implemented in the RG transformations would be similar to what occurs when the magnification of a microscope is decreased. Let us imagine that we look at image made of small pixels of definite colours (say blue, green or red). At a mesoscopic scale, the pixels are no longer seen and only a smearing of the colors of blocks of pixels can be observed. As a result, the "physics" observed would depend on the scale as in the RG transformations. This analogy has several virtues but also several drawbacks. Let us mention some. First, our brain plays a crucial role for the color vision. From the three colors blue, green and red the cones in the retina are sensitive to, our brain is smart enough to reconstruct the impression of a continuous spectrum of colors. Although the analogy leads us to believe that our perception of the colors at a mesoscopic scale is a linear combination at this scale of the elementary colors of the pixels, this is not so. Second, in a RG transformation, there are two main steps (not to mention the final change of scale to go back to the original lattice spacing). The first one is to build a stochastic variable for the block.[10] The second is to build an effective hamiltonian for this block variable by integration over short distance fluctuations. We can imagine that the first step is analogous to the superimposition of the electromagnetic field produced by the different pixels. But then what is the analog of the second step? The laws of classical electrodynamics for the propagation of light do not change from one scale to the other. Let us repeat here that the effective hamiltonians for the block variables in the Ising model are extremely complicated: they involve all powers of the fields and not only interactions among nearest neighbors.[11] There is no analog for this step in the microscope analogy although it is the crucial one from the RG point of view. In fact, things go almost the other way around. Whereas the electromagnetic field emitted by several pixels is the linear superposition of the field produced by each of them, the β-function in quantum field theory that gives the evolution of the coupling constant with the scale is a measure of the deviation to the trivial rescaling invariance (in the case of quantum electrodynamics). Thus, although the microscope analogy can be useful it should be employed with some care (and a grain of salt).

[10] This can be performed either by a majority rule as in Eq. (2.14) or by a linear relation as in Eq. (2.34).

[11] Once the continuum limit has been taken and continuous RG transformations are implemented this means that the effective hamiltonians involve all powers of the field and of its derivatives.

Let us now show how linear RG transformations can be implemented. This will allow us to prove a simple relation about the behavior under RG transformations of the two-point correlation function $\langle S_i S_j \rangle$ and thus on the correlation length [1, 9].

Linear RG Transformations and Behavior of the Correlation Length

Instead of the majority rule, we consider a linear transformation between the spins of a plaquette and the block-spin [1]. The simplest idea is to take a *spatial* average (not a thermodynamic one)

$$S'_I = \frac{1}{s^d} \sum_{i \in I} S_i^I \tag{2.33}$$

where s is the "linear size" of the block, that is s^d is the number of spins per block. In our example of the triangular lattice, $d = 2$ and $s = \sqrt{3}$. As we already said, we shall also need to perform a rescaling of all dimensionful quantities (in order to find fixed points). Thus we take:

$$\mathcal{S}_I = \frac{\lambda(s)}{s^d} \sum_{i \in I} S_i^I \tag{2.34}$$

where $\lambda(s)$ is a function that will be determined in such a way that we find a fixed point. This relation among the stochastic variables S_i^I and $\mathcal{S}_I$ leads to relations among their thermodynamic averages. The most important one is the two-point correlation function:

$$\begin{aligned} \langle \mathcal{S}_I \mathcal{S}_J \rangle &= \frac{1}{Z} \sum_{\{\mathcal{S}_I\}} \mathcal{S}_I \mathcal{S}_J e^{-H[K', \mathcal{S}_L]} = \frac{1}{Z} \sum_{\{\mathcal{S}_I\}} \mathcal{S}_I \mathcal{S}_J \sum_{\{\sigma_L^\alpha\}} e^{-H[K, \mathcal{S}_L, \sigma_L^\alpha]} \\ &= \frac{\lambda^2(s)}{s^{2d}} \frac{1}{Z} \sum_{\{\mathcal{S}_I\}} \sum_{\{\sigma_I^\alpha\}} \sum_{i \in I, j \in J} S_i^I S_j^J e^{-H[K, S_i]} = \frac{\lambda^2(s)}{s^{2d}} \sum_{i \in I, j \in J} G^{(2)}(\boldsymbol{x}_i, \boldsymbol{x}_j) \end{aligned} \tag{2.35}$$

is the two-point correlation function of the spins S_i. If the correlation length is large compared with the size of the plaquettes and if we consider two plaquettes I and J such that their distance is very large compared to a: $r_{ij} = |\boldsymbol{x}_{i \in I} - \boldsymbol{x}_{j \in J}| \gg a$, then $G^{(2)}(\boldsymbol{x}_i, \boldsymbol{x}_j)$ does not vary much for $i \in I$ and $j \in J$. Thus, in this case:

$$\sum_{i \in I, j \in J} G^{(2)}(\boldsymbol{x}_i, \boldsymbol{x}_j) \simeq s^{2d} G^{(2)}(\boldsymbol{x}_i, \boldsymbol{x}_j). \tag{2.36}$$

Therefore, close to the critical temperature and for distant plaquettes:

$$\langle \mathcal{S}_I \mathcal{S}_J \rangle \simeq \lambda^2(s) \langle S_i^I S_j^J \rangle. \tag{2.37}$$

The important point is that $\langle \mathcal{S}_I \mathcal{S}_J \rangle$ is also the two-point correlation function of a $\mathbb{Z}_2$-invariant magnetic system. The only difference with $\langle S_i S_j \rangle$ is that it is computed with the set of couplings $\boldsymbol{K}'$ instead of $\boldsymbol{K}$. We thus obtain:

$$\langle \mathcal{S}_I \mathcal{S}_J \rangle = G^{(2)}(r_{IJ}, \boldsymbol{K}') \tag{2.38}$$

where $G^{(2)}$ is the same function as in Eq. (2.35) and r_{IJ} is the distance between the plaquettes I and J.[12] We thus deduce:

$$G^{(2)}(r_{IJ}, \boldsymbol{K}') \simeq \lambda^2(s)\, G^{(2)}(r_{ij}, \boldsymbol{K}) \tag{2.39}$$

for sufficiently distant plaquettes.

Let us now explain the precise meaning of this relation. Let us suppose that we are given a new model on a triangular lattice with a set of couplings $\boldsymbol{K}'$. In principle, we can compute the correlation function $G^{(2)}$ of the spins of this new system. Our claim is that this correlation function is identical to the correlation function of the block-spins $\mathcal{S}_I$ of the original system. Of course, to compare the two functions we have to say how to compare the distances r_{ij} between spins in the two lattices. Our calculation shows that what we have to do is to measure all distances in the length unit intrinsic to the system, that is in units of the lattice spacing of each system. This means that the quantities r_{ij} and r_{IJ} that appear in Eq. (2.39) are pure numbers that must be *numerically different*

$$r_{IJ} = r_{ij}/s \tag{2.40}$$

since they correspond to the same "distance" but measured in two different units: a for the original system and $a' = sa = a\sqrt{3}$ for the coarse-grained system. It is important to understand that measured in an extrinsic length unit, like meters, these distances are indeed the same: $r_{IJ}a' = (r_{ij}/s).(sa) = r_{ij}a$ whereas they are "different"—they correspond to different numbers—when they are measured in the length unit intrinsic to each system. Put it differently, the dimensionful distances $r_{IJ}a'$ and $r_{ij}a$, measured in a common length unit, are equal whereas the dimensionless distances r_{IJ} and r_{ij}, measured in terms of the lattice spacing of each system, are different. We shall put a bar or a tilda on dimensionless quantities to distinguish them.

Let us emphasize that the value in an extrinsic unit like meters of, say, the correlation length is almost meaningless. From a physical point of view, the only relevant measure of the correlation length is in units of the lattice spacing. The difficulty in our case is two-fold. First, to write down a field theory, it is necessary to perform the continuum limit $a \to 0$. It is therefore convenient to rescale the position vectors by a factor a before performing this limit (thus, as usual, $[x]$ = length). This is consistent with the fact that the vectors $\boldsymbol{x}_i$ of the lattice have integer components

[12] Let us point out here a subtlety. This statement is not fully rigorous since the original spins are Ising spins whereas the block spins $\mathcal{S}_I$ are not. The correlation functions $\langle S_i^I S_j^J \rangle$ and $\langle \mathcal{S}_I \mathcal{S}_J \rangle$ are therefore not computed exactly in the same way since the summation over the configurations of S_i and of $\mathcal{S}_I$ do not run on the same values. In fact, after several blocking iterations, the spins that are summed over become almost continuous variables and the aforementioned difficulty disappears.

that label the position of the sites of the lattice and are therefore also dimensionless whereas, in field theory, the modulus of $\boldsymbol{x}$ represents the distance to the origin.[13] Second, since we have to consider several systems with different lattice spacings, it will be convenient to work in the continuum with lengths measured in units of the *running lattice spacings*.

Let us therefore define dimensionless quantities as

$$\bar{r} = \frac{r}{a}, \quad \bar{\xi} = \frac{\xi}{a}, \quad \bar{r}' = \frac{r}{sa}. \tag{2.41}$$

where r and ξ are the dimensionful quantities (measured in meters) that will be convenient once the continuum limit will be taken, that is in the field theory formalism. Equation (2.39) that involves only dimensionless quantities can then be rewritten:

$$G^{(2)}\left(\frac{\bar{r}}{s}, \boldsymbol{K}'\right) \simeq \lambda^2(s) G^{(2)}(\bar{r}, \boldsymbol{K}). \tag{2.42}$$

Let us give a concrete example of the meaning of this relation. In three dimensions and at large distances: $r_{ij} \gg 1$, a typical form of the two-point correlation function is:

$$\langle S_i S_j \rangle = G^{(2)}(\bar{r}, \bar{\xi}) \sim \frac{e^{-\bar{r}/\bar{\xi}}}{\bar{r}^{\theta}} \tag{2.43}$$

with $\bar{\xi}$ the correlation length in units of the lattice spacing a and $\bar{r} = \bar{r}_{ij}$. We can use the same formula as in Eq. (2.43) for the correlation function of the block-spin system:

$$\langle S_I S_J \rangle \sim \frac{e^{-\bar{r}'/\bar{\xi}'}}{\bar{r}'^{\theta}}. \tag{2.44}$$

We thus obtain

$$G^{(2)}(\bar{r}', \bar{\xi}') \sim \frac{e^{-\frac{\bar{r}/s}{\bar{\xi}'}}}{\bar{r}^{\theta}} s^{\theta} \tag{2.45}$$

and, by comparing with Eq. (2.39), we find that

$$\bar{\xi}' = \frac{\bar{\xi}}{s} \text{ and } \lambda(s) = s^{\theta/2}. \tag{2.46}$$

---

[13] It will also be convenient to rescale the spin-field by the appropriate power of a: $S_i \to \phi(\boldsymbol{x})$ so that the gradient term $(\nabla\phi)^2$ comes in the hamiltonian of the field theory with a dimensionless pre-factor (chosen to be 1/2 for convenience): $H = \int d^d x \left(\frac{1}{2}(\nabla\phi)^2 + U(\phi)\right)$. We find from this equation that $[\phi(x)] = [x^{-\frac{d-2}{2}}]$ so that the rescaling involves a factor $a^{-\frac{d-2}{2}}$. Note that the original variables S_i are dimensionless since $S_i = \pm 1$. The function $G^{(2)}$ in Eq. (2.39) is therefore also dimensionless.

By comparing all these relations we find that

- the *dimensionful* correlation lengths of the original and of the block-spin systems are identical: it is a RG-invariant. Thus, the dimensionless correlation lengths decrease as the coarse-graining scale s increases. This means that the coarse-grained systems are less correlated than the initial one and that the correlation length decreases linearly with the scale. Since the correlation length behaves as a power law close to the critical temperature

$$\xi \sim (T - T_c)^{-\nu} , \tag{2.47}$$

parametrizing $G^{(2)}$ in terms of the (reduced) temperature

$$t = \frac{T - T_c}{T_c} \tag{2.48}$$

or in terms of the correlation length is "equivalent". Thus, saying that $\bar{\xi}$ decreases with the block size s is equivalent to saying that the running reduced temperature $t(s)$ increases with s: the coarse-grained system is "less critical" than the original one. We call relevant a parameter that increases with the scale s.
- the reduced temperature t is one particular coupling among all the couplings $\boldsymbol{K}$. We shall explain in the following why the form of $G^{(2)}$ given in Eq. (2.43) and in which only the correlation length appears is valid at large distance that is why all other couplings in $\boldsymbol{K}$ play no role at large distances.
- if we combine two RG transformations of scale s_1 and s_2 we must obtain the same result as a unique transformation of scale $s_1 s_2$ (this is self-similarity). This clearly implies that

$$\lambda(s_1)\lambda(s_2) = \lambda(s_1 s_2). \tag{2.49}$$

It is straightforward to show that the only solution of this equation is a power law. The example above shows that the exponent of this power law is directly related to the power law behavior of $G^{(2)}(\bar{r})$ at $T = T_c$, Eq. (2.46).

2.1.5 Properties of the RG Flow: Fixed Points, Critical Surface, Relevant Directions

The RG flow takes place in the space of hamiltonians, that is in the space of coupling constants $\boldsymbol{K}$. We now study this flow. One of its nice properties is that, without specifying any particular statistical system, very general informations on second order transitions can be obtained from it by only making very natural assumptions.

At T_c, $\xi = \infty$ ($\bar{\xi} = \infty$) at a second order phase transition. Thus the point $\boldsymbol{K}^{(0)}$ is mapped onto $\boldsymbol{K}^{(1)}$ under a RG transformation for which $\bar{\xi}' = \infty$ again (the block-spin system is also critical). We define the *critical surface* as the set of points $\boldsymbol{K}$ in the coupling constant space for which $\bar{\xi} = \infty$. For a second order phase transition only

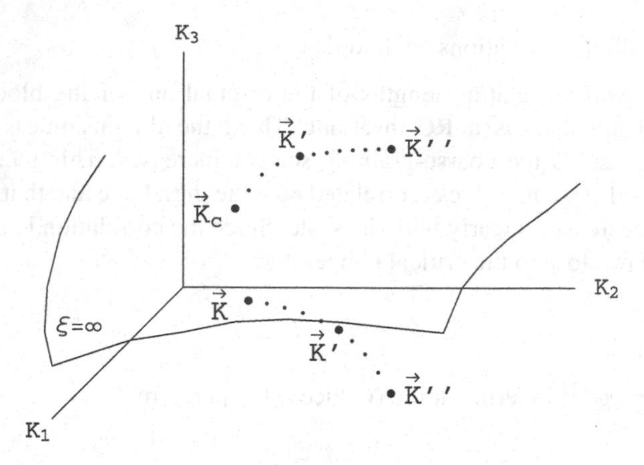

Fig. 2.2 Schematic representation of the RG flow in the space of couplings, $K_1, K_2, \ldots$. This space is infinite dimensional and the critical surface, defined by the set of points for which the correlation length is infinite is of co-dimension one. Under RG transformations $\boldsymbol{K} \to \boldsymbol{K}' \to \boldsymbol{K}'' \to \ldots$, the critical surface is stable: if $\boldsymbol{K} = \boldsymbol{K}_c$ is on the critical surface, then $\boldsymbol{K}'$, $\boldsymbol{K}''$, etc. are also on the critical surface. A point away from the critical surface is mapped onto another one "further away" from it

one parameter needs to be fine-tuned to make the system critical (the temperature for instance). Thus, the critical surface is of co-dimension one. It is stable under RG transformations (Fig. 2.2).

If we now consider a system described by a point $\boldsymbol{K}^{(0)}$ such that $\xi < \infty$ then $\xi' = \xi/s$ and the block-spin system, being "less critical", is described by $\boldsymbol{K}^{(1)}$ which is "further away" from the critical surface than $\boldsymbol{K}^{(0)}$. If we iterate the blocking process, we obtain points $\boldsymbol{K}^{(2)}, \boldsymbol{K}^{(3)}, \ldots$ that will be further and further away from the critical surface.

We shall consider in the following the continuum limit of the Ising model and this will allow us to perform continuous RG transformations. We call in this case the set of points $\boldsymbol{K}_s$, $s \in \mathbb{R}$, a RG trajectory. To different $\boldsymbol{K}_s$ on the same RG trajectory correspond systems that are microscopically different (this means at the scale of their own lattice spacing) but that lead to the *same long-distance physics* since they all have the same partition function. Let us now make the fundamental hypothesis that must be checked on each example [9]:

Hypothesis: *For points in a (finite or infinite) domain on the critical surface, the RG flow converges to a fixed point* $\boldsymbol{K}^*$: $\boldsymbol{K}^* = \boldsymbol{T}(\boldsymbol{K}^*, s)$.

This domain is called the basin of attraction of the fixed point $\boldsymbol{K}^*$. Under this hypothesis, the typical topology of the flow on the critical surface is summarized in Fig. 2.3. We have called "physical line" in this figure the line on which the temperature alone is varied. It is *not* a RG trajectory. Reciprocally, a RG trajectory does not, in general, correspond to any transformation doable on a physical system by a human being. It is only a mapping that preserves the partition function without any connection to a physical transformation.

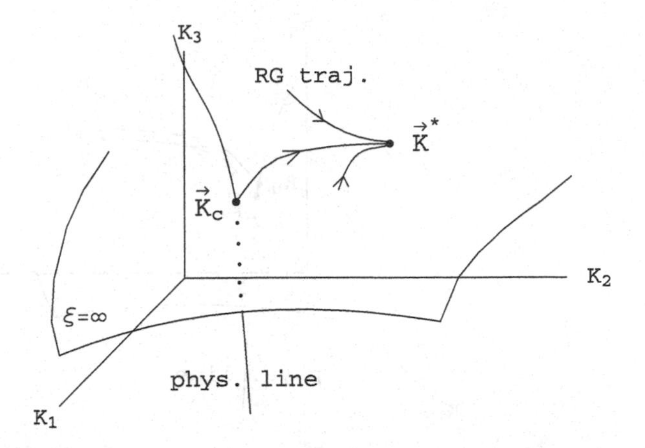

Fig. 2.3 Schematic representation of continuous RG trajectories on the critical surface. The flow converges to the fixed point $\boldsymbol{K}^*$. For a given model, the "physical line" corresponds to a change of the temperature. It is not a RG trajectory. For a more precise description of the RG flow, see Fig. 2.18

All systems that belong to the basin of attraction of $\boldsymbol{K}^*$ belong to the same universality class since they all have the same long-distance physics. Note that $\boldsymbol{K}^*$ depends on the choice of RG transformations $\boldsymbol{T}(.,s)$. Apart from being fixed for this particular choice of RG transformations, this point is nothing else than a particular critical point.

Scaling Relations: Linearization of the Flow Around the Fixed Point

The existence of an attractive (in the critical surface) fixed point is sufficient to explain universality since, independently of the starting point $\boldsymbol{K}_c$ on the critical surface, all RG trajectories end at the same point $\boldsymbol{K}^*$. However, universality holds also for systems that are not right at T_c (which is anyway impossible to reach experimentally) but close to T_c. It is thus natural to assume that the flow is continuous in the vicinity of $\boldsymbol{K}^*$. In this case, starting at a point $\boldsymbol{K}$ close to the point $\boldsymbol{K}_c$ and on the same physical line, the RG trajectory emanating from $\boldsymbol{K}$ remains close to the one emanating from $\boldsymbol{K}_c$ during many RG steps before it diverges from the critical surface, see Fig. 2.4. It is easy to estimate the typical value of s for which the RG trajectory diverges from the critical surface.

As long as the running (dimensionless) correlation length $\xi(s) = \bar{\xi}/s$ remains large, the system behaves as if it was critical and the representative point $\boldsymbol{K}(s)$ must be close to the critical surface. When the running correlation length becomes of order 1, the coarse grained system is no longer critical and $\boldsymbol{K}(s)$ must be at a distance of order 1 of the critical surface. More precisely, at the beginning of the flow, $\boldsymbol{K}(s)$ moves towards $\boldsymbol{K}^*$ (by continuity). It remains close to it as long as $\bar{\xi}(s)$ remains large so that the memory of the initial point $\boldsymbol{K}$ is largely lost. Finally, it departs from $\boldsymbol{K}^*$ when $s \sim \bar{\xi}$. Another and more precise way to state the same result is to say that the running reduced temperature $t(s)$ is of order 1 when $s \sim \bar{\xi}$:

$$t(s) \sim 1 \quad \text{for} \quad s \sim \bar{\xi}. \tag{2.50}$$

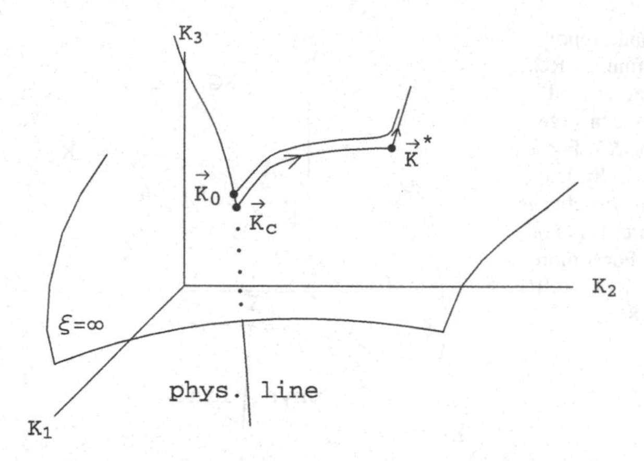

Fig. 2.4 Schematic representation of two continuous RG trajectories corresponding to the same model for two different temperatures. The trajectory starting at $\boldsymbol{K}_c$ is on the critical surface ($T = T_c$) and the other, starting at $\boldsymbol{K}_0$, is slightly away from it. There exists a RG trajectory emanating from the fixed point $\boldsymbol{K}^*$ and which is not on the critical surface. It is an eigendirection of the RG flow corresponding to the relevant direction. Note that we could parametrize the coupling constant space in such a way that the axis denoted K_3 represents the temperature. It is not necessary that this axis coincides with the relevant eigendirection of the RG flow at $\boldsymbol{K}^*$ but it is necessary that it has a non vanishing projection onto this axis since the temperature is for sure a relevant parameter. Note that this is non trivial in an infinite dimensional space

We shall see in the following that the hypothesis of the existence of a fixed point together with this relation are sufficient to predict the existence of power law behaviors for many thermodynamical quantities with critical exponents that are universal and that satisfy "scaling relations" *independently of any specific microscopic model* [9]. Clearly, to obtain these relations, only the vicinity of $\boldsymbol{K}^*$ is important since after a few RG steps, all RG trajectories emanating from points close to criticality are in the vicinity of this point, see Fig. 2.4. This will allow us to linearize the RG flow around $\boldsymbol{K}^*$.

For the sake of simplicity, we assume in the following that s can take continuous values (we work in the continuum, that is the continuous limit, $a \to 0$, has been taken). We also suppose that the RG transformation with $s = 1$ is the identity (no block of spins) and that the composition law is such that a RG transformation of parameter s_1 followed by another one of parameter s_2 is equivalent to a transformation of parameter $s_1 s_2$.

As already emphasized, to obtain approximations of the RG flow that are under control, it is preferable to perform a series of infinitesimal RG transformations rather than directly a transformation with s large. It is thus necessary to study the differential form of these transformations. There are two ways to do so:

(i) The first way consists in comparing two RG transformations of parameters s and $s + \epsilon$ performed from an initial set of couplings $\boldsymbol{K}_{\text{in}}$:

$$\boldsymbol{K}_{s+\epsilon} - \boldsymbol{K}_s = \boldsymbol{T}(\boldsymbol{K}_{\text{in}}, s+\epsilon) - \boldsymbol{T}(\boldsymbol{K}_{\text{in}}, s) \tag{2.51}$$

and thus

$$\frac{\partial \boldsymbol{K}_s}{\partial s} = \frac{\partial \boldsymbol{T}}{\partial s}|_{(\boldsymbol{K}_{\text{in}},s)}. \tag{2.52}$$

This formula is not convenient since it is expressed in terms of the initial couplings $\boldsymbol{K}_{\text{in}}$ and, as a consequence, in terms of a parameter s that can be large (and thus for RG transformations that will not be under control once approximations will be implemented). Using approximate expressions of $\boldsymbol{T}$ together with this expression would lead to violations of the composition law of RG transformations.

(ii) The second way consists in comparing two sets of running couplings differing by an infinitesimal RG transformation: $\boldsymbol{K}_s$ and $\boldsymbol{K}_{s(1+\epsilon)}$ and in *expressing the result as a function of* $\boldsymbol{K}_s$:

$$\boldsymbol{K}_{s(1+\epsilon)} - \boldsymbol{K}_s = \boldsymbol{T}(\boldsymbol{K}_s, 1+\epsilon) - \boldsymbol{T}(\boldsymbol{K}_s, 1) \tag{2.53}$$

and thus

$$s\,\frac{\partial \boldsymbol{K}_s}{\partial s} = \frac{\partial \boldsymbol{T}}{\partial s}|_{(\boldsymbol{K}_s,1)} \tag{2.54}$$

In this second formulation, the evolution at one step—that is between s and $s+\mathrm{d}s$—depends only on the couplings at this step—that is at s—and not on those at the previous steps: the evolution of the couplings is local in the space of couplings.

It is important to notice here several things.

- The composition law of RG transformations will be automatically satisfied if $\boldsymbol{K}(s)$ is obtained by integration of this expression, *even if an approximate expression of* $\boldsymbol{T}$ *is used.* This comes from the fact that, by construction, $\boldsymbol{K}(s')$ is computed in terms of $\boldsymbol{K}(s)$ by composing the infinitesimal group law between s and s': this is the meaning of integrating Eq. (2.54).
- Equation (2.54) leads naturally to the logarithmic evolution of the couplings with s, contrarily to Eq. (2.52).
- In Eq. (2.52), the variations of $\boldsymbol{T}$ had to be known for only one set of couplings $\boldsymbol{K}_{\text{in}}$ but for all s. This was the real problem of this approach since we are interested in large values of s. In Eq. (2.54) only the variation of $\boldsymbol{T}$ for $s = 1$ have to be known. The price to pay is that this must be known for all values of $\boldsymbol{K}_s$. In perturbation theory this is not problematic as long as the running coupling used in the perturbation expansion remains small *all along the flow.* Of course, if $\boldsymbol{K}_s$ converges to $\boldsymbol{K}^*$ such that the coupling(s) of the perturbation expansion at the fixed point is large, perturbation theory runs into trouble.

We define

$$\beta(\boldsymbol{K}_s) = \frac{\partial \boldsymbol{T}}{\partial s}|_{(\boldsymbol{K}_s,1)} \tag{2.55}$$

which gives the evolution of the couplings of the model with the scale. Note that $\beta(\boldsymbol{K}_s)$ does not depend explicitly on s since the right hand side of Eq. (2.55) is evaluated at $s = 1$ and thus does not depend on this variable.[14] The only dependence on s of this function is implicit and comes through the dependence of the couplings $\boldsymbol{K}_s$ on s. By definition of the fixed point $\boldsymbol{K}^*$

$$\beta\left(\boldsymbol{K}^*\right) = \boldsymbol{0}. \tag{2.56}$$

We thus obtain

$$\begin{aligned} s\,\frac{\mathrm{d}\boldsymbol{K}_s}{\mathrm{d}s} - s\,\frac{\mathrm{d}\boldsymbol{K}^*}{\mathrm{d}s} &= \beta(\boldsymbol{K}_s) - \beta\left(\boldsymbol{K}^*\right) \\ &= \frac{\mathrm{d}\beta}{\mathrm{d}\boldsymbol{K}_s}\Big|_{\boldsymbol{K}^*} . \,\delta\boldsymbol{K}_s + O\left(\delta\boldsymbol{K}_s^2\right) \end{aligned} \tag{2.57}$$

where, by definition

$$\delta\boldsymbol{K}_s = \boldsymbol{K}_s - \boldsymbol{K}^* \tag{2.58}$$

and

$$\frac{\mathrm{d}\beta}{\mathrm{d}\boldsymbol{K}_s}\Big|_{\boldsymbol{K}^*} \quad \text{is a matrix} \quad \mathcal{M}_{ij} = \frac{\mathrm{d}\beta_i}{\mathrm{d}K_{s,j}}\Big|_{\boldsymbol{K}^*}. \tag{2.59}$$

Thus, in the neighborhood of $\boldsymbol{K}^*$:

$$s\,\frac{\mathrm{d}\delta\boldsymbol{K}_s}{\mathrm{d}s} = \mathcal{M}\,\delta\boldsymbol{K}_s. \tag{2.60}$$

$\mathcal{M}$ is not symmetric in general. We suppose in the following that it can be diagonalized and that its eigenvalues are real[15] (this must be checked on each example). We moreover suppose that the set of eigenvectors $\{\boldsymbol{e}_i\}$ is a complete basis:

$$\mathcal{M}\boldsymbol{e}_i = \lambda_i\,\boldsymbol{e}_i \qquad \text{with} \qquad \lambda_i \in \mathbb{R} \tag{2.61}$$

and

$$\delta\boldsymbol{K}_s = \sum_i v_i(s)\,\boldsymbol{e}_i. \tag{2.62}$$

Under these hypotheses, we obtain:

$$s\,\frac{\mathrm{d}\delta\boldsymbol{K}_s}{\mathrm{d}s} = \sum_i v_i(s)\,\mathcal{M}\,\boldsymbol{e}_i = \sum_i \lambda_i\,v_i(s)\,\boldsymbol{e}_i \tag{2.63}$$

---

[14] In perturbation theory, the fact that the β function of the marginal coupling is cut-off independent and thus scale independent is a consequence of the perturbative renormalizability of the model.

[15] It can happen that some eigenvalues are complex. In this case, the RG flow around the fixed point is spiral-like (focus) in the corresponding eigendirections.

and thus

$$s\,\frac{dv_i(s)}{ds} = \lambda_i\, v_i(s) \quad \Rightarrow \quad v_i(s) = v_i(1)\, s^{\lambda_i}. \tag{2.64}$$

We conclude that around the fixed point, the RG flow behaves as power laws along its eigendirections (but if $\lambda_i = 0$ in which case it is logarithmic). There are three possibilities:

- $\lambda_i > 0$ and $v_i(s) \nearrow$ when $s \nearrow$ which means that the flow in the direction $\boldsymbol{e}_i$ goes away from $\boldsymbol{K}^*$. $\boldsymbol{e}_i$ is called a relevant direction and v_i a relevant coupling. As we have already seen, the reduced temperature is relevant since the system is less and less critical along the RG flow which means that $t(s)$ increases with s.
- $\lambda_i < 0$ and $v_i(s) \searrow$ when $s \nearrow$ which means that the flow in the direction $\boldsymbol{e}_i$ approaches $\boldsymbol{K}^*$. $\boldsymbol{e}_i$ is called an irrelevant direction and v_i an irrelevant coupling.
- $\lambda_i = 0$ and v_i is said to be marginal. It is necessary to go beyond the linear approximation to know whether it is relevant or irrelevant. The flow in this direction is slow: it is logarithmic instead of being a power law. This is important since it is the case of the renormalizable couplings in the critical dimension (that is $d = 4$ for the Ising model).

Physically, we expect a small number of relevant directions since, clearly, there are as many such directions as there are parameters to be fine-tuned to be on the critical surface (the co-dimension of this surface). For second order phase transitions there is one coupling to be fine-tuned to make the system critical: the temperature (for instance). There is actually one more parameter which is relevant but which is also $\mathbb{Z}_2$ non-invariant: the magnetic field. All the other directions of the RG flow are supposed to be irrelevant (this can be checked explicitly once a specific model is given).

In the following, we shall use the language of magnetic systems although our discussion will be completely general. We shall suppose that together with the magnetic field, there is only one other relevant direction. We shall show that this implies scaling laws for all thermodynamical quantities with only two independent critical exponents.

The Correlation Length and the Spin-Spin Correlation Function

Let us start by studying the critical physics of a model at zero external magnetic field having only one relevant coupling. We call v_1 this coupling and λ_1 the eigenvalue of the flow at $\boldsymbol{K}^*$ associated with v_1. We order the other eigen-couplings $v_2, v_3, \ldots$ in such a way that $0 > \lambda_2 > \lambda_3 \ldots$. Physically, the reduced temperature is expected to be a relevant parameter. It does not necessarily correspond to the relevant eigendirection of the RG flow but its projection onto this eigendirection is non vanishing. For the sake of simplicity we ignore this sublety that does not play any role in the following and we assume that $v_1 = t$. We have seen, Eq. (2.39), that for large $\bar{r}$:

$$G^{(2)}(\bar{r}, \boldsymbol{K}) \simeq \lambda^{-2}(s)\, G^{(2)}\left(\frac{\bar{r}}{s}, \boldsymbol{K}_s\right) \quad \text{with} \quad \lambda(s) = s^{\theta/2}. \tag{2.65}$$

From these relations we can deduce two others, one at T_c and another one away from T_c:

• For $t = 0$ and in the neighborhood of $\boldsymbol{K}^*$:

$$G^{(2)}(\bar{r}, 0, v_2, v_3, \dots) \simeq s^{-\theta} G^{(2)}\left(\frac{\bar{r}}{s}, 0, s^{\lambda_2} v_2, s^{\lambda_3} v_3, \dots\right). \tag{2.66}$$

Let us now suppose that we integrate out all fluctuations between scales a and r. This amounts to taking $s = \bar{r}$:

$$G^{(2)}(\bar{r}, 0, v_2, \dots) \simeq \bar{r}^{-\theta} G^{(2)}\left(1, 0, \bar{r}^{\lambda_2} v_2, \dots\right). \tag{2.67}$$

Since $\lambda_2 < 0$, we obtain that for $\bar{r}$ sufficiently large: $\bar{r}^{\lambda_2} v_2 \ll 1$ and thus

$$G^{(2)}(\bar{r}, 0, v_2, \dots) \simeq \bar{r}^{-\theta} G^{(2)}(1, 0, 0, \dots). \tag{2.68}$$

From the definition of the anomalous dimension

$$G^{(2)}(r) \underset{T=T_c}{\propto} \frac{1}{r^{d-2+\eta}}. \tag{2.69}$$

we find that

$$\theta = d - 2 + \eta. \tag{2.70}$$

Note that the $d-2$ part of θ is purely dimensional: it corresponds to the engineering dimension of the spin field. The η part corresponds to a dynamical contribution. It can be proven rigorously and for many field theories (as the ϕ^4 theory) that $\eta \geq 0$. This means that the fluctuations contribute to decrease the correlations of the system with the distance.

• For $t \neq 0$, $\xi < \infty$ and we can therefore integrate all fluctuations between scales a and ξ by taking $s = \bar{\xi}$. As explained above, Eq. (2.50), the running temperature at this scale must be of order 1 since the coarse-grained system at this scale is no longer critical. We thus deduce for large correlation lengths

$$t(s = \bar{\xi}) \simeq t\, \bar{\xi}^{\lambda_1} \sim 1. \tag{2.71}$$

From the definition of the critical exponent ν:

$$\xi \sim t^{-\nu} \tag{2.72}$$

we find that

$$\nu = \frac{1}{\lambda_1}. \tag{2.73}$$

The behavior of the correlation function close to the critical temperature follows from Eqs. (2.65, 2.71) and from $\bar{\xi} \gg 1$:

$$G^{(2)}\,(\bar{r}, t, v_2, \ldots) \simeq \bar{\xi}^{-\theta} G^{(2)} \left(\frac{\bar{r}}{\bar{\xi}}, t\, \bar{\xi}^{\lambda_1}, v_2\, \bar{\xi}^{\lambda_2} \ldots \right) \tag{2.74}$$

$$\simeq \bar{\xi}^{-\theta} G^{(2)} \left(\frac{\bar{r}}{\bar{\xi}}, 1, 0 \ldots \right)$$

$$\simeq \bar{r}^{-\theta} f \left(\frac{\bar{r}}{\bar{\xi}} \right). \tag{2.75}$$

We can see on this relation that close to T_c and at large distance the correlation function is no longer a function of infinitely many coupling constants but of only one parameter, the correlation length. One can also see that the Yukawa-like form of the correlation function that we have considered in Eq. (2.43) has the right form.

Scaling of the Correlation Function: Relations Among Exponents

We now couple the system to a uniform magnetic field. In a RG transformation:

$$B \sum_i S_i = B \sum_I \frac{s^d}{\lambda(s)} S_I \overset{\text{def}}{=} B_s \sum_I S_I. \tag{2.76}$$

Thus

$$B_s = s^{d-\theta/2}\, B. \tag{2.77}$$

We call

$$\lambda_B = d - \theta/2 = \frac{d + 2 - \eta}{2} \tag{2.78}$$

the magnetic eigenvalue. Since η is always smaller than $d + 2$ (even for $d = 1$), $\lambda_B > 0$ so that the magnetic field is a relevant variable.

Relation (2.65) can be trivially generalized in the presence of a magnetic field and to any correlation function. We now consider the magnetization per spin which is the one-point function: $G^{(1)} = m = \langle S \rangle$. Clearly, it behaves as:

$$m(t, B, \ldots) \simeq s^{-\theta/2} m(s^{\lambda_1}\, t, s^{\lambda_B}\, B, \ldots). \tag{2.79}$$

Once again, we can obtain several relations among exponents by considering the scaling of physical quantities at and away from T_c.

• For $t = 0$ and $B \neq 0$

$$m(0, B, \dots) \simeq s^{-\theta/2} m(0, s^{\lambda_B} B, \dots). \tag{2.80}$$

and, by taking s such that $s^{\lambda_B} B \simeq 1$, we obtain

$$\begin{aligned} m(0, B, \dots) &\simeq B^{\theta/2\lambda_B} m(0, 1, \dots) \\ &\propto B^{1/\delta} \end{aligned} \tag{2.81}$$

by definition of the exponent δ. We thus find:

$$\delta = \frac{d+2-\eta}{d-2+\eta} \tag{2.82}$$

• For $t < 0$ and $B = 0$ we find:

$$m(t, 0, \dots) \simeq s^{-\theta/2} m(s^{\lambda_1} t, 0, \dots). \tag{2.83}$$

and by taking $s = \bar{\xi}$:

$$\begin{aligned} m(t, 0, \dots) &\simeq \bar{\xi}^{-\theta/2} m(1, 0 \dots) \\ &\propto (T_c - T)^{\beta} \end{aligned} \tag{2.84}$$

by definition of the exponent β. Thus we find

$$\beta = \nu \frac{d-2+\eta}{2}. \tag{2.85}$$

• For the susceptibility $\chi = \partial m / \partial B$, at $t \neq 0$, we obtain:

$$\begin{aligned} \chi(t, B, \dots) &\simeq s^{-\theta/2} \frac{\partial}{\partial B} m(s^{\lambda_1} t, s^{\lambda_B} B, \dots) \\ &\simeq s^{\lambda_B - \theta/2} \chi(s^{\lambda_1} t, s^{\lambda_B} B, \dots). \end{aligned} \tag{2.86}$$

By taking $B = 0$ and $s = \bar{\xi}$, we find:

$$\begin{aligned} \chi(t, 0, \dots) &\simeq \bar{\xi}^{\lambda_B - \theta/2} \chi(1, 0 \dots) \\ &\sim (T - T_c)^{-\gamma}. \end{aligned} \tag{2.87}$$

Thus, by definition of γ

$$\gamma = \nu(2 - \eta). \tag{2.88}$$

Finally, for the exponent α of the specific heat, the calculation is subtler and requires to consider the free energy. One finds:

$$\alpha = 2 - \nu d. \tag{2.89}$$

To conclude, we have found that the hypothesis of the existence of a fixed point in the RG flow is sufficient to explain:

- universality since the critical exponents depend only on the RG flow around the fixed point and not on the point $\boldsymbol{K}_c$ representing the system when it is critical;
- the scaling behaviors of the thermodynamical quantities such as the magnetization, the susceptibility, the correlation function, etc.;
- the relations existing between critical exponents;
- the irrelevance of infinitely many couplings and the fact that the scaling of the correlation length with the temperature drives the scaling of many other thermodynamical quantities.

Two remarks are in order here. First, for second order transitions, only two exponents are independent, ν and η for instance. Second, universality is a much more general concept than what we have seen here on critical exponents. It is possible to show in particular that the RG theory enables to understand why it is possible to keep track of only a small number of coupling constants in field theory while these theories involves infinitely many degrees of freedom and thus, *a priori*, infinitely many couplings.

The Example of the Two-Dimensional Ising Model on the Triangular Lattice

This model is very famous as an example where the block-spin method *à la* Kadanoff can be implemented rather easily [16]. We shall not repeat the explicit calculation of the RG flow that can be found in most textbooks on this subject [1]. We shall only give the main ideas that will be relevant for our purpose.

As already explained above, the idea is to partition the lattice in triangular plaquettes, to build an Ising block-spin $\mathcal{S}_I$ for each plaquette by a majority rule, Eq. (2.14), and to integrate out the fluctuations inside the plaquettes compatible with a given value of $\mathcal{S}_I$. We have already said that the implementation of this idea imposes to take into account all possible $\mathbb{Z}_2$-invariant couplings among the spins. From a practical point of view it is thus necessary to perform truncations of this infinite dimensional space of coupling constants. The approximations usually performed consists in keeping only some couplings and projecting the RG flow onto this restricted space of couplings. We shall see that almost the same idea is used in the field theoretical implementation of the NPRG. The simplest such truncation consists in keeping only the nearest neighbor interaction, that is K_1, in hamiltonian (2.21) as well as the magnetic field B. For concreteness, let us give here the result of a RG transformation with $s = \sqrt{3}$ on the triangular lattice for a small magnetic field when only K_1 and B are retained:

$$K_{1,s} = 2K_1 \frac{e^{-K_1} + e^{3K_1}}{3e^{-K_1} + e^{3K_1}} \tag{2.90}$$

$$B_s = 3B \frac{e^{-K_1} + e^{3K_1}}{3e^{-K_1} + e^{3K_1}}. \tag{2.91}$$

A fixed point is trivially found

$$K_1^* = 0.336 \tag{2.92}$$

$$B^* = 0. \tag{2.93}$$

The exponent ν can be computed as well as δ and the result is:

$$\nu = 1.118, \qquad \delta = 2.17. \tag{2.94}$$

All the other exponents can be found using the scaling relations among them. The exact values, given by Onsager's solution are

$$\nu = 1, \qquad \delta = 15. \tag{2.95}$$

Let us notice that these RG results can be systematically improved by keeping more and more couplings. They drastically and rapidly improve with the next orders of approximations [16].

2.2 The Non-Perturbative Renormalization Group

"Ô insensé qui croit que je ne suis pas toi!"
V. Hugo

2.2.1 Introduction

All the different implementations of the non-perturbative renormalization group (NPRG) rely on Kadanoff-Wilson's ideas of block spins, coarse-graining and *effective long-distance theories*. However, they can substantially differ as for the way they are implemented. In the framework of field theory, there exists two main formulations: the Wilson (also called Wilson-Polchinski) approach [8, 9, 12, 17, 18] and the "effective average action" approach [19–33]. We shall deal with the second one which is not the best known, probably for historical reasons. Since it is nevertheless interesting to have an idea of the Wilson-Polchinski formulation, we start by this approach although we shall not study it in detail.

The Wilson-Polchinski Approach

We shall work in the context of statistical field theory (at equilibrium). This means that we shall not deal with a minkowskian metric (this brings its own difficulties) and that we suppose a continuum description of the systems under study. The microscopic physics is supposed to correspond to a scale Λ in momentum space which is—up to a factor of unity—the inverse of a microscopic length (a lattice spacing, an intermolecular distance, etc.). The partition function is thus given by a functional integral:

$$Z[B] = \int d\mu_{C_\Lambda}(\phi) \exp\left(-\int V(\phi) + \int B\phi\right) \tag{2.96}$$

where $d\mu_{C_\Lambda}$ is a (functional) gaussian measure with a cut-off at scale Λ:

$$d\mu_{C_\Lambda} = \mathcal{D}\phi(x) \exp\left(-\frac{1}{2}\int_{x,y} \phi(x)\, C_\Lambda^{-1}(x-y)\, \phi(y)\right) \tag{2.97}$$

with (in momentum space)

$$C_\Lambda(p) = (1 - \theta_\epsilon(p, \Lambda))\, C(p) \tag{2.98}$$

and C is the usual free propagator:

$$C(p) = \frac{1}{p^2 + r}. \tag{2.99}$$

The "cut-off" function θ_ϵ is a step function in p-space starting at Λ and smoothened around Λ on an interval of typical width ϵ, see Fig. 2.5. If $\epsilon = 0$, the quadratic part of the hamiltonian becomes the usual gradient and "mass" term, cut-off at scale Λ:

$$\frac{1}{2}\int_0^\Lambda \frac{d^dq}{(2\pi)^d}\, \phi(-p)(p^2 + r)\phi(p). \tag{2.100}$$

The role of ϵ is to smoothen the sharp cut-off at Λ which is conceptually simple but technically unpleasant.

We want to implement the block-spin idea in our field theory framework, that is to separate $\phi_p = \phi(p)$ into "rapid" and "slow" modes (with respect to a scale k). The slow modes will play the role of block-spins while the rapid ones will correspond to fluctuations inside the blocks. It is convenient to work in Fourier space where the derivative operators are diagonalized (and the cut-off simple).

We define

$$\phi_p = \phi_{p,<} + \phi_{p,>} \tag{2.101}$$

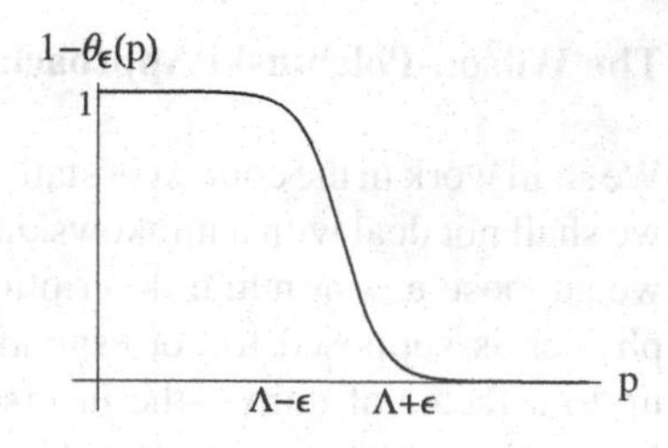

Fig. 2.5 A typical cut-off function in the Wilson-Polchinski approach

and associate

$$\phi_p \to C_\Lambda(p), \quad \phi_{p,<} \to C_k(p) \quad \phi_{p,>} \to C_\Lambda(p) - C_k(p) \tag{2.102}$$

It is important to notice that ϕ is the sum of $\phi_{p,<}$ and $\phi_{p,>}$ for all p: it does not coincide in general with $\phi_{p,<}$ on $[0,k]$ and with $\phi_{p,>}$ on $[k,\Lambda]$. The meaning of $\phi_<$ and $\phi_>$ comes from their propagator $C_k(p)$ and $C_\Lambda(p) - C_k(p)$ respectively. A beautiful identity allows us to rewrite the original partition function in terms of $\phi_<$ and $\phi_>$, $C_k(p)$ and $C_\Lambda(p) - C_k(p)$ and to perform (at least formally) the integration on the rapid modes:

$$d\mu_{C_\Lambda}(\phi) = d\mu_{C_k}(\phi_<)\, d\mu_{C_\Lambda - C_k}(\phi_>) \tag{2.103}$$

Let us show it on a one dimensional integral:

$$I = \int_{-\infty}^{+\infty} dx\, e^{-x^2/2\gamma} \tag{2.104}$$

where x is the analogue of ϕ and γ of C_Λ. We now define

$$\begin{cases} x = y + z \\ \gamma = \alpha + \beta \end{cases} \tag{2.105}$$

y and z are the analogues of $\phi_>$ and $\phi_<$ respectively and α and β the analogues of $C_\Lambda - C_k$ and of C_k. Then

$$I \propto \int_{-\infty}^{+\infty} dy\, dz\, e^{-y^2/2\alpha} e^{-z^2/2\beta}. \tag{2.106}$$

We prove this identity in the Appendix, Subsect. 2.3.2. Actually, we are not only interested in gaussian integrals and our result can be trivially generalized:

$$\int_{-\infty}^{+\infty} dx\, e^{-x^2/2\gamma - V(x)} \propto \int_{-\infty}^{+\infty} dy\, dz\, e^{-y^2/2\alpha - z^2/2\beta - V(y+z)}. \tag{2.107}$$

This result can be generalized straightforwardly to functional integrals since it is a property of the gaussian integrals.[16] It becomes

$$\int d\mu_{C_\Lambda}(\phi)\, e^{-\int V(\phi)} \propto \int d\mu_{C_k}(\phi_<)\;\; d\mu_{C_\Lambda - C_k}(\phi_>)\, e^{-\int V(\phi_< + \phi_>)} \tag{2.108}$$

Thus, by performing formally the integration on $\phi_>$, we define a running "potential" V_k at scale k:

$$e^{-\int V_k(\phi_<)} = \int d\mu_{C_\Lambda - C_k}(\phi_>) e^{-\int V(\phi_< + \phi_>)} \tag{2.109}$$

with, by definition, $V_\Lambda = V$, the initial potential. This definition leads to:

$$Z = \int d\mu_{C_k}(\phi_<)\, e^{-\int V_k(\phi_<)}. \tag{2.110}$$

Let us emphasize that V_k is called a potential because we do not have included in it the quadratic derivative term. However, generically, as soon as $k < \Lambda$, V_k involves derivative terms with, moreover, any power of the derivatives of $\phi_<$.

It is possible to write down a differential equation for the evolution of V_k with k: this is the Wilson-Polchinski equation derived in the Appendix, Subsect. 2.3.3 [8, 9, 12]. It is a possible starting point for a non-perturbative formulation of the RG. The one we shall use is mathematically equivalent to this one although more convenient in many respects when approximations are used. Before going to this second formulation, let us make some remarks here.

- $\phi_<$ at scale k represents approximately a spatial average of ϕ over a volume of order k^{-d}. It is not a thermal average. For $k = 0$, $\phi_{p=0,<}$ represents only what is (improperly) called the magnetization mode: $\int d^d x\, \phi(x)$ which is *not* the magnetization. It has a highly non trivial probability distribution and the true magnetization is the thermal average of this mode. Thus $\phi_<$ at scale $k \neq 0$ *is not a precursor* of the order parameter: it is still a stochastic variable whose physical interpretation is not so trivial.
- The flow of "potentials" $V_k(\phi_<)$ does not contain all the informations on the initial theory: the correlation functions of the rapid modes cannot be computed from this flow. It is necessary to first couple the system to an arbitrary "source" $B(x)$ and to follow the flow of this term to reconstruct the correlations of the initial fields in the whole momentum range $p \in [0, \Lambda]$. Thus the information is splitted into two different kinds of terms: on one hand, the k-dependent hamiltonians which give rise to a flow for all the couplings involved in these hamiltonians and, on the other hand, the source term. Fortunately, much informations about the theory (e.g. critical behavior) can be obtained from the flow of hamiltonians alone. It is nevertheless an open question to know if the difficulties encountered within

[16] We first consider an N-dimensional gaussian integral and then take the limit $N \to \infty$.

Wilson-Polchinski's method are related to the fact that the informations on the Green functions are not contained in the flow of hamiltonians.

- The effective hamiltonians V_k (and the corresponding Boltzmann weights) are highly abstract objects! One should remember in particular that the RG transformations do not correspond to any transformations that a human being can perform on the system. This is a purely theoretical idea that moreover will be useful only when approximations are used.
- The flow equation on V_k was written more than 30 years ago but was not much used in actual physical problems before the mid-1990s (for exceptions see, Golner, Newman, Riedel, Bagnuls, Bervillier, Zumbach, etc.) [17, 34–40]. There are three main reasons for this strange matter of fact. First, perturbation theory was extremely successful for $O(N)$ models (as well as in particle physics) and the need for the NPRG was not obvious in many situations. Second, (renormalized) perturbation theory was believed to correspond to a controlled approximation whereas approximations performed in the NPRG framework seemed uncontrolled. However, this is *completely wrong*. Renormalized perturbation series are *not* convergent [5, 41]. They are asymptotic series, at best.[17] For the $O(N)$ model at $l-1$ loops, the coefficient in front of u^l of the 4-point correlation function behaves at large l as $l!(-a)^l$ with a a real number. Thus, even in cases where many orders of the perturbation expansion have been computed, they are useless as such and resummation methods of the renormalized series are required (Padé-Borel, conformal-Borel, etc.). In many cases, these resummation techniques fail to produce converged results. As for the NPRG, there is no general theorem about the convergence of the series of approximations that are used. However, from the few results yet obtained, it seems that this method has good convergence properties [32, 42–46]. It is however too early to draw any firm conclusion on this question. Third, it seemed that the anomalous dimension was crucially depending on the choice of cut-off function θ_ϵ that separates the rapid and the slow modes whereas it should be independent of it. This was especially important when studying the $O(N)$ models in two dimensions where it seemed impossible to reproduce the perturbative results obtained from the non linear sigma model. This difficulty is very simply overcome in the "effective average action" approach [32, 33, 47].

Let us finally mention two other "psychological" difficulties related to NPRG.

- The NPRG equation on V_k can be truncated in perturbation theory. This, of course, enables to recover the usual perturbation expansion (what else could it lead to?). However, the way it was implemented most of the time in the 1970s did not allow

[17] The $O(N)$ models are completely exceptional in this respect since they are the only ones for which it has been proven that the series of the β-function is Borel-summable in $d=3$ (in the so-called zero momentum massive scheme). In all other cases, either this is not known or it is known that the series are not Borel-summable. For QED, this is not yet a problem because the smallness of the coupling constant, the fine structure constant, ensures up to now an apparent convergence of the perturbative results.

to retrieve the two-loop results.[18] It is still (erroneously) widely believed for this reason, even by "specialists", that Wilson's method does not work at two-loop order and beyond!

- $V_k(\phi)$ involves infinitely many couplings contrarily to perturbation theory that involves only the renormalizable ones. For this reason, it is widely believed that the recourse to numerical methods is unavoidable in the NPRG approach whereas it is not in the perturbative one. This is not fully correct for two reasons. First, even in perturbation theory the RG flow cannot be integrated analytically in general. Second, even in the NPRG approach, very crude approximations, involving only very few couplings, often lead both to analytically tractable computations and to highly non-trivial non-perturbative results [32].

Let us now turn to the other implementation of the NPRG formalism.

The Effective Average Action Method

Many formal results about the NPRG method as well as some "physical" results have been obtained within the Wilson-Polchinski approach.[19] However, the revival of Wilson's ideas as well as their concrete implementation in the last 15 years is largely linked with the development of an alternative, although formally equivalent, formulation first promoted by C. Wetterich at the end of the 1980s under the name of effective average action method [23–32]. In practice, this has allowed to compute in a reasonable way the anomalous dimension η and, more importantly, to study the physics of the $O(N)$ models and of many others, in all dimensions, including two [32, 33]. Moreover, the whole scheme is more intuitive, allows to retrieve very easily the one-loop results both in $4-\epsilon$ and $2+\epsilon$ dimensions and in the large N limit. This has convinced many specialists of the subject to work with this formalism.

Block-spins, coarse-graining, Legendre transform, etc.

The original Kadanoff-Wilson's idea is to perform a coarse-graining and thus to map hamiltonians onto other hamiltonians at larger scales. The hamiltonians thus obtained are the hamiltonians of the modes *that have not yet been integrated out* in the partition function, that is $\phi_<$. As already emphasized, these hamiltonians are very abstract objects.[20] Instead of computing this sequence of hamiltonians, we can compute the Gibbs free energy $\Gamma[M]$ of the rapid modes (that is $\phi_>$) that have *already*

[18] These calculations did not correspond to a series expansion of the exact NPRG equation on V_k. They enabled to retrieve the one-loop results easily but became very cumbersome beyond one-loop.

[19] See the impressive and inspiring works of Bagnuls and Bervillier about the formal aspects of Wilson's RG, as well as their criticisms of the perturbative approach [48–50].

[20] It is impossible to get easily any physical information from it except at "mean field-like" level: a functional integral has still to be performed.

been integrated out.[21] Thus, the idea is to build a one-parameter family of models, indexed by a scale k such that [32]

- when $k = \Lambda$, that is when no fluctuation has been integrated out, the Gibbs free energy $\Gamma_k[M]$ is equal to the microscopic hamiltonian[22]:

$$\Gamma_{k=\Lambda}[M] = H[\phi = M]. \tag{2.111}$$

- when $k = 0$, that is when all fluctuations are integrated out, $\Gamma_{k=0}$ is nothing but the Gibbs free energy of the original model:

$$\Gamma_{k=0}[M] = \Gamma[M]. \tag{2.112}$$

Thus, as k decreases more and more fluctuations are integrated out. The magnetization at scale k is therefore a precursor of the true magnetization (obtained at $k = 0$) and the free energy Γ_k, also called the effective average action, a precursor of the true free energy Γ (also called the effective action).

Let us notice two points. First, k plays the role of an *ultra-violet* cut-off for the slow modes $\phi_<$ in the Wilson-Polchinski formulation (analogous to Λ in the original model). It plays the role of an *infrared* cut-off in the effective average action method since Γ_k is the free energy of the rapid modes. Second, the slow modes play a fundamental role in the Wilson-Polchinski approach whereas they are absent of the effective average action method which involves only the (free energy of the) rapid modes. As a by-product, we shall see that all the informations on the model (RG flow, existence of a fixed point, computations of correlation functions, etc.) are contained in a single object: $\Gamma_k[M]$. This is a rather important advantage of this method compared to the Wilson-Polchinski one.

Now the question is to build explicitly this one-parameter family of Γ_k. The idea is to decouple the slow modes of the model in the partition function. A very convenient implementation of this idea is to give them a large mass [32]. In the language of particle physics, a large mass corresponds to a small Compton wavelength ($\hbar/mc$) and thus to a small range of distances where quantum fluctuations are important. A very heavy particle decouples from the low energy (compared to its mass) physics since it can play a role at energies below its mass threshold only through virtual processes. These processes are themselves suppressed by inverse powers of the mass of the heavy particle coming from its propagator. In the language of critical phenomena, the "mass" term $r\phi^2/2$ in the hamiltonian corresponds to the deviation from the critical temperature: $r \propto T - T_c$ (at the mean field level, at least). Thus a large

[21] The Helmoltz free energy is $F = -k_B T \log Z[B]$. It is a functional of the source $B(x)$. The Gibbs free energy is obtained by a Legendre transform from F and is a functional of the magnetization $M(x)$. It is the generating functional of the one-particle irreducible (1PI) correlation functions.

[22] Let us emphasize that at the mean-field approximation, the Gibbs free energy of the system is identical to the hamiltonian. Equation (2.111) is an exact version of this statement (remember that no fluctuation is taken into account at the mean-field level).

"mass" r corresponds to a theory which is far from criticality ($\xi \sim a$), that is where thermal fluctuations are small.

Therefore, the idea is to build a one-parameter family of models for which a "momentum-dependent mass term" has been added to the original hamiltonian [32]:

$$Z_k[B] = \int \mathcal{D}\phi(x) \exp\left(-H[\phi] - \Delta H_k[\phi] + \int B\phi\right) \tag{2.113}$$

with

$$\Delta H_k[\phi] = \frac{1}{2}\int_q R_k(q)\,\phi_q\,\phi_{-q}. \tag{2.114}$$

The function $R_k(q)$—called the cut-off function from now on—must be chosen in such a way that

- when $k = 0$, $R_{k=0}(q) = 0$ identically ($\forall q$) so that:

$$Z_{k=0}[B] = Z[B]. \tag{2.115}$$

  This will ensure that the original model is recovered when all fluctuations are integrated out, see Eq. (2.112).
- when $k = \Lambda$, all fluctuations are frozen. This will ensure that relation (2.111) is satisfied. Giving an infinite mass to all modes $q \in [0, \Lambda]$ freezes their propagation completely and we must therefore take

$$R_{k=\Lambda}(q) = \infty \qquad \forall q. \tag{2.116}$$

  An approximate, but convenient way to achieve this goal is to choose a function $R_{k=\Lambda}$ not infinite but of the order of Λ^2 for all momenta.
- when $0 < k < \Lambda$, the rapid modes (those for which $|q| > k$) must be almost unaffected by $R_k(q)$ which must therefore almost vanish for these modes:

$$R_k\,(|q| > k) \simeq 0 \tag{2.117}$$

  On the contrary, the slow modes must have a mass that almost decouple them from the long distance physics.

Remembering that R_k is homogeneous to a mass square, it is not difficult to guess its generic shape, at least if we require that it is an analytic function of q^2, see Fig. 2.6.

We shall discuss in the following some convenient choices for R_k but let us first define precisely what Γ_k is. In principle, having defined Z_k by Eq. (2.113), the Legendre transform of log Z_k should be unambiguous and should lead to Γ_k. Let us follow this program and see that there is a subtlety. We thus define

$$W_k[B] = \log Z_k[B] \tag{2.118}$$

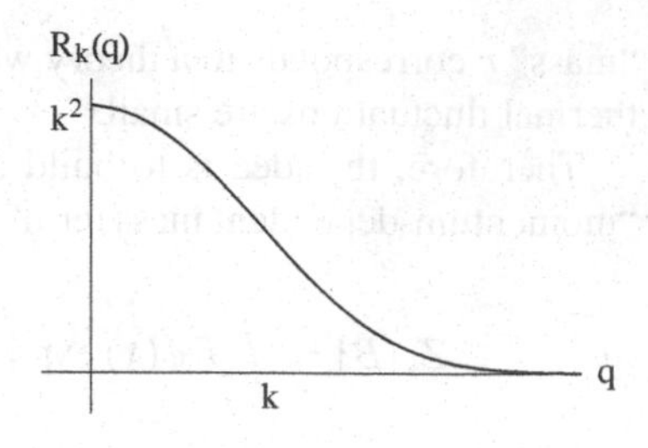

Fig. 2.6 A typical cut-off function in the effective average action approach

which is the Helmoltz free energy, up to the $-k_B T$ term that plays no role in what follows. The Legendre transform of W_k is defined by

$$\Gamma'_k[M] + W_k[B] = \int BM \tag{2.119}$$

where the magnetization $M(x)$ is, by definition the average of $\phi(x)$ and is therefore:

$$M(x) = \frac{\delta W_k}{\delta B(x)}. \tag{2.120}$$

Of course, for $k \to 0$, $R_k \to 0$, $W_k \to W$ and thus $\Gamma'_k \to \Gamma$= Gibbs free energy of the system. However, it is easy to show that $\Gamma'_\Lambda[M] \neq H[M]$ contrarily to what is expected, see Eq. (2.111). This comes from the $\Delta H_{k=\Lambda}$ term, which is large. Thus, we prefer to work with a modified free energy where the R_k term has been subtracted in Γ_k [32]. We define

$$\Gamma_k[M] + W_k[B] = \int BM - \frac{1}{2}\int_q R_k(q) M_q M_{-q}. \tag{2.121}$$

The R_k term in Eq. (2.121) does not spoil the limit $k \to 0$ since in this limit it vanishes $\forall q$. Let us now show that Eq. (2.121) is the correct definition of Γ_k leading to the limit $\Gamma_{k=\Lambda}[M] = H(M)$, Eq. (2.112).

An Integral Representation of Γ_k and the Limit $k \to \Lambda$

We start from the definition of Z_k, Eq. (2.113) and from the definition of Γ_k, Eq. (2.121). We deduce by differentiation (see, 2.3.3)

$$B_x = \frac{\delta \Gamma_k}{\delta M_x} + \int_y R_k(x-y) M_y. \tag{2.122}$$

Thus, by substituting Eq. (2.121) and Eq. (2.122) into the definition of W_k we obtain:

$$e^{-\Gamma_k[M]} = \int \mathcal{D}\phi \exp\left(-H[\phi] + \int_x \frac{\delta \Gamma_k}{\delta M_x}(\phi_x - M_x)\right).$$
$$\cdot \exp\left(-\frac{1}{2}\int_{x,y} (\phi_x - M_x)\, R_k(x-y)\, (\phi_y - M_y)\right) \tag{2.123}$$

If we choose a function $R_k(q)$ that diverges for all q as $k \to \Lambda$ then, in this limit:

$$\exp\left(-\frac{1}{2}\int (\phi_x - M_x)\, R_k(x-y)\, (\phi_y - M_y)\right) \sim \delta(\phi - M) \tag{2.124}$$

that is, it behaves as a functional Dirac delta. Therefore,

$$\Gamma_k[M] \to H[\phi = M] \quad \text{as} \quad k \to \Lambda \tag{2.125}$$

if the cut-off R_k is such that it diverges in this limit. If R_k does not diverge and is only very large,

$$\Gamma_{k=\Lambda} \sim H. \tag{2.126}$$

2.2.2 The Exact RG Equation and its Properties

The RG equation on Γ_k, sometimes called Wetterich's equation, that is the differential equation $\partial_k \Gamma_k = f(\Gamma_k)$ is derived in detail in the Appendix, Subsect. 2.3.3. The strategy is to obtain first an evolution equation for Z_k, then to deduce the equation on Γ_k. It writes

$$\partial_k \Gamma_k = \frac{1}{2}\int_q \partial_k R_k(q) \left(\Gamma_k^{(2)}[M] + \mathcal{R}_k\right)^{-1}_{q,-q} \tag{2.127}$$

where $\mathcal{R}_k(x,y) = R_k(x-y)$. The inverse $\left(\Gamma_k^{(2)}[M] + \mathcal{R}_k\right)^{-1}_{q,-q}$ has to be understood in the operator sense. It is convenient in practice to rewrite Eq. (2.127) as [32]

$$\partial_k \Gamma_k = \frac{1}{2}\tilde{\partial}_k \operatorname{Tr}\log\left(\Gamma_k^{(2)} + R_k\right) \tag{2.128}$$

where $\tilde{\partial}_k$ acts only on the k-dependence of R_k and not on $\Gamma_k^{(2)}$:

$$\tilde{\partial}_k = \frac{\partial R_k}{\partial k}\frac{\partial}{\partial R_k} \tag{2.129}$$

and the trace means integral over q (and for more complex theories, summation over any internal index).

Some General Properties of the Effective Average Action Method

Let us mention some important properties of Γ_k.

- If the microscopic hamiltonian H and the functional measure are symmetric under a group G and if there exists a cut-off function R_k such that the mass term ΔH_k respects this symmetry, then Γ_k is symmetric under G for all k and thus so is $\Gamma = \Gamma_{k=0}$. It can happen that there is no mass-like term that respects the symmetry whereas the theory, that is Γ, is invariant under G. This means that the symmetry is broken for all finite k and that the symmetry is recovered only for $k \to 0$. This is the case of gauge symmetry. This symmetry breaking term can be controlled by modified Ward identities that become, in the limit $k \to 0$, the true Ward identities. It remains nevertheless difficult up to now to compute RG flows in a completely controlled way in gauge theories.
- An exact RG equation for theories involving fermions can also be derived along the same line.
- Equation (2.128) looks very much like (the derivative of) a one-loop result since at one-loop:
$$\Gamma_k = H + \frac{1}{2}\mathrm{Tr}\log\left(H^{(2)} + R_k\right). \tag{2.130}$$
Thus, substituting $H^{(2)}$ by the full $\Gamma_k^{(2)}[M]$ functional in the derivative with respect to k of Eq. (2.130) changes the one-loop result into an exact one! There exists a diagrammatic representation of the RG equation written as in Eq. (2.127) and that emphasizes its one-loop structure, see Figs. 2.7 and 2.8.
We define
$$G_k[M] = \left(\Gamma_k^{(2)}[M] + \mathcal{R}_k\right)^{-1}, \tag{2.131}$$
which is the "full", that is M-dependent, propagator, see Fig. 2.7.
- This one-loop structure has a very important practical consequence: only one integral has to be computed and thanks to rotational invariance, it is one-dimensional. This is very different from perturbation theory where l-loop diagrams require l d-dimensional integrals. This is a tremendous simplification of the NPRG method compared to perturbation theory.
- The perturbation expansion can be retrieved from the NPRG equation (and the all-order proof of renormalizability can be simpler in this formalism).
- Because of the $\partial_k R_k$ term in the NPRG equation (2.127), only momenta q^2 of the order of k^2 or less contribute to the flow at scale k (we come back in detail to this point in the following). Thus the RG flow is regular both in the ultra-violet and in the infra-red. All the divergences of perturbation theory are avoided: we compute directly the RG flow and not first the relationship between bare and renormalized quantities from which is computed, in a second step, the RG flow.
- k acts as an infrared regulator (for $k \neq 0$) somewhat similar to a box of finite size $\sim k^{-1}$. Thus, for $k > 0$, there is no phase transition and thus no singularity in

$G_{k\, q,-q'} =$ q ——— $-q'$

Fig. 2.7 Diagrammatic representation of the "full" propagator that is the M-dependent function $G_k[M] = \left(\Gamma_k^{(2)}[M] + \mathcal{R}_k\right)^{-1}_{q,-q}$

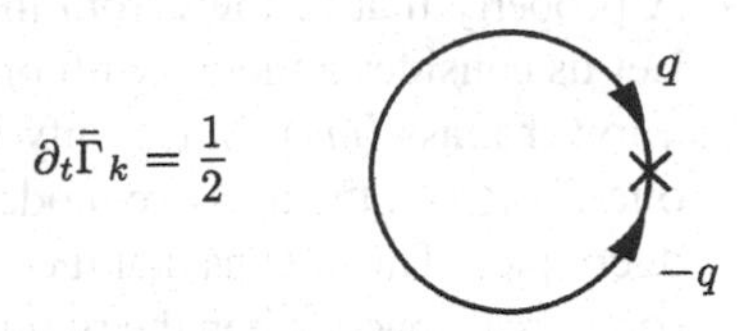

Fig. 2.8 Diagrammatic representation of the RG equation on the effective average action. The *line* represents the "full" propagator, Fig. 2.7, the cross $\partial_k R_k(q)$ and the loop, the integral over q

the free energy Γ_k. At finite k, everything is regular and can be power-expanded safely. We can therefore conclude that

(i) the singularities of Γ build up as k is lowered and are thus smoothened by k in Γ_k,

(ii) the precursor of the critical behavior should already show up at finite k for $|q| \gg k$.

- An important consequence of the regularity of Γ_k at $k > 0$ is that it can be expanded in a power series of $\nabla M(x)$. For slowly varying fields $M(x)$ this expansion is expected to be well-behaved. This is the basis of the *derivative expansion* that consists in proposing an *ansatz* for Γ_k involving only a finite number of derivatives of the field [32]. Two of the most used approximations, based on the derivative expansion, are

$$\Gamma_k = \int d^d x \left(U_k(M(x)) + \frac{1}{2} (\nabla M)^2 \right) \tag{2.132}$$

called the local potential approximation (LPA) since no field renormalization in front of the derivative term is included and

$$\Gamma_k = \int d^d x \left(U_k(M(x)) + \frac{1}{2} Z_k(M) (\nabla M)^2 \right) \tag{2.133}$$

often called the $O(\partial^2)$ approximation or the leading approximation.[23] Of course, in principle, Γ_k involves all powers of ∇M compatible with rotational and $\mathbb{Z}_2$ invariance. We shall come back in detail on these approximations in the following.
- We shall see in the following that by working with dimensionless (and renormalized) quantities, the NPRG equation can be rewritten in a way that makes no explicit reference to the scales k and Λ. This will allow us to find fixed points from which quantities like critical exponents can be computed. The fact that Λ disappears from the flow equation also ensures that the "group law" of composition

[23] The function $Z_k(M)$ has, of course, nothing to do with the partition function $Z_k[B]$ introduced in Eq. (2.113) although it is customary to use the same name for both functions.

of RG transformations is satisfied. This self-similarity property of the flow will be automatic *even if Γ_k is truncated* [as in Eqs. (2.132, 2.133) for instance]. This is a major advantage of this method compared to many others where renormalizability, that is self-similarity, is spoiled by approximations (Schwinger-Dyson approximations for instance).

- A property that follows from the last point is the decoupling of massive modes. Let us consider a theory with one very massive mode (m_1) and another one with a lower mass (m_2). Since only the modes $|q| \leq k$ contribute to the flow of Γ_k, once $k \ll m_1$, the massive mode does (almost) no longer contribute to the flow: it decouples. This means that it contributes to the flow when k is between Λ and m_1 (or, in real space, when the running lattice spacing is between a and its correlation length m_1^{-1}) and almost not below m_1. This also means that if we are given a model at a scale $k_0 \ll m_1$, we cannot know whether the underlying "fundamental theory" (at scales much larger than m_1) involves or not a massive particle since below this scale there exists almost no signal of the presence of this mode in the theory (that is, in the RG flow). Except for very precise measurements, the high energy sector of the theory has been decoupled from the low energy one and its contribution to the theory at low energy amounts to renormalization of the couplings occurring during the part of the RG flow corresponding roughly to $m_1 < k < \Lambda$.
- Self-similarity and decoupling of massive modes will have universality (in the vicinity of a second order transition) as a consequence. We shall see that universality is not just the consequence of the existence of a fixed point. It is the consequence of a very peculiar geometry of the RG flow, see Subsect. 2.2.6 [49, 51].

2.2.3 Approximation Procedures

The NPRG equation (2.127) is an extremely complicated equation. It is a functional partial differential equation since it involves the functionals $\Gamma_k[M]$ and $\Gamma_k^{(2)}[M]$. Needless to say that we do not know how to solve it in general. Some approximations are thus required. Two main kinds of approximations are usually considered: the Green function approach and the derivative expansion. In both cases, the strategy consists in solving the RG equation in *a restricted functional space* and not as a series expansion in a small parameter. This is why we can hope to obtain nonperturbative results. Both methods need to project consistently the exact RG equation in the functional space that has been chosen. Of course, the quality of the result will depend crucially on the choice of space in which we search for a solution. Depending on the problem, one choice can be drastically better than another. In all cases, it is impossible to know whether we have missed some physically crucial ingredient by making one choice rather than another one. But this problem is not specific to the NPRG. It is generic in physics…

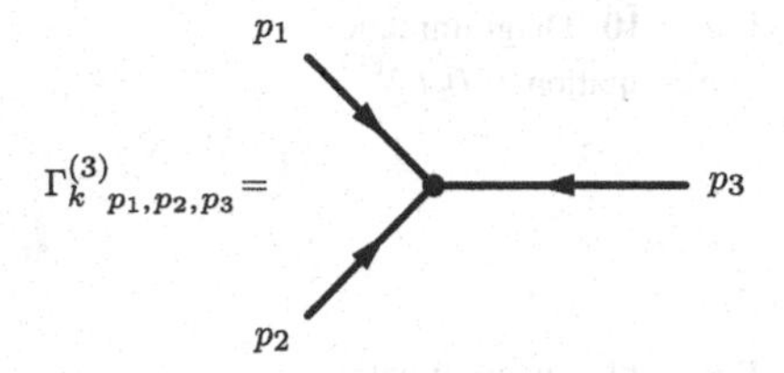

Fig. 2.9 Diagrammatic representation of $\Gamma^{(3)}_{k\ p_1,p_2,p_3}[M]$

Let us review briefly the Green function approach and then explain in some details the derivative expansion which is the most employed method in statistical mechanics.

The Green Function Approach

From Eq. (2.127) we can deduce the infinite hierarchy of RG equations on the correlation functions defined by

$$\bar{\Gamma}^{(n)}_{k,p_1\ldots p_n} = \frac{\delta^n \Gamma_k}{\delta M_{p_1}\ldots\delta M_{p_n}} \tag{2.134}$$

taken in a particular field configuration (the zero and the uniform field configurations being the most useful ones).
We define $t = \log k/\Lambda$ which is often called the RG "time". From Eq. (2.127), we obtain

$$\partial_t \frac{\delta\Gamma_k}{\delta M_p} = -\frac{1}{2}\int_{q_i} \dot{R}_k(q_1)\, G_{k\,q_1,-q_2}\, \frac{\delta\Gamma^{(2)}_{k\ q_2,-q_3}}{\delta M_p}\, G_{k\,q_3,-q_1}. \tag{2.135}$$

Therefore, setting $\dot{R}_k = \partial_t R_k = k\partial_k R_k$ we obtain

$$\begin{aligned}\partial_t \frac{\delta^2\Gamma_k}{\delta M_p\,\delta M_{p'}} &= \int_{\{q_i\}} \dot{R}_{k,q_1} G_{k\,q_1,-q_2}\,\Gamma^{(3)}_{k\ p,q_2,-q_3}\, G_{k\,q_3,-q_4}\,\Gamma^{(3)}_{k\ p',q_4,-q_5}\, G_{k\,q_5,-q_1}\\ &\quad -\frac{1}{2}\int_{\{q_i\}} \dot{R}_{k,q_1}\, G_{k\,q_1,-q_2}\,\Gamma^{(4)}_{k\ q_2,-q_3,p,p'}\, G_{k\,q_3,-q_1}\end{aligned} \tag{2.136}$$

where both the left and the right hand sides are functions of M_q since they have not yet been evaluated in a particular configuration. These equations look terrible but in fact they are not since there exists a diagrammatic way to obtain them automatically. We represent $\Gamma^{(3)}_{k\ p_1,p_2,p_3}$ as in Fig. 2.9.

We obtain for $\partial_t\Gamma^{(1)}_k$ the graph in Fig. 2.10.

Of course, if we evaluate $\partial_t\Gamma^{(1)}_k$ in a uniform field configuration, the momentum is conserved at each vertex and for each "propagator". The function is thus nonvanishing only at zero momentum and we obtain the equation in Fig. 2.11.

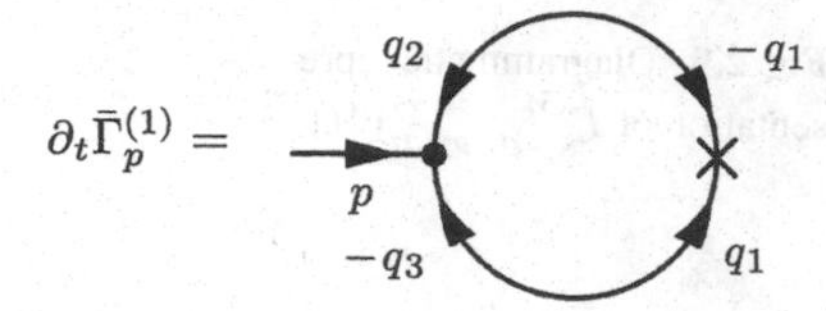

Fig. 2.10 Diagrammatic representation of $\partial_t \Gamma_k^{(1)}$

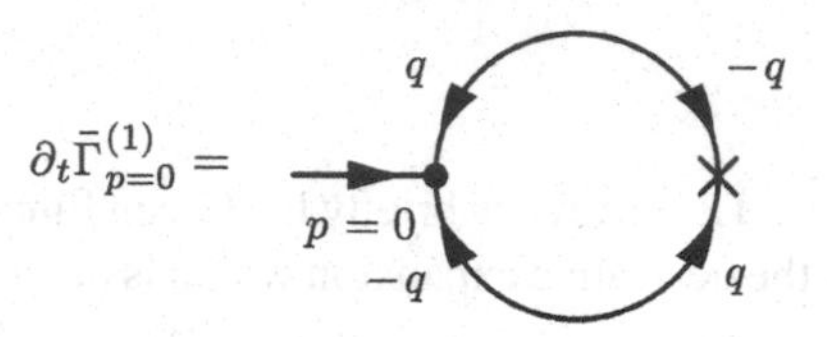

Fig. 2.11 Diagrammatic representation of $\partial_t \Gamma_k^{(1)}$ evaluated in a uniform field configuration

It is clear on the diagrammatic representation that only one momentum integral remains. The RG equation for $\partial_t \Gamma_k^{(2)}$ evaluated in a uniform field configuration is given in Fig. 2.12.

It is clear from these diagrammatic representations that $\partial_t \Gamma_k^{(n)}$ involves $\Gamma_k^{(n+1)}$ and $\Gamma_k^{(n+2)}$. If we want to solve this infinite tower of equations, we have to truncate it. A possible truncation consists, for instance, in keeping $\Gamma_k^{(2)}$ and $\Gamma_k^{(4)}$ evaluated at $M = 0$ and to neglect the contribution of $\Gamma_k^{(6)}$ in the equation on $\partial_t \Gamma_k^{(4)}$ ($\Gamma_k^{(3)}[M = 0] = 0$ in a $\mathbb{Z}_2$-invariant model). A better method is to find an ansatz for $\Gamma_k^{(6)}$ in terms of $\Gamma_k^{(2)}$ and $\Gamma_k^{(4)}$. In both cases, the systems of equations become closed and can, at least in principle, be solved. Let us finally notice that this method consists in truncating the field-dependence of $\Gamma_k[M]$ while keeping the momentum dependence of $\Gamma_k^{(2)}$ and $\Gamma_k^{(4)}$. Tremendous improvements of this type of approximation has been performed these last years [52–54].

We now study another truncation method which is somewhat the reverse.

The Derivative Expansion

The principle of this approximation has already been introduced previously, see Eqs. (2.132, 2.133). The underlying idea is that we are mostly interested (for the study of critical phenomena) in the long distance physics, that is the $|q| \to 0$ region of the correlation functions.[24] Thus, we keep only the lowest orders of the expansion of Γ_k in ∇M while we keep all orders in the field M

$$\Gamma_k = \int d^d x \left(U_k(M(x)) + \frac{1}{2} Z_k(M) \, (\nabla M)^2 \right) + O(\nabla^4) \tag{2.137}$$

[24] The computation of quantities like the total magnetization or the susceptibilities require only the knowledge of the spin-spin correlation function at zero momentum. The same thing holds for the correlation length.

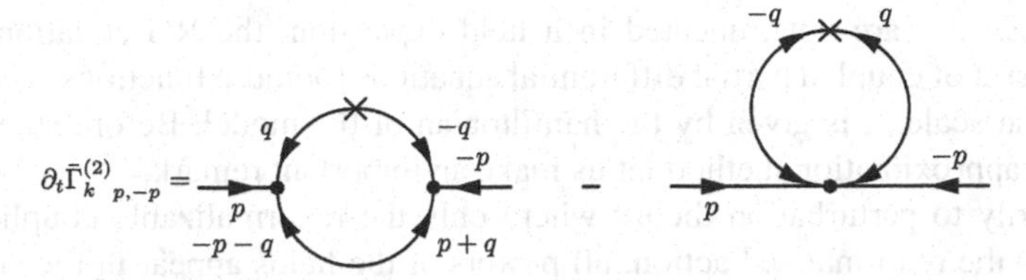

Fig. 2.12 Diagrammatic representation of $\partial_t \Gamma_k^{(2)}$ evaluated in a uniform field configuration. Note that the factor $\frac{1}{2}$ has not been represented on the figure for the tadpole because it can be retrieved from the topology of the graph itself (in fact the *minus sign* can also be retrieved from the graph)

This approximation is based on a somewhat opposite philosophy as the one that prevailed in the Green function approach. However, it should be clear that for statistical mechanics, the most important informations—e.g. the equation of state—are hidden in the effective potential $U_{k=0}$ of the theory that, therefore, needs to be computed as accurately as possible (see however [53]).

It is in fact remarkable that we can combine both methods by making a field expansion of $U_k, Z_k, \ldots$ on top of the derivative expansion while preserving many non-trivial results. The simplest such truncation consists in using the LPA and in keeping only the first two terms of the expansion of U_k in powers of M[25]:

$$\Gamma_k[M] = \int \mathrm{d}^d x \big(g_{2,k}\, M^2 + g_{4,k}\, M^4 + \frac{1}{2}(\nabla M)^2\big) \tag{2.138}$$

With this kind of ansatz, the RG equation on Γ_k becomes a set a ordinary differential equations for the couplings retained in the *ansatz*:

$$\partial_t g_{n,k} = \beta_n \left(\{g_{p,k}\}\right). \tag{2.139}$$

For instance, if the truncation above is considered we find[26]

$$\partial_t \kappa_k = -(d-2)\kappa_k + 6v_d\, l_1^d(2\kappa_k\lambda_k) \tag{2.140}$$

$$\partial_t \lambda_k = (d-4)\lambda_k + 6v_d\, \lambda_k^2\, l_2^d(2\kappa_k\lambda_k) \tag{2.141}$$

where l_1^d, l_2^d are defined in the Appendix, Subsect. 2.3.1. These equations are already non perturbative since the functions l_1^d, l_2^d are non polynomial.

[25] Let us notice that if the k-dependence of the couplings was neglected, this ansatz would exactly coincide with the *ansatz* chosen by Landau to study second order phase transitions. We know that it would lead to the mean field approximation. It is remarkable that keeping the scale dependence of the couplings and substituting precisely this *ansatz* into the RG equation of Γ_k is sufficient to capture almost all the qualitative features of the critical physics of the Ising and $O(N)$ models in all dimensions (see the following).

[26] The derivation of these equations is outlined in Subsect. 2.2.7.

If $U_k, Z_k, \ldots$ are not truncated in a field expansion, the RG equation on Γ_k becomes a set of coupled partial differential equations for these functions. The initial condition at scale Λ is given by the hamiltonian of the model. Before studying in detail this approximation method let us make an important remark.

Contrarily to perturbation theory where only the renormalizable couplings are retained in the renormalized action, all powers of the fields appear in the ansatz of Eq. (2.137). There is no longer any distinction—at this level at least—between the two kinds of couplings, renormalizable and non renormalizable. This point together with a discussion of the notion of perturbative and nonperturbative renormalizability will be discussed in the following.

2.2.4 The Local Potential Approximation for the Ising Model

The Flow Equation of the Potential

We now consider the *ansatz* Eq. (2.132) for a $\mathbb{Z}_2$-symmetric theory. As already mentioned, the problem is to project the RG equation (2.127) on the potential U_k. This is naturally performed by *defining* the potential as Γ_k computed for *uniform* field configurations:

$$U_k(M_{\text{unif.}}) = \frac{1}{\Omega}\Gamma_k[M_{\text{unif.}}] \tag{2.142}$$

where Ω is the volume of the system. To compute the RG flow of U_k we act on both sides of this equation with ∂_t and we evaluate the right hand side thanks to Eq. (2.127). The only "difficulty" of this calculation is to invert $\Gamma_k^{(2)}[M] + R_k$ for uniform field configurations. This is where truncations of Γ_k are useful: they enable explicit calculations. For the LPA *ansatz* this is performed in detail in the Appendix, Subsect. 2.3.3. The final result reads:

$$\partial_k U_k(\rho) = \frac{1}{2}\int_q \frac{\partial_k R_k(q)}{q^2 + R_k(q) + U_k'(\rho) + 2\rho\, U_k''(\rho)} \tag{2.143}$$

where

$$\rho = \frac{1}{2}M^2 \tag{2.144}$$

is the $\mathbb{Z}_2$-invariant and $U_k'(\rho)$ and $U_k''(\rho)$ are derivatives of U_k with respect to ρ. An important remark is in order here. Once the angular integral has been performed, the integrand of Eq. (2.143) is of the form (up to a constant factor)

$$f(q^2, w) = |q|^{d-1}\frac{\partial_k R_k(q^2)}{q^2 + R_k(q^2) + w}. \tag{2.145}$$

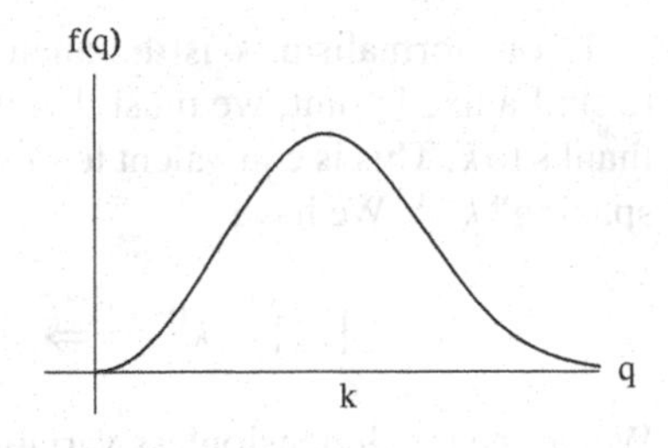

Fig. 2.13 The typical shape of the integrand $f(q)$

Let us consider a typical cut-off function:

$$R_k(q^2) = \frac{q^2}{e^{q^2/k^2} - 1}. \tag{2.146}$$

Then, for generic values of w, the typical shape of f at fixed k is given in Fig. 2.13 (for $d > 1$).

As expected, only a window of momenta around k contributes to the flow at scale k. We see in particular that the rapid modes are efficiently integrated out by this kind of cut-off function. This explains the decoupling of massive modes already discussed Subsect. 2.2.2. The cut-off function (2.146) has been used many times in the literature but it turns out that another one is sometimes more convenient because it allows to perform analytically the integral in Eq. (2.143). It writes [55]

$$R_k(q^2) = (k^2 - q^2)\,\theta(k^2 - q^2). \tag{2.147}$$

where θ is the step function. With this choice of R_k we easily find

$$\partial_t U_k(\rho) = \frac{4v_d}{d}\,\frac{k^{d+2}}{k^2 + U_k'(\rho) + 2\rho\, U_k''(\rho)} \tag{2.148}$$

where v_d is a numerical factor defined in Appendix, Subsect. 2.3.1. All we can learn about the model at this approximation is contained in the solution of this equation. We can already see that, as expected, this equation does not admit a fixed point potential U^*. We have already seen that it is necessary to go first to the dimensionless variables to find a fixed point. This is what we now do.

The Scaling Form of the RG Equation of the Dimensionless Potential

We have already emphasized when studying block-spins that it is necessary to measure all lengths in units of the running lattice spacing to find a fixed point of the RG flow. Looking for a fixed point is a convenient method to study the critical behavior of a model and we shall now spend some time deriving the RG equation on the dimensionless potential.

In our formalism, k is the analogue of the inverse running lattice spacing and, to find a fixed point, we must therefore "de-dimension" all dimensionful quantities thanks to k. This is equivalent to measuring all lengths in units of the running "lattice spacing" k^{-1}. We have

$$[\Gamma_k] = k^0 \quad \Longrightarrow \quad [M] = k^{\frac{d-2}{2}} \quad \text{and} \quad [U_k] = k^d. \tag{2.149}$$

We define the dimensionless variables by

$$y = \frac{q^2}{k^2} \tag{2.150}$$

$$R_k(q^2) = q^2 r\,(y) = k^2 y\, r(y) \tag{2.151}$$

$$\tilde{x} = k\, x \tag{2.152}$$

$$\tilde{M}(\tilde{x}) = k^{\frac{2-d}{2}}\, M(x) \tag{2.153}$$

$$\tilde{U}_t\big(\tilde{M}(\tilde{x})\big) = k^{-d}\, U_k\big(M(x)\big). \tag{2.154}$$

To derive the RG equation on $\tilde{U}_t$ we must keep in mind that the derivative ∂_t in Eq. (2.148) is computed at fixed ρ whereas we want now to compute it at fixed $\tilde{\rho}$. The detailed calculation is performed in the Appendix, Subsect. 2.3.3 and the result is:

$$\partial_t \tilde{U}_t = -d\tilde{U}_t + (d-2)\tilde{\rho}\,\tilde{U}_t' - 2v_d \int_0^\infty \mathrm{d}y\, y^{d/2+1}\, \frac{r'(y)}{y\big(1+r(y)\big) + \tilde{U}_t' + 2\tilde{\rho}\,\tilde{U}_t''}. \tag{2.155}$$

Once again, with the cut-off (2.147) the integral can be performed analytically and we find

$$\partial_t \tilde{U}_t = -d\tilde{U}_t + (d-2)\tilde{\rho}\,\tilde{U}_t' + \frac{4v_d}{d}\,\frac{1}{1+\tilde{U}_t' + 2\tilde{\rho}\,\tilde{U}_t''}. \tag{2.156}$$

We clearly see on this equation that the flow of $\tilde{U}_t$ has two parts, one that comes from the dimensions of U_k and ρ and one that comes from the dynamics of the model. This RG equation on $\tilde{U}_t$ is a rather simple partial differential equation that can be easily integrated numerically. We are thus in a position to discuss the critical behavior of the Ising model and to look for fixed points [32].

2.2.5 The Critical and Non-Critical Behavior of the Ising Model Within the LPA

Having derived the RG flow of the potential, we can relate a "microscopic" model defined at scale Λ by an hamiltonian (or directly by Γ_Λ) with the free energy $\Gamma = \Gamma_{k=0}$. Let us emphasize that there is no reason why the hamiltonian H should involve only ϕ^2 and ϕ^4 terms and not ϕ^6, ϕ^8, ... terms. In fact, the Hubbard-Stratonovich transformation enables to obtain the exact potential at the scale of the lattice spacing of a magnetic system. In the Ising case, the potential thus obtained is

$$U_\Lambda(\phi) \propto \log\cosh\phi \tag{2.157}$$

and is thus non-polynomial. In the NPRG framework this is not a problem since, even if $U_\Lambda(\phi)$ was a polynomial at scale Λ, it would become non-polynomial *at any other scale* (all $\mathbb{Z}_2$-invariant couplings are generated by RG transformations, see Subsect. 2.2.6 for a discussion).

Let us anyway, for the sake of simplicity, consider a dimensionless potential at scale Λ of the form[27]:

$$\tilde{U}_\Lambda(\tilde{\rho}) = \frac{\lambda_\Lambda}{2}(\tilde{\rho} - \kappa_\Lambda)^2 \tag{2.158}$$

with $\kappa_\Lambda > 0$. At the mean-field level and for uniform fields, $\Gamma^{\text{MF}} = \Omega\, U_\Lambda$ and we would deduce at this approximation from Eq. (2.158) that the system is in its broken phase with a spontaneous magnetization per unit volume $M_{\text{sp}} = \Lambda^{\frac{d-2}{2}}\sqrt{2\kappa_\Lambda}$.[28] However, the integration of the fluctuations can drastically change this picture: the minimum of the potential has a non-trivial RG flow that can drive it to zero. If this occurs, the system is in fact in its symmetric (high temperature) phase.

Let us call $\kappa(k)$ the running minimum of the dimensionless potential at scale k (more appropriately at "time" $t = \log k/\Lambda$):

$$\partial_{\tilde{\rho}}\tilde{U}_k\,_{|\kappa_k} = 0 \tag{2.159}$$

It is physically clear, and this can be checked on the flow of U_k, that the spontaneous magnetization decreases in comparison with its mean-field value because of the fluctuations. This means that the true spontaneous magnetization is always less than the mean-field spontaneous magnetization $\Lambda^{\frac{d-2}{2}}\sqrt{2\kappa_\Lambda}$. There are thus three possibilities.

(i) The system is in its broken phase (low temperature) and the (dimensionful) spontaneous magnetization density is given by

---

[27] The two models corresponding to the two initial conditions Eq. 2.157 and Eq. 2.158 belong to the same universality class. Thus they both have the same set of critical exponents. However, they differ as for their non universal quantities.

[28] At vanishing external magnetic field, the magnetization is given by $B = 0 = \partial U/\partial M$ and corresponds therefore to the location of the minimum of the potential.

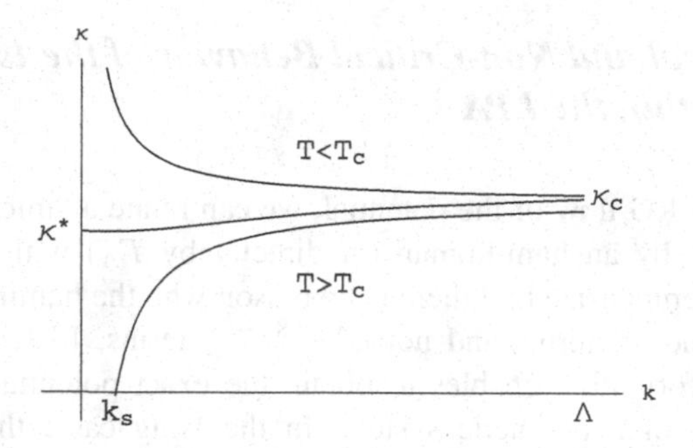

Fig. 2.14 Behavior of the running minimum $\kappa(k)$ of the dimensionless potential. From *top* to *bottom*: $T < T_c$, $T = T_c$ and $T > T_c$. κ_c is the critical initial value of κ_Λ for which the system is critical. This value should not be confused with κ^* which is the fixed point value of $\kappa(k)$. In the low temperature phase the dimensionless running minimum diverges whereas the dimensionful minimum converges to the value of the spontaneous magnetization since it is multiplied by a positive power of the scale k that compensates exactly the divergence of $\kappa(k)$. For the high temperature phase $k_s \simeq \xi^{-1}$

$$M_{\text{sp}} = \sqrt{2\rho_0(k=0)} \tag{2.160}$$

where $\rho_0(k)$ is the location of the minimum of the (dimensionful) potential $U_k(\rho)$.[29] The relation between $\kappa(k)$ and $\rho_0(k)$ is

$$\rho_0(k) = k^{d-2}\kappa(k) \tag{2.161}$$

from Eq. (2.153). Thus, if $\rho_0(k) \to M_{\text{sp}}^2/2$ when $k \to 0$, $\kappa(k)$ has to flow to infinity (for $d > 2$) as k^{2-d} in this limit, see Figs. 2.15 and 2.17.

(ii) The system is in the high temperature phase and the spontaneous magnetization is vanishing. Thus, as k decreases the minimum $\kappa(k)$ must decrease and, at a finite scale k_s, must hit the origin. It is easy to guess that k_s must be of the order of the inverse correlation length since as long as $k^{-1} \ll \xi$ the coarse-grained system remains strongly correlated and still behaves as if it was critical. It is only when $k^{-1} \sim \xi$ that "the system can feel" that its correlation length is finite and that its magnetization is vanishing. Thus $k_s \simeq \xi^{-1}$.

(iii) The system is critical. The spontaneous magnetization is vanishing which means that $\rho_0(k) = k^{d-2}\kappa(k) \to 0$ as $k \to 0$. Note that this does not require that $\kappa(0) = 0$ since $\kappa(k)$ is multiplied by a positive power of k, at least for $d > 2$.[30] In fact, $\kappa(k)$ reaches a finite fixed point value κ^*.

These three cases are summarized in Fig. 2.14.

---

[29] We shall see in the following that there is a subtlety here because of the convexity of the potential in the limit $k \to 0$.

[30] For $d \leq 2$ it is necessary to take into account the field renormalization, that is the anomalous dimension, to obtain a coherent picture of the physics. This requires to go beyond the LPA, see Subsect. 2.2.7.

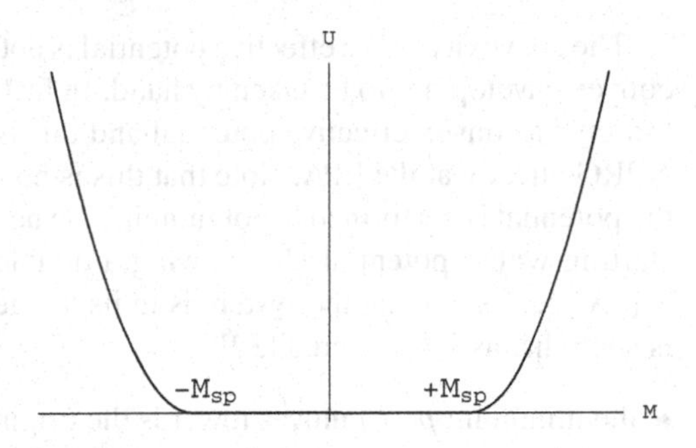

Fig. 2.15 Shape of the effective potential $U = U_{k=0}$ for $T < T_c$

For a generic initial potential

$$\tilde{U}_\Lambda(M) = \frac{\lambda_\Lambda}{2}(\tilde{\rho} - \kappa_\Lambda)^2 + \frac{u_{3,\Lambda}}{3!}(\tilde{\rho} - \kappa_\Lambda)^3 + \dots \qquad (2.162)$$

the critical value κ_c of κ_Λ is a function of all the other couplings $\lambda_\Lambda, u_{3,\Lambda}, \dots$ This means that in the space of dimensionless coupling constants there exists a "critical hypersurface" of co-dimension 1 that corresponds to systems that are critical. Generically, κ_Λ is a regular function of the temperature around T_c and at first order we can assume that

$$\kappa_\Lambda - \kappa_c \propto T_c - T. \qquad (2.163)$$

This allows us to relate the coefficients of the microscopic hamiltonian to the temperature. Note that contrarily to the mean-field analysis for which criticality is reached when the coefficient r_0 of the quadratic term of the potential (that is $r_0\phi^2/2$) is vanishing: $r_0 \propto T - T_c$, this is not true here since criticality does not correspond to the vanishing of the bare mass term. The mean-field analysis is wrong in this respect.

Let us now show what we expect for $U = U_{k=0}$ and for $\tilde{U}_{k=0}$.

The Low and the High Temperature Phases

In the low temperature phase, we expect a spontaneous magnetization, either up or down. More precisely, the equation

$$B = \frac{\partial U}{\partial M} \qquad (2.164)$$

is expected to have a solution $+M_{\text{sp}}$ for $B \to 0^+$ and $-M_{\text{sp}}$ for $B \to 0^-$. Moreover, U must be a *convex function* of M since it is obtained from a Legendre transform of W which is convex. Thus, U must have a very peculiar shape since it must have two minima at $\pm M_{\text{sp}}$ and no maximum in between (otherwise it should not be convex). The only possibility is that it is flat in between $-M_{\text{sp}}$ and $+M_{\text{sp}}$.

The convexity of the effective potential is not preserved by perturbation theory: the convex envelop has to be taken by hand. In fact, it is notoriously difficult to compute "safely" a convex effective potential and this is a property that is reproduced by the NPRG already at the LPA. Note that this is no longer the case if a field expansion of the potential is performed (a polynomial can never be flat on a whole interval). In fact, starting with a potential U_Λ showing a double-well structure and with parameters $\kappa_\Lambda, \lambda_\Lambda, \ldots$ such that the system is in its low temperature phase, the RG flow of U_k is such that as k is lowered [32]

- the minimum $\rho_0(k)$ moves towards the origin and eventually reaches a limit equal to the spontaneous magnetization $M_{\rm sp}^2/2$;
- the maximum of the potential located between the two minima decreases and goes to 0 as $k \to 0$ (the "inner part" of the potential flattens). At $k = 0$, $U_{k=0}$ looks like the potential of Fig. 2.15.

As $T \to T_c$, that is as $\kappa_\Lambda \to \kappa_c$, $M_{\rm sp}$ moves towards the origin and at $T = T_c$ coincides exactly (by definition of T_c) with the origin. In the high temperature phase, that is $\kappa_\Lambda < \kappa_c$, the spontaneous magnetization vanishes. Thus U has a single minimum at the origin in this case.

Let us finally notice that when $k \neq 0$, U_k is not convex in general. This is normal since Γ_k is not the Legendre transform of W_k (W_k is convex) because of the term $1/2 \int R_k M M$ that has been subtracted, Eq. (2.121).

The Critical Point

For $\kappa_\Lambda = \kappa_c$, the minimum of the potential U_k (or $\tilde{U}_k$) for $k > 0$ is non vanishing. It is only for $k \to 0$ that the minimum $\rho_0(k)$ of U_k reaches the origin. This means that the minimum of the potential never stops running at $T = T_c$ and that fluctuations of all wavelengths are required to make $\rho_0(k)$ vanish.

To characterize the critical point it is interesting to work with $\tilde{U}_k$ instead of U_k since, at this point, the long distance physics (compared with Λ^{-1}) is scaleless. This means that the potential, properly rescaled thanks to k, should be k-independent at $T = T_c$ and for sufficiently small k: it must be a fixed potential $\tilde{U}^*(\tilde{\rho})$

$$\partial_t \tilde{U}^*(\tilde{\rho}) = 0. \tag{2.165}$$

We deduce from this equation and from Eqs. (2.147, 2.156) that $\tilde{U}^*(\tilde{\rho})$ is solution of

$$0 = -d\tilde{U}^* + (d-2)\tilde{\rho}\tilde{U}^{*\,\prime} + \frac{4v_d}{d}\frac{1}{1+\tilde{U}^{*\,\prime} + 2\tilde{\rho}\,\tilde{U}^{*\,\prime\prime}}. \tag{2.166}$$

At first sight, the situation looks paradoxical since this is a second order differential equation that should admit infinitely many solutions indexed by two numbers whereas we expect only one fixed point in $d = 3$ corresponding to the universality class of

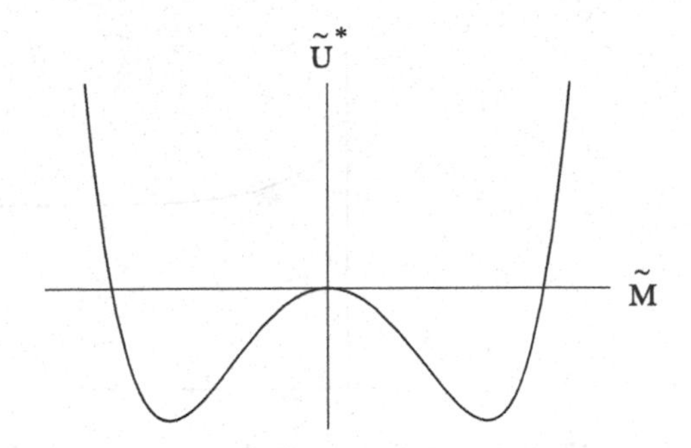

Fig. 2.16 Dimensionless fixed point potential of the Ising model in $d = 3$. The minima of this potential are located at $\pm\sqrt{2\kappa^*}$. The spontaneous magnetization is vanishing although $\kappa^* \neq 0$ (see the text)

the Ising model. In fact, it can be shown that among all these solutions, *only one* is well defined for all $\tilde{\rho} \in [0, \infty[$ [18].

An easy way to show this is the "shooting method". We first write down the fixed point equation for the derivative of $\tilde{U}^*$ with respect to $\tilde{M}$: $\tilde{U}^{*\,\prime}(\tilde{M})$ analogous to Eq. (2.166) and which is more convenient. Then, one initial condition is given by the $\mathbb{Z}_2$ symmetry: $\tilde{U}^{*\,\prime}(\tilde{M} = 0) = 0$. The other initial condition: $\tilde{U}^{*\,\prime\prime}(\tilde{M} = 0)$ is then adjusted so that $\tilde{U}^{*\,\prime}(\tilde{M})$ exists for all $\tilde{M}$. For a generic $\tilde{U}^{*\,\prime\prime}(\tilde{M} = 0)$ this is not the case: at finite $\tilde{M}$, $\tilde{U}^{*\,\prime}$ either blows up at $+\infty$ or at $-\infty$. By dichotomy, the value $\tilde{U}^{*\,\prime\prime}(\tilde{M} = 0)$ can be fine-tuned so that this occurs for larger and larger $\tilde{M}$. The fixed point potential thus obtained is shown in Fig. 2.16.

The same fixed point potential can be found "dynamically" by integrating the RG flow and by fine-tuning the value of κ_Λ to get closer and closer to κ_c. We find that for an initial κ_Λ very close to κ_c, the potential $\tilde{U}_t$ spends a very long RG time close to $\tilde{U}^*$ before either departing in the high or low temperature phase. Thus, all dimensionless couplings reach a plateau before blowing up. On this plateau, we obtain a very good approximation of $\tilde{U}^*(\tilde{M})$ very close to the one found by the shooting method. For $\kappa(k)$ very close to κ_c this is represented in Fig. 2.17.

One can observe that during a transient regime, $\kappa(k)$ for $k \simeq \Lambda$ is not stationary even at the critical temperature. This regime simply corresponds to the RG "time" necessary to approach the fixed point. It is non-universal since it depends on the starting point κ_c on the critical surface.

Let us finally point out a subtlety. When $k > 0$, it is possible to reconstruct $U_k(M)$ from $\tilde{U}_k(\tilde{M})$ by a somewhat trivial rescaling:

$$U_k(M) = k^d\, \tilde{U}_k\left(k^{-\frac{d-2}{2}} M\right). \tag{2.167}$$

The k factor acts as a magnification scale when we go from M to $\tilde{M}$ (at least for $d > 2$). For $k \to 0$ an infinitesimal range of M around the origin is mapped onto a finite range of $\tilde{M}$. It is therefore possible—and this is what indeed occurs at $T = T_c$—that $U_{k=0}$ has a trivial shape around $M = 0$ showing a minimum only at 0 (see Fig. 2.15 for $M_{\text{sp}} \to 0$) whereas $\tilde{U}_{k=0}$ shows a double-well structure (see Fig. 2.16)!

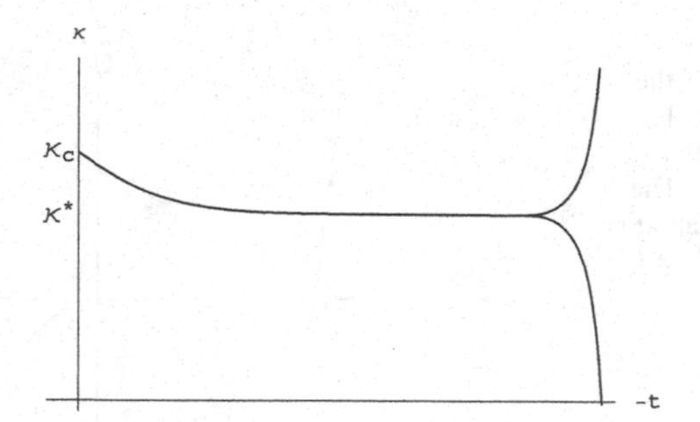

Fig. 2.17 Behavior of $\kappa(k)$ very close to κ_c. During the transient regime at small $-t$ the representative point of the system approaches the fixed point. Then, this point remains close to the fixed point and κ remains almost constant and very close to its fixed point value $\kappa*$. Then, if the initial condition of the RG flow was not right on the critical surface, it departs either in the low or in the high temperature phase where κ either blows up or goes to 0

The Critical Exponents

Within the LPA, there is no possibility of modifying the mean-field like q^2-dependence of $\Gamma_k^{(2)}(q)$ (evaluated at $M = 0$) since there is no renormalization of the derivative term. The exponent η is therefore vanishing at this order of the derivative expansion (see Eq. (2.69) and $\Gamma^{(2)}(q) = 1/G^{(2)}(q)$).

$$\eta^{\text{LPA}} = 0. \tag{2.168}$$

The exponent β defined in Eq. (2.84) can be calculated directly by fitting the behavior of M_{sp} defined in Fig. 2.15 as a function of $\kappa_\Lambda - \kappa_c$ which is itself proportional to $T_c - T$. Of course, this is not very convenient although feasible.

As for the exponent ν, we have seen that it is related to the behavior of the RG flow around the fixed point, Eq. (2.73) and, more precisely, is the inverse of the positive eigenvalue of the flow at the fixed point. This means that very close to the fixed point and away from the critical surface, the potential is such that[31]

$$\tilde{U}(\tilde{M}, t) = \tilde{U}^*(\tilde{M}) + \epsilon\, e^{-t/\nu} u(\tilde{M}). \tag{2.169}$$

It can be shown that $u(\tilde{M})$ must behave as a power law at large $\tilde{M}$ and that this ensures that there is a unique value of ν such that (2.169) holds. With the θ-cut-off function, Eq. (2.147), one finds at $d = 3$ [32, 44, 45]:

---

[31] The following relation is not general since if we choose a point on the critical surface, the corresponding potential is attracted towards $\tilde{U}^*(\tilde{M})$. This approach is governed by the so-called critical exponent ω corresponding to correction to scaling. Thus, in general, infinitesimally close to the fixed point, the evolution of the potential is given by the sum of two terms, one describing the approach to $\tilde{U}^*$ on the critical surface and one describing the way the RG flow escapes the fixed point if one starts close but away from the critical surface [18].

$$\nu^{\mathrm{LPA}} = 0.65 \tag{2.170}$$

to be compared with $\nu = 0.6297(5)$ obtained by Monte Carlo simulations.

Several remarks are in order here.

- In principle, from ν and η all the other critical exponents can be calculated thanks to the scaling relations derived in Subsect. 2.1.5. It is interesting to verify that these relations are not spoiled by the LPA. Indeed, by computing separately all the exponents, it is found that the scaling relations among them are very well satisfied.
- In the exact theory, no physical result should depend on the choice of cut-off function R_k. However, once approximations are performed a spurious dependence on the choice of R_k is observed. As a consequence, the whole scheme makes sense only if this dependence is weak. It is of course very difficult to obtain general results on this point since the space of cut-off functions is infinite dimensional and that we cannot sample it efficiently. In practice, we should choose a space of "reasonable" cut-off functions and study the variations of physical quantities like critical exponents in this space. This has been done in some details and it is observed that the dependence of ν on R_k is indeed weak [44–46]. Let us emphasize that this problem is not specific to the NPRG. Even in the perturbative schemes the critical exponents acquire a spurious dependence on several choices made during their calculations.
- The LPA is certainly not appropriate for the computation of critical exponents in $d = 2$. We have seen several times that this dimension plays a special role in the formalism although nothing spectacular is expected in $d = 2$ for the critical behavior, at least for the Ising model. This comes from the fact that $\eta = 0$ starts to be a bad approximation at and below $d = 2$ for this model [it is worse for the $O(N)$ models with $N \geq 2$ because of Mermin-Wagner's theorem]. The exact value of η is known in $d = 2$ from Onsager's solution of the Ising model: $\eta = 0.25$. At low dimensions, going to the next order of the derivative expansion cures most of the problems encountered at the level of the LPA [32, 56].
- As explained in the following for the $O(N)$ models, it is possible to go beyond the LPA and to compute with greater accuracy the critical exponents, η in particular. This kind of calculations has been performed by several authors [57] at order $O(\partial^2)$ of the derivative expansion for the Ising and $O(N)$ models in $d = 3$ and $d = 2$ [58] and also at $O(\partial^4)$ for the Ising model in $d = 3$. Let us quote the best results obtained for the Ising model in $d = 3$:

Let us now review some conceptual aspects of renormalization and in particular what Wilson's RG teaches us about perturbative and non perturabtive renormalizability (Table 2.1).

Table 2.1 Critical exponents of the three dimensional Ising model

Order	ν	η
∂^0	0.6506	0
∂^2	0.6281	0.044
∂^4	0.632	0.033
7-loops	0.6304(13)	0.0335(25)

∂^0, ∂^2 and ∂^4 correspond to the order of the truncation of the derivative expansion (the NPRG method) [46]. For completeness, we have recalled in the last line, the results obtained perturbatively [5].

2.2.6 Perturbative Renormalizability, RG Flows, Continuum Limit, Asymptotic Freedom and all That...

Our understanding of the physical meaning of renormalizability has drastically changed with the works of Wilson and Polchinski [12] (and many others).[32] It is even astonishing to see that while the technical aspects of perturbative renormalization have not changed since the 1950s, the interpretation of the notion of renormalizability has been deeply modified these last 20 years. Let us mention just a few points that illustrate this change of viewpoint on renormalization:

(i) The regularization scale of a field theory—the cut-off of the Feynman integrals or the inverse lattice spacing for instance—was considered as an unphysical scale introduced during intermediate calculations and that had to be sent to infinity at the end of the calculation. It is now thought to be the scale where new physics takes place and is thus a fundamental scale of the model. Most of the time, it is not necessary to send this scale to infinity (but perhaps for technical convenience).

(ii) For the same reason the "bare theory" (associated with the scale of the cut-off) was considered as unphysical whereas the renormalized theory was supposed to be fundamental. The bare theory is now often considered as the fundamental theory whereas the renormalized theory is only an effective theory valid at long-distance [long-distance meaning distances that are large compared to the inverse ultra-violet (UV) cut-off].

(iii) The only valid theories were thought to be built with renormalizable couplings only while nonrenormalizable couplings were believed to be forbidden (for reasons that looked rather mysterious). Nonrenormalizable couplings were thus considered as terrible monsters that nobody was able to master while super-renormalizable couplings were considered as very kind beasts. In this zoo, renormalizable couplings were classified just in between these two kinds of animals. We now know that it is useless to incorporate nonrenormalizable couplings in the "bare" action—that is in the UV regime—at least if one is interested only in the "universal" behavior of the model under study (see the following for a

[32] Much of what follows in this subsection comes from the works of Bagnuls and Bervillier to whom I owe much of my understanding of this subject [49, 51].

precise explanation of this statement). Put it differently, the contributions of the nonrenormalizable couplings included in the bare action die out at large distances and are thus under control. Reciprocally, super-renormalizable couplings of the bare action (when they exist) do not make any problem in the ultra-violet regime but are those that dominate the large distance physics of the theory and are thus the most important terms to control in this regime.

(iv) Removing the cut-off, that is taking the "infinite cut-off limit" order by order of perturbation expansions, was supposed to be the real issue of renormalizability. It turns out that the possibility of taking the infinite cut-off limit at all orders is *not* equivalent in general to the possibility of taking the "continuum limit", that is of defining the field theory at the limit where the *nonperturbative* regulator (the lattice for instance) is removed. Moreover, taking the continuum limit of a model is not in general a physically relevant issue since at very short distances the relevant physical model can well be something else than a field theory. This is already the case for the Ising model. The fundamental theory can also be a field theory but involving many other terms than the renormalizable ones and even infinitely many. The ϕ^4 model for instance is a good effective theory of the Ising model at large distances but the continuum limit of this field theory teaches us nothing about the small distance behavior of the Ising model.

Let us now discuss in more details these structural aspects of renormalization. As a starting point let us consider the different status of the couplings in the perturbative and nonperturbative frameworks. In the former some couplings are called renormalizable while the others are called nonrenormalizable whereas in the latter they are treated on the same footing. We shall mainly study the massless case in what follows although it is possible to generalize almost all arguments to the massive case.

Renormalizable Couplings and Large River Effect

To understand the origin of the difference between renormalizable and nonrenormalizable couplings, it is necessary to remember that in perturbative field theory, the action (bare or renormalized) is *not* supposed to be a physical quantity. It is only the "mathematical tool" that generates the vertices and the propagators of the theory from which Feynman diagrams are computed. As for the Feynman diagrams they are the building blocks of the Green functions that are physical (in particle physics, the S-matrix is physical). The effective action $\Gamma[M]$ is also physical. The action S or the hamiltonian H of perturbative field theory can well be polynomial and involve only a few couplings, $\Gamma[M]$ always involve all powers of the field in a non-trivial way. In the particle physics language and for the ϕ^4 theory, this means that the connected Green functions $G^{(n)}$ with $n = 6, 8, \ldots$, corresponding to the scattering of n particles are non vanishing and cannot be factorized into products of $G^{(p)}$ with $p < n$ corresponding to the scattering of fewer particles. This is a signature of the fact that a field theory involves infinitely many degrees of freedom that require infinitely many independent correlation functions to be faithfully described.[33] Thus *all* couplings g_n—defined as the values of the $\Gamma^{(n)}$'s for some configuration of the external momenta—are non-trivial, even those corresponding to an arbitrarily large n, that is to an arbitrarily large number of fields.

In Wilson's framework, all couplings g_n have a non-trivial RG evolution that, *a priori*, needs to be taken into account. This is the reason why, for instance, we must keep a complete function of M for the potential U_k and not just the first terms of its expansion in M, that is the M^2 and M^4 terms.[34] If at scale Λ we had retained only these couplings in the initial condition of the RG flow on U_k, we would have anyway found that all others would have been generated at any scale k lower than Λ.[35] Thus, at first sight, all couplings should be treated on the same footing and the distinction between renormalizable and nonrenormalizable couplings seems to have disappeared in Wilson's approach.

This is paradoxical because this seems to be in conflict with what we know from perturbation theory. In the perturbative scheme, all $\Gamma^{(n)}$ are also nontrivial and thus all (renormalized) couplings corresponding to the values of the renormalized $\Gamma_R^{(n)}$ (at zero momentum for instance) are nontrivial. However, since all renormalized functions $\Gamma_R^{(n)}$ can be expanded in powers of the renormalized coupling $g_{4,R}$, it is possible to compute all renormalized couplings $g_{6,R}, g_{8,R}$, etc. in terms of this unique coupling constant. This is what makes a renormalizable field theory predictive: once a finite number of (renormalized) masses and couplings have been determined by external means (e.g. experiments) the whole theory is entirely determined and infinitely many quantities can be computed out of it.

The paradox is resolved by explicitly computing the RG flows of the couplings g_n, for instance within the local potential approximation.[36] At sufficiently long distance, that is for k sufficiently small, all RG trajectories—that take place within the infinite dimensional space of coupling constants—are found to be attracted

---

[33] At first sight, this statement could seem incorrect in particle physics if one considers a given diffusion process. For instance the reaction $e^+e^- \to \gamma\gamma$ seems to involve four bodies only. This is actually wrong since, as virtual states involved in the loop expansion, an arbitrarily large number of particles can be exchanged during this reaction. The full Fock space structure is thus necessary to describe any kind of diffusion in the quantum and relativistic framework. The same is true in statistical mechanics. The infinite number of degrees of freedom of the system is not the relevant point. A perfect gas for instance can well involve infinitely many degrees of freedom, we all know that the whole machinery of field theory is not necessary to study it. The important point is the number of degrees of freedom that effectively interact together, that is the value of the correlation length. As long as $\xi \sim a$ field theory is not necessary since the system breaks down in small sub-systems of size ξ that are almost independent of each other. This is the reason why the law of large numbers is valid in this case and the fluctuations gaussian. Field theory is relevant only when $\xi \gg a$ in which case, for length scales l such that $a \ll l \ll \xi$ field theory is relevant.

[34] Remember that U_k corresponds to the zero momentum configuration of all $\Gamma_k^{(n)}$.

[35] Truncating the field dependence of the running potential U_k by keeping only the M^2 and M^4 terms at all scales k is nothing but a very crude approximation that eventually leads to neglecting all functions $\Gamma^{(n)}$ with $n \geq 6$ in the effective action $\Gamma[M] = \Gamma_{k=0}[M]$.

[36] All qualitative features that are explained below do not depend on this approximation.

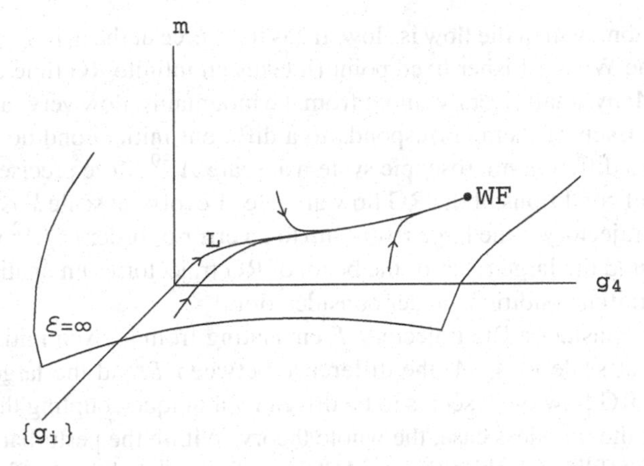

Fig. 2.18 Flow on the critical surface. The line L, called the large river, is an attractive submanifold of the RG flow of dimension one (in the Ising case). It starts at the Gaussian fixed point and ends at the Wilson-Fisher fixed point (WF). All RG trajectories approach L very rapidly while the flow on L is "slow". All the starting points of the RG trajectories correspond to different initial conditions, that is, at microscopic scales, to different $\mathbb{Z}_2$-invariant systems. The part of the RG trajectories that is far away from L is non-universal and corresponds to a transient regime. The axis $\{g_i\}$ represents the infinite number of axes different from g_4 which is the coupling of the ϕ^4 term. The axis m represents the direction of the mass which is a relevant direction of the RG flow

towards a submanifold of dimension two, see Fig. 2.18 for the critical case where the attractor is of dimension one since the massis eliminated.[37]

The RG time necessary to reach this submanifold is very short whereas the flow within it is slow. This is particularly clear, and has been studied in detail by Bagnuls and Bervillier, for the $\mathbb{Z}_2$-invariant theories in $d = 3$ when they are critical [49, 51]. In this case, the RG trajectories belong to the critical surface. All of them, after a transient regime, almost collapse on a line, that we call L, joining the gaussian fixed point to the non-trivial fixed point describing the phase transition of the Ising model (called the Wilson-Fisher fixed point). Thus, in the long-distance physics compared to Λ^{-1}, that is beyond the transient regime, the RG flow behaves as if it was driven by a unique "coupling". As long as L has a non singular projection on the axis g_4 corresponding to the M^4-coupling, it is possible to describe the flow along the line L from the flow of the g_4-coupling alone.[38] This is what perturbation theory does. This explains why in this framework all couplings have on one hand a non-trivial RG flow and on the other hand a flow determined by g_4 alone.

Bagnuls and Bervillier have used the following metaphor [49, 51]: the RG trajectories on the critical surface are like rivers in the mountains. In the valley, there is a

[37] In general, the attractive submanifold is of dimension the number of (perturbatively) renormalizable couplings including the masses. In the critical case, the mass is vanishing by definition and the attractor is of dimension one for models for which only one coupling is renormalizable.

[38] One should remember that the space of coupling constants is infinite dimensional and that it is therefore non-trivial that the projection of L onto the g_4-axis be non vanishing. A randomly chosen vector in an infinite dimensional space has in general a vanishing projection onto a given direction.

large river L along which the flow is slow. It has its source at the gaussian fixed point and stops at the Wilson-Fisher fixed point (it takes an infinite RG time to reach this fixed point). Many small rivers, coming from the mountains, flow very rapidly "into" the large one. Each of them corresponds to a different initial condition of the RG flow, that is to a different microscopic system at scale Λ.[39] More precisely, different "natural" initial conditions of the RG flow at scale Λ evolve at scale $k \ll \Lambda$ towards the same RG trajectory—the large river—up to an error of order k^2/Λ^2 which is the thickness, around the large river, of the beam of RG trajectories emanating from the set of natural initial conditions under consideration.

Let us now consider a RG trajectory T emanating from a given initial condition at scale Λ. If at scale $k \ll \Lambda$, the difference between T and the large river L is neglected, the RG flow on T seems to be driven by a unique coupling that therefore determines, in the massless case, the whole theory. Within the perturbative scheme, the infinite cut-off limit enables to get rid of any reference to the cut-off scale, that is to Λ, and leads to a unique RG flow which is valid at any scale, at least in principle. It must therefore be the flow on L and the infinite cut-off limit has thus removed the transient regime where T and L are far away. The theory has thus been "projected" onto the line L (in the massless case). This is of course technically convenient since the transient regime depends on the initial condition of the RG flow at scale Λ and thus on all couplings either renormalizable or nonrenormalizable involved in the initial condition. However, by taking the infinite cut-off limit, that is by neglecting the difference between T and L—which is a very good approximation in the IR regime—, one has given up the possibility of reversing the flow in the UV direction to go back on T to the initial condition at scale Λ. This means that it is no longer possible to initialize the flow at Λ from a "microscopic" action once T has been replaced by L (in the IR). To be consistent, the RG flow must therefore be initialized at a finite scale, say μ, which is infrared-like with respect to Λ.[40] This is the role of the "renormalization prescriptions" of perturbation theory that enable to parametrize

[39] For the Ising model, different natural initial conditions could correspond for instance to different kinds of lattices or to different types of couplings among the spins: next nearest neighbor couplings, anisotropic couplings, etc.

[40] Let us mention here that within dimensional regularization, for instance, there is no explicit ultraviolet regularization scale. The only scale introduced in this regularization scheme is the scale, often called μ, necessary to preserve dimensional analysis when, for loop-integrals, $\int d^4q$ is replaced by $\int d^dq$: $\int d^4q \to \mu^{4-d} \int d^dq$. This scale is also used most of the time as the scale of the renormalization prescriptions that are either explicit, see Eq. (2.171), or implicit as in the $\overline{\text{MS}}$ scheme (the fact that they can be implicit does not change anything to our discussion). However, any regularization consists in modifying the short distance behavior of the theory because this is where the divergences come from. Thus the ultraviolet cut-off, although not explicit must be built from μ and $\epsilon = 4 - d$. From a comparison of the divergences obtained in dimensional regularization and in the cut-off regularization, it is easy to get a qualitative correspondence between the two schemes and thus an estimate of the UV cut-off scale of dimensional regularization. When a logarithmic divergence occurs in the cut-off regularization scheme, a pole in ϵ occurs in dimensional regularization. Thus it is reasonable to imagine that $\log \Lambda \sim 1/\epsilon$. To make this correspondence dimensionally valid we must have $\log \Lambda/\mu \sim 1/\epsilon$. We therefore find that in dimensional regularization, the UV scale behaves as $\Lambda \sim \mu \exp(1/\epsilon)$.

the theory in terms of a renormalized coupling (and mass) defined as the value of the renormalized correlation function at scale μ:

$$\Gamma_R^{(4)}(\{p_i^N\}) = g_{4,R}(\mu) \tag{2.171}$$

with $\{p_i^N\}$ that are functions of μ. For instance, at a symmetric point

$$p_i^N \cdot p_j^N = \mu^2(\delta_{ij} - 1/4). \tag{2.172}$$

In the perturbative scheme, the renormalization prescriptions come from the necessity of fixing the arbitrariness of the finite parts of the subtraction procedure of the divergences. Since ∞ + anything finite = ∞, subtracting a divergence is a well-defined procedure up to a finite part. To finally get a unique and well-defined renormalized theory, it is necessary to specify this finite part. Technically, this is what renormalization prescriptions do. From a RG point of view, they also enable to initialize the RG flow at scale μ.[41] Of course, and this can be checked on each example, there are as many independent parameters in a field theory as primitive divergences—and thus as renormalization prescriptions, let apart the field renormalization—and this is also the number of dimensions of the attractive submanifold of the Wilsonian RG flow.

Let us finally mention that the back-reaction of the nonrenormalizable couplings on the flow of the renormalizable ones can be analyzed within perturbation theory. In dimension four for instance, a $g_6\phi^6$-term in the action contributes to the flows of the mass and of g_4 in the following way. Since a new vertex exists, new graphs appear in the loop-expansion of $\Gamma^{(2)}$ and $\Gamma^{(4)}$. From the power counting point of view, that is by dimensional analysis, it is clear that these new graphs contribute to $\Gamma^{(2)}$ and $\Gamma^{(4)}$ as the ϕ^4-coupling does, that is by factors Λ^2 and $\log \Lambda$ for $\Gamma^{(2)}$ and by factors $\log \Lambda$ to $\Gamma^{(4)}$. This can be checked on the graph contributing to $\Gamma^{(2)}$ shown in Fig. 2.19.

Let us notice that the "divergences" coming from the integrals are more severe for the graphs involving a ϕ^6-vertex than for the others but that the overall Λ-dependence is indeed what is expected since $g_6 \sim \Lambda^{-2}$ from dimensional analysis. For graph 2.19 for instance the two tadpoles contribute by a factor Λ^4 and the vertex by a factor Λ^{-2}. These new divergences, being of the same type than the ones coming from

---

[41] Initializing this flow in the IR has several advantages. First, as we already mentioned, this is anyway obligatory once the infinite cut-off limit has been taken since, in this case, any reference to a microscopic model defined at an UV scale has been lost. Second, if the model under study is not derived from a more fundamental model at scale Λ—in which case the analytical form of Γ_Λ is not known—, its initialization at scale Λ would require in general infinitely many phenomenological input parameters since Γ_Λ is in general not polynomial. This is of course impossible and should be compared to what is done in the IR: only the values of the renormalizable couplings have to be fixed since the dimension of the submanifold L is the number of renormalizable terms. For the Ising model for instance, it is possible to initialize the flow at scale Λ since already at this scale it is an effective model derived from a more fundamental model (coming for instance from the Hubbard model). The same occurs for fluids for instance: microscopic models can be derived from other models that are more fundamental.

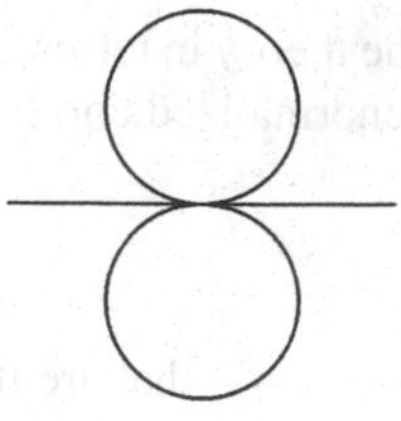

Fig. 2.19 The double tadpole diagram contributing to $\Gamma^{(2)}$ and that involves at first order the g_6 coupling. Each tadpole contributes to a factor Λ^2 and the vertex (that is g_6) to Λ^{-2}

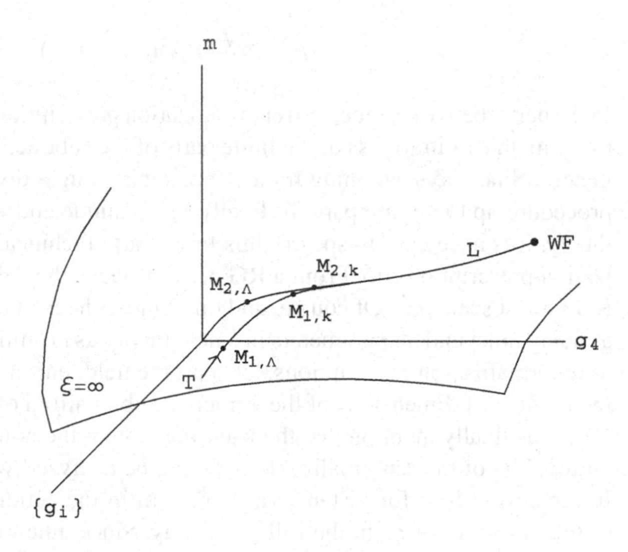

Fig. 2.20 Flows on the critical surface for two different initial conditions $M_{1,\Lambda}$ and $M_{2,\Lambda} \in L$ differing only by the initial value of an irrelevant coupling. During the transient regime on T, that is before T is close to L, the RG flows differ between the two trajectories so that at scale k the points $M_{1,k}$ and $M_{2,k}$ are not close to each other although $k \ll \Lambda$ and $M_{1,k}$ is almost on L

the ϕ^4-coupling, can be eliminated through a renormalization of the field, the mass and of g_4. The flow of the renormalized ϕ^4-coupling, $g_{4,R}$, is not affected by these contributions. From the RG point of view, this comes simply from the fact that once T and L are very close, the flow on both trajectories are almost identical: the difference between both flows that comes from a difference in the nonrenormalizable couplings at scale Λ, shows up only in the first part of the flow when T and L are far away from each other. More precisely, let us consider two initial conditions at the same scale Λ differing only by nonrenormalizable terms, for instance two points $M_{1,\Lambda}$ and $M_{2,\Lambda}$ on T and L corresponding to two different values of $g_{6,\Lambda}$. At scale $k \ll \Lambda$ these points have evolved in $M_{1,k}$ and $M_{2,k}$, see Fig. 2.20. Although $M_{1,k}$ is close to the trajectory L, the two points are in general not close to each other since the beginning of the two flows are very different. This is why non universal quantities are in general not correctly computed if nonrenormalizable terms are neglected. However, one can consider the point $M'_{2,k}$ on L which is the closest to $M_{1,k}$ and $M'_{2,\Lambda}$ the ancestor at scale Λ of this point. The two models described respectively by $M_{1,\Lambda}$ and $M'_{2,\Lambda}$ differ at short distance but are (almost) identical at all scales larger than the scale k_0 where T and L are very close since all couplings in both models are almost identical for $k < k_0$. In particular, all quantities—universal or non-universal—that can be computed from correlation functions at vanishing external momenta are found identical in both models. This shows that, at least in principle and for sufficiently long-distance physics, both universal and non-universal quantities could be computed from the flow on L—and therefore perturbatively—provided that one knows where to initialize the flow. Of course, the problem is that in general we do not have such an information and this is the very reason why non-universal quantities cannot be accurately computed within perturbation theory.

Universality

As explained previously, all the universal features of a given microscopic model can be computed from the flow on the large river. However, once the true RG trajectory T of the model has been approximated by the large river L in the long distance regime it is impossible to reverse the flow in the short distance direction to go back to the microscopic model we started with. Thus, once close to the large river, it is almost impossible to know from which small river comes the water, that is from which initial condition does the flow come from (this would require high precision measurements). This is universality. We thus observe that universality is a much more general concept than the statement that some critical exponents are independent of the microscopic details. Critical exponents are properties attached to the flow around the infrared attractive fixed point, if it exists. The attractive character of the submanifold that can be parameterized by renormalized couplings only is independent of the existence of such a fixed point. It exists for theories that are not critical and it takes place far before the theory is in the deep infrared regime. It is this property that makes field theory—or at least perturbative field theory—interesting since this is what allows us to focus on a small number of couplings. The price to pay is that nonuniversal features are not (accurately) computable within perturbative field theory.

Let us emphasize at this stage that the NPRG, because it is functional in essence, can follow any RG trajectory in the whole space of coupling constants contrarily to perturbation theory. The infinite cut-off limit $\Lambda \to \infty$ performed in perturbation theory and which is the shortcut that enables to eliminate any transient regime and to pick up the flow on the large river is not necessary in the NPRG framework. It is therefore possible to start at scale Λ with a given initial condition—polynomial or not in the field—and to integrate the RG equation down to $k = 0$. In principle, the full free energy of the model can be obtained in this limit and the only limitation of the method comes from the truncation of $\Gamma[M]$ that has been chosen to integrate the RG equation. Thus, contrarily to perturbation theory, there is no intrinsic limitation of the NPRG approach as for the computation of nonuniversal quantities. Several computations of nonuniversal quantities such as critical temperatures in fluids [59] or phase diagrams for reaction-diffusion systems (the so-called branching and annihilating random walks) [60] have been obtained accurately by NPRG methods.

Continuum Limit and Asymptotic Freedom

Taking the continuum limit of a field theory consists in removing any reference to an ultra-violet cut-off while maintaining the long-distance physics fixed. At first sight, this question seems equivalent to that of the infinite cut-off limit addressed in the perturbative context. The difference comes from the fact that the question of the existence of the continuum limit is addressed beyond perturbation theory. We shall only study the critical theory in the following although the massive case could be studied along the same line. Wilson's RG is the right framework to study this question.

We shall study in the following the particular case where the "continuum limit", $\Lambda \to \infty$, is taken on a particular RG trajectory, that is for a given physical system. For the massless $\mathbb{Z}_2$-invariant model in $d = 3$, there is only one trajectory for which this limit exists: the large river joining the gaussian to the Wilson-Fisher fixed point. The limit on this trajectory is taken in the following way. We imagine that an external information is known about the system at a scale k (an experiment has been performed for instance) that allows us to determine the value, at this scale, of at least one coupling constant. We are thus able to associate a point on the line L with the scale k. Keeping fixed this point, we can determine which ancestor at scale Λ corresponds to this point by inverting the RG flow in the ultraviolet direction. Since we have chosen the large river as RG trajectory, the flow slows down as Λ is increased since the point associated with this scale ($M_{2,\Lambda}$ in Fig. 2.20) gets closer and closer to the gaussian fixed point. When $\Lambda \to \infty$ the point associated with this scale tends to the gaussian and thus the continuum limit exists on this trajectory. The existence of a continuum limit is therefore a consequence of the fact that the theory is asymptotically free in the ultraviolet regime. One should be careful that, of course, this procedure should be taken as a limit: if one starts the RG evolution right on the gaussian fixed point, the representative point of the model remains gaussian.

Let us emphasize that for any RG trajectory other than L, the continuum limit does not exist (in the massless case) since the flow blows up at infinity in the ultraviolet direction for these trajectories. Let us remind the reader that the flow is extremely rapid away from the large river. This means that as soon as the RG trajectory is not "close" to L, the flow in the ultraviolet direction diverges very rapidly from L. This does not mean that the corresponding field theory does not exist. This simply means that the ultra-violet cut-off cannot be removed since the flow is not controlled in the UV direction. However, as an effective (cut-off) field theory it is valid in the IR and its universal behavior is the same as the field theory defined on L.

Let us emphasize that the continuum limit is probably irrelevant from a physical viewpoint: who cares about the physics at asymptotically small distances or, in particle physics, at asymptotically large energies? This question goes beyond physics since, by definition, there will never exist any experiment able to test the infinitesimally short distance regime of any model. Field theory and more precisely renormalizable field theories are most probably only effective long-distance models for the underlying and more fundamental theories that probably involve nonrenormalizable terms and/or are not even field theories.

Perturbative (Non-) Renormalizability Again

It is important at this stage to realize that the infinite cut-off limit that can be taken order by order of the perturbation expansion of a renormalizable theory does not make reference to any RG trajectory and thus neither to the existence of fixed points nor to asymptotic freedom. It depends only on the possibility of removing recursively the divergences of the theory by the redefinition of the parameters of the model. We could thus question the meaning of this infinite cut-off limit. Let us first make a remark here.

The existence of the infinite cut-off limit does not imply that of the continuum limit. The example of the $\mathbb{Z}_2$-invariant theories in $d = 4$ is illuminating in this respect. When d is increased from 3 to 4, the Wilson-Fisher fixed point moves towards the gaussian and, in $d = 4$, coincides with it. The line L does no longer exist in this dimension. The gaussian fixed point becomes infrared attractive and ultraviolet repulsive. Any point on the critical surface in $d = 4$ is attracted towards the gaussian in the infrared and pushed away from it in the ultraviolet. There is no continuum limit of any model apart from the gaussian model itself.[42] However, the infinite cut-off limit exists at all orders for the ϕ^4 theory even in $d = 4$! This means that order by order, it is possible to construct finite renormalized Green functions defined by their series expansion in the (renormalized) ϕ^4 coupling without any reference to an UV cut-off. Of course, such an expansion does not guarantee that the sum of the perturbative series of the Green functions exist. Let us remind the reader that, as usual in physics, the perturbative series are not absolutely convergent. *They are asymptotic series, at best* [5, 41, 61]. For the three-dimensional ϕ^4 theory, it has been rigorously proven that these series are Borel-summable (in the "massive zero momentum scheme" their Borel-transforms are convergent). It is very probable that in $d = 4$ the renormalized perturbative series are not summable for any finite value of $g_{4,R}$. Thus, the renormalized series of the Green functions can well exist at all orders, this does not mean that the functions themselves exist.

It could seem strange to the reader that the perturbative RG is in a better position than the perturbation expansion of the Green functions to study the problem of the existence of the continuum limit. It is important to have in mind that, even at one-loop, the β-function of the ϕ^4 coupling constant: $\beta(g_{4,R})$ takes into account the leading-logarithms *at all orders* of the perturbative series [6]. The RG goes, in this respect, beyond perturbation expansion even if the β function is computed peturbatively since it is sensitive to informations on the behavior of the sum of the series of the $\Gamma^{(n)}$'s that are unreachable at any finite order of their perturbation expansion. This is the reason why, even at one-loop, the behavior of the β function of the ϕ^4 theory leads to the conclusion (probably valid beyond perturbation theory) that the ϕ^4 theory has no non-trivial continuum limit in $d = 4$ whereas the perturbation expansion of the Green functions exists at all orders at the infinite cut-off limit.

[42] Reciprocally, any model defined at an UV scale Λ on the critical surface behaves at large distance as if it was free (this is the so-called "triviality problem" of the ϕ^4 theory in $d = 4$). This is the reason why $d = 4$ is the upper critical dimension of the $\mathbb{Z}_2$-invariant models.

Having in mind this subtlety, we could thus re-ask the question: What does mean perturbative renormalizability? We have already seen above that the infinite cut-off limit removes all finite cut-off effects from perturbation theory (the transient regime before reaching the large river L) and thus enables to pick up the RG flow on the line L. It thus allows us to select the (renormalizable) coupling(s) that drives the flow of all the others on this line. But then the relevant question is: what does mean perturbative non-renormalizability? After all, we have just learnt that all couplings are allowed (they are all non vanishing on the line L) and that they are all functions of the renormalizable one(s) once the RG trajectory has (almost) collapsed with L.[43] Thus, all couplings are present even in perturbation theory: they are simply not all independent. What does mean in this case that a coupling is perturbatively non-renormalizable? To answer this question we have to go back to what occurs in perturbation theory when a non-renormalizable coupling is present in the action of the model. In this case, at one-loop order, other non-renormalizable terms must be introduced in the action to cancel the divergences coming from this term. Then, at the next orders, these new terms generate new diagrams and new divergences that themselves require new non-renormalizable terms in the action to cancel them. This process never stops and the total action involves infinitely many terms. This is not in itself sufficient to make the theory useless: we have already seen that in fact all couplings are allowed and are likely to be there at the scale of the cut-off Λ. The very point is that, as we already mentioned above, there must be as many renormalization prescriptions as there are independent divergences to fix the arbitrariness of the finite parts of the counter-terms. For a non-renormalizable theory, this means that there are infinitely many prescriptions and the theory is thus non predictive (each prescription requires an input, an experimental result for instance). This is the very reason why non renormalizable terms are supposed to be forbidden perturbatively. We can reformulate and understand the non-renormalizability criterium from a RG point of view. Imposing a renormalization prescription on a Green function $\Gamma^{(n)}$ means imposing the value of this function for a given configuration of its momenta, that is specifying the value of the coupling g_n at an infrared scale, say μ. Let us now recall that the renormalization prescriptions are imposed independently of the RG flow within perturbation theory. The non-renormalizability of a theory simply means that it is "impossible" to impose that the RG flow of the theory goes at the infrared scale μ through a point arbitrarily chosen in the space of coupling constants: the flow must be on (or extremely close to) the large river L in the infrared. If we nevertheless impose that the flow goes through such an arbitrarily chosen point, we thus find that (i) the theory is non-predictive even in the infrared since the flow is very rapid away from L and that therefore a small change of scale below μ—or a small error in the determination of the renormalization prescriptions—leads to very large changes of the running couplings, and (ii) the ancestor in the ultra-violet of the point chosen for the prescriptions is extremely far away from the gaussian: at Λ,

[43] Once again we deal only with a massless theory. The presence of a mass does not change qualitatively the discussion below.

the initial condition of the RG flow is thus *extremely unnatural* in this case.[44] Thus, what perturbation theory calls non-renormalizable is a theory that is non-predictive in the infrared and non-natural in the ultraviolet. The subtlety here is that the reason why non-renormalizable theories should be discarded is not that it is forbidden to consider non-renormalizable couplings at scale Λ (they are in general present at this scale) but that it is impossible to choose at will their value in the infrared! The only possible choice to avoid the above-mentioned problem would be to pick up with the (infinite number of) renormalization prescriptions a point on L (or very close to L). In this case, the ancestor in the ultraviolet of this point would be also on L and we would have thus built at scale Λ a theory that perturbatively would involve infinitely many couplings, would remain predictive in the infrared and well controlled in the UV since it would become gaussian as $\Lambda \to \infty$. In fact this theory would not really involve infinitely many couplings since as already emphasized, on L the flows of all couplings are driven by those of the renormalizable ones. This program has been initiated by Symanzik and, in Wilson's words, this is the best that we can do within perturbation theory.

Let us now come back to physics and study the $O(N)$ models at order $O(\partial^2)$ of the derivative expansion.

2.2.7 The $O(N)$ Models at $O(\partial^2)$ of the Derivative Expansion

Although the Ising and $O(N)$ models have much in common, there are several non-trivial points specific to the $O(N)$ models that are worth studying. Among them is the presence of Goldstone modes in the low temperature phase, the Mermin-Wagner theorem in $d = 2$ and the Kosterlitz-Thouless transition for $N = 2$ in $d = 2$. We shall use again the derivative expansion that writes at $O(\partial^2)$:

$$\begin{aligned}\Gamma_k = \int d^d x \Big(& U_k(\boldsymbol{M}^2(x)) + \frac{1}{2} Z_k(\boldsymbol{M}^2)\,(\nabla \boldsymbol{M})^2 \\ & + \frac{1}{4} Y_k(\boldsymbol{M}^2)\,(\boldsymbol{M}.\nabla \boldsymbol{M})^2 + O(\nabla^4)\Big)\end{aligned} \tag{2.173}$$

where $\boldsymbol{M}$ is a N-component vector. In fact, we shall mainly study the LPA' that consists in neglecting $Y_k(\boldsymbol{M}^2)$ and keeping only the first term of the field-expansion of $Z_k(\boldsymbol{M}^2)$ that we call Z_k (not to be confused with the partition function):

$$\Gamma_k = \int d^d x \left(U_k(\boldsymbol{M}^2(x)) + \frac{1}{2} Z_k\,(\nabla \boldsymbol{M})^2 \right). \tag{2.174}$$

[44] One should remember that we are dealing here with dimensionless coupling constants. In a natural theory, we expect all dimensionful coupling constants, at scale Λ for instance, to be of order of Λ. This precisely means that the dimensionless coupling constants must all be of order one. Of course, the continuum limit does not exist for a non-renormalizable model.

The RG equation on Γ_k is obtained along the same line as in the Ising case. We start by constructing the partition function $Z_k[\boldsymbol{B}]$

$$Z_k[\boldsymbol{B}] = \int \mathcal{D}\phi(x) \exp\left(-H[\phi] - \Delta H_k[\phi] + \int \boldsymbol{B}.\phi\right) \tag{2.175}$$

with

$$\Delta H_k[\phi] = \frac{1}{2}\int_q R_k(q)\, \phi_q.\phi_{-q}. \tag{2.176}$$

Since $\Gamma_k[\boldsymbol{M}]$ is an $O(N)$-scalar, the RG equation involves now a trace on the $O(N)$ indices

$$\partial_k \Gamma_k = \frac{1}{2}\,\mathrm{Tr}\int_{x,y} \partial_k R_k(x-y)\, W_k^{(2)}(x,y) \tag{2.177}$$

where $W_k^{(2)}(x,y)$ is now a $N \times N$ matrix:

$$W_{k,ij}^{(2)}(x,y) = \frac{\delta^2 W_k}{\delta B_i(x)\,\delta B_j(y)}. \tag{2.178}$$

$\Gamma_k^{(2)} + \mathcal{R}_k$ is again the inverse of $W_k^{(2)}$

$$\delta(x-z)\delta_{ik} = \int_y W_{k,ij}^{(2)}(x,y)\left(\Gamma_k^{(2)} + \mathcal{R}_k\right)_{jk}(y,z). \tag{2.179}$$

where repeated indices are summed over (Eintein's convention). Thus the RG equation writes:

$$\partial_k \Gamma_k = \frac{1}{2}\,\mathrm{Tr}\int_{x,y} \partial_k R_k(x-y)\,\left(\Gamma_k^{(2)} + \mathcal{R}_k\right)^{-1}_{x,y}. \tag{2.180}$$

The RG Equation for the Potential

Once again we define the potential as $\Gamma_k[\boldsymbol{M}]$ evaluated in a uniform field configuration $\boldsymbol{M}$. By symmetry, we can choose any direction for $\boldsymbol{M}$. We take

$$\boldsymbol{M} = \begin{pmatrix} M \\ 0 \\ \vdots \\ 0 \end{pmatrix}. \tag{2.181}$$

The RG equation on the potential writes

$$\partial_t U_k = \frac{1}{2}\,\mathrm{Tr}\left(\dot{R}_k(q)\left(\frac{\partial^2 U_k}{\partial M_i\,\partial M_j} + (Z_k\,q^2 + R_k)\delta_{ij}\right)^{-1}\right). \tag{2.182}$$

where the trace means summation over $O(N)$ indices and integration on q. Since

$$\frac{\partial^2 U_k}{\partial M_i\,\partial M_j} = \frac{\partial U_k}{\partial \rho}\,\delta_{ij} + \frac{\partial^2 U_k}{\partial \rho^2}\,M_i\,M_j \tag{2.183}$$

where $\rho = 1/2M^2$, we obtain

$$\frac{\partial^2 U_k}{\partial M_i\,\partial M_j} + (Z_k\,q^2 + R_k)\delta_{ij} = \begin{pmatrix} Z_k q^2 + R_k + U_k' + 2\rho\,U_k'' & & & \\ & Z_k q^2 + R_k + U_k' & & \\ & & \ddots & \\ & & & Z_k q^2 + R_k + U_k' \end{pmatrix}. \tag{2.184}$$

It is trivial to invert this matrix and to compute the trace. We find:

$$\partial_k U_k = \frac{1}{2}\int_q \partial_k R_k(q)\left(\frac{1}{Z_k q^2 + R_k + U_k' + 2\rho\,U_k''} + \frac{N-1}{Z_k q^2 + R_k + U_k'}\right). \tag{2.185}$$

We can see two differences with what we have found in the Ising case within the LPA:

- Within the LPA there is no "field renormalization" Z_k in front of the q^2 term. Its presence will have important consequences both from the technical and physical points of view.
- There is a new term proportional to $N-1$ in Eq. (2.185). It will take care of the physics of the Goldstone bosons in the low temperature phase.

The RG Equation for the Dimensionless Potential $\tilde{U}_k$

Once again, Eq. (2.185) is not well suited for the search of fixed point since k appears explicitly in the right hand side through R_k and Z_k. We have first to go to dimensionless quantities. But now, even the change of variables to dimensionless quantities, Eq. (2.153), is not sufficient to get rid of the k-dependence since Z_k depends on k. It can be shown that in the scaling regime [52]

$$Z_{k\to 0} \sim \left(\frac{k}{\Lambda}\right)^{-\eta}. \qquad (2.186)$$

where η is the anomalous dimension. Thus, Z_k never reaches a fixed point value and it is therefore necessary to get rid of it to find the fixed point potential. We therefore introduce dimensionless and "renormalized" quantities defined by

$$y = \frac{q^2}{k^2} \qquad (2.187)$$

$$R_k(q^2) = Z_k\, q^2 r\,(y) = Z_k\, k^2 y\, r(y) \qquad (2.188)$$

$$\tilde{x} = k\, x \qquad (2.189)$$

$$\tilde{M}(\tilde{x}) = \sqrt{Z_k}\, k^{\frac{2-d}{2}}\, M(x) \qquad (2.190)$$

$$\tilde{U}_t\big(\tilde{M}(\tilde{x})\big) = k^{-d}\, U_k\big(M(x)\big). \qquad (2.191)$$

Note that a factor Z_k has been included in R_k. We can now repeat all the different steps leading to the RG equation on $\tilde{U}_k$. It is useful to define first

$$k\,\partial_k Z_k = -\eta_k Z_k \qquad (2.192)$$

where η_k could be called a "running" anomalous dimension. Because of the behavior of Z_k, Eq. (2.186), η_k reaches a fixed point value (the anomalous dimension) whereas Z_k does not. From Eqs. (2.187 to 2.191) we deduce that both $Z_k q^2 + R_k + U_k' + 2\rho\, U_k''$ and $\partial_t R_k(q^2)$ become proportional to $Z_k k^2$. Thus, with this rescaling, the explicit k- and Z_k-dependences disappear in the equation for $\tilde{U}_k$.[45] The RG equation on $\tilde{U}_k$ writes [32]:

$$\begin{aligned}\partial_t \tilde{U}_t = &\ -d\tilde{U}_t + (d-2+\eta_k)\tilde{\rho}\,\tilde{U}_t' \\ &-v_d \int_0^\infty \mathrm{d}y\, y^{d/2}\left(\eta_k\, r(y) + 2\, y\, r'(y)\right)\left(\frac{1}{y\big(1+r(y)\big)+\tilde{U}_t' + 2\tilde{\rho}\,\tilde{U}_t''}\right. \\ &\left. + \frac{N-1}{y\big(1+r(y)\big)+\tilde{U}_t'}\right).\end{aligned} \qquad (2.193)$$

We shall see in the following that it is sometimes convenient to consider the field-expansion of $\tilde{U}_t(\tilde{\rho})$. In this case, the RG flow on the potential is projected onto an

---

[45] In this sense, working with the dimensionless and renormalized quantities consists in going to a "co-moving frame" where the explicit k-dependence has been eliminated.

infinite hierarchy of ordinary differential equations for the evolution for the coupling constants appearing in this expansion. It is a non-trivial question to know around which field configuration $\tilde{\rho}$ one should perform the expansion. Of course, if the field expansion of $\tilde{U}_t(\tilde{\rho})$ is not truncated and if its radius of convergence is infinite, the point around which the expansion is performed does not matter. However, the radius of convergence is in general not infinite and we shall be of course interested in truncating the series expansion at orders in $\tilde{\rho}$ as low as possible. The rule of thumb is that each time a field-expansion has to be performed, the best choice is to perform it around the minimum of the (dimensionless) potential κ_k, Eq. (2.159):

$$\tilde{\boldsymbol{M}}_{|\mathrm{Min}} = \begin{pmatrix} \sqrt{2\kappa_k} \\ 0 \\ \vdots \\ 0 \end{pmatrix}. \tag{2.194}$$

with

$$\tilde{U}_k = \frac{\lambda_k}{2}(\tilde{\rho} - \kappa_k)^2 + \frac{u_{3,k}}{3!}(\tilde{\rho} - \kappa_k)^3 + \ldots \tag{2.195}$$

Let us notice that once $\tilde{U}_k$ has been truncated at a finite order in $\tilde{\rho}$, this equation is not sufficient to define completely $\kappa_k, \lambda_k, u_{3,k}$, etc. It is necessary to define them as

$$\frac{\partial \tilde{U}_k}{\partial \tilde{\rho}}_{|\tilde{\rho}=\kappa_k} = 0 \tag{2.196}$$

$$\frac{\partial^2 U_k}{\partial \tilde{\rho}^2}_{|\tilde{\rho}=\kappa_k} = \lambda_k \tag{2.197}$$

$$\vdots \tag{2.198}$$

$$\frac{\partial^n \tilde{U}_k}{\partial \tilde{\rho}^n}_{|\tilde{\rho}=\kappa_k} = u_{n,k}. \tag{2.199}$$

Note that Eq. (2.196) makes sense only if $\kappa_k \neq 0$ (for the search of the fixed point, this is not a problem since $\kappa^* \neq 0$).

The flow of all these coupling constants can be obtained trivially by acting on both sides of these equations with ∂_t and by using Eq. (2.193). We find for instance [28]

$$\begin{aligned} \partial_t \kappa_k &= -(d - 2 + \eta_k)\kappa_k + 2v_d\left(3 + 2\frac{\kappa_k u_{3,k}}{\lambda_k}\right) l_1^d(2\kappa_k\lambda_k) \\ &\quad + 2v_d(N-1)l_1^d(0) \end{aligned} \tag{2.200}$$

$$\partial_t \lambda_k = (d-4+2\eta_k)\lambda_k + 2v_d(N-1)\lambda_k^2 l_2^d(0) + 2v_d(3\lambda_k + 2\kappa_k u_{3,k})^2 l_2^d(2\kappa_k\lambda_k)$$
$$-2v_d\left(2u_{3,k} + 2\kappa_k u_{4,k} - 2\frac{\kappa_k u_{3,k}^2}{\lambda_k}\right) l_1^d(2\kappa_k\lambda_k) \quad (2.201)$$

where the so-called threshold functions l_n^d are defined in the Appendix, Subsect. 2.3.1. One remarks that the flow of λ_k involves $u_{3,k}$ and $u_{4,k}$. This is a general rule: the flow of $u_{n,k}$ involves $u_{n+1,k}$ and $u_{n+2,k}$. The non-perturbative character of these flows comes from the non-polynomial character of the threshold functions l_n^d. This, in turn, implies that the right hand side of Eqs. (2.200, 2.201) are not series expansions in the coupling constant λ_k.

The computation of the anomalous dimension η_k requires the computation of the flow of Z_k. As we did for the potential this is possible only after a definition of Z_k in terms of Γ_k has been found. It is clear that Z_k corresponds to the term in Γ_k which is quadratic in M and in q. In fact, this definition is not sufficient to completely characterize Z_k since it is the first term in the expansion of the function $Z_k(\tilde{\rho})$ and it is necessary to specify around which value of $\tilde{\rho}$ the expansion is performed. Here again, we choose the minimum κ_k of the potential, Eqs. (2.194–2.196).

A precise calculation shows that

$$Z_k = \frac{(2\pi)^d}{\delta(p=0)} \lim_{p^2\to 0} \frac{\mathrm{d}}{\mathrm{d}p^2}\left(\bar{\Gamma}^{(2)}_{(2,p),(2,-p)}|_{\mathrm{Min}}\right) \quad (2.202)$$

where $\bar{\Gamma}^{(2)}_{(2,p),(2,-p)}$ is the second derivative of Γ_k with respect to $M_2(p)$ and $M_2(-p)$. The flow of Z_k is now obtained by acting on both sides of (2.202) with ∂_t. After a straightforward although somewhat tedious calculation we obtain [28]:

$$\eta_k = \frac{16v_d}{d}\,\kappa_k\,\lambda_k^2\, m_{2,2}^d(2\kappa_k\lambda_k) \quad (2.203)$$

where $m_{2,2}^d$ is a threshold function defined in (2.220). We thus find that η_k is *not an independent quantity*. It is entirely defined by the other couplings.

The Limits $d \to 4$, $d \to 2$ and $N \to \infty$ and the Multicritical Ising Fixed Points

Let us first recall that the whole perturbative series of the correlation functions can be reproduced from the exact RG equation on Γ_k if it expanded perturbatively. Of course, our aim is not to use this approach perturbatively and it is thus interesting to understand the interplay between the derivative expansion and the perturbation expansion.

A very nice feature of the effective average action formalism is that the one-loop results obtained in $d = 4-\epsilon$ (with the ϕ^4 theory) and $d = 2+\epsilon$ (with the non linear sigma model for $N \geq 3$) are retrieved very simply while keeping in the *ansatz* for Γ_k only κ_k, λ_k and Z_k, Eqs. (2.200, 2.201, 2.203) [32, 33]. This means that the β-functions either of the coupling λ_k around $d = 4$ or of the temperature (which is related to the inverse of κ_k) around $d = 2$ reproduce at leading order the one-loop result respectively in $d = 4 + \epsilon$ and in $d = 2 + \epsilon$. The large N limit is also retrieved at leading order with the same *ansatz*.[46] This is a very interesting property since it shows that the *same* calculation leads to controlled results both in the upper and lower critical dimensions and at $N = \infty$.[47] The NPRG results are thus, at least, a clever interpolation between the perturbative results obtained in $d = 4$, $d = 2$ and at $N = \infty$! Let us emphasize that this should be enough to convince the most reluctant that the NPRG leads to results that are, at least, not less controlled than the perturbative ones. They are in fact probably much better controlled since they are one-loop exact in the two critical dimensions and since the three-dimensional critical exponents seem to converge rapidly with the order of the derivative expansion, see Subsect. 2.2.1.

Of course the $N = 1$ and $N = 2$ cases are peculiar since for $N = 1$ the lower critical dimension is one and for $N = 2$ there exists a finite temperature phase transition in $d = 2$ of indefinite order: the Kosterlitz-Thouless transition induced by the vortices. As for $N = 1$, the order $O(\partial^2)$ approximation leads to results in $d = 2$ that are qualitatively correct and quantitatively not so bad (error on the critical exponents around 25% [58]). In $d = 1$ the anomalous dimension η is 1 and it is thus difficult to reproduce it within the derivative expansion.[48]

As for $N = 2$ in $d = 2$, it has been shown—still with the *ansatz* involving only κ_k, λ_k and Z_k—that the Kosterlitz-Thouless transition is qualitatively well reproduced. Stricto sensu, no line of fixed points is found but rather a line of quasi-fixed points where the flow does not stop but is very slow. On this "line" the correlation length is not infinite but very large. With the complete $O(\partial^2)$ approximation, a fairly good quantitative agreement is obtained [57]. The remarkable point here is that these results have been obtained without introducing by hand the vortices as is usually done otherwise through the Villain's trick. This is very encouraging for the study of systems where the vortices play an important role but where there is no spin waves/vortices decoupling and thus no simple analytical approach available.

Morris has also shown that the infinite sequence of multicritical fixed points of the Ising model in $d = 2$ can be retrieved using the NPRG. This is also a non-trivial result since it would be very complicated to obtain them perturbatively [56].

---

[46] Let us emphasize that this is not the case with the Wilson-Polchinski approach: even the one loop result in $d = 4 - \epsilon$ is not reproduced at any finite order of the derivative expansion [62].

[47] Needless to say that this is completely out of reach of the perturbative expansions performed either from the ϕ^4 theory (around $d = 4$) or from the non-linear sigma model (around $d = 2$).

[48] Although not proven, it is reasonable to assume that the smaller the anomalous dimension the better the convergence of the derivative expansion ($\eta = 0$ within the LPA). Thus η is perhaps the "small parameter" of the derivative expansion.

2.2.8 Other Fields of Application of the NPRG in Statistical Mechanics

Apart from the $O(N)$ model, the NPRG has been used in the study of many systems both in particle physics [63] and in statistical mechanics. Let us mention a few of them in this latter field.

- *The magnetic frustrated systems.* For some antiferromagnetic systems the spins in the ground state are not collinear. The system is then said to be frustrated. For instance, continuous spins on a triangular lattice adopt a 120° structure at $T = 0$ when they interact antiferromagnetically since the spins on a triangular plaquette cannot be all anti-aligned. For three-component spins this 120° structure implies that the rotational $SO(3)$ symmetry is completely broken down in the ground state contrary to the ferromagnetic case. Both the number of Goldstone modes in the low temperature phase and the critical physics are thus different for these systems and for non frustrated ones. The frustrated systems belong to the class of systems for which the symmetry breaking scheme is $O(N) \to O(N-2)$. It is amazing that their critical physics in $d = 3$ is not yet fully elucidated: the NPRG approach predicts that for $N = 2$ and $N = 3$ in $d = 3$ generically weak first order transitions will be observed [33] whereas a fixed point is found perturbatively at five and six loops which implies a second order behavior [64, 65]. Neither the experiments nor the numerical simulations enable to discriminate up to now between the two scenarios.
- *Out of equilibrium phase transitions.* For systems that do not satisfy detailed balance the probability distribution of equilibrium states is not known a priori: there is no analogue of the Boltzmann weights. It is however a matter of fact that many characteristic features of systems at thermal equilibrium exist also out of equilibrium. For instance many such systems undergo continuous phase transitions governed by power law behaviors exhibiting universality. The whole machinery of field theory and renormalization group can be used with the subtlety that the "static" properties can be computed only through the large time limit of the underlying dynamics. For many systems this dynamics is given at a mesoscopic scale by a Langevin equation, the formal solution of which is given through a field theory. NPRG techniques have been adapted to this type of models [66]. Let us mention two non trivial results found this way in the so-called reaction-diffusion systems. One of the simplest system studied consists of particles A on a lattice that can diffuse (with a rate D), spontaneously decay (with a rate μ), annihilate by pairs when they meet on the same site (with a rate λ) and give birth spontaneously to a daughter particle (with a rate σ): $A \to 0, 2A \to 0, A \to 2A$. The physics of the system consists in the competition between the creation and the annihilation of particles. Depending on the magnitudes of μ, λ and σ the system either goes at large time into an absorbing state where all particles have disappeared or to an active state where the average density reaches a non vanishing asymptotic value. In between these two situations exists a continuous phase transition, the universality class of which is famous: it is the so-called directed percolation universality class.

However, for $\mu = 0$ (no spontaneous decay) and $\sigma \neq 0$, the system is predicted (i) at the mean field level to always reach the active phase, (ii) from perturbative RG to undergo a continuous phase transition between an absorbing and an active phase in $d = 1$ and $d = 2$ and only in these dimensions and (iii) from the NPRG to undergo a continuous phase transition in all dimensions. Numerical simulations have shown that a phase transition indeed exists in all dimensions and that the phase diagram found thanks to the NPRG is in quantitative agreement with the numerical data [60]. This is of course possible only because the NPRG is able to compute non universal quantities such as a phase diagram. For an other kind of reaction-diffusion system for which the parity of the number of particles is conserved by the dynamics ($2A \to 0$, $A \to 3A$) the universality class is different and the fixed point was not satisfactorily found within perturbation theory. Thanks to the NPRG this has been achieved and the determination of the critical exponents is in good agreement with the numerical data [67].

- *Disordered systems.* Disorder can be relevant for the critical physics of a system. Two kinds of disorder have been studied for the Ising model: disorder due either to randomness in the strength of the magnetic couplings J_{ij} between spins (random mass Ising model) or to the coupling to an external random magnetic field (random field Ising model). The first type of disorder can be studied rather easily by perturbative means and by NPRG methods. The net result of this type of disorder in three dimensions is to slightly modify the value of the critical exponents [68]. The other type of disorder has a much drastic effect and is far subtler to study. Perturbatively, it has been proven at all orders that the critical physics of the disordered system in dimension d is identical to that of the pure system in dimension $d - 2$ (this is the so-called dimensional reduction). Although true at all orders of perturbation theory, this result is wrong. The recourse to functional methods is unavoidable in this case since it can be shown that the effective potential develops non-analyticities (a cusp) along the RG flow at a scale k_0 called the (inverse) Larkin's length. It is therefore necessary to renormalize this function itself and not only the coupling constants that are its Taylor coefficients since the Taylor expansion ceases to exist below a scale k_0 of the RG flow. The functional RG as well as its nonperturbative version, the NPRG, has been successfully employed in this context [69–74].
- *Lifshitz point, bubble nucleation, Bose-Einstein condensation.*
  Let us finally mention other results obtained in statistical mechanics with the NPRG.
  The study of Lifshitz critical points that is of critical systems exhibiting modulated phases (in space) has been achieved by Bervillier [36]. This requires to take into account anisotropic derivative terms as well as terms of order four in the derivatives. This kind of critical phenomena exhibits also genuinely non perturbative phenomena.
  Bubble nucleation is one of the most important phenomena occurring at a first order phase transition. The calculation of the nucleation rate is far from trivial and has been computed using the LPA. It compares well with other approaches [75, 76].

Recently, the problem of the cross-over between BCS superfluidity and Bose-Einstein condensation in fermionic systems has attracted much attention in particular because of experimental breakthroughs. The NPRG is ideally suited to study the cross-over between these phenomena since non universal quantities can be computed out of it and since it is functional. A quantitative comparison between experimental data and theoretical predictions is now possible [77, 78].
The NPRG has also been applied to the study of simple fluids. A study on the special case of CO_2 has been performed [59] with good results for the calculation of the critical temperature for instance and a general formalism has been developed in [79, 80].

2.2.9 Conclusion

In this introduction to the NPRG we have focused on its application to statistical mechanics and on some of its relations with perturbative renormalization.[49] We have seen three important points.

First, at a conceptual level, the NPRG enables to understand how microphysics can be continuously related to macrophysics, something that is not possible in general within perturbative field theory. As a by-product, one can solve this way the paradox that a field theory involving infinitely many *interacting* degrees of freedom can be described in the infrared regime with only a finite (and small) number of coupling constants, precisely those that are called renormalizable within perturbation theory. This comes from the attractive character of the submanifold spanned by the renormalizable couplings in the space of coupling constants (the large river effect) and is the very meaning of universality.

Second, we have seen that contrary to common belief, it is possible to obtain qualitatively good results about the long distance physics with NPRG techniques from very short ansätze and even results that reproduce one-loop results around the upper and the lower critical dimensions and at large N. This is probably why the NPRG results obtained at finite N and for dimensions in between the upper and the lower critical dimension are reliable.

Third, the series obtained from the derivative expansion seem to converge rapidly, at least in dimension three for the Ising model. This makes the NPRG a quantitative tool for studying strongly correlated systems and not only, as often claimed, a qualitative one. Of course, this claim should be substantiated by calculations performed beyond the $O(\partial^4)$ and also in dimension two. It is nevertheless encouraging to see that critical exponents already converge at this order to the best known values without any resummation and that non universal quantities can be accurately computed.

---

[49] For a pedagogical introduction to the NPRG applied in gauge theories see Ref. [63] and for a discussion of some features of the NPRG see Ref. [81].

Let us finally mention that a crucial drawback of the derivative expansion is its inadequacy to the calculation of the momentum dependence of the correlation functions. In fact, it can be shown that the derivative expansion makes sense only when the external momenta of the correlation functions are less than the running scale k. Thus, when $k \to 0$ only the infrared physics can be computed with the derivative expansion. Crucial improvements in the computation of the momentum dependence of $\Gamma^{(2)}$ and $\Gamma^{(4)}$ has been performed these last years [52–54] and there is no doubt that if this method works it will be a new step in our possibility of computing new non perturbative phenomena in field theory.

Acknowledgments I want first to thank D. Mouhanna without whom the little group in Paris (with an antenna in Grenoble) working on the NPRG would have never existed. I owe my understanding of this subject to many discussions and common works with him. I also thank M. Tissier with whom I collaborated and from whom I have learned a lot. It is also a great pleasure to thank L. Canet and H. Chaté with whom I have collaborated and discussed many aspects of statistical mechanics and renormalization. I also thank I. Dornic and J. Vidal with whom I collaborated and G. Tarjus, J. Berges, C. Bervillier and C. Wetterich for many discussions about the NPRG. More recently, discussions with R. Mendez-Galain and N. Wschebor on their own version of the NPRG have greatly improved my understanding of this subject and I want to thank them for that. I also thank J-M. Caillol for several discussions. Finally, I thank Yu. Holovatch who encouraged me to write down my lecture notes, the students who helped me to improve them (P. Hosteins and G. Gurtner in particular) and A. Schwenk and J.-P. Blaizot who invited me in Trento to give a set of lectures on the NPRG.

2.3 Appendix

"Nobody ever promised you a rose garden."
J. Polchinski

2.3.1 Definitions, Conventions

- **Integrals in x and q spaces**
  In real and Fourier spaces we define

$$\int_x = \int \mathrm{d}^d x, \quad \int_q = \int \frac{\mathrm{d}^d q}{(2\pi)^d} \tag{2.204}$$

- **Fourier transform**

$$f(x) = \int_q \tilde{f}(q)\, e^{iqx}, \quad \tilde{f}(q) = \int_x f(x)\, e^{-iqx}. \tag{2.205}$$

Depending on the context, we omit or not the tilde on the Fourier transform.

• **Definition of v_d**

$$\int_q f(q^2) = 2v_d \int_0^\infty \mathrm{d}x\, x^{d/2-1} f(x). \tag{2.206}$$

with

$$v_d = \frac{1}{2^{d+1}\pi^{d/2}\Gamma\left(\frac{d}{2}\right)}. \tag{2.207}$$

• **Functional derivatives**

$$\frac{\delta}{\delta\tilde{\phi}_q} = \int_x \frac{\delta\phi(x)}{\delta\tilde{\phi}(q)}\frac{\delta}{\delta\phi(x)} = \int_x \frac{e^{iqx}}{(2\pi)^d}\frac{\delta}{\delta\phi_x} \tag{2.208}$$

• **Correlation functions**

$\Gamma[M]$ is a functional of $M(x)$. We define the 1PI correlation functions by

$$\Gamma^{(n)}[M(x); x_1, \ldots, x_n] = \frac{\delta^n \Gamma[M]}{\delta M_{x_1} \ldots \delta M_{x_n}} \tag{2.209}$$

We also define the Fourier transform of $\Gamma^{(n)}$ by

$$\tilde{\Gamma}^{(n)}[M(x); q_1, \ldots, q_n] = \int_{x_1 \ldots x_n} e^{-i\sum_i q_i x_i}\, \Gamma^{(n)}[M(x); x_1, \ldots, x_n]. \tag{2.210}$$

and also

$$\bar{\Gamma}^{(n)}[M(x); q_1, \ldots, q_n] = \frac{\delta^n \Gamma}{\delta\tilde{M}_{q_1} \ldots \delta\tilde{M}_{q_n}}. \tag{2.211}$$

The relation between $\tilde{\Gamma}^{(n)}$ and $\bar{\Gamma}^{(n)}$ follows from Eq. (2.208):

$$\bar{\Gamma}^{(n)}[M(x); q_1, \ldots, q_n] = (2\pi)^{-nd}\, \tilde{\Gamma}^{(n)}[M(x); q_1, \ldots, q_n]. \tag{2.212}$$

• **Cut-off function in x and q spaces**

$$\Delta H_k[\phi] = -\frac{1}{2}\int_q \tilde{R}_k(q^2)\, \tilde{\phi}_q \tilde{\phi}_{-q} = -\frac{1}{2}\int_{x,y} \phi_x\, R_k(x-y)\, \phi_y \tag{2.213}$$

One should be careful about the fact that R_k is sometimes considered as a function of q and sometimes as a function of q^2. It can be convenient to define a cut-off function with two entries by

$$\mathcal{R}_k(x, y) = R_k(x-y). \tag{2.214}$$

Then

$$\tilde{\mathcal{R}}_k(q, q') = (2\pi)^d \delta^d(q + q')\tilde{R}_k(q) \tag{2.215}$$

- **k-dependent anomalous dimension**
  By definition:

$$k\,\partial_k Z_k = -\eta_k Z_k. \tag{2.216}$$

- **Threshold functions l_n^d**

$$l_n^d(w, \eta) = \frac{n + \delta_{n,0}}{2} \int_0^\infty \mathrm{d}y\, y^{d/2-1} \frac{s(y)}{\big(y(1 + r(y)) + w\big)^{n+1}} \tag{2.217}$$

where

$$R_k(q^2) - Z_k q^2 r(y) \quad \text{with} \quad y = \frac{q^2}{k^2} \tag{2.218}$$

and, by definition of $s(y)$

$$k\,\partial_k R_k(q^2) = k\,\partial_k \left(Z_k q^2 r\left(\frac{q^2}{k^2}\right)\right) = Z_k k^2\big(-\eta_k\, y\, r(y) - 2y^2 r'(y)\big) = Z_k k^2 s(y). \tag{2.219}$$

- **Threshold functions m_{n_1,n_2}^d**

$$m_{n_1,n_2}^d(w) = -\frac{1}{2} Z_k^{-1} k^{d-6} \int_0^\infty \mathrm{d}x\, x^{d/2} \tilde{\partial}_t \frac{(\partial_x P)^2(x, 0)}{P^{n_1}(x, 0) P^{n_2}(x, w)} \tag{2.220}$$

with

$$P(x, w) = Z_k\, x + R_k(x) + w \tag{2.221}$$

- **Universal value of $l_n^{2n}(0, 0)$ for $n > 0$**
  For $n > 0$ and independently of the choice of cut-off function R_k:

$$l_n^{2n}(0, 0) = \frac{n}{2} \int_0^\infty \mathrm{d}y\, (-2) \frac{r'(y)}{(1 + r(y))^{n+1}} = 1 \tag{2.222}$$

- **Derivative of l_n^d**

$$\partial_w l_n^d(w, \eta) = -(n + \delta_{n,0})\, l_{n+1}^d(w, \eta) \tag{2.223}$$

- **θ-cut-off**
  A convenient cut-off function R_k that allows to compute analytically some threshold functions is

$$R_k(q) = Z_k (k^2 - q^2)\theta\left(1 - \frac{q^2}{k^2}\right). \tag{2.224}$$

With this cut-off we find

$$r(y) = \frac{1-y}{y}\,\theta(1-y) \tag{2.225}$$

• **Threshold functions l_n^d and $m_{2,2}^d$ with the θ-cut-off**

With the cut-off function, Eq. (2.224), the l_n^d threshold functions can be computed analytically

$$l_n^d(w,\eta) = \frac{2}{d}(n+\delta_{n,0})\left(1-\frac{\eta_k}{d+2}\right)\frac{1}{(1+w)^{n+1}} \tag{2.226}$$

$$m_{2,2}^d(w) = \frac{1}{(1+w)^2} \tag{2.227}$$

2.3.2 *Proof of Eq. (2.106)*

We define:

$$J = \int_{y,z} e^{-y^2/2\alpha - z^2/2\beta} \tag{2.228}$$

and we rewrite the exponent:

$$\begin{aligned} -\frac{y^2}{2\alpha} - \frac{z^2}{2\beta} &= -\frac{1}{2}\left(\frac{1}{\alpha}+\frac{1}{\beta}\right)y^2 + \frac{xy}{\beta} - \frac{x^2}{2\beta} \\ &= -\frac{1}{2}\frac{\gamma}{\alpha\beta}\left(y - \frac{\alpha}{\gamma}x\right)^2 + \frac{\alpha}{2\beta\gamma}x^2 - \frac{x^2}{2\beta}. \end{aligned} \tag{2.229}$$

We now define

$$u = y - \frac{\alpha}{\gamma}x \tag{2.230}$$

and change variables: $(y,z) \to (u,x)$. The jacobian is 1 and thus:

$$J = \int_{u,x} e^{-\gamma u^2/2\alpha\beta - x^2/2\gamma} = \sqrt{\frac{2\pi\alpha\beta}{\gamma}}\, I. \tag{2.231}$$

2.3.3 *The Exact RG Equations*

For the sake of simplicity, we consider a scalar theory (e.g. Ising). We have by definition

$$Z_k[B] = \int \mathcal{D}\phi \, \exp\left(-H[\phi] - \Delta H_k[\phi] + \int B\phi\right)$$

$$\text{with } \Delta H_k[\phi] = \frac{1}{2}\int_q R_k(q)\,\phi_q\,\phi_{-q}$$

$$W_k[B] = \log Z_k[B]$$

$$\Gamma_k[M] + W_k[B] = \int_x BM - \frac{1}{2}\int_{x,y} M_x\,R_{k,x-y}\,M_y$$

with, by definition of $M(x)$:

$$\frac{\delta W_k}{\delta B(x)} = M(x) = \langle\phi(x)\rangle. \tag{2.232}$$

When $B(x)$ is taken k-independent (as in $Z_k[B]$) then $M(x)$ computed from W_k is k-dependent. Reciprocally, if $M(x)$ is taken fixed (as in $\Gamma_k[M]$), then $B(x)$ computed from Eq. (2.236) becomes k-dependent.

RG Equation for $W_k[B]$

$$\begin{aligned}
\partial_k e^{W_k} &= -\frac{1}{2}\int \mathcal{D}\phi \left(\int_{x,y} \phi_x\,\partial_k R_k(x-y)\,\phi_y\right) \\
&\quad \cdot \exp\left(-H[\phi] - \frac{1}{2}\int_q R_k(q)\phi_q\phi_{-q} + \int B\phi\right) \\
&= \left(-\frac{1}{2}\int_{x,y} \partial_k R_k(x-y)\frac{\delta}{\delta B_x}\frac{\delta}{\delta B_y}\right) e^{W_k[B]}.
\end{aligned} \tag{2.233}$$

We therefore obtain for W_k:

$$\partial_k W_k[B] = -\frac{1}{2}\int_{x,y} \partial_k R_k(x-y)\left(\frac{\delta^2 W_k}{\delta B_x\,\delta B_y} + \frac{\delta W_k}{\delta B_x}\frac{\delta W_k}{\delta B_y}\right) \tag{2.234}$$

which is equivalent to the Polchinski equation.

RG Equation for $\Gamma_k[M]$

We first derive the reciprocal relation of Eq. (2.232). The Legendre transform is symmetric with respect to the two functions that are transformed. Here the Legendre transform of W_k is $\Gamma_k + 1/2\int R_k MM$. Thus

$$\frac{\delta}{\delta M_x}\left(\Gamma_k + \frac{1}{2}\int_{x,y} M_x\, R_k(x-y)\, M_y\right) = B_x \tag{2.235}$$

and then

$$\frac{\delta \Gamma_k}{\delta M_x} = B_x - \int_y R_k(x-y) M_y. \tag{2.236}$$

In the Polchinski equation (2.234), the k-derivative is taken at fixed B_x. We must convert it to a derivative at fixed M:

$$\partial_{k|B} = \partial_{k|M} + \int_x \partial_k M_{x|B} \frac{\delta}{\delta M_x} \tag{2.237}$$

Acting on Eq. (2.232) with $\partial_{k|B}$, we obtain:

$$\begin{aligned}\partial_k \Gamma_k[M]_{|B} + \partial_k W_k[B]_{|B} &= \int_x B\, \partial_k M_{|B} \\ -\frac{1}{2}\int_{x,y} \partial_k R_{k,x-y}\, M_x\, M_y &- \int_{x,y} R_{k,x-y} M_x \partial_k M_{y|B}\end{aligned} \tag{2.238}$$

Substituting Eqs. (2.236, 2.234, 2.237) into this equation we finally obtain

$$\partial_k \Gamma_k[M] = \frac{1}{2}\int_{x,y} \partial_k R_k(x-y)\, \frac{\delta^2 W_k}{\delta B_x\, \delta B_y} \tag{2.239}$$

The last step consists in rewriting the right hand side of this equation in terms of Γ_k only. We start from (2.232) and act on it with $\delta/\delta M_z$:

$$\delta(x-z) = \frac{\delta^2 W_k}{\delta B_x\, \delta M_z} = \int_y \frac{\delta^2 W_k}{\delta B_x\, \delta B_y}\, \frac{\delta B_y}{\delta M_z}. \tag{2.240}$$

Now, using (2.236), we obtain

$$\delta(x-z) = \int_y \frac{\delta^2 W_k}{\delta B_x\, \delta B_y}\left(\frac{\delta^2 \Gamma_k}{\delta M_y\, \delta M_z} + R_k(y-z)\right). \tag{2.241}$$

We define

$$W_k^{(2)}(x,y) = \frac{\delta^2 W_k}{\delta B_x\, \delta B_y} \tag{2.242}$$

and thus

$$\delta(x-z) = \int_y W_k^{(2)}(x,y)\left(\Gamma_k^{(2)} + \mathcal{R}_k\right)(y,z). \tag{2.243}$$

$\Gamma_k^{(2)} + \mathcal{R}_k$ is therefore the inverse of $W_k^{(2)}$ in the operator sense. Note that although we did not specify it, $W_k^{(2)}$ is a functional of $B(x)$ and $\Gamma_k^{(2)}$ a functional of $M(x)$. Relation (2.243) is valid for arbitrary M. The RG equation (2.239) can now be written in terms of Γ_k only:

$$\partial_k \Gamma_k[M] = \frac{1}{2} \int_{x,y} \partial_k R_k(x-y) \left(\Gamma_k^{(2)} + \mathcal{R}_k\right)^{-1} (x,y). \tag{2.244}$$

In Fourier space this equation becomes:

$$\partial_k \Gamma_k[M] = \frac{1}{2} \int_q \partial_k \tilde{R}_k(q) \left(\tilde{\Gamma}_k^{(2)} + \tilde{\mathcal{R}}_k\right)^{-1}_{q,-q}. \tag{2.245}$$

RG Equation for the Effective Potential

The derivative expansion consists in expanding Γ_k as

$$\Gamma_k[M(x)] = \int_x \left(U_k(M^2) + \frac{1}{2} Z_k(M^2)\, (\nabla M)^2 + \ldots\right) \tag{2.246}$$

where we have supposed that the theory is $\mathbb{Z}_2$ symmetric so that $U_k, Z_k, \ldots$ are functions of M^2 only. To compute the flow of these functions it is necessary to define them from Γ_k. The effective potential U_k coincides with Γ_k when it is evaluated for uniform field configurations $M_{\text{unif.}}$:

$$\Gamma_k[M_{\text{unif.}}] = \Omega\, U_k(M^2_{\text{unif.}}) \tag{2.247}$$

where Ω is the volume of the system. It is easy to derive an RG equation from this definition of U_k if we use the local potential approximation (LPA) that consists in truncating Γ_k as in (2.246) with $Z_k(M) = 1$:

$$\Gamma_k^{\text{LPA}}[M(x)] = \int_x \left(U_k(M^2(x)) + \frac{1}{2}\, (\nabla M)^2\right). \tag{2.248}$$

By acting on Eq. (2.247) with ∂_k we obtain:

$$\partial_k U_k(M) = \frac{1}{2\Omega} \int_q \partial_k R_k(q) \left(\Gamma_k^{(2)}{}_{|M_{\text{unif.}}} + R_k\right)^{-1}_{q,-q}. \tag{2.249}$$

Thus, we have to invert $\Gamma_k^{(2)} + R_k$ for a uniform field configuration and within the LPA. From now on, we omit the superscript LPA on Γ_k. An elementary calculation leads to

$$\tilde{\Gamma}^{(2)}_{k,q,q'}{}_{|M_{\text{unif.}}} = \left(\frac{\partial^2 U_k}{\partial M^2} + q^2\right) (2\pi)^{-d} \delta(q+q'). \tag{2.250}$$

Using $\delta(q=0)=\Omega(2\pi)^{-d}$ we find

$$\partial_k U_k = \frac{1}{2}\int_q \frac{\partial_k R_k(q)}{q^2 + R_k(q) + \dfrac{\partial^2 U_k}{\partial M^2}}. \tag{2.251}$$

It is convenient to re-express this equation in terms of

$$\rho = \frac{1}{2}M^2 \tag{2.252}$$

which is the $\mathbb{Z}_2$-invariant.

$$\partial_k U_k(\rho) = \frac{1}{2}\int_q \frac{\partial_k R_k(q)}{q^2 + R_k(q) + U_k'(\rho) + 2\rho\, U_k''(\rho)} \tag{2.253}$$

where $U_k'(\rho)$ and $U_k''(\rho)$ are derivatives of U_k with respect to ρ.

To obtain the RG equation for the dimensionless potential we have to perform the change of variables of Eq. 2.154. We find

$$\partial_{t\,|_\rho} = \partial_{t\,|_{\tilde\rho}} + (2-d-\eta)\tilde\rho\,\frac{\partial}{\partial\tilde\rho} \tag{2.254}$$

and

$$\partial_{t\,|_{q^2}} = \partial_{t\,|_y} - 2y\partial_y. \tag{2.255}$$

Inserting these relations together with Eq. (2.154) and with

$$\partial_t R_k(q^2) = -y\big(\eta_k\, r(y) + 2y\, r'(y)\big) Z_k\, k^2 \tag{2.256}$$

in Eq. (2.253) leads to the RG equation on $\tilde U_t$, Eq. (2.155).

References

1. Le Bellac, M.: Quantum and Statistical Field Theory. Clarendon Press, Oxford (1991)
2. Binney, J.J., Dowrick, N.J., Fisher, A.J., Newman, M.E.J.: The Theory of Critical Phenomena: An Introduction to the Renormalization Group. Oxford University Press, Oxford (1992)
3. Goldenfeld, N.: Lectures on Phase Transitions and the Renormalization Group. Addison-Wesley, Reading (1992)
4. Ryder, L.H.: Quantum Field Theory. Cambridge University Press, Cambridge (1985)
5. Zinn-Justin, J.: Quantum Field Theory and Critical Phenomena, 3rd edn. Oxford University Press, New York (1989)
6. Delamotte, B.: Am. J. Phys. **72**, 170 (2004)
7. Kadanoff, L.P.: Physica **2**, 263 (1966)
8. Wegner, F.J., Houghton, A.: Phys. Rev. A **8**, 401 (1973)

9. Wilson, K.G., Kogut, J.: Phys. Rep., Phys. Lett. **12C**, 75 (1974)
10. Wegner F.J.: The critical state, general aspects. In: Domb, C., Greene, M.S. (eds.) Phase Transitions and Critical Phenomena, vol. 6. Academic Press, New York (1976)
11. Weinberg, S.: Critical phenomena for field theorists. In: Zichichi, A. (ed.) Understanding the Fundamental Constituents of Matter. Plenum Press, New York (1978)
12. Polchinski, J.: Nucl. Phys. B **231**, 269 (1984)
13. Shirkov, D.V., Kovalev, V.F.: Phys. Rep. **352**, 219 (2001)
14. Fortuin, C.M., Kasteleyn, P.W.: Physica **57**, 536 (1972)
15. Swendsen, R.H., Wang, J.S.: Phys. Rev. Lett. **58**, 86 (1987)
16. Niemeijer, T., van Leuwen, J.: Renormalization: ising-like spin systems. In: Domb, C., Greene, M.S. (eds.) Phase Transitions and Critical Phenomena, vol. 6. Academic Press, New York (1975)
17. Golner, G.R., Riedel, E.K.: Phys. Rev. Lett. **34**, 856 (1975)
18. Hasenfratz, A., Hasenfratz, P.: Nucl. Phys. B [FS] **270**, 687 (1986)
19. Nicoll, J.F., Chang, T.S., Stanley, H.E.: Phys. Rev. Lett. **33**, 540 (1974). err. ibid. **33**, 1525 (1974)
20. Nicoll, J.F., Chang, T.S., Stanley, H.E.: Phys. Lett. A **57**, 7 (1976)
21. Nicoll, J.F., Chang, T.S.: Phys. Lett. A **62**, 287 (1977)
22. Ringwald, A., Wetterich, C.: Nucl. Phys. B **334**, 506 (1990)
23. Wetterich, C.: Nucl. Phys. B **352**, 529 (1991)
24. Tetradis, N., Wetterich, C.: Nucl. Phys. B **383**, 197 (1992)
25. Wetterich, C.: Phys. Lett. B **301**, 90 (1993)
26. Ellwanger, U.: Z. Phys. C **58**, 619 (1993)
27. Ellwanger, U.: Z. Phys. C **62**, 503 (1993)
28. Tetradis, N., Wetterich, C.: Nucl. Phys. B [FS] **422**, 541 (1994)
29. Ellwanger, U.: Phys. Lett. B **335**, 364 (1994)
30. Morris, T.R.: Int. J. Mod. Phys. A **9**, 2411 (1994)
31. Morris, T.R.: Phys. Lett. B **329**, 241 (1994)
32. Berges, J., Tetradis, N., Wetterich, C.: Phys. Rep. **363**, 223 (2002)
33. Delamotte, B., Mouhanna, D., Tissier, M.: Phys. Rev. B **69**, 134413 (2004)
34. Newman, K.E., Riedel, E.K.: Phys. Rev. B **30**, 6615 (1984)
35. Newman, K.E., Riedel, E.K., Muto, S.: Phys. Rev. B **29**, 302 (1984)
36. Bervillier, C.: Phys. Lett. A **331**, 110 (2004)
37. Kunz, H., Zumbach, G.: J. Phys. A **26**, 3121 (1993)
38. Zumbach, G.: Phys. Rev. Lett. **71**, 2421 (1993)
39. Zumbach, G.: Nucl. Phys. B **413**, 754 (1994)
40. Zumbach, G.: Phys. Lett. A **190**, 225 (1994)
41. Kleinert, H., Schulte-Frohlinde, V.: Critical Properties of Phi 4-Theories. World Scientific, Singapore (2001)
42. Liao, S., Polonyi, J., Strickland, M.: Nucl. Phys. B **567**, 493 (2000)
43. Litim, D.: J. High Energy Phys. **0111**, 059 (2001)
44. Litim, D.F.: Nucl. Phys. B **631**, 128 (2002)
45. Canet, L., Delamotte, B., Mouhanna, D., Vidal, J.: Phys. Rev. D **67**, 065004 (2003)
46. Canet, L., Delamotte, B., Mouhanna, D., Vidal J.: Phys. Rev. B **68**, 064421 (2003)
47. Tissier, M., Delamotte, B., Mouhanna, D.: Phys. Rev. Lett. **84**, 5208 (2000)
48. Bagnuls, C., Bervillier, C.: Phs.Rev. B **41**, 402 (1990)
49. Bagnuls, C., Bervillier, C.: Phys. Rep. **348**, 91 (2001)
50. Bervillier, C.: J. Phys. Condens. Matter **17**, S1929 (2005)
51. Bagnuls, C., Bervillier, C.: Int. J. Mod. Phys. A **16**, 1825 (2001)
52. Blaizot, J.-P., Galain, R.M., Wschebor, N.: Phys. Rev. E **74**, 051116 (2006)
53. Blaizot, J.-P., Galain, R.M., Wschebor, N.: Phys. Lett. B **632**, 571 (2006)
54. Blaizot, J.-P., Galain, R.M., Wschebor, N.: Phys. Rev. E 051117 (2006)
55. Litim, D.: Mind the gap. Int. J. Mod. Phys. A **16**, 2081 (2001)
56. Morris, T.R.: Phys. Lett. B **345**, 139 (1995)
57. Gersdorff, G.v., Wetterich, C.: Phys. Rev. B **64**, 054513 (2001)

58. Ballhausen, H., Berges, J., Wetterich, C.: Phys. Lett. B **582**, 144 (2004)
59. Seide, S., Wetterich, C.: Nucl. Phys. B **562**, 524 (1999)
60. Canet, L., Chaté, H., Delamotte, B.: Phys. Rev. Lett. **92**, 255703 (2004)
61. Itzykson, C., Drouffe, J.M.: Statistical Field Theory. Cambridge University Press, Cambridge (1989)
62. Morris, T.R., Tighe, J.F.: J. High Energy Phys. **08**, 007 (1999)
63. Gies, H.: hep-ph/0611146 (2006)
64. Pelissetto, A., Vicari, E.: Phys. Rep. **368**, 549 (2002)
65. Calabrese, P., Parruccini, P., Pelissetto, A., Vicari, E.: Phys. Rev. B **70**, 174439 (2004)
66. Canet, L., Delamotte, B., Deloubrière, O., Wschebor, N.: Phys. Rev. Lett. **92**, 195703 (2004)
67. Canet, L., Chaté, H., Delamotte, B., Dornic, I., Munoz, M.A.: Phys. Rev. Lett. **95**, 100601 (2005)
68. Tissier, M., Mouhanna, D., Vidal, J., Delamotte, B.: Phys. Rev. B **65**, 140402 (2002)
69. Fisher, D.S.: Phys. Rev. Lett. **56**, 1964 (1986)
70. Le Doussal, P., Wiese, K.J.: Phys. Rev. Lett. **89**, 125702 (2002)
71. Feldman, D.E.: Phys. Rev. Lett. **88**, 177202 (2002)
72. Le Doussal, P., Wiese, K.J., Chauve, P.: Phys. Rev. E **69**, 026112 (2004)
73. Tarjus, G., Tissier, M.: Phys. Rev. Lett. **93**, 267008 (2004)
74. Tissier, M., Tarjus, G.: Phys. Rev. Lett. **96**, 087202 (2006)
75. Strumia, A., Tetradis, N.: Nucl. Phys. B **542**, 719 (1999)
76. Strumia, A., Tetradis, N.: JHEP, **9911**, 023 (1999)
77. Diehl, S., Wetterich, C.: Phys. Rev. A **73**, 033615 (2006)
78. Diehl, S., Gies, H., Pawlowski, J.M., Wetterich, C.: Phys. Rev. A **76**, 021602 (2007)
79. Parola, A., Reatto, L.: Adv. Phys. **44**, 211 (1995)
80. Caillol, J.-M.: Mol. Phys. **104**, 1931 (2006)
81. Delamotte, B., Canet, L.: Condens. Matter Phys. **8**, 163 (2005)

Chapter 3
EFT for DFT

R. J. Furnstahl

These lectures give an overview of the ongoing application of effective field theory (EFT) and renormalization group (RG) concepts and methods to density functional theory (DFT), with special emphasis on the nuclear many-body problem. Many of the topics covered are still in their infancy, so rather than a complete review these lectures aim to provide an introduction to the developing literature.

3.1 EFT, RG, DFT for Fermion Many-Body Systems

3.1.1 Overview of Fermion Many-Body Systems

There are a wide range of many-body systems featuring fermion degrees of freedom. These can be collections of "fundamental" fermions (electrons, quarks, …) or of composites each made of *odd* number of fermions (e.g., protons). Here are some general categories and examples:

1. Isolated atoms or molecules, which contain electrons interacting via the long-range (screened) Coulomb force.
2. Bulk solid-state materials, such as metals, insulators, semiconductors, superconductors, etc.
3. Liquid ^{3}He (a superfluid!).
4. Cold fermionic atoms in (optical) traps (note that ^{6}Li is a fermion but ^{7}Li is a boson).
5. Atomic nuclei.
6. Neutron stars, which could mean neutron matter or color superconducting quark matter.

R. J. Furnstahl (✉)
Department of Physics, Ohio State University, Columbus, OH 43210, USA
e-mail: furnstahl.1@osu.edu

J. Polonyi and A. Schwenk (eds.), *Renormalization Group and Effective Field Theory Approaches to Many-Body Systems*, Lecture Notes in Physics 852,
DOI: 10.1007/978-3-642-27320-9_3,

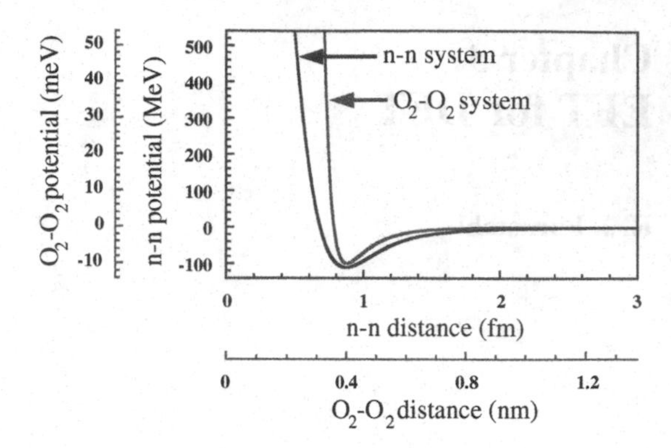

Fig. 3.1 Molecule-molecule and nucleon-nucleon potentials compared [2]

DFT has been most widely applied to systems in categories 1 and 2. We will focus in these lectures on categories 4 and 5 (and neutrons stars are treated in Thomas Schäfer's lectures).

For our purposes, it won't matter whether the fermions are structureless (as far as we know) such as quarks or electrons, or are composites such as interacting atoms. Note that an individual atom, studied as a many-body system of electrons (with external potential from the nucleus) is a fermion many-body system, while a collection of these atoms might be either a boson or fermion many-body system [1].

If we label the axes appropriately in Fig. 3.1 (note the many orders of magnitude difference!), we see qualitative similarities between the central part of a conventional nucleon-nucleon (NN) potential and potentials between atoms or molecules (a Lennard-Jones potential is shown). In particular, there is midrange attraction and strong short-range repulsion (or a "hard core"). In the atomic case, the attraction is van der Waals in nature from induced polarization (which is why it falls off as $1/r^6$), and the rapid repulsion is from when the electron clouds overlap. In the nuclear case, the long- and mid-range attraction is mostly from one- and two-pion exchange. The hard core is often described in terms of vector-meson exchange but is generally phenomenological. The potential shown is the central part of the NN interaction; there are also important spin dependences and a non-central tensor force [3].

What might one expect qualitatively from a many-body system with such a potential? Start with the equation of state of an ideal gas $PV = nRT$ (n is number of moles). Hard-core means "excluded volume" so $V \to (V - nb)$ with b constant. Attraction lowers the pressure on the container, so we find: $P = \frac{nRT}{V-nb} - \frac{an^2}{V^2}$. The end result is a van der Waals equation of state: $(P + \frac{an^2}{V^2})(V - nb) = nRT$, which has a liquid-gas phase transition. This is consistent with nuclei! The hard core keeps particles apart, leading to "short-range" correlations in the wave function. They make many-body problems difficult but might seem to be essential in saturating the nuclear liquid. Both liquid helium and nuclei can be thought of as liquid drops, whose radii scale with the number of particles A to the 1/3 power: $R \sim r_0 A^{1/3}$. If a hard core of radius c is responsible for saturation, one can estimate that $0.55c \leq r_0 \leq 2.4c$ [4, 5]. Liquid ^{3}He has $r_0 \sim 2.4\,\text{A} \sim c$, but for nuclei $r_0 \sim 1.1\,\text{fm} \sim 2.75c$.

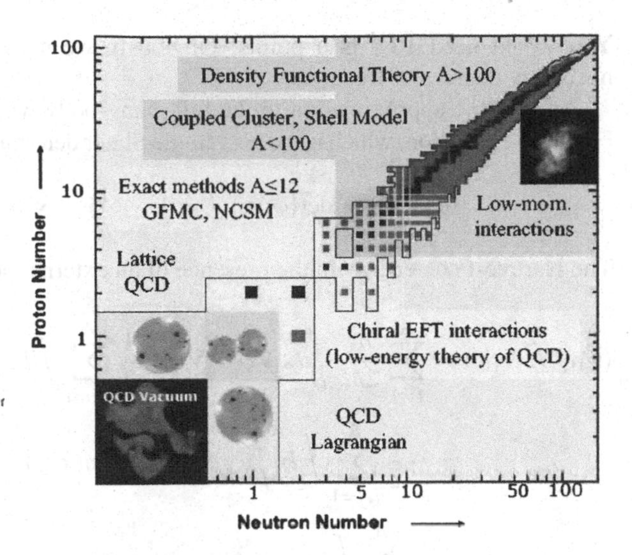

Fig. 3.2 Overview of low-energy many-body nuclear systems [6]

This implies that nuclear matter is dilute and "delicately" bound, which means that an EFT expansion may be particularly useful. Other questions we might ask are whether there are common ("universal") features of atomic and nuclear systems, and how do we relate the many-body physics to more fundamental underlying theories? As we'll see, EFT will also help to address these questions.

In Fig. 3.2, we show the "big picture" of low-energy nuclear physics, which features a specific scientific goal of predicting properties of unstable nuclei (the non-black squares), and a general scientific goal, which is connecting the whole picture from quantum chromodynamics (QCD) to superheavy nuclei in a systematic way with robust predictions. In principle, the nuclear part of the problem is simple: given internucleon potentials, just solve the many-body Schrödinger equation. This turns out to be feasible only for the smallest nuclei.

Why is this many-body problem so difficult computationally? Let's think about the many-body Schrödinger wave function [7]. How can we represent the wave function for an A-body nucleus? Consider a ^{8}Be ($Z = 4$ protons, $N = 4$ neutrons) wave function with spin, isospin, and space components:

$$|\Psi\rangle = \sum_{\sigma,\tau} \chi_\sigma \chi_\tau \phi(\mathbf{R}), \quad \text{where } \mathbf{R} \text{ are the } 3A \text{ spatial coordinates},$$

$$\chi_\sigma = \downarrow_1 \uparrow_2 \cdots \downarrow_A \ (2^A \text{ terms}) = 256 \text{ for } A = 8,$$

$$\chi_\tau = n_1 n_2 \cdots p_A \ (\tfrac{A!}{N!Z!} \text{ terms}) = 70 \text{ for } {}^8\text{Be}.$$

So for ^{8}Be there are 17,920 complex functions in $3A - 3 = 21$ dimensions! Suppose for a nucleus of size 10 fm you represent this with a mesh spacing of 0.5 fm.

You would need 10^{27} grid points! Obviously we need to drastically reduce the necessary degrees of freedom.

An extreme approximation to the full many-body wave function is the Hartree-Fock wave function, which is the best single Slater determinant in a variational sense:

$$|\Psi_{\mathrm{HF}}\rangle = \det\{\phi_i(\mathbf{x}), i = 1 \cdots A\}, \quad \mathbf{x} = (\mathbf{r}, \sigma, \tau). \tag{3.1}$$

The Hartree-Fock energy in the presence of an external potential is [8]

$$\begin{aligned}\langle\Psi_{\mathrm{HF}}|\widehat{H}|\Psi_{\mathrm{HF}}\rangle = &\sum_{i=1}^{A}\frac{\hbar^2}{2M}\int d\mathbf{x}\, \nabla\phi_i^* \cdot \nabla\phi_i + \frac{1}{2}\sum_{i,j=1}^{A}\int d\mathbf{x}\int d\mathbf{y}\, |\phi_i(\mathbf{x})|^2 v(\mathbf{x},\mathbf{y})|\phi_j(\mathbf{y})|^2 \\ &-\frac{1}{2}\sum_{i,j=1}^{A}\int d\mathbf{x}\int d\mathbf{y}\, \phi_i^*(\mathbf{x})\phi_i(\mathbf{y})v(\mathbf{x},\mathbf{y})\phi_j^*(\mathbf{y})\phi_j(\mathbf{x}) \\ &+\sum_{i=1}^{A}\int d\mathbf{y}\, v_{\mathrm{ext}}(\mathbf{y})|\phi_j(\mathbf{y})|^2. \end{aligned} \tag{3.2}$$

We determine the ϕ_i by varying with fixed normalization:

$$\frac{\delta}{\delta\phi_i^*(\mathbf{x})}\Big(\langle\Psi_{\mathrm{HF}}|\widehat{H}|\Psi_{\mathrm{HF}}\rangle - \sum_{j=1}^{A}\epsilon_j\int d\mathbf{y}\, |\phi_j(\mathbf{y})|^2\Big) = 0. \tag{3.3}$$

We solve this self-consistently, which is non-trivial because the potential is non-local, but drastically simpler than solving for the full wave function. However, while Hartree-Fock is a reasonable starting point for atoms, it is not for nuclear potentials of the form in Fig. 3.1 (e.g., note that the Argonne v_{18} "1st order" curve in Fig. 3.5 is not even bound).

3.1.2 Density Functional Theory

An alternative to working with the many-body wave function is density functional theory (DFT) [9–11], which as the name implies, has fermion densities as the fundamental "variables". To date, the dominant application of DFT has been to the inhomogeneous electron gas, which means interacting point electrons in the static potentials of atomic nuclei. This has led to "ab initio" calculations of atoms, molecules, crystals, surfaces, and more [12]. DFT is founded on a theorem of Hohenberg and Kohn (HK): There *exists* an energy functional $E_{v_{\mathrm{ext}}}[\rho]$ of the density ρ such that

$$E_{v_{\mathrm{ext}}}[\rho] = F_{\mathrm{HK}}[\rho] + \int d^3x\, v_{\mathrm{ext}}(\mathbf{x})\rho(\mathbf{x}), \tag{3.4}$$

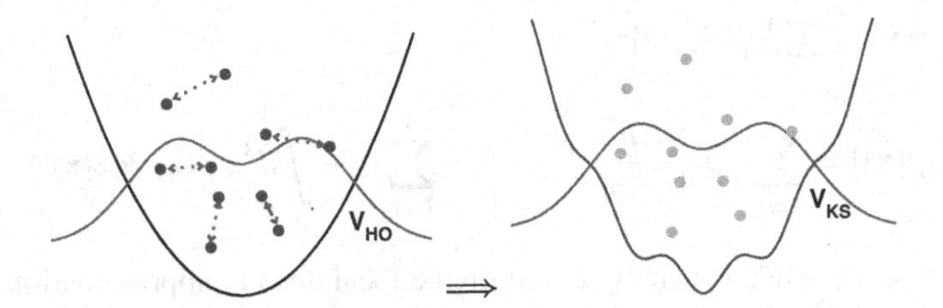

Fig. 3.3 Kohn-Sham DFT for a $v_{\text{ext}} = V_{\text{HO}}$ harmonic trap. On the *left* is the interacting system and on the *right* the Kohn-Sham system. The density profile is the same in each

where F_{HK} is *universal* (the same for any external potential v_{ext}), the same for H_2 to DNA! This is useful *if* you can approximate the energy functional.

The general procedure is to introduce single-particle orbitals and to minimize the energy functional to obtain the ground-state energy E_{gs} and density ρ_{gs}. This is called Kohn-Sham DFT, and is illustrated schematically in Fig. 3.3. Here the interacting density for A fermions in the external potential V_{HO} is equal (by construction) to the non-interacting density in V_{KS}. Orbitals $\{\psi_i(\mathbf{x})\}$ in the local potential $V_{\text{KS}}([\rho], \mathbf{x})$ are solutions to

$$[-\nabla^2/2m + V_{\text{KS}}(\mathbf{x})]\psi_i = \varepsilon_i \psi_i \tag{3.5}$$

and determine the density

$$\rho(\mathbf{x}) = \sum_{i=1}^{A} |\psi_i(\mathbf{x})|^2 \tag{3.6}$$

(the sum is over the lowest A states). The magical Kohn-Sham potential $V_{\text{KS}}([\rho], \mathbf{x})$ is in turn determined from $\delta E_{v_{\text{ext}}}[\rho]/\delta\rho(\mathbf{x})$ (see below for an example). Thus the Kohn-Sham orbitals depend on the potential, which depends on the density, which depends on the orbitals, so we must solve self-consistently (for example, by iterating until convergence).

DFT for solid-state or molecular systems starts with the HK free energy for an inhomogeneous electron gas [10]:

$$F_{\text{HK}}[\rho(\mathbf{x})] = T_{\text{KS}}[\rho(\mathbf{x})] + \frac{e^2}{2}\int d^3x\, d^3x' \frac{\rho(\mathbf{x})\rho(\mathbf{x}')}{|\mathbf{x} - \mathbf{x}'|} + E_{\text{xc}}[\rho(\mathbf{x})]. \tag{3.7}$$

Then $V_{\text{KS}} = v_{\text{ext}} - e\phi + v_{\text{xc}}$ with $v_{\text{xc}}(\mathbf{x}) = \delta E_{\text{xc}}/\delta\rho(\mathbf{x})$. To calculate the Kohn-Sham kinetic energy $T_{\text{KS}}[\rho(\mathbf{x})]$, find the normalized $\{\psi_i, \epsilon_i\}$ from

$$\left(-\frac{\hbar^2}{2m}\nabla^2 + V_{\text{KS}}(\mathbf{x})\right)\psi_i(\mathbf{x}) = \epsilon_i \psi_i(\mathbf{x}) \tag{3.8}$$

and, with $\rho(\mathbf{x}) = \sum_{i=1}^{A} |\psi_i(\mathbf{x})|^2$,

$$T_{\mathrm{KS}}[\rho(\mathbf{x})] = \sum_{i=1}^{A} \left\langle \psi_i \left| -\frac{\hbar^2}{2m}\nabla_i^2 \right| \psi_i \right\rangle = \sum_{i=1}^{A} \epsilon_i - \int d^3x\, \rho(\mathbf{x}) V_{\mathrm{KS}}(\mathbf{x}). \tag{3.9}$$

In practice, the DFT is usually based on the local density approximation (LDA): $E_{\mathrm{xc}}[\rho(\mathbf{x})] \approx \int d^3x\, \mathcal{E}_{\mathrm{xc}}(\rho(\mathbf{x}))$ with $\mathcal{E}_{\mathrm{xc}}(\rho)$ fit to a Monte Carlo calculation of the uniform electron gas. For example, one parametric formula for the energy density is [10]

$$\mathcal{E}_{\mathrm{xc}}(\rho)/\rho = -0.458/r_s - 0.0666G(r_s/11.4), \tag{3.10}$$

with

$$G(x) = \frac{1}{2}\left\{(1+x)^3 \log(1+x^{-1}) - x^2 + \frac{1}{2}x - \frac{1}{3}\right\}. \tag{3.11}$$

This is just like a simple Hartree approach with the additional potential:

$$v_{\mathrm{xc}}(\mathbf{x}) = \left.\frac{d[\mathcal{E}_{\mathrm{xc}}(\rho)]}{d\rho}\right|_{\rho=\rho(\mathbf{x})}. \tag{3.12}$$

The LDA is improved with the Generalized Gradient Approximation (GGA), such as the van Leeuwen–Baerends GGA [10],

$$v_{\mathrm{xc}}(\mathbf{r}) = -\beta\rho^{1/3}(\mathbf{r}) \frac{x^2(\mathbf{r})}{1 + 3\beta x(\mathbf{r}) \sinh^{-1}(x(\mathbf{r}))} \tag{3.13}$$

with $x = |\nabla\rho|/\rho^{4/3}$. For these Coulomb systems, Hartree-Fock is generally a good starting point, DFT/LDA is better, and DFT/GGA is better still.

There are some concerns, however, about DFT. Here are some quotes from the DFT literature that help motivate the application of EFT to DFT:

> From *A Chemist's Guide to DFT* [13]: "To many, the success of DFT appeared somewhat miraculous, and maybe even unjust and unjustified. Unjust in view of the easy achievement of accuracy that was so hard to come by in the wave function based methods. And unjustified it appeared to those who doubted the soundness of the theoretical foundations. "

> From *Density Functional Theory* [10]: "It is important to stress that all practical applications of DFT rest on essentially uncontrolled approximations, such as the LDA ..."

> From *Meta-Generalized Gradient Approximation* [14] "Some say that 'there is no systematic way to construct density functional approximations.' But there are more or less systematic ways, and the approach taken ... here is one of the former."

Thus, a microscopic, controlled, and systematic approach to DFT would be welcome.

We end this section with a preview of DFT as an effective action [15]. Recall ordinary thermodynamics with N particles at $T = 0$. The thermodynamic potential is related to the partition function, with the chemical potential μ acting as a source to change $N = \langle \widehat{N} \rangle$,

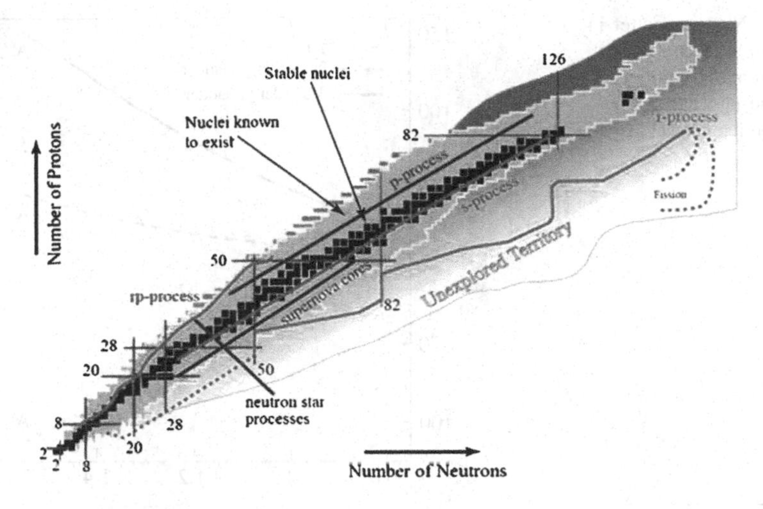

Fig. 3.4 Table of the nuclides

$$\Omega(\mu) = -kT \ln Z(\mu) \quad \text{and} \quad N = -\left(\frac{\partial \Omega}{\partial \mu}\right)_{TV}. \tag{3.14}$$

If we *invert* to find $\mu(N)$ and apply a Legendre transform, we obtain

$$F(N) = \Omega(\mu(N)) + \mu(N)N. \tag{3.15}$$

This is our (free) energy function of the particle number, which is analogous to the DFT energy functional of the density. Indeed, if we generalize to a spatially dependent chemical potential $J(\mathbf{x})$, then

$$Z(\mu) \longrightarrow Z[J(\mathbf{x})] \quad \text{and} \quad \mu N = \mu \int \psi^\dagger \psi \longrightarrow \int J(\mathbf{x})\psi^\dagger \psi(\mathbf{x}). \tag{3.16}$$

Now Legendre transform from $\ln Z[J(\mathbf{x})]$ to $\Gamma[\rho(\mathbf{x})]$, where $\rho = \langle \psi^\dagger \psi \rangle_J$, and we have DFT with Γ simply proportional to the energy functional!

3.1.3 DFT for Nuclei: EFT and RG Approaches

Figure 3.4 shows the table of the nuclides, labeled by the total numbers of protons and neutrons. For example, ^{208}Pb is found at the intersection of the horizontal line labeled "82" (protons) and the vertical line labeled "126" (neutrons). Stable nuclei are in black, known in pink. Here are some basic nuclear physics questions (which *everyone* should know how to answer):

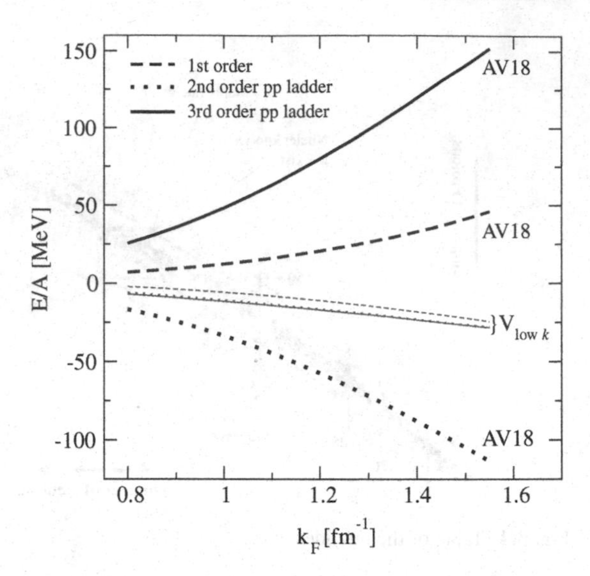

Fig. 3.5 Nuclear matter in perturbation theory for a conventional NN potential (Argonne v_{18}) and a low-momentum potential [16]

- Why is the slope of the black region less than a 45° angle once it is past $Z = N = 20$ or so?
- How do the binding energies of the nuclei in black compare? (e.g., do they vary over a wide range? Do they have a regular pattern?)
- What happens to the binding energy as you move perpendicular to the black line?
- What is the difference between being unstable and unbound? What are the drip lines?

(See [3, 8] for nuclear physics background.) The nucleosynthesis r-process lives largely in "Unexplored Territory." Radioactive beam facilities (existing and proposed) address this physics, as well as exotic nuclei such as halo nuclei and many other phenomena. As one gets further from stability, the importance of pairing grows, which highlights a difference between nuclear many-body physics and the physics of Coulomb systems, like atoms and molecules.

Let's try solving nuclear matter in low-order perturbation theory. One expects a minimum in the energy/particle (E/A) versus density (here the Fermi momentum $k_F \propto \rho^{1/3}$) at about $k_F \sim 1.35\,\text{fm}^{-1}$ and $E/A \sim -16\,\text{MeV}$. The standard Argonne v_{18} potential [17] is used in Fig. 3.5, with "Brueckner ladder" contributions shown order-by-order. First order is Hartree-Fock and it is not even bound! The repulsive core means that the series diverges badly.

But there is an energy density functional approach that seems to be based on Hartree-Fock (HF), which works quite well throughout Fig. 3.4. This is the phenomenological Skyrme HF approach [8, 18–20]. Recall from our earlier Hartree-Fock discussion that solving self-consistently is somewhat tricky because the potential is non-local. It would be much simpler if $v(\mathbf{x}, \mathbf{y}) \propto \delta(\mathbf{x} - \mathbf{y})$. This is the case with the

Skyrme interaction $V_2^{\text{Skyrme}} + V_3^{\text{Skyrme}}$, where $\langle \mathbf{k}|V_2^{\text{Skyrme}}|\mathbf{k}'\rangle = t_0 + \frac{1}{2}t_1(\mathbf{k}^2 + \mathbf{k}'^2) + t_2\mathbf{k}\cdot\mathbf{k}' + i W_0(\sigma_1+\sigma_2)\cdot\mathbf{k}\times\mathbf{k}'$. This motivates the Skyrme energy density functional (for $N = Z$) [8]:

$$\begin{aligned}\mathcal{E}[\rho,\tau,\mathbf{J}] &= \frac{1}{2M}\tau + \frac{3}{8}t_0\rho^2 + \frac{1}{16}t_3\rho^{2+\alpha} + \frac{1}{16}(3t_1+5t_2)\rho\tau \\ &\quad + \frac{1}{64}(9t_1-5t_2)(\nabla\rho)^2 - \frac{3}{4}W_0\rho\nabla\cdot\mathbf{J} + \frac{1}{32}(t_1-t_2)\mathbf{J}^2, \end{aligned} \tag{3.17}$$

where $\rho(\mathbf{x}) = \sum_i |\phi_i(\mathbf{x})|^2$ and $\tau(\mathbf{x}) = \sum_i |\nabla\phi_i(\mathbf{x})|^2$ (see [8] for the $\mathbf{J}(\mathbf{x})$ formula). We minimize $E = \int d\mathbf{x}\, \mathcal{E}[\rho,\tau,\mathbf{J}]$ by varying the (normalized) ϕ_i's,

$$\left(-\nabla\frac{1}{2M^*(\mathbf{x})}\nabla + U(\mathbf{x}) + \frac{3}{4}W_0\nabla\rho\cdot\frac{1}{i}\nabla\times\sigma\right)\phi_i(\mathbf{x}) = \epsilon_i\,\phi_i(\mathbf{x}), \tag{3.18}$$

with $U = \frac{3}{4}t_0\rho + (\frac{3}{16}t_1 + \frac{5}{16}t_2)\tau + \cdots$ and $\frac{1}{2M^*(\mathbf{x})} = \frac{1}{2M} + (\frac{3}{16}t_1 + \frac{5}{16}t_2)\rho$. Onc iterates until the ϕ_i's and ϵ_i's are self-consistent.

While phenomenologically successful, there are many questions and possible criticisms of Skyrme HF. Typical [e.g., SkIII] model parameters are: $t_0 = -1129$, $t_1 = 395$, $t_2 = -95$, $t_3 = 14000$, $W_0 = 120$ (in units of MeV-fm^n). These seem large; is there an expansion parameter? Where does $\rho^{2+\alpha}$ come from? Why not $\rho^{2+\beta}$? Is this just parameter fitting? A famous quote from von Neumann (via Fermi via Dyson) says: "*With four parameters I can fit an elephant, and with five I can make him wiggle his trunk.*" One might also say that Skyrme HF is only a mean-field model, which is too simple to accommodate NN correlations. How do we improve the approach? Is pairing treated correctly? How does Skyrme HF relate to NN (and NNN) forces? And so on. There is also the observation that Skyrme functionals works well where there is already data, but elsewhere fails to give consistent predictions (and the theoretical error bar is unknown).

Rather than focus on the Skyrme *interaction*, we consider the Skyrme energy functional as an approximate DFT functional. (Note: this is the viewpoint of modern practitioners.) Our master plan is to use EFT and renormalization group (RG) methods to elevate something close to Skyrme HF to a full DFT treatment. We want to use EFT and RG to provide a systematic input potential, including three-body potentials, and to generate systematically improved energy functionals. At the same time, we want to be able to estimate theoretical errors, so that extrapolation is under control.

This is a relatively new and different approach. In Table 3.1 we summarize aspects of the "traditional" approach to the (nuclear) many-body problem that are being challenged by the EFT approach. There are many continuing successes of conventional many-body approaches. The idea is not to prove standard methods wrong but to highlight where new insight can be provided. For each "old" item in this table (see endnotes for further explanations), we'll have a "new" perspective from EFT (see Table 3.2).

Table 3.1 (Nuclear) Many-body physics: "Old" approach

One Hamiltonian for all problems and energy/length scales (not QCD!)	For nuclear structure, protons and neutrons with a *local* potential [21] fit to NN data
Find the "best" potential	NN potential with $\chi^2/\text{dof} \approx 1$ up to $\sim 300\,\text{MeV}$ energy [22]
Two-body data may be sufficient; many-body forces as last resort	Let phenomenology dictate whether three-body forces are needed (answer: yes! [23])
Avoid (hide) divergences	Add "form factor" to suppress high-energy intermediate states; do not consider cutoff dependence [24]
Choose approximations by "art"	Use physical arguments (often handwaving) to justify the subset of diagrams used [25]

Table 3.2 (Nuclear) Many-body physics: "Old" vs. "New"

One Hamiltonian for all problems and energy/length scales	*Infinite* # of low-energy potentials; different resolutions $\Longrightarrow$ different dof's and Hamiltonians
Find the "best" potential	There *is* no best potential $\Longrightarrow$ use a convenient one!
Two-body data may be sufficient; many-body forces as last resort	Many-body data needed and many-body forces *inevitable*
Avoid divergences	Exploit divergences (cutoff dependence as tool)
Choose diagrams by "art"	Power counting determines diagrams and truncation error

3.1.4 EFT Analogies

From an effective field theory we desire systematic calculations with error estimates. We want them to be robust and model independent, which will enable reliable extrapolation. To help understand how this is accomplished, we can explore useful analogies between EFT and sophisticated numerical analysis.

- Naive error analysis: pick a method and reduce the mesh size (e.g., increase grid points) until the error is "acceptable". Sophisticated error analysis: understand scaling and sources of error (e.g., algorithm vs. round-off errors). *Does it work as well as it should?*
- Representation dependence means that not all ways of calculating are equally effective!

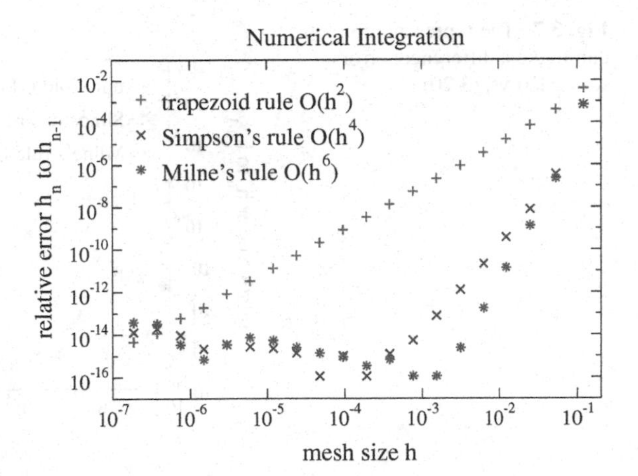

Fig. 3.6 Error plots in numerical integration

- Reliable extrapolation requires completeness of an expansion basis. An EFT lagrangian provides the analog of a complete basis.

Note: quantum mechanics makes EFT trickier than "classical" numerical analysis (see Lepage's lectures)!

Consider the numerical calculation of an integral using equal-spaced integration rules of increasing sophistication. How do the *numerical* errors behave? Log-log plots of the relative error against a parameter such as the mesh size are very helpful; a straight line indicates a power law and the exponent is read off from the slope. An example is shown in Fig. 3.6. Similar plots can be made for order-by-order EFT calculations, as described in Peter Lepage's lectures. Just like computer math is not equivalent to ordinary math, EFT is not the same as a theory applicable at all energies. It breaks down at high resolution, as the numerical calculations break down and degrade because of round-off errors at small mesh sizes. (Note: Don't carry this analogy to extremes!) Next consider an elliptic integral:

$$\int_0^1 \sqrt{(1-x^2)(2-x)}\, dx \tag{3.19}$$

with errors plotted in Fig. 3.7. Something is wrong; the errors do not behave as expected. However, after a simple transformation:

$$\int_0^{\pi/2} \sin^2 y \sqrt{2-\cos y}\, dy. \tag{3.20}$$

As seen in Fig. 3.7, the transformation makes a big difference! If you have freedom to change the representation, you can make a big difference in the ability to calculate accurately and easily. This is a moral we will apply when using EFT and RG methods to many-body problems.

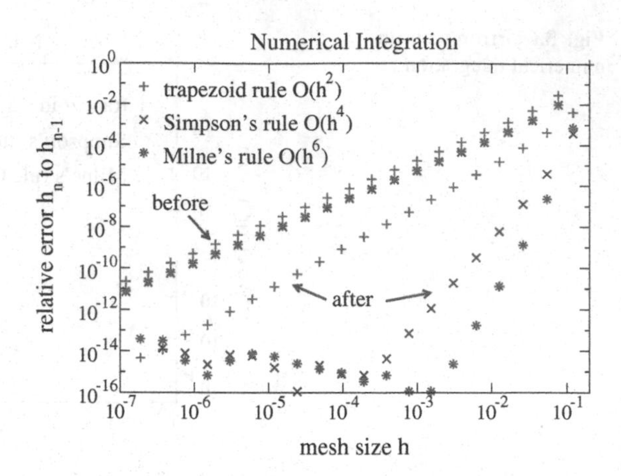

Fig. 3.7 The representation can make a difference: errors for (3.19) vs. (3.20)

3.1.5 Principles of Effective Low-Energy Theories

There are some basic physics principles underlying *any* low-energy effective model or theory. A high-energy, short-wavelength probe sees detail down to scales comparable to the wavelength. Thus, high-energy electron scattering at Jefferson Lab resolves the quark substructure of protons and neutrons in a nucleus. But at lower energies, details are not resolved, and one can replace short-distance structure with something simpler, as in a multipole expansion of a complicated charge or current distribution. So it is not necessary to do full QCD to do strong interaction physics at low energies, we can replace quarks and gluons by neutrons and protons (and pions). Chiral effective field theory is a systematic approach to carrying out this program using a local lagrangian framework.

It is not obvious that working at low resolution will work in quantum mechanics as it does for pixels or point dots or the classical multipole expansion, because *virtual* states can have high energies that are not, in reality, simple. But renormalization theory says it can be done! (See Lepage's lectures and [26, 27].) Note that this doesn't mean that we are *insensitive* to all short-distance details, only that their effects at low energies can be accounted for in a simple way.

We can use the possibility of changing the resolution scale to change the "perturbativeness" of nuclei. There are several sources of nonperturbative physics for nucleon-nucleon interactions:

1. Strong short-range repulsion ("hard core").
2. Iterated tensor (S_{12}) interactions (e.g., from pion exchange).
3. Near zero-energy bound states.

In Coulomb DFT, Hartree-Fock gives the dominate contribution to the energy, and correlations are small corrections. This may be why DFT works. In contrast, for NN interactions, correlations $\gg$ HF; does this mean DFT fails?? However, the first two sources depend on the resolution (i.e., the cutoff of high-energy physics), and the

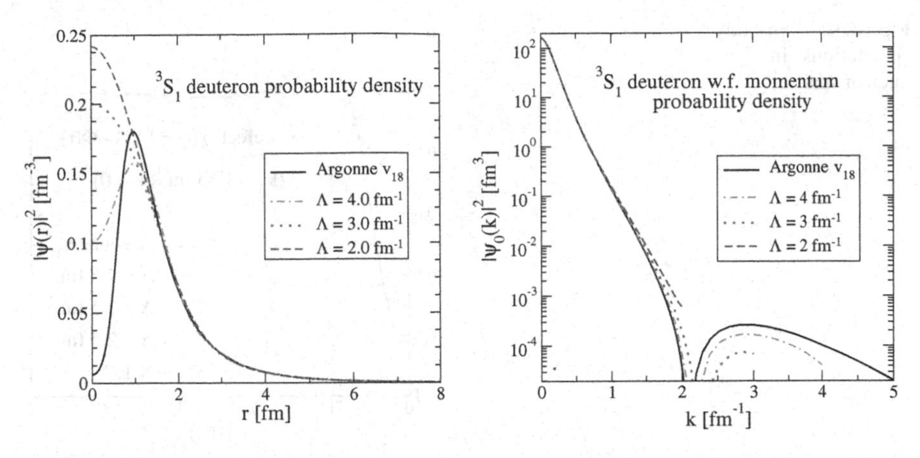

Fig. 3.8 The deuteron probability density at different resolutions (as indicated by the sharp momentum cutoff Λ)

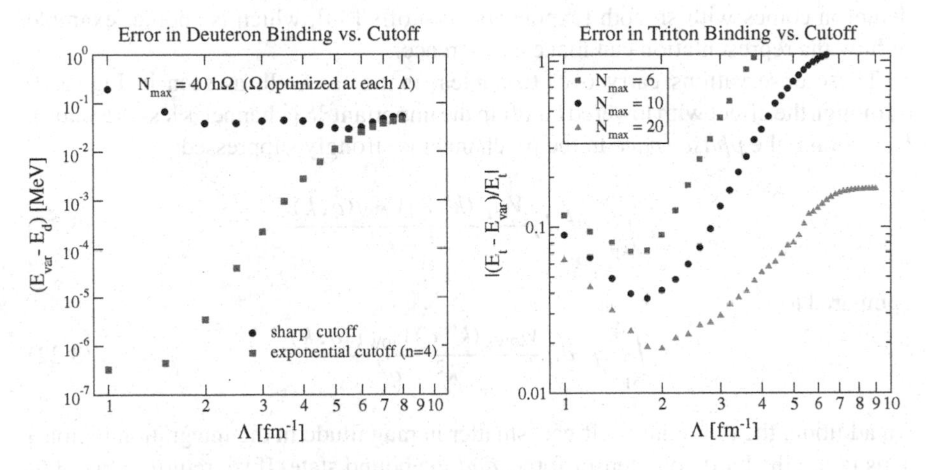

Fig. 3.9 Variational calculations at different resolutions [28, 29]

third one is affected by Pauli blocking. Thus we might use the freedom of low-energy theories to simplify calculations.

We can see the impact of different resolutions on the deuteron wave function in Fig. 3.8. The repulsive core leads to short-distance suppression and important high-momentum (small λ) components. This makes the many-body problem complicated. In contrast, potentials evolved by renormalization group (RG) methods to low momentum, generically called $V_{\mathrm{low}\,k}$ [30–32], have much simpler wave functions! (See Andreas Nogga's lectures for more detail on such potentials.) We note that chiral EFT potentials are naturally low-momentum potentials, but lowering their cutoffs further is generally advantageous. The consequence of lower resolution for variational calculations is illustrated in Fig. 3.9 [28, 29]. Note that the improvement for the

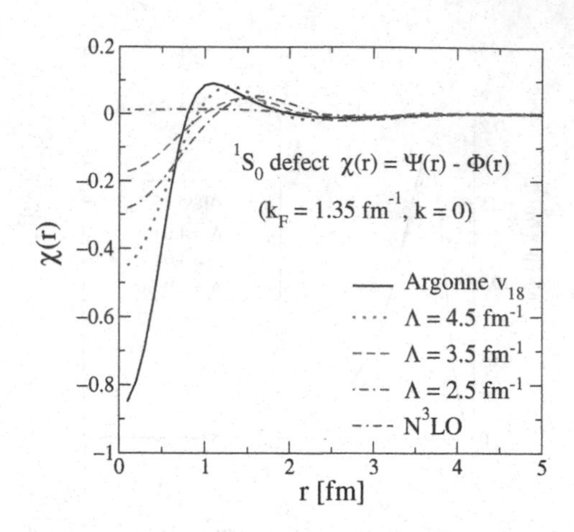

Fig. 3.10 Two-body correlations in nuclear matter

deuteron comes with smooth (exponential) cutoffs [33], which is another example of how the representation can make a difference.

These observations carry over to nuclear matter as well, as seen in Fig. 3.10 (although the effect with lowered cutoff in the important $^3S_1$ channel is less dramatic). In medium, the *phase space* in the pp-channel is strongly suppressed:

$$\int_{k_F}^{\infty} q^2\, dq \frac{V_{NN}(k', q) V_{NN}(q, k)}{k^2 - q^2} \tag{3.21}$$

compared to

$$\int_{k_F}^{\Lambda} q^2\, dq \frac{V_{\mathrm{low}\,k}(k', q) V_{\mathrm{low}\,k}(q, k)}{k^2 - q^2}. \tag{3.22}$$

(In addition, the potential itself gets smaller in magnitude in the integration region.) This tames the hard core, tensor force, *and* the bound state. If we return to Fig. 3.5, we see the consequence, which is very rapid convergence (by 2nd order) of the in-medium T-matrix for $V_{\mathrm{low}\,k}$ [16]. But there is no saturation in sight!

There were active attempts to transform away hard cores and soften the tensor interaction in the late 1960s and early 1970s [34–36]. But the requiem for soft potentials was given by Bethe (1971):

> "Very soft potentials must be excluded because they do not give saturation; they give too much binding and too high density. In particular, a substantial tensor force is required."

The next 30+ years were spent trying to solve accurately with "hard" potentials. But the story is not complete: three-nucleon forces (3NF)! When they are added consistently, we have the advantages of soft potentials while answering Bethe's criticism.

Ideally we would start with chiral NN + 3 NF EFT interactions and then evolve downward in Λ. What is possible now is to run the NN and *fit* 3NF EFT at each Λ [37]. The consequence is shown in Fig. 3.11. There is saturation even at

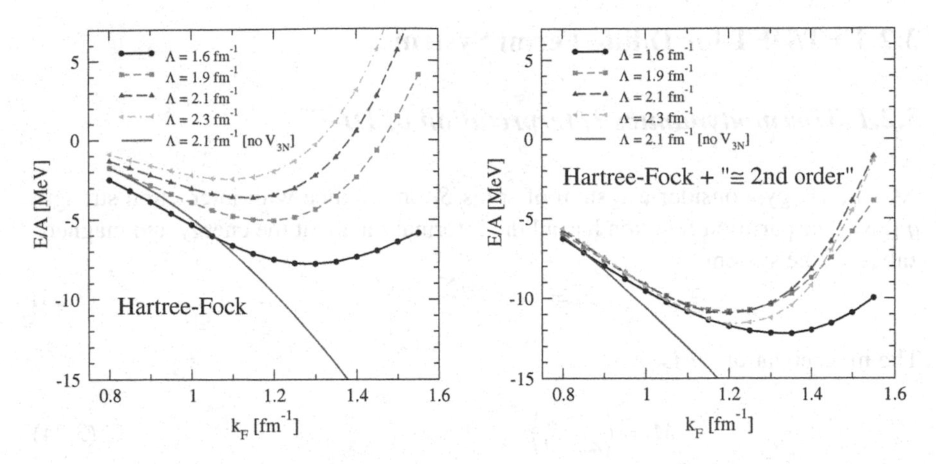

Fig. 3.11 Nuclear matter energy per particle [16]

the Hartree-Fock level, which is now driven by the three-body force. At second order, the cutoff dependence is greatly reduced and the minimum moves closer to the empirical point. One might worry that the three-body force is unnaturally large, but it is consistent with EFT power counting. Excellent results are also found for neutron matter. While there is much to be done, these results are very encouraging and motivate a microscopic DFT for nuclei.

3.1.6 Summary

In summary, there is a new attitude for many-body physics inspired by effective field theory. Bethe wrote about the nuclear case:

> "The theory must be such that it can deal with any nucleon-nucleon (NN) force, including hard or 'soft' core, tensor forces, and other complications. It ought not to be necessary to tailor the NN force for the sake of making the computation of nuclear matter (or finite nuclei) easier, but the force should be chosen on the basis of NN experiments (and possibly subsidiary experimental evidence, like the binding energy of H^3)."

The new attitude is to instead seek to make the problem easier. It's like the old vaudeville joke about a doctor and his patient:
Patient: Doctor, doctor, it hurts when I do this!
Doctor: Then don't do that.
We also follow Weinberg's Third Law of Progress in Theoretical Physics: "*You may use any degrees of freedom you like to describe a physical system, but if you use the wrong ones, you'll be sorry!*" We conclude with a new table (Table 3.2) of many-body physics, contrasting the old and the new approaches.

3.2 EFT/DFT for Dilute Fermi Systems

3.2.1 Thermodynamics Interpretation of DFT

As an analogy, consider a system of spins S_i on a lattice with interaction strength g [38]. The partition function has all the information about the energy and magnetization of the system:

$$\mathcal{Z} = \mathrm{Tr}\, e^{-\beta g \sum_{\{i,j\}} S_i S_j} . \tag{3.23}$$

The magnetization M is

$$M = \Big\langle \sum_i S_i \Big\rangle \tag{3.24}$$

$$= \frac{1}{\mathcal{Z}} \mathrm{Tr} \Big[\Big(\sum_i S_i \Big) e^{-\beta g \sum_{\{i,j\}} S_i S_j} \Big] . \tag{3.25}$$

Now add an external magnetic probe source H. The source adjusts the spin configurations near the ground state,

$$\mathcal{Z}[H] = e^{-\beta F[H]} = \mathrm{Tr}\, e^{-\beta (g \sum_{\{i,j\}} S_i S_j - H \sum_i S_i)} . \tag{3.26}$$

Variations of the source yield the magnetization

$$M = \Big\langle \sum_i S_i \Big\rangle_H = -\frac{\partial F[H]}{\partial H} , \tag{3.27}$$

where $F[H]$ is the Helmholtz free energy. For the ground state, we set $H = 0$ (or equal to a real external source) at the end.

Now if we find $H[M]$ by inverting $M[H]$,

$$M = \Big\langle \sum_i S_i \Big\rangle_H = -\frac{\partial F[H]}{\partial H} , \tag{3.28}$$

we can Legendre transform to the Gibbs free energy

$$\Gamma[M] = F[H] + H\, M . \tag{3.29}$$

Then the ground-state magnetization M_{gs} follows by minimizing $\Gamma[M]$:

$$H = \frac{\partial \Gamma[M]}{\partial M} \longrightarrow \frac{\partial \Gamma[M]}{\partial M}\bigg|_{M_{\mathrm{gs}}} = 0 . \tag{3.30}$$

Thus, we have a function of M (or functional if H is inhomogeneous) that is minimized at the ground-state free energy and magnetization.

DFT has an analogous structure as an effective action [15]. An effective action is generically the Legendre transform of a generating functional with an external source (or sources). For DFT, we use a source to adjust the density instead of the magnetization. The partition function in the presence of $J(x)$ coupled to the density is (we'll use a schematic notation here):

$$\mathcal{Z}[J] = e^{-W[J]} \sim \operatorname{Tr} e^{-\beta(\widehat{H}+J\widehat{\rho})} \longrightarrow \int \mathcal{D}[\psi^\dagger]\mathcal{D}[\psi]\, e^{-\int[\mathcal{L}+J\,\psi^\dagger\psi]}. \tag{3.31}$$

The density $\rho(x)$ in the presence of $J(x)$ is (keep in mind that we want $J = 0$ eventually),

$$\rho(x) = \langle\widehat{\rho}(x)\rangle_J = \frac{\delta W[J]}{\delta J(x)}. \tag{3.32}$$

After inverting to find $J[\rho]$, we Legendre transform from J to ρ:

$$\Gamma[\rho] = W[J] - \int J\,\rho \quad \text{and} \quad J(x) = -\frac{\delta\Gamma[\rho]}{\delta\rho(x)}. \tag{3.33}$$

Now consider the partition function in the zero-temperature limit of a Hamiltonian with time-independent source $J(\mathbf{x})$ [39, 40]:

$$\widehat{H}(J) = \widehat{H} + \int J\,\psi^\dagger\psi. \tag{3.34}$$

If the ground state is isolated (and bounded from below),

$$e^{-\beta\widehat{H}} = e^{-\beta E_0}\left[|0\rangle\langle 0| + \mathcal{O}(e^{-\beta(E_1-E_0)})\right]. \tag{3.35}$$

As $\beta \to \infty$, $\mathcal{Z}[J]$ yields the ground state of $\widehat{H}(J)$ with energy $E_0(J)$:

$$E_0(J) = \lim_{\beta\to\infty} -\frac{1}{\beta}\log\mathcal{Z}[J] = \frac{1}{\beta}W[J]. \tag{3.36}$$

Substitute and separate out the pieces:

$$E_0(J) = \langle\widehat{H}(J)\rangle_J = \langle\widehat{H}\rangle_J + \int J\,\langle\psi^\dagger\psi\rangle_J = \langle\widehat{H}\rangle_J + \int J\,\rho(J). \tag{3.37}$$

Rearranging, the expectation value of $\widehat{H}$ in the ground state generated by $J[\rho]$ is

$$\langle\widehat{H}\rangle_J = E_0(J) - \int J\,\rho = \frac{1}{\beta}\Gamma[\rho]. \tag{3.38}$$

Let's put it all together:

$$\frac{1}{\beta}\Gamma[\rho] = \langle \widehat{H} \rangle_J \xrightarrow{J\to 0} E_0 \quad \text{and} \quad J(x) = -\frac{\delta\Gamma[\rho]}{\delta\rho(x)} \xrightarrow{J\to 0} \left.\frac{\delta\Gamma[\rho]}{\delta\rho(x)}\right|_{\rho_{gs}(\mathbf{x})} = 0. \tag{3.39}$$

So for static $\rho(\mathbf{x})$, $\Gamma[\rho]$ is proportional to the DFT energy functional F_{HK}! Furthermore, the true ground state (with $J = 0$) is a variational minimum, so additional sources should be better than just one source coupling to the density (as we'll consider below). The simple, universal dependence on external potential follows directly in this formalism:

$$\Gamma_v[\rho] = W_v[J] - \int J\,\rho = W_{v=0}[J+v] - \int [(J+v) - v]\,\rho = \Gamma_{v=0}[\rho] + \int v\,\rho. \tag{3.40}$$

[Note: the functionals will change with resolution or field redefinitions; only stationary points are observables.]

There are a number of paths to the DFT effective action:

1. Follow the usual Coulomb Kohn-Sham DFT by calculating the uniform system as function of density, which yields an LDA ("local density approximation") functional and a standard Kohn-Sham procedure. Improve the functional with a semi-empirical gradient expansion.
2. Derive the functional with an RG approach [41].
3. Use the auxiliary field method [42, 43]. Couple $\psi^\dagger\psi$ to an auxiliary field φ, and eliminate all or part of $(\psi^\dagger\psi)^2$. Add a source term $J\varphi$ and perform a loop expansion about the expectation value $\phi = \langle\varphi\rangle$. A Kohn-Sham version uses the freedom of the expansion to require the density be unchanged at each order.
4. The inversion method [44–47] yields a systematic Kohn-Sham DFT, based on an order-by-order expansion. For example, we can apply the EFT power counting for a dilute system.

We'll expand here on the last path.

3.2.2 *EFT for Dilute Fermi Systems*

We consider first one of the simplest many-body systems: a collection of "hard spheres," which means that the potential is infinitely repulsive at a separation R of the fermions. Since the potential is zero outside of R and the wave function must vanish in the interior of the potential (so that the energy is finite), we can trivially write down the S-wave scattering solution for momentum k: it is just a sine function shifted by kR from the origin (see Fig. 3.12). Our problem will be to find the energy per particle (and other observables) of a system of particles interacting with this potential at $T = 0$, given the density.

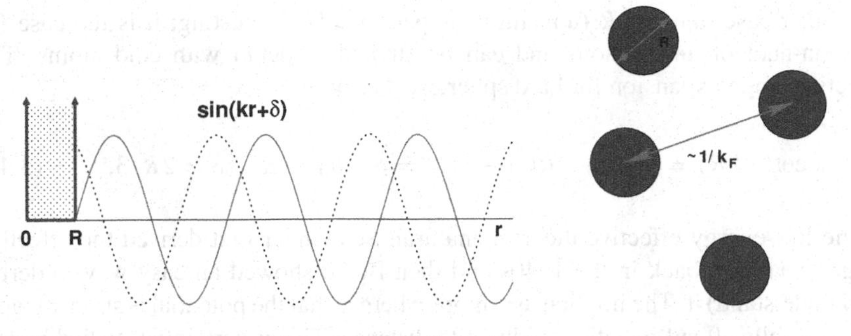

Fig. 3.12 Hard-sphere phase shifts and scales at finite density

Let's do a quick review of scattering. (More details on scattering at this level can be found in practically any first-year graduate quantum mechanics text. For a more specialized but very readable account of nonrelativistic scattering, check out "Scattering Theory" by Taylor.) Consider relative motion with total momentum $P = 0$:

$$\psi(r) \overset{r\to\infty}{\longrightarrow} e^{i\mathbf{k}\cdot\mathbf{r}} + f(k,\theta)\frac{e^{ikr}}{r}, \tag{3.41}$$

where $k^2 = k'^2 = ME_k$ and $\cos\theta = \hat{k}\cdot\hat{k}'$. The differential cross section is $d\sigma/d\Omega = |f(k,\theta)|^2$. For a central potential, we use partial waves:

$$f(k,\theta) = \sum_l (2l+1) f_l(k) P_l(\cos\theta), \tag{3.42}$$

where [48]

$$f_l(k) = \frac{e^{i\delta_l(k)}\sin\delta_l(k)}{k} = \frac{1}{k\cot\delta_l(k) - ik} \tag{3.43}$$

and the S-wave phase shift is defined by

$$u_0(r) \overset{r\to\infty}{\longrightarrow} \sin[kr + \delta_0(k)] \quad \Longrightarrow \quad \delta_0(k) = -kR \text{ for hard sphere.} \tag{3.44}$$

Note: we can do a partial wave expansion even if the potential is not central (as in the nuclear case!); it merely means that different l's will mix. The more important question is how many total l's do we need to include to ensure convergence.

As first shown by Schwinger, $k^{l+1}\cot\delta_l(k)$ has a power series expansion. For $l = 0$,

$$k\cot\delta_0 = -\frac{1}{a_0} + \frac{1}{2}r_0k^2 - Pr_0^3k^4 + \cdots, \tag{3.45}$$

which defines the *scattering length* a_0 and the *effective range* r_0. While $r_0 \sim R$, the range of the potential, a_0 can be anything; if $a_0 \sim R$, it is called "natural".

The other case $|a_0| \gg R$ (unnatural) is particularly interesting; it is the case for nucleon-nucleon interactions and can be studied in detail with cold atoms. The effective range expansion for hard sphere scattering is:

$$k\cot(-kR) = -\frac{1}{R} + \frac{1}{3}Rk^2 + \cdots \quad \Longrightarrow \quad a_0 = R \quad r_0 = 2R/3, \tag{3.46}$$

so the low-energy effective theory is natural. Schwinger first derived the effective range expansion back in the 1940s and then Bethe showed an easy way to derive (and understand) it. The implicit assumption here is that the potential is short-ranged; that is, it falls off sufficiently rapidly with distance. This is certainly satisfied by any potential that actually vanishes beyond a certain distance. Long-range potentials like the Coulomb potential must be treated differently (but a Yukawa potential with finite range is ok).

So now consider the EFT for a natural, short-ranged interaction [49]. A simple, general interaction is a sum of delta functions and derivatives of delta functions. In momentum space,

$$\langle \mathbf{k}|V_{\text{eft}}|\mathbf{k}'\rangle = C_0 + \frac{1}{2}C_2(\mathbf{k}^2 + \mathbf{k}'^2) + C_2'\,\mathbf{k}\cdot\mathbf{k}' + \cdots \tag{3.47}$$

Or, we construct the effective lagrangian $\mathcal{L}_{\text{eft}}$ from the most general local (contact) interactions:

$$\mathcal{L}_{\text{eft}} = \psi^\dagger\Big[i\frac{\partial}{\partial t} + \frac{\overrightarrow{\nabla}^2}{2M}\Big]\psi - \frac{C_0}{2}(\psi^\dagger\psi)^2 + \frac{C_2}{16}[(\psi\psi)^\dagger(\psi\overleftrightarrow{\nabla}^2\psi) + \text{h.c.}]$$
$$+ \frac{C_2'}{8}(\psi\overleftrightarrow{\nabla}\psi)^\dagger\cdot(\psi\overleftrightarrow{\nabla}\psi) - \frac{D_0}{6}(\psi^\dagger\psi)^3 + \ldots \tag{3.48}$$

Dimensional analysis (with a bit of additional insight to give us the 4π's) implies

$$C_{2i} \sim \frac{4\pi}{M}R^{2i+1}, \quad D_{2i} \sim \frac{4\pi}{M}R^{2i+4}, \tag{3.49}$$

which will enable us to make quantitative power-counting estimates.

The ingredients for an effective field theory are nicely summarized in the "Crossing the Border" review [50]:

1. *Use the most general $\mathcal{L}$ with low-energy degrees-of-freedom consistent with global and local symmetries of underlying theory.* Here,

$$\mathcal{L}_{\text{eft}} = \psi^\dagger\left[i\frac{\partial}{\partial t} + \frac{\nabla^2}{2M}\right]\psi - \frac{C_0}{2}(\psi^\dagger\psi)^2 - \frac{D_0}{6}(\psi^\dagger\psi)^3 + \ldots \tag{3.50}$$

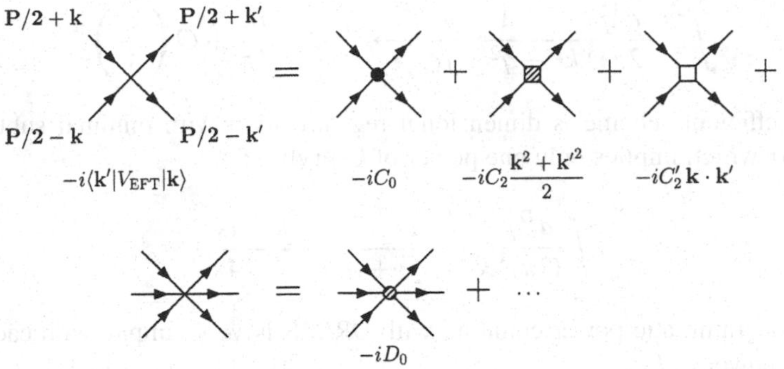

Fig. 3.13 Feynman rules in free space [49]

2. *Declare a regularization and renormalization scheme.* For a natural a_0, using dimensional regularization and minimal subtraction is particularly convenient and efficient.
3. *Establish a well-defined power counting*, which means identifying small expansion parameters, typically using the separation of scales. Here, k_F/Λ with $\Lambda \sim 1/R$, which implies $k_F a_0$, $k_F r_0$, etc. are expansion parameters. In the end, this will be manifest in the energy density:

$$\mathcal{E} = \rho \frac{k_F^2}{2M}\left[\frac{3}{5} + \frac{2}{3\pi}(k_F a_0) + \frac{4}{35\pi^2}(11 - 2\ln 2)(k_F a_0)^2 + \cdots\right]. \tag{3.51}$$

The Feynman rules for the EFT lagrangian are summarized in Fig. 3.13 [49]. We need to reproduce $f_0(k)$ in perturbation theory (the Born series):

$$f_0(k) \propto a_0 - i a_0^2 k - (a_0^3 - a_0^2 r_0/2)k^2 + \mathcal{O}(k^3 a_0^4). \tag{3.52}$$

The leading potential $V_{\text{EFT}}^{(0)}(\mathbf{x}) = C_0 \delta(\mathbf{x})$ or

$$\langle \mathbf{k}|V_{\text{eft}}^{(0)}|\mathbf{k}'\rangle \Longrightarrow$$

$$\Longrightarrow C_0. \tag{3.53}$$

Thus, choosing $C_0 \propto a_0$ gets the first term. Next is $\langle \mathbf{k}|V G_0 V|\mathbf{k}'\rangle$:

$$\Longrightarrow C_0 M \int \frac{d^3q}{(2\pi)^3} \frac{1}{k^2 - q^2 + i\epsilon} C_0 \longrightarrow \infty! \tag{3.54}$$

This is a linear divergence. If the integral is cutoff at Λ_c, we can absorb the linear dependence on Λ_c into C_0, but we'll have all powers of k^2:

$$\int^{\Lambda_c} \frac{d^3q}{(2\pi)^3} \frac{1}{k^2 - q^2 + i\epsilon} \longrightarrow \frac{\Lambda_c}{2\pi^2} - \frac{ik}{4\pi} + \mathcal{O}\left(\frac{k^2}{\Lambda_c}\right). \tag{3.55}$$

A more efficient scheme is dimensional regularization with minimal subtraction (DR/MS), which implies only one power of k survives:

$$\int \frac{d^Dq}{(2\pi)^3} \frac{1}{k^2 - q^2 + i\epsilon} \stackrel{D\to 3}{\longrightarrow} -\frac{ik}{4\pi}. \tag{3.56}$$

The diagrammatic power counting with DR/MS is very simple, with each loop adding a power of k:

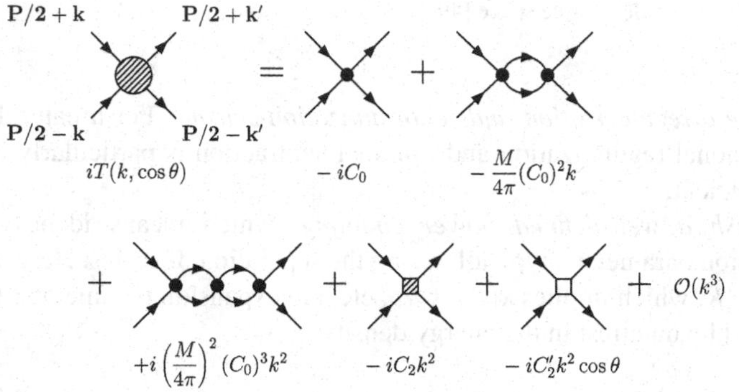

P/2 + k P/2 + k′

P/2 − k P/2 − k′

$iT(k, \cos\theta) = -iC_0 + -\frac{M}{4\pi}(C_0)^2 k$

$+ \; +i\left(\frac{M}{4\pi}\right)^2 (C_0)^3 k^2 \; + \; -iC_2 k^2 \; + \; -iC_2' k^2 \cos\theta \; + \; \mathcal{O}(k^3)$

After matching to the scattering amplitude,

$$C_0 = \frac{4\pi}{M} a_0 = \frac{4\pi}{M} R, \quad C_2 = \frac{4\pi}{M} \frac{a_0^2 r_0}{2} = \frac{4\pi}{M} \frac{R^3}{3}, \quad C_2' = \frac{4\pi}{M} a_p^3, \quad \cdots \tag{3.57}$$

recovers the effective range expansion order-by-order with diagrams:

$$\frac{4\pi}{M}(a_0 - i a_0^2 p - a_0^3 p^2 + a_0^2 r_0 p^2 + \cdots), \tag{3.58}$$

with one power of k per diagram and *natural* coefficients, so we can estimate truncation errors from simple dimensional analysis.

3.2.3 Apply at Finite Density

Consider a noninteracting Fermi sea of particles at $T = 0$. Put the system in a large box ($V = L^3$) with periodic boundary conditions and spin-isospin degeneracy ν (e.g., for nuclei, $\nu = 4$). Fill momentum states up to Fermi momentum k_F, so that

$$N = \nu \sum_{\mathbf{k}}^{k_{\mathrm{F}}} 1, \qquad E = \nu \sum_{\mathbf{k}}^{k_{\mathrm{F}}} \frac{\hbar^2 k^2}{2M}. \tag{3.59}$$

We can evaluate the sums using

$$\int F(k)\, dk \approx \sum_i F(k_i)\Delta k_i = \sum_i F(k_i)\frac{2\pi}{L}\Delta n_i = \frac{2\pi}{L}\sum_i F(k_i). \tag{3.60}$$

In one dimension (try finding the 1-D analogs of the following results!),

$$N = \nu \frac{L}{2\pi}\int_{-k_{\mathrm{F}}}^{+k_{\mathrm{F}}} dk = \frac{\nu k_{\mathrm{F}}}{\pi} L \Longrightarrow \rho = \frac{N}{L} = \frac{\nu k_{\mathrm{F}}}{\pi}; \quad \frac{E}{L} = \frac{1}{3}\frac{\hbar^2 k_{\mathrm{F}}^2}{2M}\rho, \tag{3.61}$$

while in three dimensions:

$$N = \nu \frac{V}{(2\pi)^3}\int^{k_{\mathrm{F}}} d^3k = \frac{\nu k_{\mathrm{F}}^3}{6\pi^2} V \Longrightarrow \rho = \frac{N}{V} = \frac{\nu k_{\mathrm{F}}^3}{6\pi^2}; \quad \frac{E}{V} = \frac{3}{5}\frac{\hbar^2 k_{\mathrm{F}}^2}{2M}\rho. \tag{3.62}$$

The volume/particle $V/N = 1/\rho \sim 1/k_{\mathrm{F}}^3$, so the spacing $\sim 1/k_{\mathrm{F}}$, as implied by Fig. 3.12.

We find the energy density by summing over the Fermi sea. In leading order, we found $V_{\mathrm{EFT}}^{(0)}(\mathbf{x}) = C_0\delta(\mathbf{x})$, so that $V_{\mathrm{EFT}}^{(0)}(\mathbf{k}, \mathbf{k}') = C_0$, and

$$\longrightarrow \qquad \mathcal{E}_{\mathrm{LO}} = \frac{C_0}{2}\nu(\nu - 1)\left(\sum_{\mathbf{k}}^{k_{\mathrm{F}}} 1\right)^2 \propto a_0 k_{\mathrm{F}}^6. \tag{3.63}$$

At the next order, we get a linear divergence again:

$$\Longrightarrow \qquad \mathcal{E}_{\mathrm{NLO}} \propto \int_{k_{\mathrm{F}}}^{\infty} \frac{d^3q}{(2\pi)^3}\frac{C_0^2}{k^2 - q^2}. \tag{3.64}$$

The *same* renormalization fixes it!

$$\int_{k_{\mathrm{F}}}^{\infty} \frac{1}{k^2 - q^2} = \int_0^{\infty} \frac{1}{k^2 - q^2} - \int_0^{k_{\mathrm{F}}} \frac{1}{k^2 - q^2} \xrightarrow{D \to 3} -\int_0^{k_{\mathrm{F}}} \frac{1}{k^2 - q^2} \propto a_0^2 k_{\mathrm{F}}^7. \tag{3.65}$$

We also note that particles $\longrightarrow$ holes through the renormalization.

The Feynman rules for the energy density $\mathcal{E}$ at $T = 0$, which is the sum of *Hugenholtz* diagrams [38] (closed, connected Feynman diagrams with symmetry factors) with the same vertices as free space (and the same renormalization!), are:

1. Each line is assigned conserved $\widetilde{k} \equiv (k_0, \boldsymbol{k})$ and $[\omega_{\boldsymbol{k}} \equiv k^2/2M]$,

$$iG_0(\widetilde{k})_{\alpha\beta} = i\delta_{\alpha\beta}\left(\frac{\theta(k-k_{\rm F})}{k_0-\omega_{\boldsymbol{k}}+i\epsilon}+\frac{\theta(k_{\rm F}-k)}{k_0-\omega_{\boldsymbol{k}}-i\epsilon}\right). \quad (3.66)$$

2.  $\longrightarrow (\delta_{\alpha\gamma}\delta_{\beta\delta}+\delta_{\alpha\delta}\delta_{\beta\gamma})$ (if spin-independent).
3. After spin summations, $\delta_{\alpha\alpha} \to -\nu$ in every closed fermion loop.
4. Integrate $\int d^4k/(2\pi)^4$ with $e^{ik_0 0^+}$ for tadpoles
5. The symmetry factor $i/(S\prod_{l=2}^{l_{\max}}(l!)^k)$ counts vertex permutations and equivalent l-tuples of lines (see [49] for examples).

These Feynman rules in turn lead to power counting rules:

1. For every propagator (line): $M/k_{\rm F}^2$
2. For every loop integration: $k_{\rm F}^5/M$
3. For every n-body vertex with $2i$ derivatives: $k_{\rm F}^{2i}/M\Lambda^{2i+3n-5}$

Thus, a diagram with V_{2i}^n n-body vertices scales as $(k_{\rm F})^\beta$ *only*:

$$\beta = 5+\sum_{n=2}^{\infty}\sum_{i=0}^{\infty}(3n+2i-5)V_{2i}^n. \quad (3.67)$$

For example, at leading order,

$$\Longrightarrow V_0^2 = 1 \Longrightarrow \beta = 5+(3\cdot 2+2\cdot 0-5)\cdot 1 = 6 \Longrightarrow \mathcal{O}(k_{\rm F}^6), \quad (3.68)$$

and at next-to-leading order,

$$\Longrightarrow V_0^2 = 2 \Longrightarrow \beta = 5+(3\cdot 2+2\cdot 0-5)\cdot 2 = 7 \Longrightarrow \mathcal{O}(k_{\rm F}^7). \quad (3.69)$$

We emphasize that Pauli blocking doesn't change the free-space ultraviolet (short distance) renormalization, since the density is a long-distance effect. As noted before, particles become holes:

$$\int_{k_{\rm F}}^{\infty} = \int_0^{\infty} - \int_0^{k_{\rm F}} \longrightarrow -\int_0^{k_{\rm F}}. \quad (3.70)$$

The power counting is exceptionally clean, with a separation of vertex factors $\propto a_0, r_0, \ldots$ and a dimensionless geometric integral times $k_{\rm F}^n$, with each diagram contributing to exactly one order in the expansion. This is a systematic expansion: the ratio of successive terms is $\sim k_{\rm F}R$, so you can *estimate* excluded contributions.

The full result for the energy density through $\mathcal{O}(k_\mathrm{F}^8)$ is [49]:

$$\frac{E}{V} = \rho\frac{k_\mathrm{F}^2}{2M}\bigg[\frac{3}{5} + (\nu-1)\frac{2}{3\pi}(k_\mathrm{F}a_0) + (\nu-1)\frac{4}{35\pi^2}(11-2\ln 2)(k_\mathrm{F}a_0)^2$$
$$+ (\nu-1)\big(0.076+0.057(\nu-3)\big)(k_\mathrm{F}a_0)^3 + (\nu-1)\frac{1}{10\pi}(k_\mathrm{F}r_0)(k_\mathrm{F}a_0)^2$$
$$+ (\nu+1)\frac{1}{5\pi}(k_\mathrm{F}a_p)^3 + \cdots\bigg]. \tag{3.71}$$

This looks like a power series in k_F, but it's not! There are new *logarithmic* divergences in 3–3 scattering, in these diagrams:

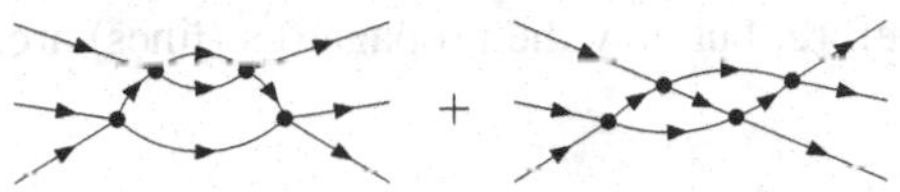

$$\propto (C_0)^4 \ln(k/\Lambda_c). \tag{3.72}$$

Changes in Λ_c must be absorbed by the *3-body* coupling $D_0(\Lambda_c)$, so [51]

$$D_0(\Lambda_c) \propto (C_0)^4 \ln(a_0\Lambda_c) + \text{const.} \tag{3.73}$$

Then requiring the sum to be independent of Λ_c,

$$\frac{d}{d\Lambda_c}\Big[\ \cdots\ \Big] = 0 \tag{3.74}$$

fixes the coefficient! This implies for the energy density,

$\mathcal{O}(k_\mathrm{F}^9 \ln(k_\mathrm{F}))$:

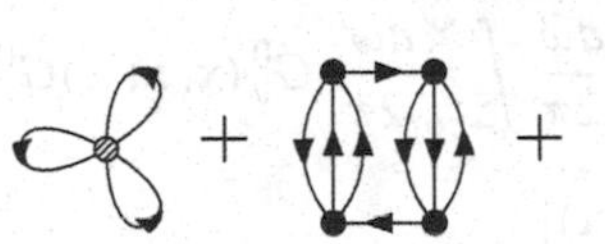

$$\propto (\nu-2)(\nu-1)\,(k_\mathrm{F}a_0)^4\ln(k_\mathrm{F}a_0) \tag{3.75}$$

without actually carrying out the calculation. Similar analyses can identify the higher logarithmic terms in the expansion of the energy density [49, 51].

Divergences indicate sensitivity to short-distance behavior. The cutoff Λ_c here serves as a resolution scale; as we increase Λ_c, we see more of the short-distance details. Observables (such as scattering amplitudes) must not change when Λ_c changes, so they must be absorbed in a coupling. But it can't be a coupling from 2–2 scattering, because we already took care of all the divergences there. So there must be a point-like three-body force included, whose coupling D_0 can absorb the changes.

Let's summarize the dilute Fermi system with natural a_0. We find that the many-body energy density is perturbative (but not analytic) in $k_\mathrm{F}a_0$, and is efficiently reproduced by the EFT approach. Power counting gives us error estimates from omitted diagrams. Three-body forces are *inevitable* in the low-energy effective theory and not unique; they depend on the two-body potential. The case of a natural scattering length is under control for a uniform system, but what if the scattering length is not natural? We'll come back to that situation in the last lecture. First we consider a non-uniform system: a finite number of fermions in a trap, which takes us back to DFT.

3.2.4 DFT via EFT

Return to the thermodynamic version of DFT through the effective action and ask: What can EFT do for DFT? We can construct the effective action as a path integral by finding $W[J]$ order-by-order in an EFT expansion. For a dilute short-range system, this means the same diagrams as before, but now the propagators (lines) are in the background field $J(\mathbf{x})$ [52–54]:

$$G_J^0(\mathbf{x},\mathbf{x}';\omega) = \sum_\alpha \psi_\alpha(\mathbf{x})\psi_\alpha^*(\mathbf{x}')\left[\frac{\theta(\epsilon_\alpha-\epsilon_F)}{\omega-\epsilon_\alpha+i\eta}+\frac{\theta(\epsilon_F-\epsilon_\alpha)}{\omega-\epsilon_\alpha-i\eta}\right], \tag{3.76}$$

where $\psi_\alpha(\mathbf{x})$ satisfies:

$$\left[-\frac{\nabla^2}{2M}+v_{\text{ext}}(\mathbf{x})-J(\mathbf{x})\right]\psi_\alpha(\mathbf{x}) = \epsilon_\alpha\psi_\alpha(\mathbf{x}). \tag{3.77}$$

Applying this to the leading-order (LO) contribution $W_1[J]$, which is Hartree-Fock, yields

$$\begin{aligned} W_1[J] &= \frac{1}{2}\nu(\nu-1)C_0\int d^3\mathbf{x}\int_{-\infty}^{\infty}\frac{d\omega}{2\pi}\int_{-\infty}^{\infty}\frac{d\omega'}{2\pi}\,G_J^0(\mathbf{x},\mathbf{x};\omega)G_J^0(\mathbf{x},\mathbf{x};\omega') \\ &= -\frac{1}{2}\frac{(\nu-1)}{\nu}C_0\int d^3\mathbf{x}\,[\rho_J(\mathbf{x})]^2, \end{aligned} \tag{3.78}$$

where $\rho_J(\mathbf{x}) \equiv \nu\sum_\alpha^{\epsilon_F}|\psi_\alpha(\mathbf{x})|^2$. Expressions for the other W_i's proceed directly from the Feynman rules using the new propagator.

Given $W[J]$ as an EFT expansion, how do we perform the Legendre transformation,

$$\Gamma[\rho] = W[J] - \int J\rho, \tag{3.79}$$

in a systematic way? The EFT power counting gives us a means to invert to find $J[\rho]$. In particular, the "inversion method" provides an order-by-order inversion from $W[J]$ to $\Gamma[\rho]$ [44–46]. It proceeds by decomposing $J(x) = J_0(x)+J_{\text{LO}}(x)+J_{\text{NLO}}(x)+\ldots$ with two conditions on J_0:

$$\rho(x) = \frac{\delta W_0[J_0]}{\delta J_0(x)} \quad \text{and} \quad J_0(x)|_{\rho=\rho_{gs}} = \left.\frac{\delta \Gamma_{\text{interacting}}[\rho]}{\delta \rho(x)}\right|_{\rho=\rho_{gs}}. \tag{3.80}$$

We are using the freedom to split J into J_0 and the rest, in the same way that one adds and subtracts a single-particle potential U to a Hamiltonian: $H = T + V = (T + U) + (V - U)$ and then uses the freedom to choose U to improve many-body convergence. In our case, we choose J_0 so that there are no corrections to the zeroth order density at each order in the expansion. The interpretation is that J_0 is the external potential that yields for a noninteracting system the exact density. This is the Kohn-Sham potential! The two conditions involving J_0 imply a self-consistent procedure.

The inversion method for effective action DFT [44–46] is an order-by-order matching in a counting parameter λ (e.g., an EFT expansion):

$$\text{diagrams} \Longrightarrow W[J,\lambda] = W_0[J] + \lambda W_1[J] + \lambda^2 W_2[J] + \cdots \tag{3.81}$$

$$\text{assume} \Longrightarrow J[\rho,\lambda] = J_0[\rho] + \lambda J_1[\rho] + \lambda^2 J_2[\rho] + \cdots \tag{3.82}$$

$$\text{derive} \Longrightarrow \Gamma[\rho,\lambda] = \Gamma_0[\rho] + \lambda \Gamma_1[\rho] + \lambda^2 \Gamma_2[\rho] + \cdots \tag{3.83}$$

We start with the exact expressions for Γ and ρ [note: β and $T = 1$ here],

$$\Gamma[\rho] = W[J] - \int d^4x\, J(x)\rho(x), \tag{3.84}$$

$$\rho(x) = \frac{\delta W[J]}{\delta J(x)}, \quad J(x) = -\frac{\delta \Gamma[J]}{\delta \rho(x)}. \tag{3.85}$$

Then plug in each of the expansions, with ρ treated as order unity. Zeroth order is the noninteracting system with potential $J_0(x)$,

$$\Gamma_0[\rho] = W_0[J_0] - \int d^4x\, J_0(x)\rho(x) \quad \Longrightarrow \quad \rho(x) = \frac{\delta W_0[J_0]}{\delta J_0(x)}, \tag{3.86}$$

which is the Kohn-Sham system with the exact density! (Note: $J_0 \equiv V_{KS}$ here.) To evaluate $W_0[J_0]$, we introduce the orbitals from (3.77), which diagonalize W_0, so that it yields a sum of ε_i's for the occupied states. Then we find J_0 for the ground state via a self-consistency loop:

$$J_0 \to W_1 \to \Gamma_1 \to J_1 \to W_2 \to \Gamma_2 \to \cdots \Longrightarrow J_0(x) = \sum_{i>0} \frac{\delta \Gamma_i[\rho]}{\delta \rho(x)}, \tag{3.87}$$

which is the second of our two conditions on J_0.

We note that the Kohn-Sham potential is local:

$$J_0(\mathbf{x}) = \frac{\delta \Gamma_{\text{int}}[\rho]}{\delta \rho(\mathbf{x})}, \tag{3.88}$$

in stark contrast to the non-local and state-dependent self-energy $\Sigma^*(\mathbf{x}, \mathbf{x}'; \omega)$. Evaluating the functional derivatives is easiest if Γ is approximated so that the dependence on the density is explicit, as with the LDA or DME (see below). Otherwise we need to use a chain rule with the "inverse density-density correlator" [45]

$$J_0(\mathbf{R}) = \frac{\delta \Gamma_{\text{int}}[\rho]}{\delta \rho(\mathbf{R})} = \int \left(\frac{\delta \rho(\mathbf{R})}{\delta J_0(\mathbf{y})} \right)^{-1} \frac{\delta \Gamma_{\text{int}}[\rho]}{\delta J_0(\mathbf{y})} = - \underset{\mathbf{R}}{(\ldots)} - \underset{\mathbf{R}}{(\ldots)} + \cdots$$

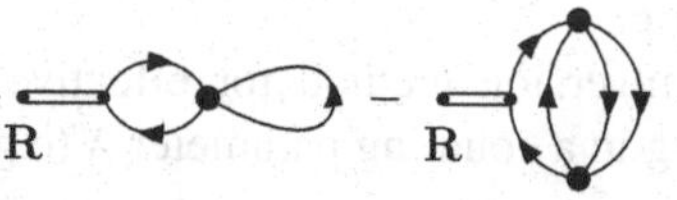

There are new Feynman rules for Γ_{int} for evaluating such diagrams [45]. (A related approach is the OEP method [11, 55–57].)

In constructing the diagrams for $W[J]$ and new diagrams for $\Gamma[\rho]$ order by order in the expansion (e.g., EFT power counting), the source $J_0(\mathbf{x})$ is now the background field (rather than the full $J(\mathbf{x})$). Propagators (lines) in the background field $J_0(\mathbf{x})$ are

$$G^0_{\text{KS}}(\mathbf{x}, \mathbf{x}'; \omega) = \sum_\alpha \psi_\alpha(\mathbf{x}) \psi^*_\alpha(\mathbf{x}') \left[\frac{\theta(\epsilon_\alpha - \epsilon_\text{F})}{\omega - \epsilon_\alpha + i\eta} + \frac{\theta(\epsilon_\text{F} - \epsilon_\alpha)}{\omega - \epsilon_\alpha - i\eta} \right], \tag{3.89}$$

where $\psi_\alpha(\mathbf{x})$ satisfies:

$$\left[-\frac{\nabla^2}{2M} + v(\mathbf{x}) - J_0(\mathbf{x}) \right] \psi_\alpha(\mathbf{x}) = \epsilon_\alpha \psi_\alpha(\mathbf{x}). \tag{3.90}$$

For example, if we apply this prescription to the short-range LO contribution (i.e., Hartree-Fock), we obtain

$$\begin{aligned} W_1[J_0] &= \frac{1}{2} \nu(\nu - 1) C_0 \int d^3\mathbf{x} \int_{-\infty}^{\infty} \frac{d\omega}{2\pi} \int_{-\infty}^{\infty} \frac{d\omega'}{2\pi}\, G^0_{\text{KS}}(\mathbf{x}, \mathbf{x}; \omega) G^0_{\text{KS}}(\mathbf{x}, \mathbf{x}; \omega') \\ &= -\frac{1}{2} \frac{(\nu - 1)}{\nu} C_0 \int d^3\mathbf{x}\, [\rho_{J_0}(\mathbf{x})]^2, \end{aligned} \tag{3.91}$$

where

$$\rho_{J_0}(\mathbf{x}) \equiv \nu \sum_\alpha^{\epsilon_\text{F}} |\psi_\alpha(\mathbf{x})|^2. \tag{3.92}$$

Let us construct the $T = 0$ local density approximation (LDA). In a uniform system, each line is a non-interacting propagator. The energy density in the uniform system evaluates to:

$$\frac{E}{V} = \rho\frac{k_F^2}{2M}\left[\frac{3}{5} + (\nu-1)\frac{2}{3\pi}(k_F a_0) + (\nu-1)\frac{4}{35\pi^2}(11 - 2\ln 2)(k_F a_0)^2 \right.$$
$$+ (\nu-1)\big(0.076 + 0.057(\nu-3)\big)(k_F a_0)^3$$
$$\left. + (\nu-1)\frac{1}{10\pi}(k_F r_0)(k_F a_0)^2 + (\nu+1)\frac{1}{5\pi}(k_F a_p)^3 + \cdots \right]. \tag{3.93}$$

with $k_F = (6\pi^2\rho/\nu)^{1/3}$. Using this relation to replace k_F everywhere by $\rho(\mathbf{x})$, we directly obtain the LDA expression for $\Gamma[\rho]$,

$$\Gamma[\rho] = \int d^3x \left[T_{KS}(\mathbf{x}) + \frac{1}{2}\frac{(\nu-1)}{\nu}\frac{4\pi a_0}{M}[\rho(\mathbf{x})]^2 + d_1\frac{a_0^2}{2M}[\rho(\mathbf{x})]^{7/3} \right.$$
$$+ d_2\, a_0^3[\rho(\mathbf{x})]^{8/3} + d_3\, a_0^2\, r_0[\rho(\mathbf{x})]^{8/3}$$
$$\left. + d_4\, a_p^3[\rho(\mathbf{x})]^{8/3} + \cdots \right]. \tag{3.94}$$

The Kohn-Sham J_0 according to the EFT expansion follows immediately in the LDA from (3.88):

$$J_0(\mathbf{x}) = \left[-\frac{(\nu-1)}{\nu}\frac{4\pi a_0}{M}\rho(\mathbf{x}) - c_1\frac{a_0^2}{2M}[\rho(\mathbf{x})]^{4/3} - c_2\, a_0^3[\rho(\mathbf{x})]^{5/3} \right.$$
$$\left. - c_3\, a_0^2\, r_0[\rho(\mathbf{x})]^{5/3} - c_4\, a_p^3[\rho(\mathbf{x})]^{5/3} + \cdots \right]. \tag{3.95}$$

(Finding the $\{d_i\}$'s and $\{c_i\}$'s is left as an exercise for the reader.)

Given (3.94) and (3.95), the iteration procedure is:

1. Guess an initial density profile $\rho(r)$ (e.g., the Thomas-Fermi density).
2. Evaluate the local single-particle potential $V_{KS}(r) \equiv v_{ext}(r) - J_0(r)$.
3. Find the wave functions and energies $\{\psi_\alpha, \epsilon_\alpha\}$ of the lowest A states (including degeneracies) by solving:

$$\left[-\frac{\nabla^2}{2M} + V_{KS}(r) \right] \psi_\alpha(\mathbf{x}) = \epsilon_\alpha \psi_\alpha(\mathbf{x}). \tag{3.96}$$

4. Compute a new density $\rho(r) = \sum_{\alpha=1}^{A} |\psi_\alpha(\mathbf{x})|^2$. Other observables are simple functionals of $\{\psi_\alpha, \epsilon_\alpha\}$.
5. Repeat 2.–4. until changes are small ("self-consistent")

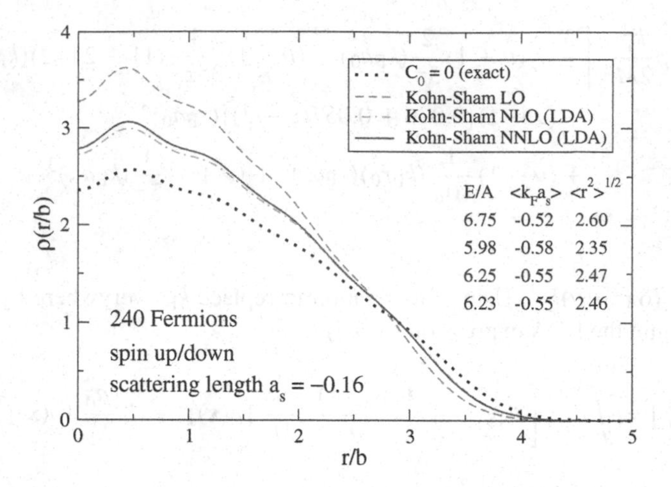

Fig. 3.14 Density profile of a dilute system of fermions in a trap [52]

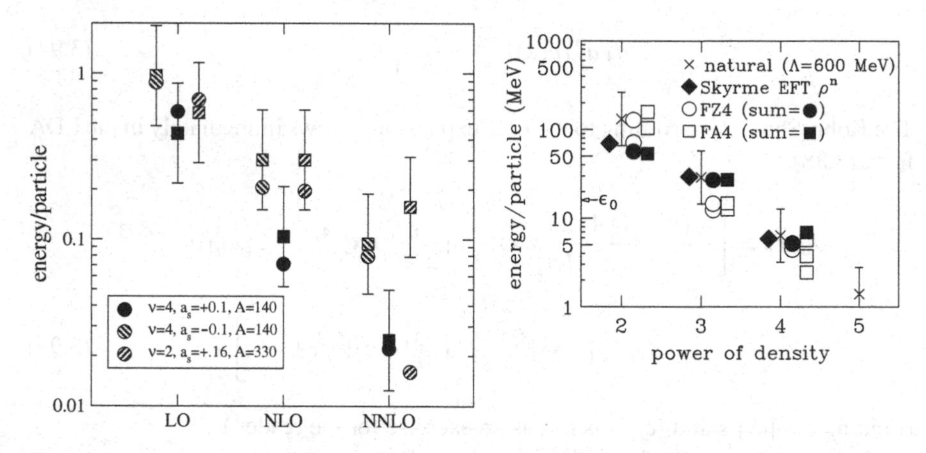

Fig. 3.15 Estimates for energy functionals for a dilute fermions in a harmonic trap (*left*) and for phenomenological energy functionals for nuclei (*right*)

This sounds like a simple Hartree calculation! Results at different EFT orders for a dilute Fermi gas in a harmonic oscillator trap is given in Fig. 3.14. Note the systematic progression from order to order.

An important consequence of the systematic EFT approach is that we can also estimate individual terms in energy functionals. If we scale contributions to the energy per particle according to the average density or $\langle k_F \rangle$, we can make estimates [52, 53]. This is shown in Fig. 3.15 for both the dilute trapped fermions, which is under complete control, and for phenomenological energy functionals for nuclei, to which a postulated QCD power counting is applied. In both cases, the estimates agree well with the actual numbers (sometimes overestimating the contribution because of accidental cancellations), which means that truncation errors are understood.

Conventional DFT is one example of using effective actions, which feature sources coupled to composite operators. It's possible that for some applications a different type of effective action may be better. There are many outstanding questions from the present discussion, particularly as we try to adapt it to real nuclei. We'll address some of them in the next lecture.

3.3 Refinements: Toward EFT/DFT for Nuclei

Let's enumerate some questions about DFT and nuclear structure.

- How is Kohn-Sham DFT more than mean field? That is, where are the approximations and how do we truncate? How do we include long-range effects (correlations)?
- What can you calculate in a DFT approach? Can we calculate single-particle properties? Or excited states?
- How does pairing work in DFT? Can we (should we) decouple *pp* and *ph*? Are higher-order contributions important?
- The Skyrme functional depends on multiple densities: $\rho(\mathbf{x})$, $\tau(\mathbf{x})$, and $\mathbf{J}(\mathbf{x})$; how does that work?
- What about broken symmetries that arise with self-bound systems? (translation, rotation, ...)
- How do we connect to the free, microscopic NN$\cdots$N interactions? Can we use chiral EFT or low-momentum interactions/RG?

We'll explore some answers to these questions (and note which ones are open) in this lecture.

Consider Kohn-Sham DFT compared to the Thomas-Fermi energy functional, for which the entire functional is treated in the local density approximation (LDA). In Kohn-Sham DFT, treating kinetic energy non-locally leads to the shell structure of electrons in atoms and in trapped atoms, as seen in Fig. 3.16. This motivates going even further beyond the LDA.

As a simple step beyond Kohn-Sham LDA, we consider functionals of the kinetic energy density in addition to the usual fermion density. The phenomenological Skyrme E is a functional of ρ *and* $\tau \equiv \langle \nabla\psi^\dagger \cdot \nabla\psi \rangle$ (and $\mathbf{J}$):

$$\begin{aligned}E[\rho,\tau,\mathbf{J}] &= \int d^3x \left\{ \frac{1}{2M}\tau + \frac{3}{8}t_0\rho^2 + \frac{1}{16}t_3\rho^{2+\alpha} + \frac{1}{16}(3t_1+5t_2)\rho\tau \right. \\ &\quad \left. + \frac{1}{64}(9t_1-5t_2)(\nabla\rho)^2 - \frac{3}{4}W_0\rho\nabla\cdot\mathbf{J} + \frac{1}{32}(t_1-t_2)\mathbf{J}^2 \right\}.\end{aligned} \tag{3.97}$$

To do this in DFT/EFT, add to the Lagrangian $\eta(\mathbf{x})\,\nabla\psi^\dagger\nabla\psi$ and generalize our Legendre transformation and inversion to $\Gamma[\rho,\tau]$,

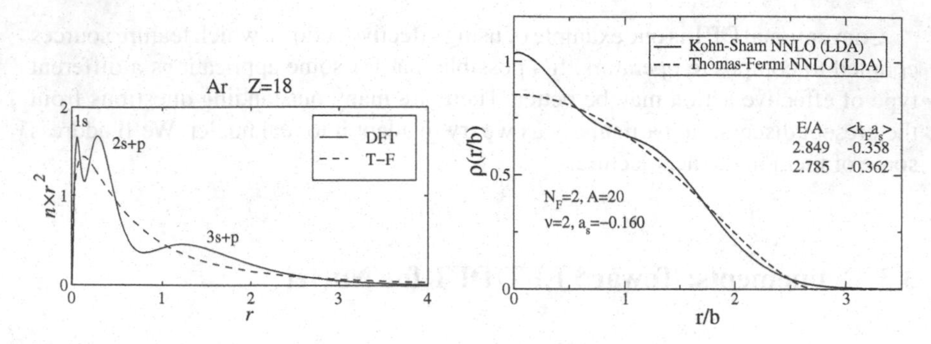

Fig. 3.16 Thomas-Fermi vs. DFT for atoms [10] (*left*) and trapped fermions (*right*)

$$\Gamma[\rho,\tau] = W[J,\eta] - \int J(x)\rho(x) - \int \eta(x)\tau(x). \tag{3.98}$$

Now there are two Kohn-Sham potentials:

$$J_0(\mathbf{x}) = \frac{\delta \Gamma_{\text{int}}[\rho,\tau]}{\delta\rho(\mathbf{x})} \quad \text{and} \quad \eta_0(\mathbf{x}) = \frac{\delta \Gamma_{\text{int}}[\rho,\tau]}{\delta\tau(\mathbf{x})}. \tag{3.99}$$

The Kohn-Sham equation defines $1/M^*(\mathbf{x}) \equiv 1/M - 2\eta_0(\mathbf{x})$:

$$\left(-\nabla\cdot\frac{1}{2M^*(\mathbf{x})}\nabla + v_{\text{ext}}(\mathbf{x}) - J_0(\mathbf{x})\right)\phi_\alpha(\mathbf{x}) = \epsilon_\alpha\,\phi_\alpha(\mathbf{x}). \tag{3.100}$$

A simple first application is to evaluate Hartree-Fock diagrams including the quadratic gradient terms [53]. Consider the HF "bowtie diagrams"

that have vertices with derivatives:

$$\mathcal{L}_{\text{eft}} = \ldots + \frac{C_2}{16}\left[(\psi\psi)^\dagger(\psi\overleftrightarrow{\nabla}^2\psi) + \text{ h.c.}\right] + \frac{C_2'}{8}(\psi\overleftrightarrow{\nabla}\psi)^\dagger\cdot(\psi\overleftrightarrow{\nabla}\psi) + \ldots \tag{3.101}$$

The energy density in Kohn-Sham LDA is

$$\mathcal{E}_{\text{int}}[\rho] = \ldots + \frac{C_2}{8}\left[\frac{3}{5}\left(\frac{6\pi^2}{\nu}\right)^{2/3}\rho^{8/3}\right] + \frac{3C_2'}{8}\left[\frac{3}{5}\left(\frac{6\pi^2}{\nu}\right)^{2/3}\rho^{8/3}\right] + \ldots \tag{3.102}$$

while the energy density in Kohn-Sham with τ ($\nu = 2$) is

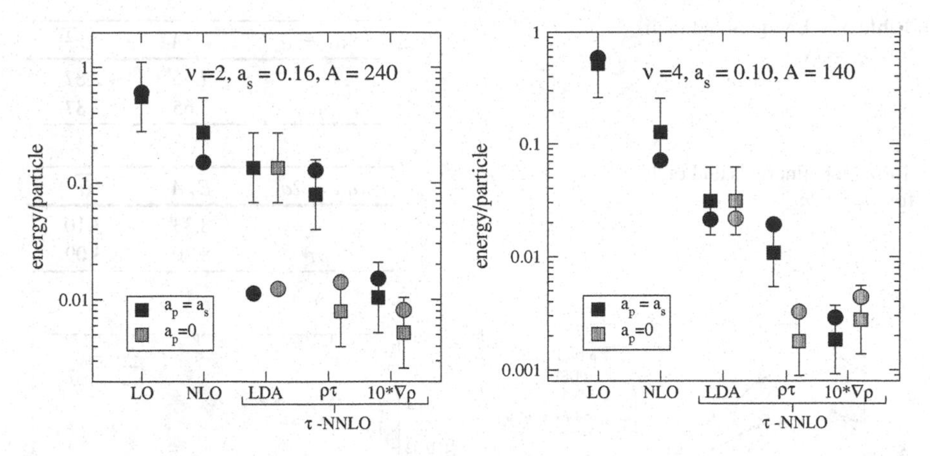

Fig. 3.17 Estimates of terms in the energy functional, including those with gradients, compared to actual values [53]

$$\mathcal{E}_{\rm int}[\rho,\tau] = \ldots + \frac{C_2}{8}\left[\rho\tau + \frac{3}{4}(\nabla\rho)^2\right] + \frac{3C_2'}{8}\left[\rho\tau - \frac{1}{4}(\nabla\rho)^2\right] + \ldots \qquad (3.103)$$

We find that power counting estimates for terms in the energy functional also work with gradient terms (see Fig. 3.17).

Now let's compare the dilute fermion functional to the phenomenological Skyrme functional. The Skyrme energy density functional (for $N = Z$) is given in (3.97) while the corresponding dilute energy density functional for $\nu = 4$ (and $V_{\rm external} = 0$) is

$$E[\rho,\tau,\mathbf{J}] = \int d^3x \left\{ \frac{\tau}{2M} + \frac{3}{8}C_0\rho^2 + \frac{1}{16}(3C_2 + 5C_2')\rho\tau + \frac{1}{64}(9C_2 - 5C_2')(\nabla\rho)^2 \right.$$
$$\left. - \frac{3}{4}C_2''\rho\nabla\cdot\mathbf{J} + \frac{c_1}{2M}C_0^2\rho^{7/3} + \frac{c_2}{2M}C_0^3\rho^{8/3} + \frac{1}{16}D_0\rho^3 + \cdots \right\}. \qquad (3.104)$$

They have the same terms after the association $t_i \leftrightarrow C_i$, except that the Skyrme functional is missing non-analytic terms, the three-body contribution, and other features. We note after matching to empirical Skyrme coefficients that the "effective" scattering length from C_0 is $a_0 \approx -2$–3 fm, but that $|k_{\rm F}a_p|$, $|k_{\rm F}r_0| < 1$ (with $a_p < 0$). However, we want the Skyrme functional to account for the finite-ranged pion, so it *should not* be equivalent to a short-distance expansion. Thus, the close correspondence suggests that the Skyrme functional is lacking and should be generalized.

It is useful to compare results from the ρ only functional compared to the ρ and τ functional, as in Fig. 3.18 and these Tables 3.3 and 3.4.
We see very little difference in the Kohn-Sham observables, which are the binding energy and the density distribution. However, the single-particle Kohn-Sham

Table 3.3 Energies and radii for $a_p = a_s$

$a_p = a_s$	E/A	$\sqrt{\langle r^2 \rangle}$
ρ	7.66	2.87
$\rho\tau$	7.65	2.87

Table 3.4 Energies and radii for $a_p = 2a_s$

$a_p = 2a_s$	E/A	$\sqrt{\langle r^2 \rangle}$
ρ	8.33	3.10
$\rho\tau$	8.30	3.09

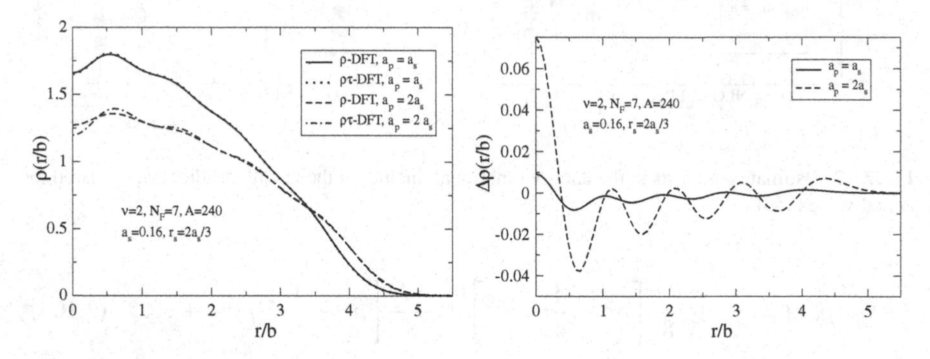

Fig. 3.18 Comparing densities from energy functionals of ρ only and ρ, τ [53]

spectrum, which is *not* an observable, shows significant differences, as evidenced by Fig. 3.19. (Note: The effective mass M^* is closely related to single-particle levels.) We can show for a uniform system that the HF single-particle levels satisfy

$$\varepsilon_{\mathbf{k}}^{\rho} - \varepsilon_{\mathbf{k}}^{\rho\tau} = \frac{\pi}{\nu}[(\nu-1)a_s^2 r_s + 2(\nu+1)a_p^3]\,\frac{k_F^2 - \mathbf{k}^2}{2M}\rho, \qquad (3.105)$$

and the $\rho\tau$ result is the one corresponding to the spectrum from the full Hartree-Fock propagator. So the issue becomes how the full G (as shown below with the self-energy) is related to the Kohn-Sham G_{ks} [54]; the closer they are, the better approximation G_{ks} will provide for single-particle properties.

$$+ \quad + \quad + \quad + \cdots \Longrightarrow \quad = \quad + \quad \mathbf{x}' \; \mathbf{x} \Longrightarrow \Sigma^*(\mathbf{x}, \mathbf{x}'; \omega)$$

To explore this connection, we add a non-local source $\xi(x', x)$ coupled to $\psi(x)\psi^{\dagger}(x')$ [we're back in Minkowski space here for no particular reason!]:

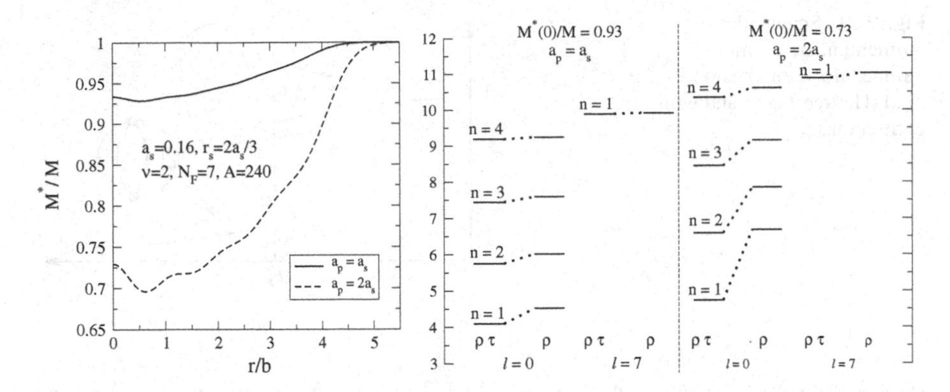

Fig. 3.19 Comparing effective masses and single-particle spectra from energy functionals of ρ only and ρ, τ [53]

Fig. 3.20 Full Green's function G in terms of the Kohn-Sham Green's function G_{ks}

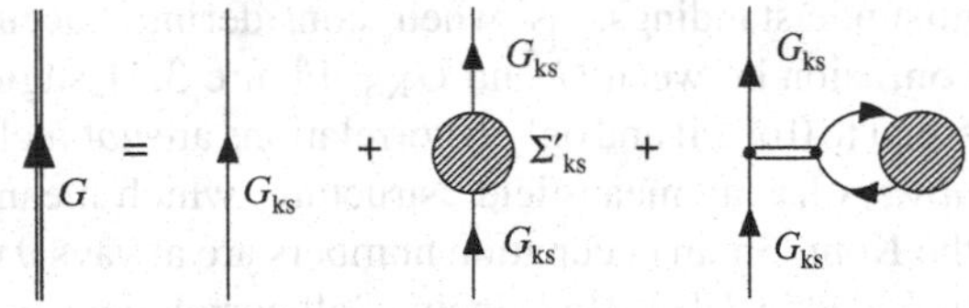

$$Z[J,\xi] = e^{iW[J,\xi]} = \int D\psi D\psi^\dagger \, e^{i\int d^4x\,[\mathcal{L} + J(x)\psi^\dagger(x)\psi(x) + \int d^4x'\,\psi(x)\xi(x,x')\psi^\dagger(x')]} . \tag{3.106}$$

Writing $\Gamma[\rho,\xi] = \Gamma_0[\rho,\xi] + \Gamma_{\text{int}}[\rho,\xi]$,

$$G(x,x') = \frac{\delta W}{\delta \xi}\bigg|_J = \frac{\delta \Gamma}{\delta \xi}\bigg|_\rho = G_{\text{ks}}(x,x') + G_{\text{ks}}\Big[\frac{1}{i}\frac{\delta \Gamma_{\text{int}}}{\delta G_{\text{ks}}} + \frac{\delta \Gamma_{\text{int}}}{\delta \rho}\Big]G_{\text{ks}}, \tag{3.107}$$

which is represented diagrammatically in Fig. 3.20. Now G and G_{ks} yield the same density by *construction*; that is, $\rho_{\text{ks}}(\mathbf{x}) = -i\nu G^0_{\text{KS}}(x,x^+)$ equals $\rho(\mathbf{x}) = -i\nu G(x,x^+)$. Here is a simple diagrammatic demonstration (the double line is minus the inverse of a single ph ring):

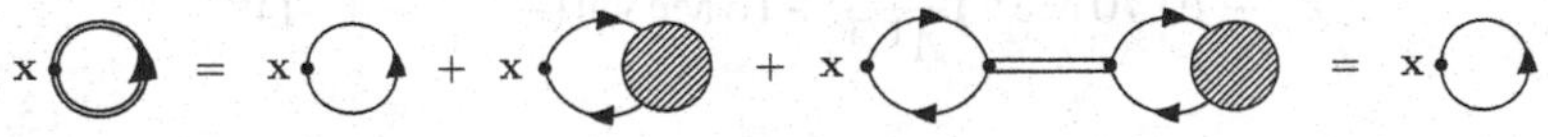

But other single-particle properties (e.g., the spectrum) are generally different, since the last two terms in Fig. 3.20 will not cancel.

We can ask whether the Kohn-Sham basis is a useful one for G. Or, more simply, ask how close is G_{KS} to G. We find that it depends on what sources are used, as

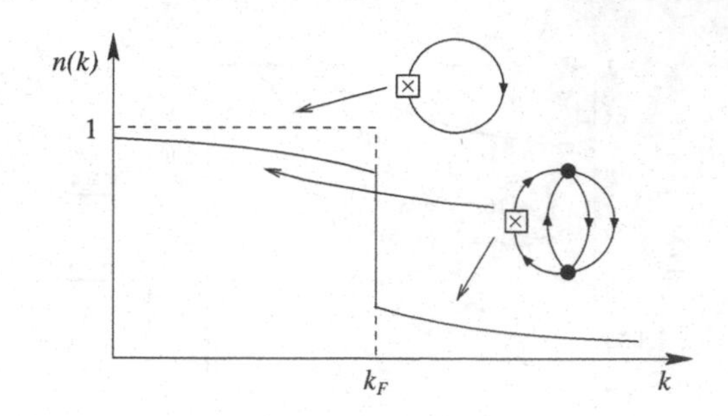

Fig. 3.21 Schematic momentum occupation number $n(k)$ for mean-field (Hartree-Fock) and with correlations

shown by the comparison of single-particle spectra in (3.105), with more sources implying less difference. This is a topic that merits further investigation.

The comparison of Kohn-Sham DFT and "mean-field" models often leads to misunderstandings, as when considering "occupation numbers", because of a confusion between G and G_{KS}. Figure 3.21 suggests that occupation numbers are equal to 0 or 1 if and only if correlations are *not* included. The Kohn-Sham propagator *always* has a "mean-field" structure, which means that (in the absence of pairing) the Kohn-Sham occupation numbers are always 0 or 1. But correlations are certainly included in $\Gamma[\rho]$! (In principle, all correlations can be included; in practice, certain types like long-range particle-hole correlations may be largely omitted.) Further, $n(\mathbf{k}) = \langle a^\dagger_{\mathbf{k}} a_{\mathbf{k}} \rangle$ is resolution dependent (not an observable!); the operator related to experiment is more complicated. Additional discussion on these issues can be found in [58].

3.3.1 Pairing in Kohn-Sham DFT

There is abundant evidence for pairing in nuclei. The semi-empirical mass formula reproduces nuclear masses only with a pairing term (the last one):

$$\begin{aligned}
B(N,Z) = (15.6\,\mathrm{MeV})\left[1 - 1.5\left(\frac{N-Z}{A}\right)^2\right] A - (17.2\,\mathrm{MeV})A^{2/3} \\
- (0.70\,\mathrm{MeV})\frac{Z^2}{A^{1/3}} + (6\,\mathrm{MeV})[(-1)^N + (-1)^Z]/A^{1/2},
\end{aligned} \tag{3.108}$$

which implies an odd-even staggering of binding energies (left panel of Fig. 3.22). Other evidence is the energy gap in the spectra of deformed nuclei, low-lying 2^+ states in even nuclei (right panel of Fig. 3.22), and deformations and moments of inertia (the theory requires pairing to reproduce data).

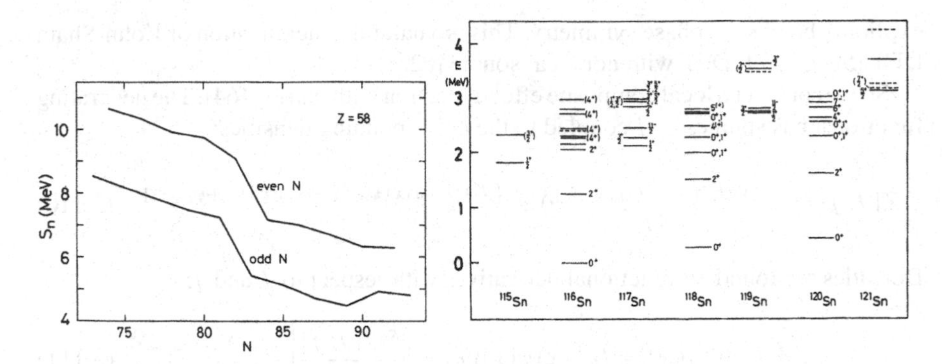

Fig. 3.22 Evidence for pairing in nuclei (see text and [8])

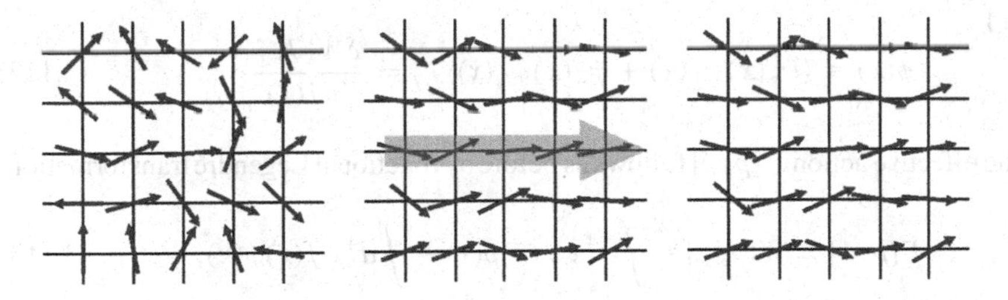

Fig. 3.23 Spontaneous symmetry breaking analogy with spins

Pairing is an example of spontaneous symmetry breaking, which is naturally accommodated in an effective action framework. For example, consider testing for zero-field magnetization M in a spin system by introducing an external field H to break the rotational symmetry (Fig. 3.23). Legendre transform the Helmholtz free energy $F(H)$:

$$\text{invert } M = -\partial F(H)/\partial H \quad \Longrightarrow \quad \Gamma[M] = F[H(M)] + MH(M). \tag{3.109}$$

Since $H = \partial \Gamma/\partial M \longrightarrow 0$, we look for the stationary points of Γ to identify possible ground states, including whether the symmetry broken state is lowest.

For pairing, the broken symmetry is a $U(1)$ [phase] symmetry. The textbook effective action treatment in condensed matter is to introduce a contact interaction [42, 43]: $g\psi^\dagger\psi^\dagger\psi\psi$, and perform a Hubbard-Stratonovich transformation with an auxiliary pairing field $\hat{\Delta}(x)$ coupled to $\psi^\dagger\psi^\dagger$, which eliminates the contact interaction. Then one constructs the 1PI effective action $\Gamma[\Delta]$ with $\Delta = \langle\hat{\Delta}\rangle$, and looks for values for which $\delta\Gamma/\delta\Delta = 0$. To leading order in the loop expansion (mean field), this yields the BCS weak-coupling gap equation with gap Δ.

The natural alternative here is to combine an expansion (e.g., EFT power counting) and the *inversion* method for effective actions [44–46]. Thus we introduce another external current $j(x)$, which is coupled to the fermion pair density in order to explicitly breaks the phase symmetry. This is a natural generalization of Kohn-Sham DFT [59–61] (cf. DFT with nonlocal source [62, 63]).

So we consider a local composite effective action with pairing [64]. The generating functional has sources J, j coupled to the corresponding densities:

$$Z[J,j] = e^{-W[J,j]} = \int D(\psi^\dagger\psi)\, e^{-\int d^4x[\mathcal{L}+J(x)\,\psi_\alpha^\dagger\psi_\alpha+j(x)(\psi_\uparrow^\dagger\psi_\downarrow^\dagger+\psi_\downarrow\psi_\uparrow)]}. \quad (3.110)$$

Densities are found by functional derivatives with respect to J and j:

$$\rho(x) \equiv \langle\psi^\dagger(x)\psi(x)\rangle_{J,j} = \left.\frac{\delta W[J,j]}{\delta J(x)}\right|_j, \quad (3.111)$$

and

$$\phi(x) \equiv \langle\psi_\uparrow^\dagger(x)\psi_\downarrow^\dagger(x)+\psi_\downarrow(x)\psi_\uparrow(x)\rangle_{J,j} = \left.\frac{\delta W[J,j]}{\delta j(x)}\right|_J. \quad (3.112)$$

The effective action $\Gamma[\rho,\phi]$ follows as before by functional Legendre transformation:

$$\Gamma[\rho,\phi] = W[J,j] - \int d^4x\, J(x)\rho(x) - \int d^4x\, j(x)\phi(x), \quad (3.113)$$

and is proportional to the (free) energy functional $E[\rho,\phi]$; at finite temperature, the proportionality constant is β. The sources are given by functional derivatives wrt ρ and ϕ:

$$\frac{\delta E[\rho,\phi]}{\delta\rho(\mathbf{x})} = J(\mathbf{x}) \quad \text{and} \quad \frac{\delta E[\rho,\phi]}{\delta\phi(\mathbf{x})} = j(\mathbf{x}). \quad (3.114)$$

But the sources are zero in the ground state, so we determine the ground-state $\rho(\mathbf{x})$ and $\phi(\mathbf{x})$ by stationarity:

$$\left.\frac{\delta E[\rho,\phi]}{\delta\rho(\mathbf{x})}\right|_{\rho=\rho_{\rm gs},\phi=\phi_{\rm gs}} = \left.\frac{\delta E[\rho,\phi]}{\delta\phi(\mathbf{x})}\right|_{\rho=\rho_{\rm gs},\phi=\phi_{\rm gs}} = 0. \quad (3.115)$$

This is Hohenberg-Kohn DFT extended to pairing!

We need a method to carry out the Legendre transforms to get Kohn-Sham DFT; an obvious choice is to apply the Kohn-Sham inversion method again, with order-by-order matching in the counting parameter λ. Once again,

$$\begin{aligned}
\text{diagrams} &\Longrightarrow W[J,j,\lambda] = W_0[J,j] + \lambda W_1[J,j] + \lambda^2 W_2[J,j] + \cdots \\
\text{assume} &\Longrightarrow J[\rho,\phi,\lambda] = J_0[\rho,\phi] + \lambda J_1[\rho,\phi] + \lambda^2 J_2[\rho,\phi] + \cdots \\
\text{assume} &\Longrightarrow j[\rho,\phi,\lambda] = j_0[\rho,\phi] + \lambda j_1[\rho,\phi] + \lambda^2 j_2[\rho,\phi] + \cdots \\
\text{derive} &\Longrightarrow \Gamma[\rho,\phi,\lambda] = \Gamma_0[\rho,\phi] + \lambda\Gamma_1[\rho,\phi] + \lambda^2\Gamma_2[\rho,\phi] + \cdots
\end{aligned}$$

Start with the exact expressions for Γ and ρ

$$\Gamma[\rho, \phi] = W[J, j] - \int J\,\rho - \int j\,\phi, \tag{3.116}$$

and

$$\rho(x) = \frac{\delta W[J, j]}{\delta J(x)}, \quad \phi(x) = \frac{\delta W[J, j]}{\delta j(x)}, \tag{3.117}$$

and plug in the expansions, with ρ, ϕ treated as order unity. Zeroth order is the Kohn-Sham system with potentials $J_0(\mathbf{x})$ and $j_0(\mathbf{x})$,

$$\Gamma_0[\rho, \phi] = W_0[J_0, j_0] - \int J_0\,\rho - \int j_0\,\phi, \tag{3.118}$$

so the *exact* densities $\rho(\mathbf{x})$ and $\phi(\mathbf{x})$ are by *construction*

$$\rho(x) = \frac{\delta W_0[J_0, j_0]}{\delta J_0(x)}, \quad \phi(x) = \frac{\delta W_0[J_0, j_0]}{\delta j_0(x)}. \tag{3.119}$$

Now introduce single-particle orbitals and solve

$$\begin{pmatrix} h_0(\mathbf{x}) - \mu_0 & j_0(\mathbf{x}) \\ j_0(\mathbf{x}) & -h_0(\mathbf{x}) + \mu_0 \end{pmatrix} \begin{pmatrix} u_i(\mathbf{x}) \\ v_i(\mathbf{x}) \end{pmatrix} = E_i \begin{pmatrix} u_i(\mathbf{x}) \\ v_i(\mathbf{x}) \end{pmatrix} \tag{3.120}$$

where

$$h_0(\mathbf{x}) \equiv -\frac{\nabla^2}{2M} + V_{\rm trap}(\mathbf{x}) - J_0(\mathbf{x}). \tag{3.121}$$

This is just like the Skyrme Hartree-Fock Bogliubov approach [8].

The diagrammatic expansion of the W_i's is the same as without pairing, except now lines in diagrams are KS Nambu-Gor'kov Green's functions,

$\Gamma_{\rm int} =$ 

$$\mathbf{G} = \begin{pmatrix} \langle T_\tau \psi_\uparrow(x)\psi_\uparrow^\dagger(x')\rangle_0 & \langle T_\tau \psi_\uparrow(x)\psi_\downarrow(x')\rangle_0 \\ \langle T_\tau \psi_\downarrow^\dagger(x)\psi_\uparrow^\dagger(x')\rangle_0 & \langle T_\tau \psi_\downarrow^\dagger(x)\psi_\downarrow(x')\rangle_0 \end{pmatrix} \equiv \begin{pmatrix} G^0_{\rm ks} & F^0_{\rm ks} \\ F^{0\,\dagger}_{\rm ks} & -\widetilde{G}^0_{\rm ks} \end{pmatrix}. \tag{3.122}$$

The extra diagram shown follows from the inversion (here it removes anomalous diagrams). In frequency space, the Kohn-Sham Green's functions are

$$G^0_{\text{ks}}(\mathbf{x},\mathbf{x}';\omega) = \sum_j \left[\frac{u_j(\mathbf{x})\,u_j^*(\mathbf{x}')}{i\omega - E_j} + \frac{v_j(\mathbf{x}')\,v_j^*(\mathbf{x})}{i\omega + E_j}\right], \tag{3.123}$$

$$F^0_{\text{ks}}(\mathbf{x},\mathbf{x}';\omega) = -\sum_j \left[\frac{u_j(\mathbf{x})\,v_j^*(\mathbf{x}')}{i\omega - E_j} - \frac{u_j(\mathbf{x}')\,v_j^*(\mathbf{x})}{i\omega + E_j}\right]. \tag{3.124}$$

The Kohn-Sham self-consistency procedure involves the same iterations as in phenomenological Skyrme HF (or relativistic mean-field) when pairing is included. In terms of the orbitals, the fermion density is

$$\rho(\mathbf{x}) = 2\sum_i |v_i(\mathbf{x})|^2, \tag{3.125}$$

and the pair density is (warning: this is unrenormalized!)

$$\phi(\mathbf{x}) = \sum_i [u_i^*(\mathbf{x})v_i(\mathbf{x}) + u_i(\mathbf{x})v_i^*(\mathbf{x})]. \tag{3.126}$$

The chemical potential μ_0 is fixed by $\int \rho(\mathbf{x}) = A$. Diagrams for $\Gamma[\rho,\phi] \propto E_0[\rho,\phi] + E_{\text{int}}[\rho,\phi]$ yield the Kohn-sham potentials

$$J_0(\mathbf{x})\Big|_{\rho=\rho_{\text{gs}}} = \frac{\delta E_{\text{int}}[\rho,\phi]}{\delta\rho(\mathbf{x})}\Big|_{\rho=\rho_{\text{gs}}} \quad \text{and} \quad j_0(\mathbf{x})\Big|_{\phi=\phi_{\text{gs}}} = \frac{\delta E_{\text{int}}[\rho,\phi]}{\delta\phi(\mathbf{x})}\Big|_{\phi=\phi_{\text{gs}}}. \tag{3.127}$$

3.3.2 Renormalization of Pairing

When we carry out the DFT pairing calculation for a uniform dilute Fermi system, we find divergences almost immediately. The generating functional with constant sources μ and j is:

$$e^{-W[\mu,j]} = \int D(\psi^\dagger\psi)\exp\Big\{-\int d^4x\, \Big[\psi_\alpha^\dagger\Big(\frac{\partial}{\partial\tau} - \frac{\nabla^2}{2M} - \mu\Big)\psi_\alpha \tag{3.128}$$

$$+ \frac{C_0}{2}\psi_\uparrow^\dagger\psi_\downarrow^\dagger\psi_\downarrow\psi_\uparrow + j(\psi_\uparrow\psi_\downarrow + \psi_\downarrow^\dagger\psi_\uparrow^\dagger)\Big]\Big\} + \frac{1}{2}\zeta\, j^2\Big]\Big\} \tag{3.129}$$

(cf. adding an integration over an auxiliary field $\int D(\Delta^*,\Delta)\, e^{-\frac{1}{|C_0|}\int|\Delta|^2}$, then shifting variables to eliminate $\psi_\uparrow^\dagger\psi_\downarrow^\dagger\psi_\downarrow\psi_\uparrow$ for $\Delta^*\psi_\uparrow\psi_\downarrow$). There are new divergences because of j, e.g., expand W to $\mathcal{O}(j^2)$:

$W[\mu, j] = \cdots +$ 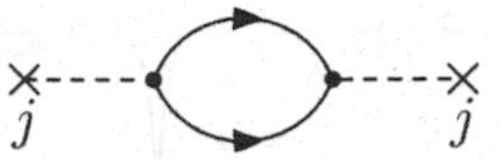 $+ \cdots$

which has the same linear divergence as in 2-to-2 scattering. To renormalize, we add the counterterm $\frac{1}{2}\zeta|j|^2$ to $\mathcal{L}$ (see [39]), which is additive to W (cf. $|\Delta|^2$), so there is no effect on scattering.

We'll use dimensional regularization again, but generalize from DR/MS (as used by Papenbrock and Bertsch [65]) to DR/PDS, which generates explicit Λ dependence to "check" renormalization (by verifying that Λ dependence cancels). The basic free-space integral in D spatial dimensions is

$$\left(\frac{\Lambda}{2}\right)^{3-D} \int \frac{d^D k}{(2\pi)^D} \frac{1}{p^2 - k^2 + i\epsilon} \xrightarrow{\text{PDS}} -\frac{1}{4\pi}(\Lambda + ip), \tag{3.130}$$

where $\int \frac{1}{\epsilon_k^0} \to \frac{M\Lambda}{2\pi}$. Renormalizing and matching free-space scattering yields for $C_0(\Lambda)$:

$$C_0(\Lambda) = \frac{4\pi a_s}{M}\frac{1}{1 - a_s\Lambda} = \frac{4\pi a_s}{M} + \frac{4\pi a_s^2}{M}\Lambda + \mathcal{O}(\Lambda^2) = C_0^{(1)} + C_0^{(2)} + \cdots \tag{3.131}$$

Note: we recover DR/MS by taking $\Lambda = 0$. As an exercise, you can verify that NLO renormalization in free space (left):

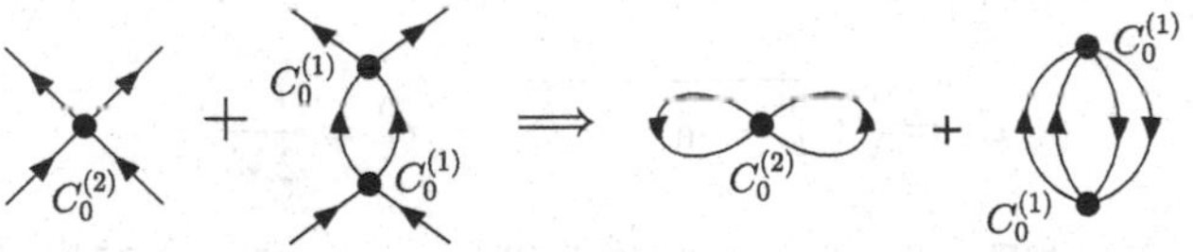

implies that the corresponding sum of diagrams at finite density (right) is independent of Λ.

Now consider the Kohn-Sham noninteracting system for a uniform system, where we have constant chemical potential μ_0 and pairing source j_0 (rather than spatially dependent sources). The bare density ρ is:

$$\rho = -\frac{1}{\beta V}\frac{\partial W_0[\mu_0, j_0]}{\partial \mu_0} = \frac{2}{V}\sum_{\mathbf{k}} v_k^2 = \int \frac{d^3k}{(2\pi)^3}\left(1 - \frac{\epsilon_k^0 - \mu_0}{E_k}\right), \tag{3.132}$$

and the bare pair density ϕ_B is:

$$\phi_B = \frac{1}{\beta V}\frac{\partial W_0[\mu_0, j_0]}{\partial j_0} = \frac{2}{V}\sum_{\mathbf{k}} u_k v_k = -\int \frac{d^3k}{(2\pi)^3}\frac{j_0}{E_k}. \tag{3.133}$$

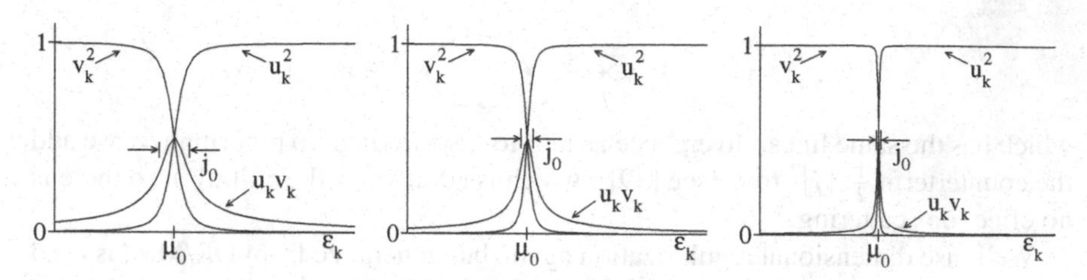

Fig. 3.24 Quasiparticle wave functions for a uniform system for several values of j_0/μ_0. As j_0/μ_0 decreases, $u_k v_k$ becomes sharply peaked at μ_0

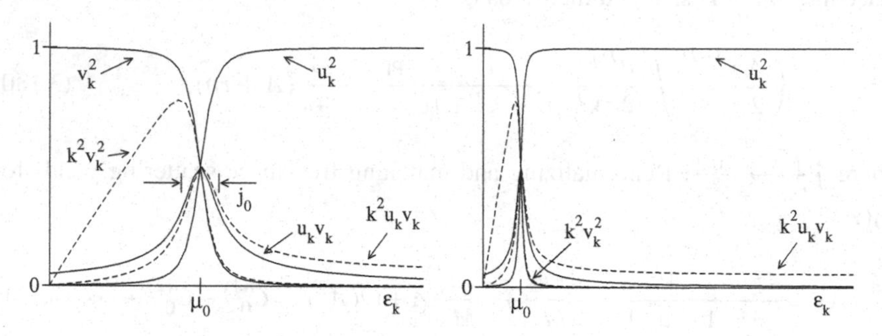

Fig. 3.25 Illustration of the divergence in $\sum_k u_k v_k$

In these expressions, j_0 plays role of constant gap; e.g., the spectrum is

$$E_k = \sqrt{(\epsilon_k^0 - \mu_0)^2 + j_0^2}, \qquad \epsilon_k^0 = \frac{k^2}{2M}. \tag{3.134}$$

(See also Fig. 3.24.) The divergence in ϕ_B is illustrated in Fig. 3.25.

The basic DR/PDS integral in D dimensions, with $x \equiv j_0/\mu_0$, is

$$\begin{aligned} I(\beta) &\equiv \left(\frac{\Lambda}{2}\right)^{3-D} \int \frac{d^D k}{(2\pi)^D} \frac{(\epsilon_k^0)^\beta}{\sqrt{(\epsilon_k^0 - \mu_0)^2 + j_0^2}} \\ &= \frac{M\Lambda}{2\pi} \mu_0^\beta \left(1 - \delta_{\beta,2}\frac{x^2}{2}\right) \\ &\quad + (-)^{\beta+1} \frac{M^{3/2}}{\sqrt{2}\pi} [\mu_0^2(1+x^2)]^{(\beta+1/2)/2} P^0_{\beta+1/2}\left(\frac{-1}{\sqrt{1+x^2}}\right). \end{aligned} \tag{3.135}$$

We can check that in the KS density equation the Λ dependence cancels:

$$\rho = -\frac{1}{\beta V}\frac{\partial W_0[\,]}{\partial \mu_0} = \int \frac{d^3k}{(2\pi)^3}\left(1 - \frac{\epsilon_k^0}{E_k} + \frac{\mu_0}{E_k}\right) \longrightarrow 0 - I(1) + \mu_0\, I(0). \tag{3.136}$$

The KS equation for the pair density ϕ fixes $\zeta^{(0)}$:

$$\phi = \frac{1}{\beta V}\frac{\partial W_0[\,]}{\partial j_0} = -\int \frac{d^3k}{(2\pi)^3}\frac{j_0}{E_k} + \zeta^{(0)} j_0 \longrightarrow -j_0\, I(0) + \zeta^{(0)} j_0 \qquad (3.137)$$

so that

$$\zeta^{(0)} = \frac{M\Lambda}{2\pi}. \qquad (3.138)$$

Calculating to nth order, we find $\Gamma_{1\le i\le n}[\rho, \phi]$ by first constructing all of the $W_{1\le i\le n}[\mu_0(\rho,\phi), j_0(\rho,\phi)]$, including additional Feynman rules [46],

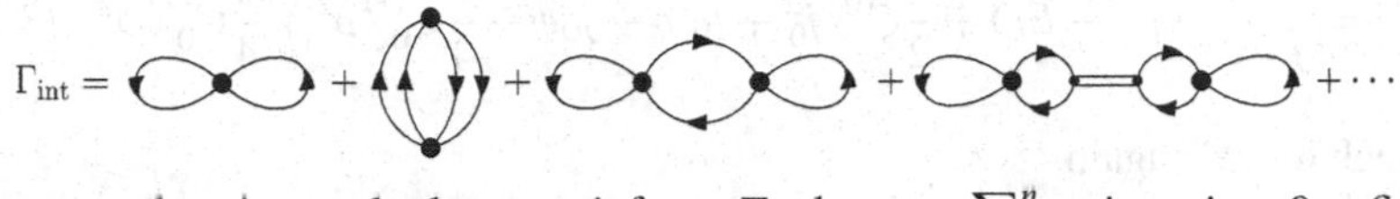

$\Gamma_{\rm int} = \; + \; + \; + \; + \cdots$

So the procedure is to calculate μ_i, j_i from Γ_i, then use $\sum_{i=0}^{n} j_i = j \to 0$ to find j_0. The renormalization conditions mean that there is no freedom in choosing $C_0(\Lambda)$, so the Λ's must cancel! In leading order, diagrams for $\Gamma_1[\rho,\phi] = W_1[\mu_0(\rho,\phi), j_0(\rho,\phi)]$ are

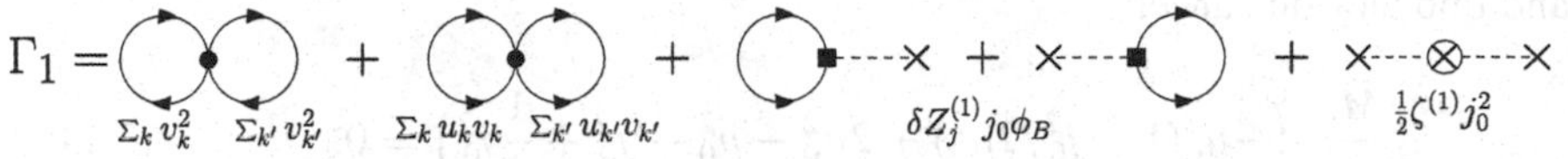

$\Gamma_1 = \; + \; + \; + \; + \;$

$\Sigma_k v_k^2 \quad \Sigma_{k'} v_{k'}^2 \qquad \Sigma_k u_k v_k \quad \Sigma_{k'} u_{k'} v_{k'} \qquad \delta Z_j^{(1)} j_0 \phi_B \qquad \frac{1}{2}\zeta^{(1)} j_0^2$

and we choose $\delta Z_j^{(n)}$ and $\zeta^{(n)}$ to convert ϕ_B to the renormalized ϕ, yielding

$$\frac{1}{\beta V}\Gamma_1[\rho,\phi] = \frac{1}{4}C_0^{(1)}\rho^2 + \frac{1}{4}C_0^{(1)}\phi^2 \;\text{ with }\; C_0^{(1)} = \frac{4\pi a_s}{M}. \qquad (3.139)$$

The Γ_1 dependence on ρ and ϕ is explicit, so it is easy to find μ_1 and j_1:

$$\mu_1 = \frac{1}{\beta V}\frac{\partial \Gamma_1}{\partial \rho} = \frac{1}{2}C_0^{(1)}\rho \quad \text{and} \quad j_1 = -\frac{1}{\beta V}\frac{\partial \Gamma_1}{\partial \phi} = -\frac{1}{2}C_0^{(1)}\phi. \qquad (3.140)$$

The "gap" equation then follows from $j = j_0 + j_1 = 0$:

$$j_0 = -j_1 = -\frac{1}{2}|C_0^{(1)}|\phi = \frac{1}{2}|C_0^{(1)}|\, j_0\left[\int \frac{d^3k}{(2\pi)^3}\frac{1}{\sqrt{(\epsilon_k^0 - \mu_0)^2 + j_0^2}} - \zeta^{(0)}\right]. \qquad (3.141)$$

DR/PDS reproduces the Papenbrock/Bertsch result [65] (with $x \equiv |j_0/\mu_0|$)

$$\begin{aligned} 1 &= -\sqrt{2M\mu_0}|a_s|(1+x^2)^{1/4} P_{1/2}^0\left(\frac{-1}{\sqrt{1+x^2}}\right) \\ &\xrightarrow{x\to 0} k_{\rm F} a_s\left[\frac{4 - 6\log 2}{\pi} + \frac{2}{\pi}\log x\right], \end{aligned} \qquad (3.142)$$

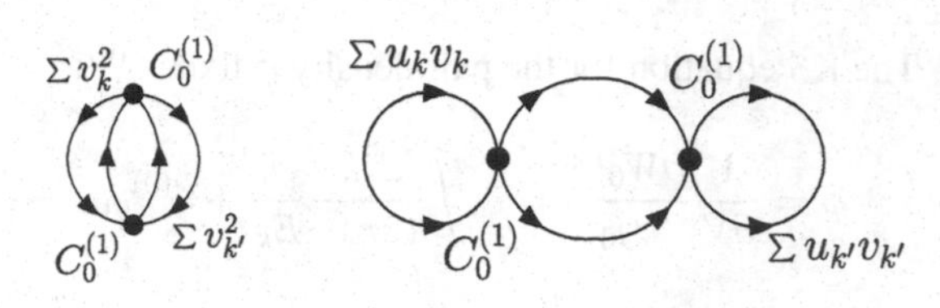

Fig. 3.26 Contributions to the NLO energy density

and if $k_F a_s < 1$, then $j_0/\mu_0 = (8/e^2)e^{-\pi/2k_F|a_s|}$ holds to very good approximation. The renormalized effective action $\Gamma = \Gamma_0 + \Gamma_1$ is

$$\frac{1}{\beta V}\Gamma = \int(\epsilon_k^0 - \mu_0 - E_k) + \frac{1}{2}\zeta^{(0)} j_0^2 + \mu_0\rho - j_0\phi + \frac{1}{4}C_0^{(1)}\rho^2 + \frac{1}{4}C_0^{(1)}\phi^2. \quad (3.143)$$

We check for Λ's again,

$$\frac{1}{\beta V}\Gamma = 0 - I(2) + 2\mu_0 I(1) - (\mu_0^2 + j_0^2)I(0) + \frac{1}{2}\frac{M\Lambda}{2\pi}j_0^2 + \cdots \quad (3.144)$$

and find they do cancel:

$$\frac{M\Lambda}{2\pi}\left(-\mu_0^2(1 - j_0^2/2\mu_0^2) + 2\mu_0^2 - \mu_0^2 - j_0^2 + \frac{1}{2}j_0^2\right) = 0. \quad (3.145)$$

To find the energy density, evaluate Γ at the stationary point $j_0 = -\frac{1}{2}|C_0^{(1)}|\phi$ with μ_0 fixed by the equation for ρ, yielding the same results as Papenbrock/Bertsch (plus an HF term) [64].

Life gets more complicated at Next-to-Leading Order (NLO), where dependence of Γ_2 on ρ and ϕ is no longer explicit and analytic formulas for DR integrals not available. Γ_2 at NLO is (see Fig. 3.26) [64]

$$\begin{aligned} -(C_0^{(1)})^2 \int\frac{d^3p}{(2\pi)^3}\int\frac{d^3k}{(2\pi)^3}\int\frac{d^3q}{(2\pi)^3}&\frac{1}{E_p + E_k + E_{p-q} + E_{k+q}} \\ \times\big[u_p^2\, u_k^2\, v_{p-q}^2\, v_{k+q}^2 &- 2u_p^2\, v_k^2\,(uv)_{p-q}\,(uv)_{k+q} \\ &+ (uv)_p\,(uv)_k\,(uv)_{p-q}\,(uv)_{k+q}\big] \end{aligned} \quad (3.146)$$

and

$$-(C_0^{(1)})^2\int\frac{d^3k}{(2\pi)^3}\frac{1}{2E_k}\big[\rho(u_k v_k)^2 + \frac{1}{2}\phi_B(u_k^2 - v_k^2)\big]^2. \quad (3.147)$$

The UV divergences can be identified from

$$v_k^2 = \frac{1}{2}\left(1 - \frac{\xi_k}{E_k}\right) \stackrel{k\to\infty}{\longrightarrow} \frac{j_0^2 M^2}{k^4}, \qquad u_k^2 = \frac{1}{2}\left(1 + \frac{\xi_k}{E_k}\right) \stackrel{k\to\infty}{\longrightarrow} 1 - \frac{j_0^2 M^2}{k^4}, \tag{3.148}$$

and

$$u_k v_k = -\frac{j_0}{2E_k} \stackrel{k\to\infty}{\longrightarrow} -\frac{j_0 M}{k^2}, \qquad \frac{1}{E_k} \stackrel{k\to\infty}{\longrightarrow} \frac{2M}{k^2}. \tag{3.149}$$

For the renormalization at NLO to work, the bowtie diagram with the $C_0^{(2)} = (4\pi a_s^2/M)\Lambda$ vertex must precisely cancel the Λ dependence from the beachball with $C_0^{(1)} = 4\pi a_s/M$ vertices:

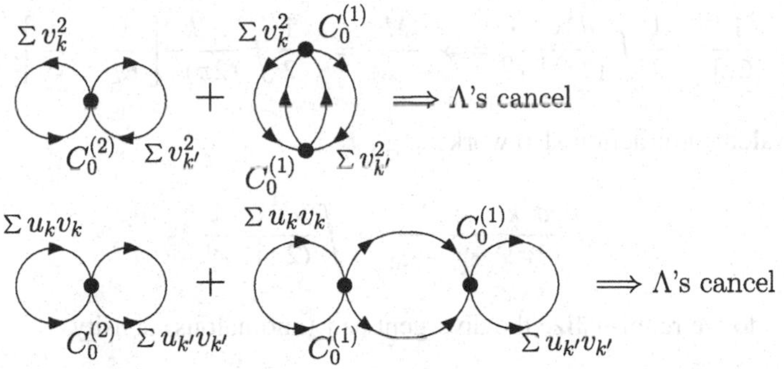

Σv_k^2 $C_0^{(2)}$ $\Sigma v_{k'}^2$ + Σv_k^2 $C_0^{(1)}$ $C_0^{(1)}$ $\Sigma v_{k'}^2$ $\Longrightarrow$ Λ's cancel

$\Sigma u_k v_k$ $C_0^{(2)}$ $\Sigma u_{k'} v_{k'}$ + $\Sigma u_k v_k$ $C_0^{(1)}$ $C_0^{(1)}$ $\Sigma u_{k'} v_{k'}$ $\Longrightarrow$ Λ's cancel

(Note that the $\delta Z_j^{(1)}$ vertex takes $\phi_B \to \phi$.) How do we see cancellation of Λ's and evaluate renormalized results without analytic formulas?

Before addressing that issue, we first see how the standard induced interaction result [66, 67] is recovered here. As $j_0 \to 0$, $u_k v_k$ peaks at μ_0 (see Fig. 3.24). At leading order (for $T = 0$),

$$\Delta_{LO}/\mu_0 = \frac{8}{e^2}\, e^{-1/N(0)|C_0|} = \frac{8}{e^2}\, e^{-\pi/2k_F|a_s|}, \tag{3.150}$$

where we make the association $j_0 \to \Delta_{LO}$. At NLO the exponent is modified, which changes the prefactor, $\Delta_{NLO} \approx \Delta_{LO}/(4e)^{1/3}$, using

$$\Gamma_1 + \Gamma_2 = \text{(bowtie)} + \text{(beachball)} \Longrightarrow j_1 + j_2 = \frac{1}{2}|C_0|\left[1 - |C_0|\langle \Pi_0 \rangle_{|\mathbf{k}|=|\mathbf{k}'|=k_F}\right]\phi$$

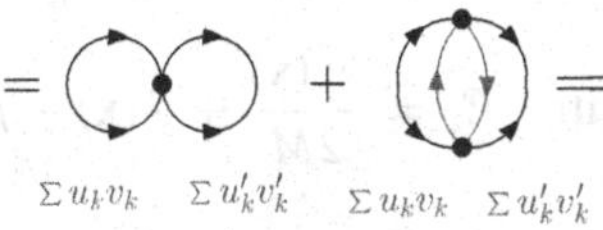

$\Sigma u_k v_k$ $\Sigma u'_k v'_k$ $\Sigma u_k v_k$ $\Sigma u'_k v'_k$

Further details can be found in [64]. An unexplored question is how does the Kohn-Sham gap compare to "real" gap?

Now we return to the question of renormalizing in practice; an alternative approach is to use subtractions. The NLO integrals with $E_k = \sqrt{(\epsilon_k - \mu_0)^2 + j_0^2}$ are intractable, but we directly obtain a renormalized result with the substitution

$$\int \frac{1}{E_1 + E_2 + E_3 + E_4} \longrightarrow \int \left[\frac{1}{E_1 + E_2 + E_3 + E_4} - \frac{\mathcal{P}}{\epsilon_1^0 + \epsilon_2^0 - \epsilon_3^0 - \epsilon_4^0} \right] \tag{3.151}$$

plus a DR/PDS integral that is proportional to Λ. When applied at LO,

$$\int \frac{1}{E_k} = \int \left[\frac{1}{E_k} - \frac{\mathcal{P}}{\epsilon_k^0} \right] + \frac{M\Lambda}{2\pi}. \tag{3.152}$$

This is the same sort of subtraction used to eliminate C_0 in the gap equation,

$$\frac{M}{4\pi a_s} + \frac{1}{|C_0|} = \frac{1}{2} \int \frac{d^3k}{(2\pi)^3} \frac{1}{\epsilon_k^0} \Longrightarrow \frac{M}{4\pi a_s} = -\frac{1}{2} \int \frac{d^3k}{(2\pi)^3} \left[\frac{1}{E_k} - \frac{1}{\epsilon_k^0} \right]. \tag{3.153}$$

Any equivalent subtraction also works, e.g.,

$$\int \frac{d^3k}{(2\pi)^3} \frac{\mathcal{P}}{\epsilon_k^0 - \mu_0} = \int \frac{d^3k}{(2\pi)^3} \frac{1}{\epsilon_k^0}. \tag{3.154}$$

So how do we renormalize the divergent pair (anomalous) density,

$$\phi(\mathbf{x}) = \sum_i [u_i^*(\mathbf{x}) v_i(\mathbf{x}) + u_i(\mathbf{x}) v_i^*(\mathbf{x})] \longrightarrow \infty, \tag{3.155}$$

in a finite system? (Cf. the scalar density $\rho_s = \sum_i \overline{\psi}(\mathbf{x})\psi(\mathbf{x})$ for relativistic Hartree theory). Answer: use the subtracted expression for ϕ in the uniform system,

$$\phi = \int^{k_c} \frac{d^3k}{(2\pi)^3} j_0 \left(\frac{1}{\sqrt{(\epsilon_k^0 - \mu_0)^2 + j_0^2}} - \frac{1}{\epsilon_k^0} \right) \stackrel{k_c \to \infty}{\longrightarrow} \text{finite}, \tag{3.156}$$

and apply this in a local density approximation (Thomas-Fermi):

$$\phi(\mathbf{x}) = 2 \sum_i^{E_c} u_i(\mathbf{x}) v_i(\mathbf{x}) - j_0(\mathbf{x}) \frac{M\, k_c(\mathbf{x})}{2\pi^2} \quad \text{with} \quad E_c = \frac{k_c^2(\mathbf{x})}{2M} + J(\mathbf{x}) - \mu_0. \tag{3.157}$$

This procedure was worked out by Bulgac and collaborators [59–61]. Convergence is very slow as the energy cutoff is increased, so Bulgac/Yu devised a different subtraction,

$$\phi = \int^{k_c} \frac{d^3k}{(2\pi)^3}\, j_0 \left(\frac{1}{\sqrt{(\epsilon_k^0 - \mu_0)^2 + j_0^2}} - \frac{\mathcal{P}}{\epsilon_k^0 - \mu_0} \right) \stackrel{k_c \to \infty}{\longrightarrow} \text{finite.} \tag{3.158}$$

A comparison of convergence in uniform system for the two subtraction schemes (3.156) and (3.158):

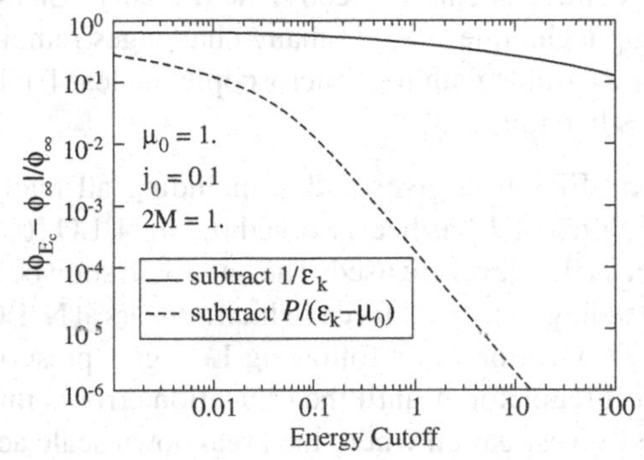

shows dramatic improvement for the Bulgac/Yu subtraction. Bulgac et al. have demonstrated that this works in finite systems, and there have been recent practical applications.

We finish this lecture with a brief mention of alternatives to a local Kohn-Sham formalism for pairing. One alternative is to couple a source to the *non-local* pair field [62, 63]:

$$\widehat{H} \longrightarrow \widehat{H} - \int dx\, dx' \,[D^*(x,x')\psi_\uparrow(x)\psi_\downarrow(x') + \text{H.c.}], \tag{3.159}$$

which yields essentially a two-particle-irreducible (2PI) effective action $\Gamma[\rho, \Delta]$ with $\Delta(x,x') = \langle \psi_\uparrow(x)\psi_\downarrow(x')\rangle$. Or one could use auxiliary fields: introduce $\hat{\Delta}^*(x)\psi(x)\psi(x) + \text{H.c.}$ via a Hubbard-Stratonovich transformation to obtain a 1PI effective action in $\Delta(x) = \langle \hat{\Delta}(x)\rangle$. By adopting a special saddle point evaluation, one can obtain Kohn-Sham DFT. Finally there is the possibility of deriving a density functional (without Kohn-Sham orbits) by direct renormalization group evolution [41].

3.4 Loose Ends and Challenges Plus Cold Atoms

In this final lecture, we touch on some loose ends raised in previous lectures, and outline some of the plans and challenges for moving forward toward a microscopic DFT for nuclei based on effective field theory and renormalization group ideas and methods [68]. We'll also briefly consider cold atom physics and some recent work on density functionals for that problem.

3.4.1 Toward a Microscopic Nuclear DFT

We have outlined a framework for generating density functional theory based on effective actions. A key ingredient is a tractable hierarchy of many-body approximations to which we can apply the inversion method. A scenario for carrying this out has emerged, which combines chiral effective field theory (EFT) interactions with renormalization group techniques. While many challenges remain, it is a plausible and systematically improvable path to a microscopic nuclear DFT.

This scenario goes like this:

1. Construct a chiral EFT to a given order, including all many-body forces. At present, the NN chiral EFT has been worked out to N^3LO [69, 70], while three-body forces at the N^2LO level are used. The latter will soon be extended to N^3LO and already the leading four-body force (which appears at N^3LO) has been tested. To minimize the truncation error following Lepage's prescription, one should increase the cutoff regulator Λ until the truncation error is minimized. (Note: it is still a matter of investigation where the breakdown scale actually lies.)
2. Evolve the Hamiltonian to lower Λ with renormalization group (RG) methods. There are choices here, including the $V_{\mathrm{low}\,k}$ approach, the Similarity Renormalization Group (SRG) [71, 72], and possibly simply a direct construction of the chiral EFT at a lower cutoff [73]. Cutoffs in the range of $\Lambda \approx 2\,\mathrm{fm}^{-1}$ appear to be appropriate for ordinary nuclei. One needs the consistent evolution of *all* interactions *and* other operators. As discussed in the first lecture, by decoupling high and low momentum the nuclear many-body problem becomes perturbative in the particle-particle channel, in stark contrast to the situation with conventional interactions [16, 72].
3. Generate the density functional in effective action form. A by-product of evolving to low momentum is that the convergence of the many-body diagrams no longer is critically dependent on the choice of single-particle potential. This opens the door to choosing it to maintain the density as in a Kohn-Sham approach. In the short term, a direct construction of the functional in the Skyrme form is possible via an adaption (and extension) of Negele and Vautherin's density matrix expansion (DME) [74, 75]. In the long term, chain-rule constructions will allow non-local effects to be included [see after (3.88)].

This program is well underway and is part of a larger project to construct and constrain a universal nuclear energy density functional (UNEDF). A detailed overview and an explanation of the DME will be available in [76].

We'll briefly describe the idea of the DME [74, 75], which starts by expressing the Hartree-Fock energy using the density matrix. Recall that we take the best single Slater determinant in a variational sense

$$|\Psi_{\mathrm{HF}}\rangle = \det\{\psi_i(\mathbf{x}), i = 1\cdots A\}, \quad \mathbf{x} = (\mathbf{r}, \sigma, \tau), \tag{3.160}$$

to find the Hartree-Fock energy (suppressing σ, τ):

$$\langle \Psi_{\mathrm{HF}}|\widehat{H}|\Psi_{\mathrm{HF}}\rangle = \cdots + \frac{1}{2}\sum_{i,j=1}^{A}\int d\mathbf{r}_1 \int d\mathbf{r}_2\, |\psi_i(\mathbf{r}_1)|^2 v(\mathbf{r}_1,\mathbf{r}_2)|\psi_j(\mathbf{r}_2)|^2 \tag{3.161}$$

$$-\frac{1}{2}\sum_{i,j=1}^{A}\int d\mathbf{r}_1 \int d\mathbf{r}_2\, \psi_i^\dagger(\mathbf{r}_1)\psi_i(\mathbf{r}_2) v(\mathbf{r}_1,\mathbf{r}_2)\psi_j^\dagger(\mathbf{r}_2)\psi_j(\mathbf{r}_1). \tag{3.162}$$

We can trivially express this in terms of the single-particle density matrix:

$$\rho(\mathbf{r}_1,\mathbf{r}_2) = \nu \sum_{\epsilon_\alpha \le \epsilon_{\mathrm{F}}} \psi_\alpha^\dagger(\mathbf{r}_1)\psi_\alpha(\mathbf{r}_2). \tag{3.163}$$

The idea is to write this in the *Kohn-Sham basis* (i.e., the ψ_α's are Kohn-Sham orbitals), so that it is compatible with the DFT diagrammatic expansion. If we change to $\mathbf{R} = \frac{1}{2}(\mathbf{r}_1 + \mathbf{r}_2)$ and $\mathbf{s} = \mathbf{r}_1 - \mathbf{r}_2$, we can expand in $\mathbf{s}$

$$\rho(\mathbf{R}+\mathbf{s}/2, \mathbf{R}-\mathbf{s}/2) = e^{\mathbf{s}\cdot(\nabla_1 - \nabla_2)/2}\, \rho(\mathbf{r}_1,\mathbf{r}_2)|_{\mathbf{s}=0}\,. \tag{3.164}$$

Negele and Vautherin obtained an expansion in terms of the fermion, kinetic energy, and other densities:

$$\rho(\mathbf{r}_1,\mathbf{r}_2) = \frac{3j_1(sk)}{sk}\rho(\mathbf{R}) + \frac{35 j_3(sk)}{2sk^3}\left(\frac{1}{4}\nabla^2\rho(\mathbf{R}) - \tau(\mathbf{R}) + \frac{3}{5}k^2\rho(\mathbf{R}) + \cdots\right), \tag{3.165}$$

which leads to functionals of these densities, for which we can take the $\delta/\delta\rho(\mathbf{R})$, $\delta/\delta\tau(\mathbf{R})$, etc. derivatives directly. (See also DME applied to ChPT in nuclear medium by Kaiser and collaborators [77–79].) This is clear at the Hartree-Fock level, but generalizations are needed for higher-order diagrams. These are also in progress [80].

There are some important open questions for this approach or any DFT treatment of finite nuclei. These include:

- For pairing, the energy interpretation, number projection, renormalization in finite systems, and efficient numerical implementation. Also, a unified microscopic treatment of particle-particle and particle-hole physics.
- DFT for self-bound systems. Self-bound systems have no external potential, which implies that the true ground-state density is uniform! More generally, how do we deal with symmetry breaking (translational, rotational invariance, particle number) and restoration. There has been little or no guidance from Coulomb DFT. There are analogous issues and methods for effective actions, namely soliton zero modes and projection methods. Work on an energy functional for the intrinsic density is in its infancy [81].
- Long-range effects, as illustrated schematically in Fig. 3.27. This includes long-range forces (i.e., pion and Coulomb exchange) but also long-range correlations.

Fig. 3.27 Long-range effects contributing to energy density functionals

$J_0(\mathbf{x}) = -$ [diagram] $+$ [diagram] $+ \cdots$

$=$ [diagram] $+$ [diagram] $+ \cdots$

The latter can be understood as non-localities from near-on-shell particle-hole excitations, as in the lower diagrams pictured in Fig. 3.27.

3.4.2 Covariant DFT

Thus far, our discussion has included only nonrelativistic EFT and DFT. However, there is a successful phenomenology based on relativistic mean field energy functionals [82–84]. Can we make a connection?

In principle we could proceed by deriving a covariant EFT. We start by observing that *all* low-energy effective theories have incorrect UV behavior. Sensitivity to short-distance physics is signalled by divergences but finiteness (e.g., with cutoff) doesn't mean there is not sensitivity! One must absorb (and correct) sensitivity by renormalization. Instances of UV divergences for low-energy nuclear physics are

nonrelativistic	covariant
scattering	scattering
pairing	pairing
	anti-nucleons

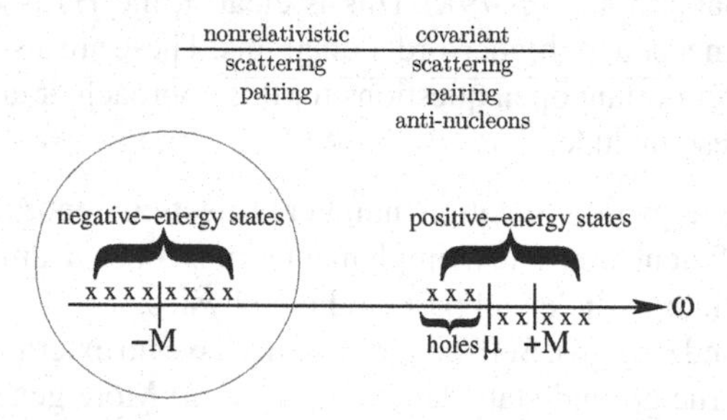

Thus, there is an additional source of divergences in the covariant case from the "Dirac sea".

Gasser et al. [85] derived chiral perturbation theory (ChPT) for πN physics using relativistic nucleon degrees of freedom. But they found that the loop and momentum expansions no longer agree (as they do in nonrelativistic ChPT), which means that systematic power counting was lost. The heavy-baryon EFT restores power counting by a $1/M$ expansion, and has been the basis for nonrelativistic NN EFT treatments [86].

However, Hua-Bin Tang [87] (and with Paul Ellis [88]) observed:

> "...EFT's permit useful low-energy expansions only if we absorb *all* of the hard-momentum effects into the parameters of the Lagrangian." "When we include the nucleons relativistically, the anti-nucleon contributions are also hard-momentum effects."

They advocated moving the "Dirac sea" physics into the coefficients, thereby absorbing the "hard" part of a diagram into parameters, while the remaining "soft" part satisfies chiral power counting. The original πN prescription by Tang and Ellis (expand, integrate term-by-term, and resum propagators) was systematized for πN by Becher and Leutwyler under the name "infrared regularization" or IR [89]. It is not unique; e.g., Fuchs et al. have used additional finite subtractions in DR [90]. The extension of IR to multiple heavy particles by Lehmann and Prézeau [91], with a convenient reformulation by Schindler et al. [92], offers the possibility of a working covariant EFT.

If we restrict our attention to purely short-ranged, natural interactions, there are tremendous simplifications. In particular, tadpoles and $N\overline{N}$ loops in free space vanish! For example, leading order (LO) has scalar, vector, etc. vertices,

$$\mathcal{L}_{\text{eft}} = \cdots - \frac{C_s}{2}(\overline{\psi}\psi)(\overline{\psi}\psi) - \frac{C_v}{2}(\overline{\psi}\gamma^{\mu}\psi)(\overline{\psi}\gamma_{\mu}\psi) + \cdots \tag{3.166}$$

which we designate as

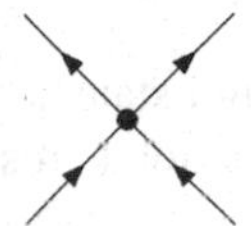

and consider all possible diagrams at NLO:

$$+ \quad + \cdots + \quad + \quad +$$

Only the particle-particle loop diagram survives IR and all of the others pictured here vanish. Since only forward-going nucleons contribute in the end, one obtains the same scattering amplitude as in nonrelativistic DR/MS for small k.

Unlike QED DFT, "no sea" for nuclear structure is a misnomer; one should include $\overline{N}N$ "vacuum physics" in coefficients via renormalization. But note that requiring renormalizability at the hadronic level corresponds to making a model for the short-distance behavior, which has proven to be a poor model phenomenologically. Fixing short-distance behavior is not the same thing as throwing away negative-energy states. For a long time, people searched for *unique* "relativistic effects"; these were largely misguided efforts.

The further investigation of covariant EFT and its extension to DFT is motivated by the successes of relativistic mean-field phenomenology and other arguments about low-energy QCD. But there is much to be done for it to be competitive with the nonrelativistic EFT.

3.4.3 DFT for Cold Atoms with Large Scattering Length

Finally we return to the large scattering length problem, which is realizable with cold atoms. The total cross section for scattering is expressed in term of partial-wave phase shifts as

$$\sigma_{\text{total}} = \frac{4\pi}{k^2}\sum_{l=0}^{\infty}(2l+1)\sin^2\delta_l(k). \tag{3.167}$$

Recall that an attractive potential pulls the asymptotic wave function (outside the potential) in, by an amount at each energy or k called the phase shift. At low energy ($\lambda = 2\pi/k \gg 1/R$), the S-wave phase shift $\delta_0(k)$ satisfies:

$$k\cot\delta_0(k) \xrightarrow{k\to 0} -\frac{1}{a_0} + \frac{1}{2}r_0k^2 + \ldots \tag{3.168}$$

where a_0 is the "scattering length" and r_0 is the "effective range". The effective range expansion for low-energy scattering goes back to Schwinger and Bethe and others. The effective range expansion typifies the general principles we have stated for EFT's: If a complicated potential produces scattering with a given a_0 and r_0, we can replace it by a simpler potential with the same values and everything agrees at low energies. In general, the effective-range expansion is reproduced and extended by EFT.

Having a bound-state or near-bound state at zero energy means large scattering lengths ($a_0 \to \pm\infty$). For $kR \to 0$, the total cross section is

$$\sigma_{\text{total}} = \sigma_{l=0} = \frac{4\pi a_0^2}{1+(ka_0)^2} = \begin{cases} 4\pi a_0^2 \text{ for } ka_0 \ll 1, \\ \frac{4\pi}{k^2} \text{ for } ka_0 \gg 1 \text{ (unitarity limit).} \end{cases} \tag{3.169}$$

We are particularly interested in cases where there is a bound-state near zero energy or there just misses being a bound state. These pictures:

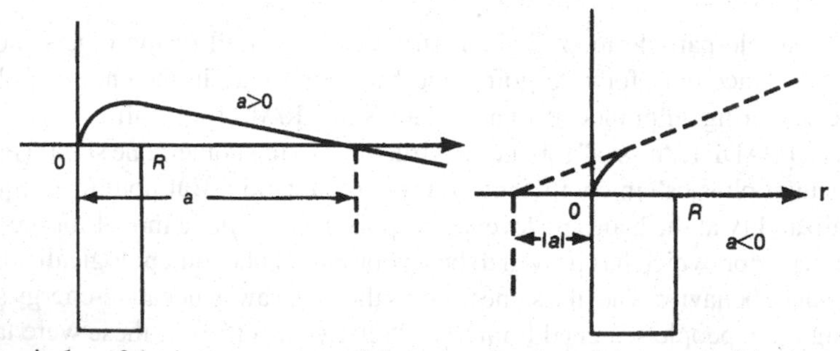

are a reminder of the interpretation of the scattering length in terms of the intercept of the zero energy scattering wave function, which is a straight line outside the potential. For potentials that just have a bound state, the wave function just turns over and a_0

is large and positive. If the potential just fails to have a bound state, it doesn't quite make it to horizontal and a_0 is large and negative.

At low energies, depending on the size of k times a_0, the cross section first goes like the square of a_0 and then saturates at the unitarity limit. So if we could adjust the depth of the bound state, we can control the "strength" of the interaction in a sense. This is possible for atoms by changing an external magnetic field to produce resonant scattering. For QCD is it possible to adjust the quark mass *theoretically* so that m_π changes and the nuclear a_0 can be tuned to $\pm\infty$!

Let's consider the large scattering length many-body problem, which means we have an attractive two-body potential with $a_0 \to \infty$. If $R \ll 1/k_F \ll a_0$, as in this figure,

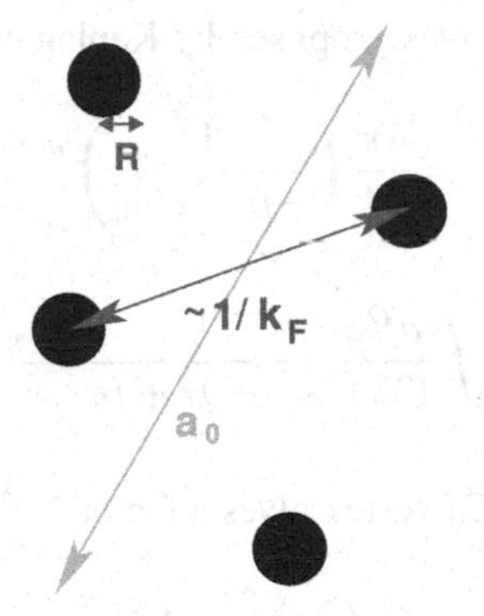

then we expect scale invariance (since we lose both R and a_0 as possible scales). This means that the energy and superfluid gap should be pure numbers times $E_{FG} = \frac{3}{5}\frac{k_F^2}{2M}$.

Recall that for the natural scattering length case, EFT power counting led to an organized perturbative expansion:

P/2 + k   P/2 + k′
P/2 − k   P/2 − k′

$$iT(k, \cos\theta) = -iC_0 + \left(-\frac{M}{4\pi}(C_0)^2 k\right) + \left(+i\left(\frac{M}{4\pi}\right)^2 (C_0)^3 k^2\right) + \left(-iC_2 k^2\right) + \left(-iC_2' k^2 \cos\theta\right) + \mathcal{O}(k^3)$$

with $C_0 = \frac{4\pi}{M}a_0$ and $ka_0 \ll 1$. But in the large scattering length limit, $ka_0 \gg 1$ so the bubble series diverges. This is not a difficulty in free space, because the geometric sum of bubbles is easily performed. This sum yields the $f_0(k)$ expansion by keeping ka_0 to all orders and expanding the rest:

$$f_0(k) \propto \frac{1}{-1/a_0 + r_0 k^2/2 - ik} \longrightarrow \frac{-1}{1/a_0 + ik}\left[1 + \frac{r_0/2}{1/a_0 + ik}k^2 + \cdots\right] \tag{3.170}$$

With a natural a_0 and a perturbative expansion, we found the DR/MS (minimal subtraction) scheme particularly convenient. With large a_0, we need a new renormalization scheme. DR/PDS was proposed by Kaplan et al. [93] and counts $\mu \sim k$:

$$\Longrightarrow\ C_0(\mu) = \frac{4\pi}{M}\left(\frac{1}{-\mu + 1/a_0}\right) \stackrel{a_0\to\infty}{\longrightarrow} -\frac{4\pi}{M\mu}, \tag{3.171}$$

$$\Longrightarrow \int \frac{d^D q}{(2\pi)^3}\frac{1}{k^2 - q^2 + i\epsilon} \stackrel{D\to 3}{\longrightarrow} -\frac{\mu+ik}{4\pi}. \tag{3.172}$$

In medium, each additional C_0 vertex gives a factor

$$C_0(k_F)\left(\frac{M}{k_F^2}\right)^2\left(\frac{k_F^5}{M}\right) \sim k_F^0, \tag{3.173}$$

which means that all C_0 diagrams are leading order! Thus, we are told to sum *all* many-body diagrams with C_0 vertices. This is only possible numerically (or possibly with an additional expansion).

So we turn to numerical calculations. GFMC results from Chang et al. [94], in which one extrapolates to large numbers of fermions, are shown in Fig. 3.28. They find the energy per particle is $E/N = 0.44(1)E_{FG}$. Diffusion Monte Carlo (DMC) results [95], with a square-well potential tuned to $a_0 \to \infty$ and an extrapolation to large numbers of fermions, find a similar result, namely an energy per particle of $E/N = 0.42(1)E_{FG}$.

Papenbrock has considered DFT for the unitary regime [96]. He assumes a simple, constrained form of the density functional,

$$\mathcal{E}[\rho] = \frac{\hbar^2}{m}\left[\frac{m}{2m_{\text{eff}}}\sum_{j=1}^{N}|\nabla\phi_j(r)|^2 + \left(\xi - \frac{m}{m_{\text{eff}}}\right)c\rho^{5/3}\right] + \frac{1}{2}m\omega^2 r^2\rho, \tag{3.174}$$

with non-localities and gradient terms via the effective mass m_{eff}. The parameters in $E[\rho] = \int d\mathbf{x}\,\mathcal{E}[\rho(\mathbf{x})]$ can be fit for $N = 2$ to exact results for two fermions in harmonic trap [Busch et al. (1998)]:

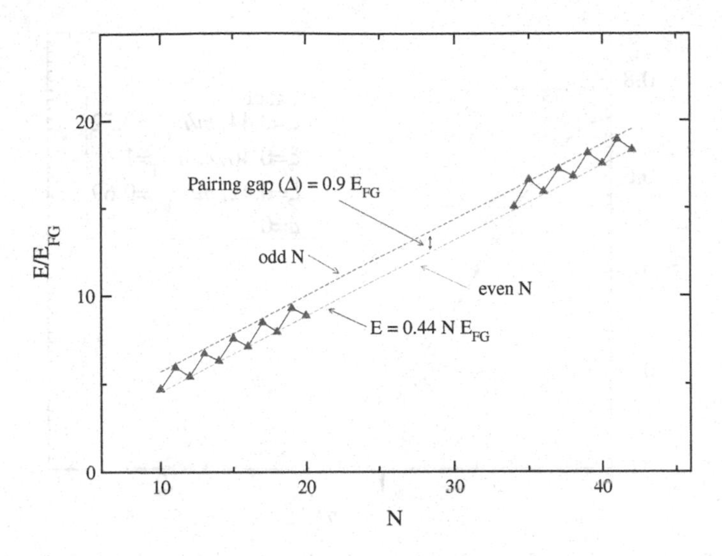

Fig. 3.28 GFMC results from Chang et al. [94] for the unitary fermion system

$$\psi_{\text{rel}}(r) = \frac{1}{\sqrt{2^{3/2}\pi l^3}} \frac{l}{r} e^{-r^2/4l^2}, \qquad l = \sqrt{\hbar/m\omega}, \quad E = 2\hbar\omega, \tag{3.175}$$

and a gaussian center-of-mass wave function. The result is

$$\rho_{\text{exact}}(r) = \frac{4}{\pi^{3/2}l^3} \frac{l}{r} e^{-2(r/l)^2} \int_0^r dx\, e^{x^2}. \tag{3.176}$$

Results of such fits are shown in Fig. 3.29. They predict $\xi = 0.42$ and $m/m_{\text{eff}} = 0.69$ from the best fit. The value for ξ is amazingly close to the Monte Carlo result. Papenbrock and Bhattacharyya [97, 98] consider corrections for an LDA density functional close to the unitary limit

$$\mathcal{E}[\rho] = \mathcal{E}_{\text{FG}} \left(\xi + \frac{c_1}{a\rho^{1/3}} + c_2 r_0 \rho^{1/3} \right). \tag{3.177}$$

Again, this is fit to the harmonically trapped two-fermion system. Results are given in Fig. 3.30 and show impressive agreement with Monte Carlo calculations.

Finally, there are interesting investigations of the constraints of general coordinate and conformal invariance by Dam Son and collaborators [99, 100]. They ask: Is there more than scale invariance for the unitary Fermi gas? The symmetries can be exposed by adding a background gauge field A_μ and curved space with metric $g_{ij}(t, \mathbf{x})$:

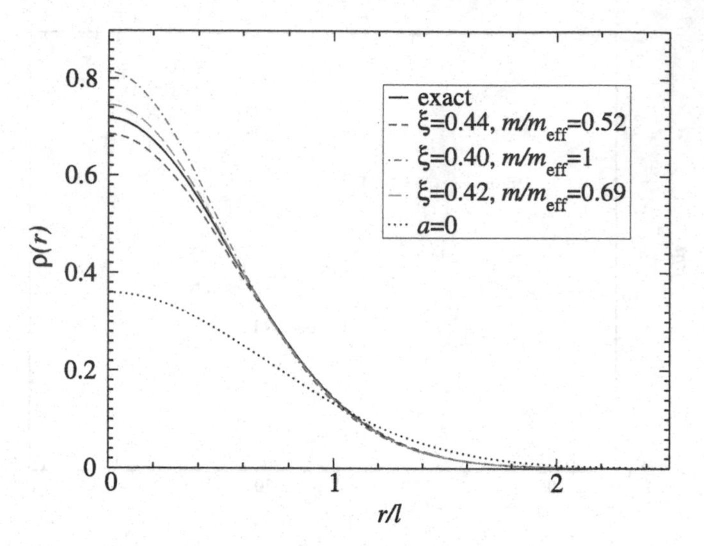

Fig. 3.29 Papenbrock results from a density functional for the unitary regime

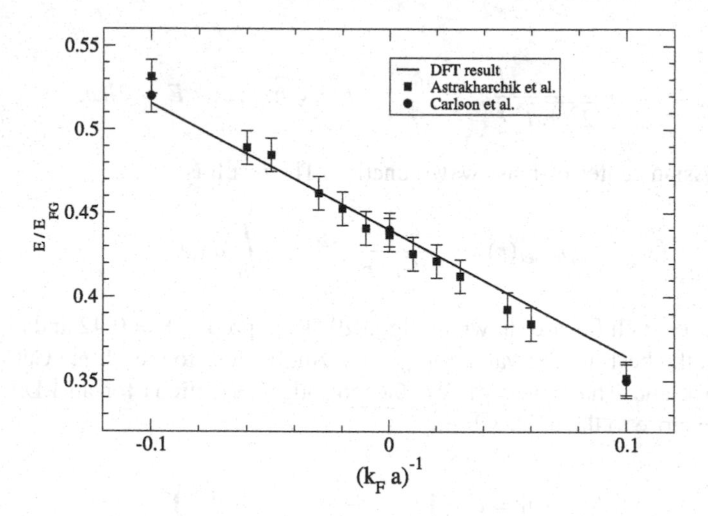

Fig. 3.30 Papenbrock/Bhattacharyya DFT results for finite scattering length a compared to Monte Carlo calculations

$$S \longrightarrow \int dt\, d\mathbf{x}\, \sqrt{g}\Big[\frac{i}{2}\psi^{\dagger}\overleftrightarrow{\partial}_t\psi - \frac{g^{ij}}{2m}(\partial_i + iA_i)\psi^{\dagger}(\partial_j - iA_j)\psi$$
$$+(q_0\sigma - A_0)\psi^{\dagger}\psi - \frac{g^{ij}}{2}\partial_i\sigma\partial_j\sigma - \frac{\sigma^2}{2r_0^2}\Big]. \qquad (3.178)$$

This is more than scale and Galilean invariance! Direct consequences include extra constraints on $\mathcal{L}_{\text{eff}}$ at NLO, which naively involve five arbitrary functions but these symmetries show there are only three. For the unitary Fermi gas, three constants from scale invariance are reduced to two constants from conformal invariance. This leads us to ask: What additional constraints can we find for the energy functional? This and the other open questions are ripe for investigation!

Acknowledgments I thank the organizers of the Trento school, Janos Polonyi and Achim Schwenk, for the opportunity to participate in an excellent lecture series, and all the student participants, who made giving the lectures a pleasure. This work was supported in part by the National Science Foundation under Grant No. PHY–0354916.

References

1. Recall that the wave function of a system of identical fermions must be antisymmetric under exchange of any two particles. This has important consequences for many-body systems, such as the development of a Fermi sea and "Pauli blocking"
2. This figure was made by Jacek Dobaczewski
3. Preson, M.A., Bhaduri, R.K.: Structure of the Nucleus. Addison-Wesley, Reading (1975)
4. Jackson, A.D.: The once and future nuclear many-body problem. In: Ainsworth, T.L., Campbell, C.E., Clements, B.E., Krotscheck, E. (eds.) Recent Progress in Many-Body Theories, vol. 3. Plenum, New York (1992)
5. Jackson, A.D., Wettig, T.: Phys. Rep. **237**, 325 (1994)
6. This figure was adapted by Achim Schwenk from a picture by A. Richter
7. This discussion is adapted from a talk by Joe Carlson
8. Ring, P., Schuck, P.: The Nuclear Many-Body Problem. Springer, Berlin (2005)
9. Dreizler, R.M., Gross, E.K.U.: Density Functional Theory. Springer, Berlin (1990)
10. Argaman, N., Makov, G.: Am. J. Phys. **68**, 69 (2000)
11. Fiolhais, C., Nogueira, F., Marques, M. (eds.): A Primer in Density Functional Theory. Springer, Berlin (2003)
12. The term "ab initio" is often used in this context, but the calculations do not in practice proceed only from the Coulomb interaction
13. Koch, W., Holthausen, M.C.: A Chemist's Guide to Density Functional Theory. Wiley, New York (2000)
14. Perdew, J.P., Kurth, S., Zupan, A., Blaha, P.: Phys. Rev. Lett. **82**, 2544 (1999)
15. Polonyi, J., Sailer, K.: Phys. Rev. B **66**, 155113 (2002) [arXiv:cond-mat/0108179]
16. Bogner, S.K., Schwenk, A., Furnstahl, R.J., Nogga, A.: Nucl. Phys. A **763**, 59 (2005)
17. Wiringa, R.B., Stoks, V.G.J., Schiavilla, R.: Phys. Rev. C **51**, 38 (1995)
18. Vautherin, D., Brink, D.M.: Phys. Rev. C **5**, 626 (1972)
19. Dobaczewski, J., Nazarewicz, W., Reinhard, P.G.: Nucl. Phys. A **693**, 361 (2001) [arXiv:nucl-th/0103001]
20. Bender, M., Heenen, P.H., Reinhard, P.-G.: Rev. Mod. Phys. **75**, 121 (2003)
21. A "local" potential is one whose action on a wavefunction in the Schrödinger equation is just $V(r)\psi(r)$; that is, it happens at one point. More generally, we expect something like $\int dr'\, V(r,r')\psi(r')$, which is not diagonal in coordinate representation. In momentum representation, this means $\langle k|V|k'\rangle$ is not a function of $q \equiv k - k'$ only
22. The goal of devising a potential that fits the scattering data as best as possible (that is, $\chi^2/\text{dof} \approx 1$) below where inelastic effects become important has been realized by several potentials, including CD-Bonn, Nijmegen I and II, and Argonne v_{18}
23. Calculations of the energy of few-body nuclei using accurate two-body potentials have demonstrated the need for a three-body force

24. All NN potentials have a cut-off on high-momentum (short-distance) contributions. Putting in a cut-off doesn't mean the high-energy physics is thrown out, however! We need to renormalize (this is one of the main points of our discussion)
25. One way of specifying a many-body approximation is to say which Feynman diagrams are included. (If all are included, we're solving the problem exactly!) Usually one needs to select an (infinite) subset (for example, the ladder or ring diagrams). In some case there are rigorous justifications for which to sum, but not always and there are seldom estimates available for what quantitative error is made by the truncation
26. Lepage, G.P.: What is renormalization? In: DeGrand, T., Toussaint, D. (eds.) From Actions to Answers (TASI-89), p. 483. World Scientific, Singapore (1989)
27. Lepage, G.P.: How to renormalize the Schrödinger equation. Lectures given at 9th Jorge Andre Swieca Summer School: Particles and Fields, Sao Paulo, Brazil (1997). arXiv:nucl-th/9706029
28. Bogner, S.K., Furnstahl, R.J.: Phys. Lett. B **632**, 501 (2006)
29. Bogner, S.K., Furnstahl, R.J.: Phys. Lett. B **639**, 237 (2006)
30. Bogner, S.K., Kuo, T.T.S., Schwenk, A., Entem, D.R., Machleidt, R.: Phys. Lett. B **576**, 265 (2003)
31. Bogner, S.K., Kuo, T.T.S., Schwenk, A.: Phys. Rept. **386**, 1 (2003)
32. Bogner, S.K., Schwenk, A., Kuo, T.T.S., Brown, G.E.: arXiv:nucl-th/0111042
33. Bogner, S.K., Furnstahl, R.J., Ramanan, S., Schwenk, A.: Nucl. Phys. A **773**, 203 (2006)
34. Kerman, A.K., Svenne, J.P., Villars, F.M.H.: Phys. Rev. **147**, 710 (1966)
35. Bassichis, W.H., Kerman, A.K., Svenne, J.P.: Phys. Rev. **160**, 746 (1967)
36. Strayer, M.R., Bassichis, W.H., Kerman, A.K.: Phys. Rev. C **8**, 1269 (1973)
37. Nogga, A., Bogner, S.K., Schwenk, A.: Phys. Rev. C **70**, 061002(R) (2004)
38. Negele, J.W., Orland, H.: Quantum Many-Particle Systems. Addison-Wesley, New York (1988)
39. Zinn-Justin, J.: Quantum Field Theory and Critical Phenomena, Chapter 12. 4th edn. Oxford University Press, New York (2002)
40. For the Minkowski-space version of this discussion, see S. Weinberg: The Quantum Theory of Fields: vol. II, Modern Applications. Cambridge University Press, Cambridge (1996)
41. Schwenk, A., Polonyi, J.: arXiv:nucl-th/0403011
42. Nagaosa, N.: Quantum Field Theory in Condensed Matter Physics. Springer, Berlin (1999)
43. Stone, M.: The Physics of Quantum Fields. Springer, Berlin (2000)
44. Fukuda, R., Komachiya, M., Yokojima, S., Suzuki, Y., Okumura, K., Inagaki, T.: Prog. Theor. Phys. Suppl. **121**, 1 (1995)
45. Valiev, M., Fernando, G.W.: arXiv:cond-mat/9702247
46. Valiev, M., Fernando, G.W.: Phys. Lett. A **227**, 265 (1997)
47. Rasamny, M., Valiev, M.M., Fernando, G.W.: Phys. Rev. B **58**, 9700 (1998)
48. To derive the second equality for $f_l(k)$, first put the $e^{i\delta_l}$ in the denominator, then replace it by $\cos + i \sin$ and divide top and bottom by sin
49. Hammer, H.W., Furnstahl, R.J.: Nucl. Phys. A **678**, 277 (2000)
50. Beane, S.R., Bedaque, P.F., Haxton, W.C., Phillips, D.R., Savage, M.J.: arXiv:nucl-th/0008064
51. Braaten, E., Nieto, A.: Phys. Rev. B **55**, 8090 (1997); **56**, 14745 (1997)
52. Puglia, S.J., Bhattacharyya, A., Furnstahl, R.J.: Nucl. Phys. A **723**, 145 (2003) [arXiv:nucl-th/0212071]
53. Bhattacharyya, A., Furnstahl, R.J.: Nucl. Phys. A **747**, 268 (2005) [arXiv:nucl-th/0408014]
54. Bhattacharyya, A., Furnstahl, R.J.: Phys. Lett. B **607**, 259 (2005) [arXiv:nucl-th/0410105]
55. Bartlett, R.J., Lotrich, V.F., Schweigert, I.V.: J. Chem. Phys. **123**, 062205 (2005)
56. Görling, A.: J. Chem. Phys. **123**, 062203 (2005)
57. Baerends, E.J., Gritsenko, O.V.: J. Chem. Phys. **123**, 062202 (2005)
58. Furnstahl, R.J., Hammer, H.W.: Phys. Lett. B **531**, 203 (2002) [arXiv:nucl-th/0108069]
59. Bulgac, A.: Phys. Rev. C **65**, 051305 (2002) [arXiv:nucl-th/0108014]
60. Bulgac, A., Yu, Y.: Phys. Rev. Lett. **88**, 042504 (2002) [arXiv:nucl-th/0106062]
61. Yu, Y., Bulgac, A.: Phys. Rev. Lett. **90**, 222501 (2003) [arXiv:nucl-th/0210047]

62. Oliveira, L.N., Gross, E.K.U., Kohn, W.: Phys. Rev. Lett. **60**, 2430 (1988)
63. Kurth, S., Marques, M., Lüders, M., Gross, E.K.U.: Phys. Rev. Lett. **83**, 2628 (1999)
64. Furnstahl, R.J., Hammer, H.W., Puglia, S.J.: Ann. Phys. **322**, 2703 (2007) [arXiv:nuclth/0612086]
65. Papenbrock, T., Bertsch, G.F.: Phys. Rev. C **59**, 2052 (1999)
66. Gor'kov, L.P., Melik-Barkhudarov, T.K.: Sov. Phys. JETP **13**, 1018 (1961)
67. Heiselberg, H., Pethick, C.J., Smith, H., Viverit, L.: Phys. Rev. Lett. **85**, 2418 (2000)
68. Furnstahl, R.J.: J. Phys. G **31**, S1357 (2005) [arXiv:nucl-th/0412093]
69. Entem, D.R., Machleidt, R.: Phys. Rev. C **68**, 041001(R) (2003)
70. Epelbaum, E., Glöckle, W., Meißner, U.G.: Nucl. Phys. A **747**, 362 (2005)
71. Bogner, S.K., Furnstahl, R.J., Perry, R.J.: Phys. Rev. C **75**, 061001 (2007) [arXiv:nucl-th/0611045]
72. Bogner, S.K., Furnstahl, R.J., Perry, R.J., Schwenk, A.: Phys. Lett. B **649**, 488 (2007) [arXiv:nucl-th/0701013]
73. Coraggio, L., Covello, A., Gargano, A., Itaco, N., Entem, D.R., Kuo, T.T.S., Machleidt, R.: Phys. Rev. C **75**, 024311 (2007) [arXiv:nucl-th/0701065]
74. Negele, J.W., Vautherin, D.: Phys. Rev. C **5**, 1472 (1972)
75. Negele, J.W., Vautherin, D.: Phys. Rev. C **11**, 1031 (1975)
76. Bogner, S.K., Furnstahl, R.J., Platter, L.: Eur. Phys. J. A **39**, 219 (2009)
77. Kaiser, N., Fritsch, S., Weise, W.: Nucl. Phys. A **724**, 47 (2003)
78. Kaiser, N.: Phys. Rev. C **68**, 014323 (2003)
79. Kaiser, N., et al.: Nucl. Phys. A **750**, 259 (2005)
80. Rotivale, V., Bogner, S.K., Duguet, T., Furnstahl, R.J.: in preparation
81. Engel, J.: Phys. Rev. C **75**, 014306 (2007) [arXiv:nucl-th/0610043]
82. Serot, B.D., Walecka, J.D.: Adv. Nucl. Phys. **16**, 1 (1986)
83. Ring, P.: Prog. Part. Nucl. Phys. **37**, 193 (1996), and references therein
84. Serot, B.D., Walecka, J.D.: Int. J. Mod. Phys. E **6**, 515 (1997), and references therein
85. Gasser, J., Sainio, M.E., Svarc, A.: Nucl. Phys. B **307**, 779 (1988)
86. Jenkins, E., Manohar, A.V.: Phys. Lett. B **255**, 558 (1991); Bernard, V., Kaiser, N., Meissner, U.-G.: Nucl. Phys. B **388**, 315 (1992)
87. Tang, H.-B.: arXiv:hep-ph/9607436
88. Ellis, P.J., Tang, H.-B.: Phys. Rev. C **57**, 643 (1998)
89. Becher, T., Leutwyler, H.: Eur. Phys. J. C **9**, 643 (1999)
90. Fuchs, T., et al.: Phys. Rev. D **68**, 056005 (2003) [arXiv:hep-ph/0302117]
91. Lehmann, D., Prezeau, G.: Phys. Rev. D **65**, 016001 (2002)
92. Schindler, M.R., Gegelia, J., Scherer, S.: Phys. Lett. B **586**, 258 (2004) [arXiv:hep-ph/0309005]
93. Kaplan, D.B., Savage, M.J., Wise, M.B.: Phys. Lett. B **424**, 390 (1998); Nucl. Phys. B **534**, 329 (1998)
94. Chang, S.Y. et al.: Nucl. Phys. A **746**, 215 (2004) [arXiv:nucl-th/0401016]
95. Astrakharchik, G.E., Boronat, J., Casulleras, J.: S. Giorgini Phys. Rev. Lett. **93**, 200404 (2004)
96. Papenbrock, T.: Phys. Rev. A **72**, 041603 (2005) [arXiv:cond-mat/0507183]
97. Bhattacharyya, A., Papenbrock, T.: Phys. Rev. A **74**, 041602 (2006) [arXiv:nucl-th/0602050]
98. Papenbrock, T., Bhattacharyya, A.: Phys. Rev. C **75**, 014304 (2007) [arXiv:nucl-th/0609084]
99. Son, D.T., Wingate, M.: Annals Phys. **321**, 197 (2006) [arXiv:cond-mat/0509786]
100. Son, D.T.: Phys. Rev. Lett. **98**, 020604 (2007) [arXiv:cond-mat/0511721]

Chapter 4
Effective Theories of Dense and Very Dense Matter

Thomas Schäfer

4.1 Introduction

The exploration of the phase diagram of dense baryonic matter is an area of intense theoretical and experimental activity. Baryonic systems, from dilute neutron matter at low density to superconducting quark matter at high density, exhibit an enormous variety of many-body effects. Despite its simplicity all these phenomena are ultimately described by the lagrangian of QCD.

In practice it is usually very difficult to describe QCD many-body systems directly in terms of the QCD lagrangian, and even in cases where this is possible it is often not the most convenient and most transparent description. Instead, it is advantageous to employ an effective field theory (EFT) formulated in terms of the relevant degrees of freedom. EFTs also provide a unified description of physical systems involving very different length scales, such as Fermi liquids in nuclear and atomic physics, or non-Fermi liquid gauge theories involving colored quarks or charged electrons.

In these lectures we shall discuss the many body physics of several effective field theories relevant to the structure of hadronic matter. We will concentrate on two regimes in the phase diagram. At low baryon density the relevant degrees of freedom are neutrons and protons, while at very high baryon density the degrees of freedom are quarks and gluons. These lectures do not provide an introduction to effective field theories (see [1–3]), nor an exhaustive treatment of many body physics (see [4–6]) or the physics of dense quark matter (see [7, 8]).

T. Schäfer (✉)
Department of Physics, North Carolina State University,
Raleigh, NC 27695, USA
e-mail: thomas_schaefer@ncsu.edu

J. Polonyi and A. Schwenk (eds.), *Renormalization Group and Effective Field Theory Approaches to Many-Body Systems*, Lecture Notes in Physics 852,
DOI: 10.1007/978-3-642-27320-9_4,

4.2 Fermi Liquids

4.2.1 Effective Field Theory for Non-Relativistic Fermions

If the relevant momenta are small neutrons and protons can be described as point-like non-relativistic fermions interacting via local forces. Effective field theories for nuclear systems have been studied extensively over the past couple of years [3, 9–11]. If the typical momenta are on the order of the pion mass pions have to be included as explicit degrees of freedoms. For simplicity we will consider neutrons only and focus on momenta small compared to m_π. The effective lagrangian is

$$\mathcal{L}_0 = \psi^\dagger\left(i\partial_0 + \frac{\nabla^2}{2m}\right)\psi - \frac{C_0}{2}\left(\psi^\dagger\psi\right)^2 + \frac{C_2}{16}\left[(\psi\psi)^\dagger(\psi\overleftrightarrow{\nabla}^2\psi) + h.c\right] + \dots, \tag{4.1}$$

where m is the neutron mass, C_0 and C_2 are dimensionful coupling constants, $\overleftrightarrow{\nabla} = \overrightarrow{\nabla} - \overleftarrow{\nabla}$ is a Galilei invariant derivative, and ... denotes interactions with more derivatives. We have only displayed terms that act in the s-channel. The coupling constant are determined by the neutron-neutron scattering amplitude. For non-relativistic scattering the amplitude is related to the scattering phase shift δ by

$$\mathcal{A} = \frac{4\pi}{m}\frac{1}{p\cot\delta - ip}. \tag{4.2}$$

For small momenta the quantity $p\cot\delta$ can be expanded as a Taylor series in p. This expansion is called the effective range expansion

$$p\cot\delta = -\frac{1}{a} + \frac{1}{2}\sum_{n=0}^{\infty} r_n p^{2(n+1)}, \tag{4.3}$$

where a is the scattering length, and r_0 is the effective range. The situation is simplest if the scattering length is small. In this case the scattering amplitude has a perturbative expansion in C_i. At tree level

$$C_0 = \frac{4\pi a}{m}, \qquad C_2 = C_0\frac{ar_0}{2}. \tag{4.4}$$

However, there are many systems of physical interest in which the scattering length is not small. This happens whenever there is a two-body bound state with a very small binding energy, or if the two-body system is very close to forming a bound state. For neutrons $a_{nn} = -17$ fm, much larger than typical strong interaction length scales.

If the scattering length is large then loop diagrams with the leading order interaction $C_0(\psi^\dagger\psi)^2$ have to be resummed. The one-loop correction involves the loop integral

$$L(E) = i\int \frac{d^{d+1}q}{(2\pi)^{d+1}} \frac{1}{(E/2+q_0-\mathbf{q}^2/(2m)+i\epsilon)(E/2-q_0-\mathbf{q}^2/(2m)+i\epsilon)}$$
$$= \int \frac{d^d q}{(2\pi)^d} \frac{1}{E-\mathbf{q}^2/m+i\epsilon}$$
$$= -\frac{m}{(4\pi)^{d/2}}\Gamma\left(\frac{2-d}{2}\right)(-mE-i\epsilon)^{\frac{d-2}{2}}, \tag{4.5}$$

where E is the center-of-mass energy. We have regularized the integral by analytic continuation to $d{+}1$ dimensions. In order to define the theory we have to specify a subtraction scheme. Here, we will employ the modified minimal subtraction $\overline{MS}$ scheme. See [12] for a discussion of different renormalization schemes. We get

$$L(E) = \frac{m}{4\pi}\sqrt{-mE-i\epsilon} = -\frac{m}{4\pi}ip\,, \tag{4.6}$$

where $p=\sqrt{mE}$ is the nucleon momentum in the center-of-momentum frame. It is now straightforward to sum all the bubble diagrams. The result is

$$\mathcal{A} = -\frac{C_0}{1+iC_0(mp/4\pi)}. \tag{4.7}$$

Higher order corrections due to the C_i terms ($i\geq 2$) can be treated perturbatively. The bubble sum can now be matched to the effective range expansion. In the $\overline{MS}$ scheme the result is particularly simple since Eq. (4.6) only contains the contribution from the unitarity cut. As a consequence, the result given in Eq. (4.4) is not modified even if C_0 is summed to all orders.

4.2.2 Dilute Fermi Liquid

The lagrangian given in Eq. (4.1) is invariant under the $U(1)$ transformation $\psi\to e^{i\phi}\psi$. The $U(1)$ symmetry implies that the fermion number

$$N = \int d^3x\,\psi^\dagger\psi \tag{4.8}$$

is conserved. As a consequence, it is meaningful to study a system of fermions at finite density $\rho = N/V$. We will do this in the grand-canonical formalism. We introduce a chemical potential μ conjugate to the fermion number N and study the partition function

$$Z(\mu,\beta) = \mathrm{Tr}\left[e^{-\beta(H-\mu N)}\right]. \tag{4.9}$$

Here, H is the Hamiltonian associated with $\mathcal{L}$ and $\beta = 1/T$ is the inverse temperature. The trace in Eq. (4.9) runs over all possible states of the system. The average number of particles for a given chemical potential μ and temperature T is given by $\langle N\rangle = T(\partial \log Z)/(\partial\mu)$. At zero temperature the chemical potential is the energy required to add one particle to the system.

There is a formal resemblance between the partition function Eq. (4.9) and the quantum mechanical time evolution operator $U = \exp(-iHt)$. In order to write the partition function as a time evolution operator we have to identify $\beta \to it$ and add the term $-\mu N$ to the Hamiltonian. Using standard techniques we can write the time evolution operators as a path integral [13, 14]

$$Z = \int D\psi D\psi^\dagger \exp\left(-\int_0^\beta d\tau \int d^3x\, \mathcal{L}_E\right). \tag{4.10}$$

Here, $\mathcal{L}_E$ is the euclidean lagrangian

$$\mathcal{L}_E = \psi^\dagger\left(\partial_\tau - \mu - \frac{\nabla^2}{2m}\right)\psi + \frac{C_0}{2}(\psi^\dagger\psi)^2 + \dots. \tag{4.11}$$

The fermion fields satisfy anti-periodic boundary conditions $\psi(\beta) = -\psi(0)$. Equation (4.11) is the starting point of the imaginary time formalism in thermal field theory. The chemical potential simply results in an extra term $-\mu\psi^\dagger\psi$ in the lagrangian. From Eq. (4.11) we can easily read off the free fermion propagator

$$S^0_{\alpha\beta}(p) = \frac{\delta_{\alpha\beta}}{ip_4 + \mu - \frac{p^2}{2m}}, \tag{4.12}$$

where α, β are spin labels. We observe that the chemical potential simply shifts the four-component of the momentum. This implies that we have to carefully analyze the boundary conditions in the path integral in order to fix the pole prescription. The correct Minkowski space propagator is

$$S^0_{\alpha\beta}(p) = \frac{\delta_{\alpha\beta}}{p_0 - \epsilon_p + i\delta\mathrm{sgn}(\epsilon_p)} = \delta_{\alpha\beta}\left\{\frac{\Theta(p-p_F)}{p_0-\epsilon_p+i\delta} + \frac{\Theta(p_F-p)}{p_0-\epsilon_p-i\delta}\right\}, \tag{4.13}$$

where $\epsilon_p = E_p - \mu$, $E_p = \boldsymbol{p}^2/(2m)$ and $\delta \to 0^+$. The quantity $p_F = \sqrt{2m\mu}$ is called the Fermi momentum. We will refer to the surface defined by the condition $|\boldsymbol{p}| = p_F$ as the Fermi surface. The two terms in Eq. (4.13) have a simple physical interpretation. At finite density and zero temperature all states with momenta below the Fermi momentum are occupied, while all states above the Fermi momentum are

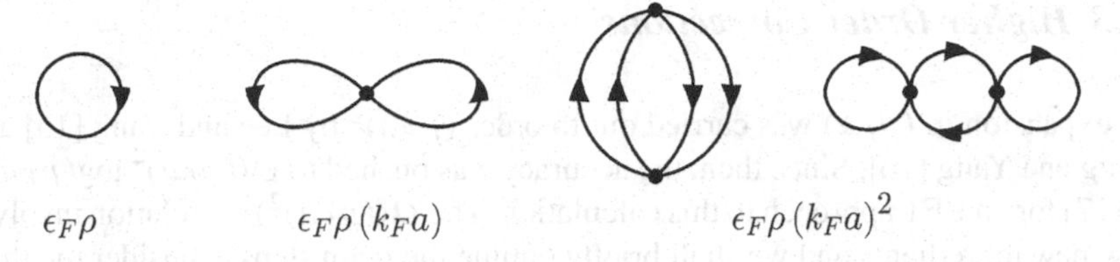

Fig. 4.1 Leading order Feynman diagrams for the ground state energy of a dilute gas of fermions interacting via a short range potential

empty. The possible excitation of the system are particles above the Fermi surface or holes below the Fermi surface, corresponding to the first and second term in Eq. (4.13). The particle density is given by

$$\rho = \langle \psi^\dagger \psi \rangle = \int \frac{d^4p}{(2\pi)^4} S^0_{\alpha\alpha}(p) e^{ip_0\delta}\Big|_{\delta\to 0^+} = 2\int \frac{d^3p}{(2\pi)^3}\Theta(p_F - p) = \frac{p_F^3}{3\pi^2}. \tag{4.14}$$

Tadpole diagrams require an extra $i\delta$ prescription which can be derived from a careful analysis of the path integral representation at $\mu \neq 0$. As a first simple application we can compute the energy density as a function of the fermion density. For free fermions, we find

$$\mathcal{E} = 2\int \frac{d^3p}{(2\pi)^3} E_p \Theta(p_F - p) = \frac{3}{5}\rho\frac{p_F^2}{2m}. \tag{4.15}$$

We can also compute the corrections to the ground state energy due to the interaction $(C_0/2)(\psi^\dagger\psi)^2$. The first term is a two-loop diagram with one insertion of C_0, see Fig. 4.1. There are two possible contractions and the spin-factor of the diagram is $(\delta_{\alpha\alpha}\delta_{\beta\beta} - \delta_{\alpha\beta}\delta_{\alpha\beta}) = g(g-1)$ where $g = (2s+1)$ is the degeneracy and s is the spin of the fermions. In the following we will always set $g = 2$. The diagram is proportional to the square of the density and we get

$$\mathcal{E}_1 = C_0\left(\frac{p_F^3}{6\pi^2}\right)^2. \tag{4.16}$$

We observe that the sum of the first two terms in the energy density can be written as

$$\mathcal{E} = \rho\frac{p_F^2}{2m}\left(\frac{3}{5} + \frac{2}{3\pi}(p_F a) + \ldots\right), \tag{4.17}$$

which shows that the C_0 term is the first term in an expansion in $p_F a$, suitable for a dilute, weakly interacting, Fermi gas.

4.2.3 Higher Order Corrections

The expansion in $(p_F a)$ was carried out to order $(p_F a)^2$ by Lee and Yang [15] and Huang and Yang [16]. Since then, the accuracy was pushed to $O((p_F a)^4 \log(p_F a))$, see [17] for an EFT approach to this calculation. The $O((p_F a)^2)$ calculation involves a few new ingredients and we shall briefly outline the main steps. Consider the third diagram in Fig. 4.1. The contribution to the vacuum energy is

$$\mathcal{E}_2 = -i\frac{C_0^2}{2}\int\frac{d^4q_1}{(2\pi)^4}\int\frac{d^4q_2}{(2\pi)^4}\int\frac{d^4q_3}{(2\pi)^4}S(q_1)S(q_2)S(q_3)S(q_1+q_2-q_3). \quad (4.18)$$

We begin by performing two of the energy integrals using contour integration. The contours can be placed in such a way that the two poles correspond to two particles or two holes (but not a particle and a hole). This allows us to write

$$\mathcal{E}_2 = \frac{C_0^2}{2}\int\frac{d^3q_1}{(2\pi)^3}\int\frac{d^3q_2}{(2\pi)^3}\,\theta_q^- \Pi_{pp}(q_1+q_2) + h.c., \quad (4.19)$$

where θ_q^- is the Pauli-blocking factor corresponding to a pair of holes

$$\theta_q^- = \theta\,(p_F - q_1)\,\theta\,(p_F - q_2), \quad (4.20)$$

and Π_{pp} is the one-loop particle-particle scattering amplitude. Since $q_{1,2}$ are on-shell we can write Π_{pp} as a function of the center-of-mass and relative momenta $\boldsymbol{P}$ and $\boldsymbol{k}$ with $\boldsymbol{q}_{1,2} = \boldsymbol{P}/2 \pm \boldsymbol{k}$. Note that because of Galilean invariance the vacuum scattering amplitude only depends on $\boldsymbol{k}$. We find

$$\Pi_{pp} = \int\frac{d^3q}{(2\pi)^3}\,\frac{m\theta_q^+}{\boldsymbol{k}^2 - \boldsymbol{q}^2 + i\epsilon} = \Pi_{pp}^{vac}(k) + \frac{mp_F}{(2\pi)^2}f_{pp}(\kappa, s), \quad (4.21)$$

where $\theta_q^+ = \theta\,(p_F + q_1)\,\theta\,(p_F + q_2)$ is defined in analogy with Eq. (4.20). The first term on the RHS is the vacuum contribution and the second term is the medium contribution which depends on the scaled momenta $\boldsymbol{\kappa} = \boldsymbol{k}/p_F$ and $\boldsymbol{s} = \boldsymbol{P}/(2p_F)$. The vacuum contribution is divergent and needs to be renormalized. In dimensional regularization Π_{pp}^{vac} is purely imaginary and does not contribute to the vacuum energy. In other regularization schemes the vacuum contributions combines with the $O(C_0)$ graph to give the correct one-loop relation between C_0 and the scattering length.

For $s < 1$ the in-medium scattering amplitude is given by

$$f_{PP}(\kappa, s) = 1+s+\kappa\log\left|\frac{1+s-\kappa}{1+s+\kappa}\right| + \frac{1-\kappa^2-s^2}{2s}\log\left|\frac{(1+s)^2-\kappa^2}{1-\kappa^2-s^2}\right|. \quad (4.22)$$

The contribution to the energy density can now be determined by integrating Eq. (4.22) over phase space. We find

$$\begin{aligned}\mathcal{E}_2 &= C_0^2 \frac{p_F m}{4\pi^2} \int \frac{d^3P}{(2\pi)^3} \frac{d^3k}{(2\pi)^3}\, \theta_k^- \, f_{PP}(\kappa, s) \\ &= \rho \frac{p_F^2}{2m} \frac{4}{35\pi^2} (11 - 2\log(2))\, (p_F a)^2 . \end{aligned} \tag{4.23}$$

The fourth diagram in Fig. 4.1 involves a particle-hole pair with zero energy and the corresponding phase space factor vanishes [5].

The effective lagrangian can also be used to study many other properties of the system, such as corrections to the fermion propagator. Near the Fermi surface the propagator can be written as

$$S_{\alpha\beta} = \frac{Z\delta_{\alpha\beta}}{p_0 - v_F(|\boldsymbol{p}| - p_F) + i\delta \mathrm{sgn}(|\boldsymbol{p}| - p_F)}, \tag{4.24}$$

where Z is the wave function renormalization and $v_F = p_F/m^*$ is the Fermi velocity. Z and m^* can be worked out order by order in $(p_F a)$, see [4, 18]. The leading order results are

$$\frac{m^*}{m} = 1 - \frac{8}{15\pi^2} (1 - 7\log(2))\, (p_F a)^2 + \ldots \tag{4.25}$$

$$Z^{-1} = 1 - \frac{4}{\pi^2} \log(2)\, (p_F a)^2 + \ldots \tag{4.26}$$

The main observation is that the structure of the propagator is unchanged even if interactions are taken into account. The low energy excitations are quasi-particles and holes, and near the Fermi surface the lifetime of a quasi-particle is infinite. This is the basis of Fermi liquid theory [19, 20].

4.2.4 Screening and Damping

An important aspect of a dilute Fermi gas of charged particles is the response to an external electromagnetic field. We consider a system in which the total charge is neutralized by a homogeneous background (such as positive ions in a metal). The response to an electric field is governed by the gauge coupling $eA_0\psi^\dagger\psi$. The medium correction to the photon propagator is determined by the polarization function

$$\Pi_{00}(q) = e^2 \int d^4x\, e^{-iqx} \langle \psi^\dagger\psi(0)\psi^\dagger\psi(x)\rangle . \tag{4.27}$$

The one-loop contribution is given by

$$\Pi_{00}(q) = -ie^2 \int \frac{d^4p}{(2\pi)^4} \frac{1}{q_0 + p_0 - \epsilon_{p+q} + i\delta \mathrm{sgn}(\epsilon_{p+q})} \frac{1}{p_0 - \epsilon_p + i\delta \mathrm{sgn}(\epsilon_p)}. \tag{4.28}$$

Performing the p_0 integral using contour integration we find

$$\Pi_{00}(q) = e^2 \int \frac{d^3p}{(2\pi)^3} \frac{n_{p+q} - n_p}{E_{p+q} - E_p}, \tag{4.29}$$

where we have introduced the Fermi distribution function $n_p = \Theta(p_F - p)$. We observe that in the limit $\boldsymbol{q} \to 0$ the polarization function only receives contributions from particle-hole pairs that are very close to the Fermi surface. On the other hand, the energy denominator diverges in this limit because the photon can excite particle-hole pairs with arbitrarily small energy. These two effects combine to give a finite contribution

$$\Pi_{00}(q_0 = 0, \boldsymbol{q} \to 0) = e^2 \int \frac{d^3p}{(2\pi)^3} \frac{\partial n_p}{\partial E_p} = e^2 \frac{p_F m}{2\pi^2}, \tag{4.30}$$

which is proportional to the density of states on the Fermi surface. Equation (4.30) implies that the static photon propagator in the limit $\boldsymbol{q} \to 0$ is modified according to $1/\boldsymbol{q}^2 \to 1/(\boldsymbol{q}^2 + m_D^2)$, where

$$m_D^2 = e^2 \left(\frac{p_F m}{2\pi^2} \right) \tag{4.31}$$

is called the Debye mass. The factor $N = (p_F m)/(2\pi^2)$ is equal to the density of states on the Fermi surface. In a relativistic theory we find the same result as in Eq. (4.31) with the density of states replaced by the correct relativistic expression $N = (p_F E_F)/(2\pi^2)$. The Coulomb potential is modified as

$$V(r) = -e \frac{e^{-r/r_D}}{r}, \tag{4.32}$$

where $r_D = 1/m_D$ is called the Debye screening length. The physics of screening is very easy to understand. A test charge can polarize virtual particle-hole pairs that act to shield the charge.

In the same fashion we can study the response to an external vector potential $\boldsymbol{A}$. The coupling of a non-relativistic fermion to the vector potential is determined in the usual way by replacing $\boldsymbol{p} \to \boldsymbol{p} + e\boldsymbol{A}$. Since the kinetic energy operator is quadratic in the momentum we find a linear and a quadratic coupling of the vector potential. The one-loop diagrams that contribute to the polarization tensor are shown in Fig. 4.2. In the limit of small external momenta we find

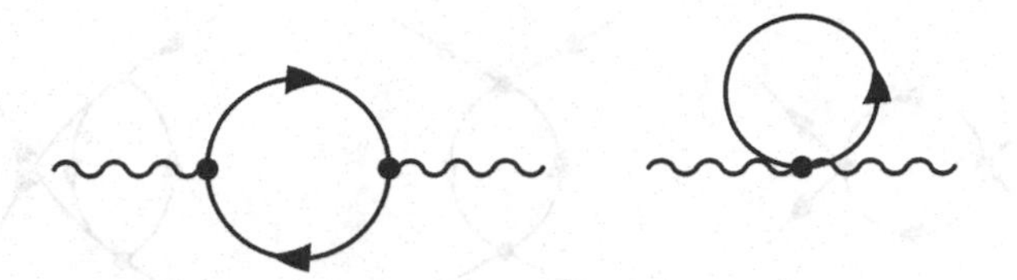

Fig. 4.2 Leading order Feynman diagrams that contribute to the photon polarization function in a non-relativistic Fermi liquid. The tadpole diagram shown in the *right panel* only appears in the spatial part of the polarization tensor

$$\Pi_{ij}(q) = -e^2 m_D^2 \int \frac{d\Omega}{4\pi} \left\{ v_i v_j \frac{\boldsymbol{v}\cdot\boldsymbol{q}}{q_0 - \boldsymbol{v}\cdot\boldsymbol{q}} + \frac{1}{3} v^2 \delta_{ij} \right\}, \tag{4.33}$$

where $\boldsymbol{v} = \boldsymbol{p}/m$ is the Fermi velocity. In the limit $q_0 = 0$ the polarization tensor vanishes. There is no screening of static magnetic fields. For non-zero q_0 the trace of the polarization tensor is given by

$$\Pi_{ii}(q) = m_D^2 \frac{vq_0}{2q} \log\left(\frac{q_0 - vq}{q_0 + vq}\right). \tag{4.34}$$

The result has an imaginary part for $vq > q_0$. This phenomenon is known as Landau damping. The physical mechanism is that the photon is loosing energy as it scatters of the electrons in the Fermi liquid, see [21] for a detailed discussion in the context of kinetic theory.

4.3 Superconductivity

4.3.1 BCS Instability

One of the most remarkable phenomena that take place in many body systems is superconductivity. Superconductivity is related to an instability of the Fermi surface in the presence of attractive interactions between fermions. Let us consider fermion-fermion scattering in the simple model introduced in Sect. 4.2. At leading order the scattering amplitude is given by

$$\Gamma_{\alpha\beta\gamma\delta}(p_1, p_2, p_3, p_4) = C_0 \left(\delta_{\alpha\gamma}\delta_{\beta\delta} - \delta_{\alpha\delta}\delta_{\beta\gamma}\right). \tag{4.35}$$

At next-to-leading order we find the corrections shown in Fig. 4.3. A detailed discussion of the role of these corrections can be found in [1, 4, 22]. The BCS diagram is special, because in the case of a spherical Fermi surface it can lead to an instability in weak coupling. The main point is that if the incoming momenta satisfy $\boldsymbol{p}_1 \simeq -\boldsymbol{p}_2$

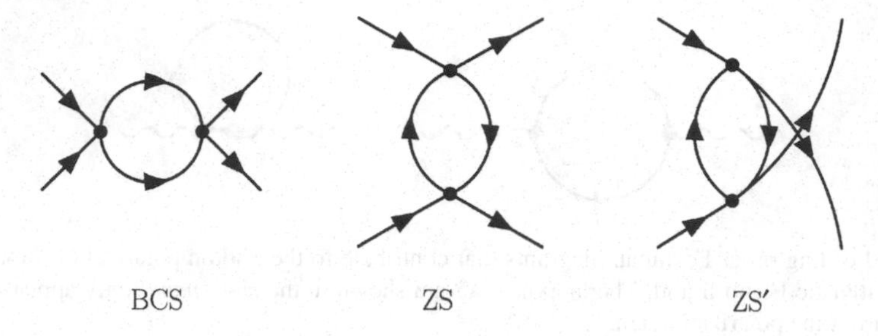

Fig. 4.3 Second order diagrams that contribute to particle-particle scattering. The three diagrams are known as the *ZS* (zero sound), ZS' and *BCS* (Bardeen-Cooper-Schrieffer) contribution

then there are no kinematic restrictions on the loop momenta. As a consequence, all back-to-back pairs can mix and there is an instability even in weak coupling.

For $\boldsymbol{p}_1 = -\boldsymbol{p}_2$ and $E_1 = E_2 = E$ the BCS diagram is given by

$$\Gamma_{\alpha\beta\gamma\delta} = C_0^2\left(\delta_{\alpha\gamma}\delta_{\beta\delta} - \delta_{\alpha\delta}\delta_{\beta\gamma}\right)\int\frac{d^4q}{(2\pi)^4}\,\frac{1}{E+q_0-\epsilon_q+i\delta\mathrm{sgn}(\epsilon_q)}\,\frac{1}{E-q_0-\epsilon_q+i\delta\mathrm{sgn}(\epsilon_q)}. \tag{4.36}$$

As $E \to 0$ the loop integral develops an infrared divergence. This divergence comes from momenta near the Fermi surface and we can approximate $d^3q \simeq p_F^2 dl$ with $l = |\boldsymbol{q}| - p_F$. The scattering amplitude is proportional to

$$\Gamma_{\alpha\beta\gamma\delta} = \left(\delta_{\alpha\gamma}\delta_{\beta\delta} - \delta_{\alpha\delta}\delta_{\beta\gamma}\right)\left\{C_0 - C_0^2\left(\frac{p_F m}{2\pi^2}\right)\log\left(\frac{E_0}{E}\right)\right\}, \tag{4.37}$$

where E_0 is an ultraviolet cutoff. The logarithmic divergence can also be seen by expanding Eq. (4.22) around $s = 0$ and $\kappa = 1$. The term in the curly brackets can be interpreted as an effective energy dependent coupling. The coupling constant satisfies the renormalization group equation [1, 22]

$$E\frac{dC_0}{dE} = C_0^2\left(\frac{p_F m}{2\pi^2}\right), \tag{4.38}$$

with the solution

$$C_0(E) = \frac{C_0(E_0)}{1 + NC_0(E_0)\log(E_0/E)}, \tag{4.39}$$

where $N = (p_F m)/(2\pi^2)$ is the density of states. Equation (4.39) shows that there are two possible scenarios. If the initial coupling is repulsive, $C_0(E_0) > 0$, then the renormalization group evolution will drive the effective coupling to zero and the Fermi liquid is stable. If, on the other hand, the initial coupling is attractive, $C_0(E_0) < 0$, then the effective coupling grows and reaches a Landau pole at

$$E_{crit} \sim E_0 \exp\left(-\frac{1}{N|C_0(E_0)|}\right). \tag{4.40}$$

At the Landau pole the Fermi liquid description has to break down. The renormalization group equation does not determine what happens at this point, but it seems natural to assume that the strong attractive interaction will lead to the formation of a fermion pair condensate. The fermion condensate $\langle \epsilon^{\alpha\beta}\psi_\alpha\psi_\beta \rangle$ signals the breakdown of the $U(1)$ symmetry and leads to a gap Δ in the single particle spectrum.

The scale of the gap is determined by the position of the Landau pole, $\Delta \sim E_{crit}$. A more quantitative estimate of the gap can be obtained in the mean field approximation. In the path integral formulation the mean field approximation is most easily introduced using the Hubbard-Stratonovich trick. For this purpose we first rewrite the four-fermion interaction as

$$\frac{C_0}{2}(\psi^\dagger\psi)^2 = \frac{C_0}{4}\{(\psi^\dagger\sigma_2\psi^\dagger)(\psi\sigma_2\psi) + (\psi^\dagger\sigma_2\boldsymbol{\sigma}\psi^\dagger)(\psi\boldsymbol{\sigma}\sigma_2\psi)\}, \tag{4.41}$$

where we have used the Fierz identity $2\delta^{\alpha\beta}\delta^{\gamma\rho} = \delta^{\alpha\rho}\delta^{\gamma\beta} + (\boldsymbol{\sigma})^{\alpha\rho}(\boldsymbol{\sigma})^{\gamma\beta}$. Note that the second term in Eq. (4.41) vanishes because $(\sigma_2\boldsymbol{\sigma})$ is a symmetric matrix. We now introduce a factor of unity into the path integral

$$1 = \frac{1}{Z_\Delta}\int D\Delta \exp\left(\frac{\Delta^*\Delta}{C_0}\right), \tag{4.42}$$

where we assume that $C_0 < 0$. We can eliminate the four-fermion term in the lagrangian by a shift in the integration variable Δ. The action is now quadratic in the fermion fields, but it involves a Majorana mass term $\psi\sigma_2\Delta\psi + h.c.$ The Majorana mass terms can be handled using the Nambu-Gorkov method. We introduce the bispinor $\Psi = (\psi, \psi^\dagger\sigma_2)$ and write the fermionic action as

$$\mathcal{S} = \frac{1}{2}\int \frac{d^4p}{(2\pi)^4}\Psi^\dagger \begin{pmatrix} p_0 - \epsilon_p & \Delta \\ \Delta^* & p_0 + \epsilon_p \end{pmatrix} \Psi. \tag{4.43}$$

Since the fermion action is quadratic we can integrate the fermion out and obtain the effective lagrangian

$$\mathcal{L} = \frac{1}{2}\mathrm{Tr}\left[\log\left(G_0^{-1}G\right)\right] + \frac{1}{C_0}|\Delta|^2, \tag{4.44}$$

where G is the fermion propagator

$$G(p) = \frac{1}{p_0^2 - \epsilon_p^2 - |\Delta|^2}\begin{pmatrix} p_0 + \epsilon_p & \Delta^* \\ \Delta & p_0 - \epsilon_p \end{pmatrix}. \tag{4.45}$$

The diagonal and off-diagonal components of $G(p)$ are sometimes referred to as normal and anomalous propagators. Note that we have not yet made any approximation. We have converted the fermionic path integral to a bosonic one, albeit with a very non-local action. The mean field approximation corresponds to evaluating the bosonic path integral using the saddle point method. Physically, this approximation means that the order parameter does not fluctuate. Formally, the mean field approximation can be justified in the large N limit, where N is the number of fermion fields. The saddle point equation for Δ gives the gap equation

$$\Delta = |C_0| \int \frac{d^4p}{(2\pi)^4} \frac{\Delta}{p_0^2 - \epsilon_p^2 - \Delta^2}. \tag{4.46}$$

Performing the p_0 integration we find

$$1 = \frac{|C_0|}{2} \int \frac{d^3p}{(2\pi)^3} \frac{1}{\sqrt{\epsilon_p^2 + \Delta^2}}. \tag{4.47}$$

Since $\epsilon_p = E_p - \mu$ the integral in Eq. (4.47) has an infrared divergence on the Fermi surface $|\boldsymbol{p}| \sim p_F$. As a result, the gap equation has a non-trivial solution even if the coupling is arbitrarily small. We can estimate the size of the gap as we did earlier by writing $d^3p \simeq p_F^2 dl$ and introducing a cutoff Λ for the integral over l. We find $\Delta = 2\Lambda \exp(-1/(N|C_0|))$. In order to obtain a more accurate result we compute the RHS of Eq. (4.47) without the approximation $d^3q \simeq p_F^2 dl$. We use dimensional regularization and

$$\int_0^\infty dz \frac{z^\alpha}{\sqrt{(z-1)^2 + x^2}} = -\frac{\pi}{\sin(\pi\alpha)} (1 + x^2)^{\alpha/2} P_\alpha\left(-\frac{1}{\sqrt{1+x^2}}\right). \tag{4.48}$$

The dimensionally regularized gap equation is [23, 24]

$$1 = \frac{\lambda\pi}{\sin(\pi\alpha)} (1 + x^2)^{\alpha/2} P_\alpha\left(-\frac{1}{\sqrt{1+x^2}}\right), \tag{4.49}$$

where $2\lambda = C_0 m p_F^{d-2} \Omega_d/(2\pi)^d$ is a dimensionless coupling constant, Ω_d is the surface area of the d-dimensional unit ball and $x = \Delta/E_F$ is the dimensionless gap. $P_\alpha(z)$ is the Legendre function of order α and $\alpha = (d-2)/2$. Dimensional regularization sets the power divergence in Eq. (4.47) to zero. As a result, we can set $d = 3$ and $C_0 = 4\pi a/m$ in Eq. (4.49). If the gap is small, $x \ll 1$, Eq. (4.49) can be solved using the asymptotic behavior of the Legendre function $P_\alpha(z)$ near the logarithmic singularity at $z = -1$,

$$P_\alpha(z) \simeq \frac{\sin(\alpha\pi)}{\pi} \left(\log\left(\frac{1+z}{2}\right) + 2\gamma + 2\psi(\alpha+1) + \pi\cot(\alpha\pi) \right). \tag{4.50}$$

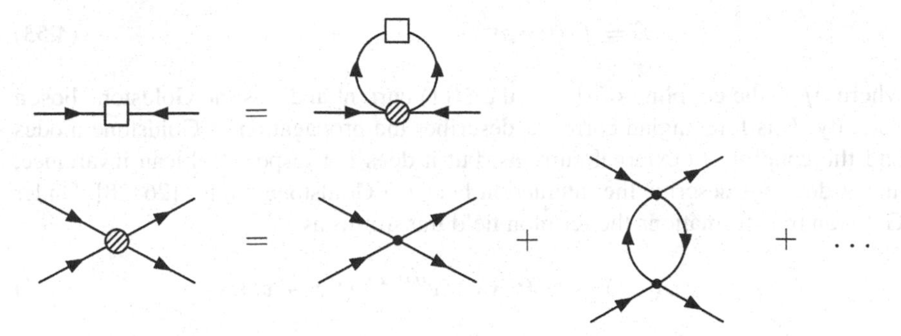

Fig. 4.4 Gap equation for the superfluid gap in a theory with short range interactions

We find

$$\Delta = \frac{8E_f}{e^2}\exp\left(-\frac{\pi}{2p_F|a|}\right). \tag{4.51}$$

The term in the exponent represents the leading term in an expansion in $p_F|a|$, see Fig. 4.4. This means that in order to determine the pre-exponent in Eq. (4.51) we have to solve the gap equation at next-to-leading order. The contribution from the second diagram in Fig. 4.4 was first computed by Gorkov and Melik-Barkhudarov [25]. The second order graph screens the leading order particle-particle scattering amplitude and suppresses the s-wave gap by a factor $(4e)^{1/3} \sim 2.2$.

For neutron matter the scattering length is large, $a = -18.8$ fm, and Eq. (4.51) is not very useful, except at very small density. At moderate density a rough estimate of the gap can be obtained by replacing $1/(p_F a)$ with $\cot(\delta(k_F))$, where $\delta(k)$ is the s-wave phase shift. This estimate gives neutron gaps on the order of 1 MeV at nuclear matter density.

4.3.2 Superfluidity

Pairing leads to important physical effects. If the fermions are charged, pairing causes superconductivity. If the fermions are neutral, pairing leads to superfluidity. We first discuss superfluidity. The superfluid order parameter $\langle\psi\psi\rangle$ breaks the $U(1)$ symmetry and leads to the appearance of a Goldstone boson. The Goldstone boson field is defined as the phase of the order parameter

$$\langle\psi\psi\rangle = |\langle\psi\psi\rangle|e^{2i\varphi}. \tag{4.52}$$

In the following we shall construct an effective lagrangian for the Goldstone field φ. The $U(1)$ symmetry implies that the lagrangian can only depend on derivatives of φ. The simplest possibility is

$$\mathcal{L} = f^2\big((\partial_0\varphi)^2 - v^2(\partial_i\varphi)^2 + \ldots\big), \tag{4.53}$$

where vf is the coupling of $\partial_i\varphi$ to the $U(1)$ current and v is the Goldstone boson velocity. This Lagrangian correctly describes the propagation of Goldstone modes and the coupling to external currents, but it does not respect Galilean invariance, and it does not describe the interaction between Goldstone modes [26–28]. Under Galilean transformations the fermion field transforms as

$$\psi(t, \boldsymbol{x}) \to \psi'(t, \boldsymbol{x}) = e^{im\boldsymbol{v}\cdot\boldsymbol{x}}\psi(t, \boldsymbol{x} - \boldsymbol{v}t). \tag{4.54}$$

This implies that φ transforms as $\varphi(t, \boldsymbol{x}) \to \varphi(t, \boldsymbol{x} - \boldsymbol{v}t) + m\boldsymbol{v}\cdot\boldsymbol{x}$. We also observe that the chemical potential enters the microscopic theory like the time component of a $U(1)$ gauge field. We can impose the constraints of Galilei invariance and $U(1)$ symmetry by constructing an effective lagrangian that only depends on the variable

$$X = \mu - \partial_0\varphi - \frac{(\partial_i\varphi)^2}{2m}. \tag{4.55}$$

In the following it will be useful to consider a low energy expansion in which $\partial_0\varphi$, $\partial_i\varphi$ are $O(1)$ but higher derivatives $\partial_i\partial_j\varphi$, etc. are suppressed. In this case the leading order lagrangian contains arbitrary powers of X, but terms with derivatives of X are suppressed. The functional form of $\mathcal{L}(X)$ can be determined using the following simple argument. For constant fields $\varphi = const$ the lagrangian $\mathcal{L}(X) = \mathcal{L}(\mu)$ is equal to minus the thermodynamic potential Ω. Since $\Omega = -P$, where P is the pressure, we conclude that $\mathcal{L}(X) = P(X)$.

As an example consider superfluidity in a weakly coupled Fermi gas. At leading order the equation of state is that of a free Fermi gas, $P = m^{3/2}(2\mu)^{5/2}/(15\pi^2)$. The effective lagrangian is given by

$$\mathcal{L} = \frac{2^{5/2}m^{3/2}}{15\pi^2}\left(\mu - \partial_0\varphi - \frac{(\partial_i\varphi)^2}{2m}\right)^{5/2}. \tag{4.56}$$

We can determine the Goldstone boson propagator as well as Goldstone boson interactions by expanding this result in powers of $\partial_0\varphi$ and $\partial_i\varphi$. There are some predictions that are independent of the equation of state. Consider the effective theory at second order in $(\partial\varphi)$,

$$\mathcal{L} = P(\mu) - n\partial_0\varphi + \frac{1}{2}\frac{\partial n}{\partial\mu}(\partial_0\varphi)^2 - \frac{n}{2m}(\partial_i\varphi)^2 + \ldots, \tag{4.57}$$

where we have used $n = (\partial P)/(\partial\mu)$. The Goldstone boson velocity is given by

$$v^2 = \frac{n}{m}\frac{\partial\mu}{\partial n} = \frac{\partial P}{\partial\rho}. \tag{4.58}$$

where $\rho = nm$ denotes the mass density. We observe that the Goldstone boson velocity is given by the same formula as the speed of sound in a normal fluid. In a weakly interacting Fermi gas $v^2 = v_F^2/3$.

It is also instructive to study the relation to fluid dynamics in more detail. The equation of motion for the field φ is given by

$$\partial_0 \bar{n} + \frac{1}{m} \nabla (\bar{n} \nabla \varphi) = 0, \tag{4.59}$$

where we have defined $\bar{n} = P'(X)$. Equation (4.59) is the continuity equation for the current $j_\mu = \bar{n}(1, \boldsymbol{v}_s)$ where we have identified the fluid velocity

$$\boldsymbol{v}_s = \frac{\nabla \varphi}{m}. \tag{4.60}$$

In the hydrodynamic description the independent variables are $\bar{n}$ and $\boldsymbol{v}_s$. We can derive a second equation by using the identity $dP = nd\mu$. We get

$$\partial_0 \boldsymbol{v}_s + \frac{1}{2} \nabla v_s^2 = -\frac{1}{m} \nabla \mu. \tag{4.61}$$

This is the Euler equation for non-viscous, irrotational fluid. The fact that the flow is irrotational follows from the definition of the velocity as the gradient of φ. We conclude that the low energy effective lagrangian is equivalent to superfluid hydrodynamics.

4.3.3 *Landau-Ginzburg Theory*

In this section we shall study the properties of a superconductor in more detail. Superconductors are characterized by the fact that the $U(1)$ symmetry is gauged. The order parameter $\Phi = \langle \epsilon^{\alpha\beta} \psi_\alpha \psi_\beta \rangle$ breaks $U(1)$ invariance. Consider a gauge transformation

$$A_\mu \to A_\mu + \partial_\mu \Lambda. \tag{4.62}$$

The order parameter transforms as

$$\Phi \to \exp(2ie\Lambda)\Phi. \tag{4.63}$$

The breaking of gauge invariance is responsible for most of the unusual properties of superconductors [29, 30]. This can be seen by constructing the low energy effective action of a superconductor. For this purpose we write the order parameter in terms of its modulus and phase

$$\Phi(x) = \exp(2ie\phi(x))\tilde{\Phi}(x). \tag{4.64}$$

The field ϕ corresponds to the Goldstone mode. Under a gauge transformation $\phi(x) \to \phi(x) + \Lambda(x)$. Gauge invariance restricts the form of the effective Lagrange function as

$$L = -\frac{1}{4}\int d^3x\, F_{\mu\nu}F_{\mu\nu} + L_s(A_\mu - \partial_\mu\phi). \tag{4.65}$$

There is a large amount of information we can extract even without knowing the explicit form of L_s. Stability implies that $A_\mu = \partial_\mu\phi$ corresponds to a minimum of the energy. This means that up to boundary effects the gauge potential is a total divergence and that the magnetic field has to vanish. This phenomenon is known as the Meissner effect.

Equation (4.65) also implies that a superconductor has zero resistance. The equations of motion relate the time dependence of the Goldstone boson field to the potential,

$$\dot{\phi}(x) = -V(x). \tag{4.66}$$

The electric current is related to the gradient of the Goldstone boson field. Equation (4.66) shows that the time dependence of the current is proportional to the gradient of the potential. In order to obtain a static current the gradient of the potential has to vanish throughout the sample, and the resistance is zero.

In order to study the properties of a superconductor in more detail we have to specify L_s. For this purpose we assume that the system is time-independent, that the spatial gradients are small, and that the order parameter is small. In this case we can write

$$L_s = \int d^3x \left\{ -\frac{1}{2}\left|(\nabla - 2ie\mathbf{A})\,\Phi\right|^2 + \frac{1}{2}m_H^2\left(\Phi^*\Phi\right)^2 - \frac{1}{4}g\left(\Phi^*\Phi\right)^4 + \ldots \right\}, \tag{4.67}$$

where m_H and g are unknown parameters that depend on the temperature. Equation (4.67) is known as the Landau-Ginzburg effective action. Strictly speaking, the assumption that the order parameter is small can only be justified in the vicinity of a second order phase transition. Nevertheless, the Landau-Ginzburg description is instructive even in the regime where $t = (T - T_c)/T_c$ is not small. It is useful to decompose $\Phi = \rho\exp(2ie\phi)$. For constant fields the effective potential,

$$V(\rho) = -\frac{1}{2}m_H^2\rho^2 + \frac{1}{4}g\rho^4, \tag{4.68}$$

is independent of ϕ. The minimum is at $\rho_0^2 = m_H^2/g$ and the energy density at the minimum is given by $\mathcal{E} = -m_H^4/(4g)$. This shows that the two parameters m_H and g can be related to the expectation value of Φ and the condensation energy. We also observe that the phase transition is characterized by $m_H(T_c) = 0$.

In terms of ϕ and ρ the Landau-Ginzburg action is given by

$$L_s = \int d^3x \left\{ -2e^2\rho^2 (\nabla\phi - \boldsymbol{A})^2 + \frac{1}{2} m_H^2 \rho^2 - \frac{1}{4} g\rho^4 - \frac{1}{2} (\nabla\rho)^2 \right\}. \quad (4.69)$$

The equations of motion for $\boldsymbol{A}$ and ρ are given by

$$\nabla \times \boldsymbol{B} = 4e^2\rho^2 (\nabla\phi - \boldsymbol{A}), \quad (4.70)$$
$$\nabla^2\rho = -m_H^2\rho^2 + g\rho^3 + 4e^2\rho(\nabla\phi - \boldsymbol{A})^2. \quad (4.71)$$

Equation (4.70) implies that $\nabla^2\boldsymbol{B} = -4e^2\rho^2\boldsymbol{B}$. This means that an external magnetic field $\boldsymbol{B}$ decays over a characteristic distance $\lambda = 1/(2e\rho)$. Equation (4.71) gives $\nabla^2\rho = -m_H^2\rho + \ldots$. As a consequence, variations in the order parameter relax over a length scale given by $\xi = 1/m_H$. The two parameters λ and ξ are known as the penetration depth and the coherence length.

The relative size of λ and ξ has important consequences for the properties of superconductors. In a type II superconductor $\xi < \lambda$. In this case magnetic flux can penetrate the system in the form of vortex lines. At the core of a vortex the order parameter vanishes, $\rho = 0$. In a type II material the core is much smaller than the region over which the magnetic field goes to zero. The magnetic flux is given by

$$\int_A \boldsymbol{B} \cdot d\boldsymbol{S} = \oint_{\partial A} \boldsymbol{A} \cdot d\boldsymbol{l} = \oint_{\partial A} \nabla\phi \cdot d\boldsymbol{l} = \frac{n\pi\hbar}{e}, \quad (4.72)$$

and quantized in units of $\pi\hbar/e$. In a type II superconductor magnetic vortices repel each other and form a regular lattice known as the Abrikosov lattice. In a type I material, on the other hand, vortices are not stable and magnetic fields can only penetrate the sample if superconductivity is destroyed.

4.3.4 Microscopic Calculation of the Screening Mass

In this section we shall study screening of gauge fields in a superconductor from a more microscopic point of view. The calculation is analogous to the one discussed in Sect. 4.2.4. The difference is that the propagators contain the gap, and that there is an extra trace over Nambu-Gorkov indices, see Fig. 4.5. The polarization functions contains normal contributions proportional to $G_{11}G_{11}$ and $G_{22}G_{22}$ as well as anomalous terms proportional to $G_{12}G_{21}$, where G_{ij} is the Nambu-Gorkov propagator give in Eq. (4.45). The sum of the normal and anomalous diagrams is given by

$$\Pi_{00}(q=0) = -ie^2 \int \frac{d^4p}{(2\pi)^4} \left\{ \frac{p_0^2 + \epsilon_p^2}{(p_0^2 - \epsilon_p^2 - \Delta^2)^2} - \frac{\Delta^2}{(p_0^2 - \epsilon_p^2 - \Delta^2)^2} \right\}. \quad (4.73)$$

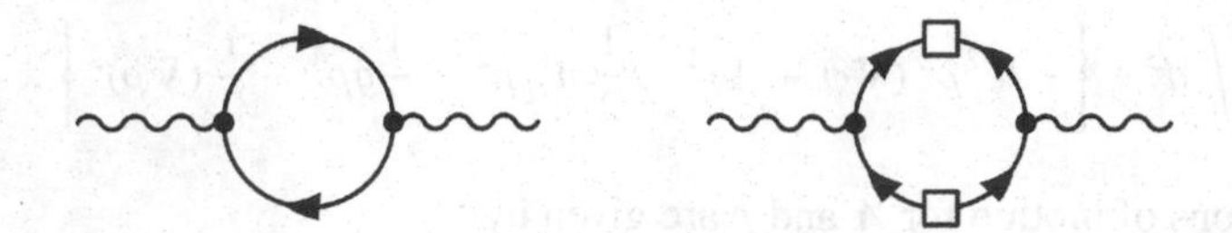

Fig. 4.5 Leading order Feynman diagrams that contribute to the photon polarization function in a superconducting Fermi gas. The figure does not show the tadpole diagram

The integral over p_0 can be done by contour integration. The two terms in Eq. (4.73) give equal contributions. We find

$$\Pi_{00}(q=0) = e^2 \int \frac{d^3p}{(2\pi)^3} \frac{\Delta^2}{(\epsilon_p^2+\Delta^2)^{3/2}}. \tag{4.74}$$

This integral is dominated by very small energies $|\epsilon_p| = |E_p - \mu| \sim \Delta$ and we can approximate $\epsilon_p = v_F(p - p_F)$. We find

$$\Pi_{00}(q=0) = e^2 \frac{p_F m}{2\pi^2}, \tag{4.75}$$

which is identical to the result in the normal phase. There are a number of subtleties that are worth commenting on. First we note that the polarization function in the superfluid phase is analytic in the external momenta and we can set $q_0 = \boldsymbol{q} = 0$ from the beginning. We also note that the normal contribution is formally ultraviolet divergent. The correct prescription to deal with this divergence is to perform the p_0 integral first [4]. Finally we observe that while the screening masses in the normal and superfluid phase are the same, only half of the result in the superfluid phase is contributed by the normal term.

The calculation of the electric polarization function is easily generalized to the magnetic case. There are three diagrams. The first is the tadpole contribution discussed in Sect. 4.2.4. This contribution is proportional to the total density and is the same in the normal and superfluid phase. The normal and anomalous one-loop diagrams are similar to the electric case, but the coupling e^2 is replaced by $e^2 v_i v_j$ in the normal contribution and $e^2 v_i(-v_j)$ in the anomalous term. As a result the two terms cancel and the polarization function is given by the tadpole term

$$\Pi_{ij}(q=0) = -e^2 v_F^2 \delta_{ij} \frac{p_F m}{6\pi^2}. \tag{4.76}$$

We find that there is a non-zero magnetic screening mass in the superfluid phase, and that the Meissner mass is controlled not by the gap, but by the density of states on the Fermi surface. This does not contradict the fact that the magnetic screening mass goes to zero as $\Delta \to 0$. We find that the photon mass term has the structure $m_D^2(A_0^2 - v_F^2 \mathbf{A}^2/3)$. This result can also be obtained by gauging the effective Lagrangian for the Goldstone boson, Eq. (4.53), together with the result $v^2 = v_F^2/3$ for the speed of sound in a weakly interacting Fermi gas.

4.4 Strongly Interacting Fermions

Up to this point we have concentrated on weakly coupled many body systems. In this section we shall consider a cold, dilute gas of fermionic atoms in which the scattering length a of the atoms can be changed continuously. This system can be realized experimentally using Feshbach resonances, see [31] for a review. A small negative scattering length corresponds to a weak attractive interaction between the atoms. This case is known as the BCS limit. As the strength of the interaction increases the scattering length becomes larger. It diverges at the point where a bound state is formed. The point $a = \infty$ is called the unitarity limit, since the scattering cross section saturates the s-wave unitarity bound $\sigma = 4\pi/k^2$. On the other side of the resonance the scattering length is positive. In the BEC limit the interaction is strongly attractive and the fermions form deeply bound molecules.

A dilute gas of fermions in the unitarity limit is a strongly coupled quantum liquid that exhibits many interesting properties. One interesting feature is universality. We are interested in the limit $(k_F a) \to \infty$ and $(k_F r) \to 0$, where k_F is the Fermi momentum, a is the scattering length and r is the effective range. From dimensional analysis it is clear that the energy per particle at zero temperature has to be proportional to energy per particle of a free Fermi gas at the same density

$$\frac{E}{A} = \xi \left(\frac{E}{A}\right)_0 = \xi \frac{3}{5}\left(\frac{k_F^2}{2m}\right). \tag{4.77}$$

The constant ξ is universal, i.e. independent of the details of the system. Similar universal constants govern the magnitude of the gap in units of the Fermi energy and the equation of state at finite temperature.

Universal behavior in the unitarity limit is relevant to the physics of dilute neutron matter. The neutron-neutron scattering length is $a_{nn} = -18$ fm and the effective range is $r_{nn} = 2.8$ fm. This means that there is a range of densities for which the inter-particle spacing is large compared to the effective range but small compared to the scattering length. It is interesting to note that the neutron scattering length depends on the quark masses in a way that is very similar to the dependence of atomic scattering lengths on the magnetic field near a Feshbach resonance [32], see Fig. 4.6.

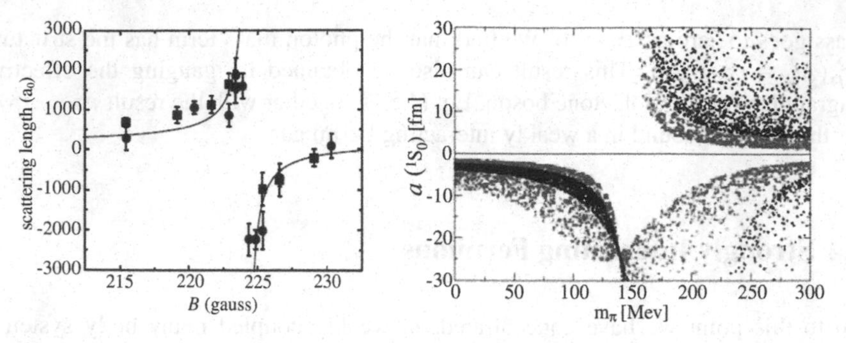

Fig. 4.6 The *left panel* shows the scattering length of ^{40}K Atoms as a function of the magnetic field near a Feshbach resonance, from Regal [31]. The *right panel* shows the nucleon-nucleon scattering length in the $^1S_0$ channel as a function of the pion mass. The scatter plot indicates the uncertainty due to higher order terms in the chiral effective lagrangian. Figure from Beane and Savage [32]

4.4.1 Numerical Calculations

The calculation of the dimensionless quantity ξ is a non-perturbative problem. In this section we shall tackle this problem using a combination of effective field theory and lattice field theory methods. We will study an analytical approach in the next section. We first observe that in the low density limit the details of the interaction are not important. The physics of the unitarity limit is captured by an effective lagrangian of point-like fermions interacting via a short-range interaction. The lagrangian is

$$\mathcal{L} = \psi^\dagger \left(i\partial_0 + \frac{\nabla^2}{2m} \right) \psi - \frac{C_0}{2} \left(\psi^\dagger \psi \right)^2 , \tag{4.78}$$

as in Eq. (4.1). The usual strategy for dealing with the four-fermion interaction is to use a Hubbard-Stratonovich transformation as in Sect. 4.3.1. The partition function can be written as [33]

$$Z = \int Ds\, Dc\, Dc^* \exp\left[-S\right], \tag{4.79}$$

where s is the Hubbard-Stratonovich field and c is a Grassmann field. S is a discretized euclidean action

$$\begin{aligned} S = \sum_{\boldsymbol{n},i} &\left[e^{-\hat{\mu}\alpha_t} c_i^*(\boldsymbol{n}) c_i(\boldsymbol{n}+\hat{0}) - e^{\sqrt{-C_0\alpha_t}s(\boldsymbol{n}) + \frac{C_0\alpha_t}{2}} (1-6h) c_i^*(\boldsymbol{n}) c_i(\boldsymbol{n}) \right] \\ &- h \sum_{\boldsymbol{n},l_s,i} \left[c_i^*(\boldsymbol{n}) c_i(\boldsymbol{n}+\hat{l}_s) + c_i^*(\boldsymbol{n}) c_i(\boldsymbol{n}-\hat{l}_s) \right] + \frac{1}{2} \sum_{\boldsymbol{n}} s^2(\boldsymbol{n}). \end{aligned} \tag{4.80}$$

Here i labels spin and $\boldsymbol{n}$ labels lattice sites. Spatial and temporal unit vectors are denoted by $\hat{l}_s$ and $\hat{0}$, respectively. The temporal and spatial lattice spacings are

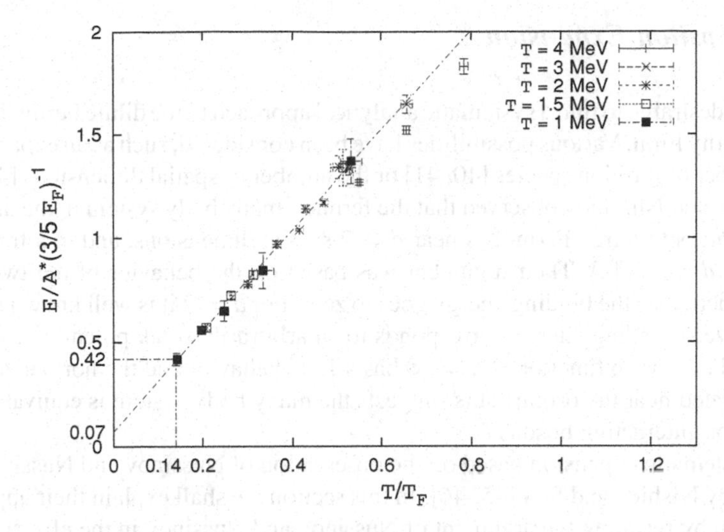

Fig. 4.7 Lattice results for the energy per particle of a dilute Fermi gas from Lee and Schäfer [33]. We show the energy per particle in units of $3E_F/5$ as a function of temperature in units of T_F

b_τ and b. The dimensionless chemical potential is given by $\hat{\mu} = \mu b_\tau$. We define α_t as the ratio of the temporal and spatial lattice spacings and $h = \alpha_t/(2\hat{m})$. Note that for $C_0 < 0$ the action is real and standard Monte Carlo simulations are possible.

The four-fermion coupling is fixed by computing the sum of all particle-particle bubbles as in Sect. 4.2.1 but with the elementary loop function regularized on the lattice. Schematically,

$$\frac{m}{4\pi a} = \frac{1}{C_0} + \frac{1}{2}\sum_p \frac{1}{E_p}, \tag{4.81}$$

where the sum runs over discrete momenta on the lattice and E_p is the lattice dispersion relation. A detailed discussion of the lattice regularized scattering amplitude can be found in [33–35]. For a given scattering length a the four-fermion coupling is a function of the lattice spacing. The continuum limit correspond to taking the temporal and spatial lattice spacings b_τ, b to zero

$$b_\tau \mu \to 0, \qquad bn^{1/3} \to 0, \tag{4.82}$$

keeping $an^{1/3}$ fixed. Here, μ is the chemical potential and n is the density. Numerical results in the unitarity limit are shown in Fig. 4.7. From these simulations we concluded that $\xi = (0.09–0.42)$. Lee performed canonical simulations at $T = 0$ and obtained [36] $\xi = 0.25$. Green Function Monte Carlo calculations give [37] $\xi = 0.44$, and finite temperature lattice simulations have been extrapolated to $T = 0$ to yield similar results [38, 39].

4.4.2 Epsilon Expansion

It is also desirable to find a systematic analytical approach to the dilute Fermi liquid in the unitarity limit. Various possibilities have been considered, such as an expansion in the number of fermion species [40, 41] or the number of spatial dimensions [42, 43]. Nussinov and Nussinov observed that the fermion many body system in the unitarity limit reduces to a free Fermi gas near $d = 2$ spatial dimensions, and to a free Bose gas near $d = 4$ [44]. Their argument was based on the behavior of the two-body wave function as the binding energy goes to zero. For $d = 2$ it is well known that the limit of zero binding energy corresponds to an arbitrarily weak potential. In $d = 4$ the two-body wave function at $a = \infty$ has a $1/r^2$ behavior and the normalization is concentrated near the origin. This suggests the many body system is equivalent to a gas of non-interacting bosons.

A systematic expansion based on the observation of Nussinov and Nussinov was studied by Nishida and Son [45, 46]. In this section we shall explain their approach. We begin by restating the argument of Nussinov and Nussinov in the effective field theory language. In dimensional regularization $a \to \infty$ corresponds to $C_0 \to \infty$. The fermion-fermion scattering amplitude (see Eq. 4.7) is given by

$$\mathcal{A}(p_0, \boldsymbol{p}) = \left(\frac{4\pi}{m}\right)^{d/2}\left[\Gamma\left(1-\frac{d}{2}\right)\right]^{-1}\frac{i}{\left(-p_0 + E_p/2 - i\delta\right)^{\frac{d}{2}-1}}, \tag{4.83}$$

where $\delta \to 0+$. As a function of d the Gamma function has poles at $d = 2, 4, \ldots$ and the scattering amplitude vanishes at these points. Near $d = 2$ the scattering amplitude is energy and momentum independent. For $d = 4 - \epsilon$ we find

$$\mathcal{A}(p_0, \boldsymbol{p}) = \frac{8\pi^2\epsilon}{m^2}\frac{i}{p_0 - E_p/2 + i\delta} + O(\epsilon^2). \tag{4.84}$$

We observe that at leading order in ϵ the scattering amplitude looks like the propagator of a boson with mass $2m$. The boson-fermion coupling is $g^2 = (8\pi^2\epsilon)/m^2$ and vanishes as $\epsilon \to 0$. This suggests that we can set up a perturbative expansion involving fermions of mass m weakly coupled to bosons of mass $2m$. In the unitarity limit the Hubbard-Stratonovich transformed lagrangian reads

$$\mathcal{L} = \Psi^\dagger\left[i\partial_0 + \sigma_3\frac{\nabla^2}{2m}\right]\Psi + \mu\Psi^\dagger\sigma_3\Psi + \left(\Psi^\dagger\sigma_+\Psi\phi + h.c.\right), \tag{4.85}$$

where $\Psi = (\psi_\uparrow, \psi_\downarrow^\dagger)^T$ is a two-component Nambu-Gorkov field, σ_i are Pauli matrices acting in the Nambu-Gorkov space and $\sigma_\pm = (\sigma_1 \pm i\sigma_2)/2$. In the superfluid phase ϕ acquires an expectation value. We write

$$\phi = \phi_0 + g\varphi, \qquad g = \frac{\sqrt{8\pi^2\epsilon}}{m}\left(\frac{m\phi_0}{2\pi}\right)^{\epsilon/4}, \tag{4.86}$$

where $\phi_0 = \langle\phi\rangle$. The scale $M^2 = m\phi_0/(2\pi)$ was introduced in order to have a correctly normalized boson field. The scale parameter is arbitrary, but this particular choice simplifies some of the loop integrals. In order to get a well defined perturbative expansion we add and subtract a kinetic term for the boson field to the lagrangian. We include the kinetic term in the free part of the lagrangian

$$\mathcal{L}_0 = \Psi^\dagger\left[i\partial_0 + \sigma_3\frac{\nabla^2}{2m} + \phi_0(\sigma_+ + \sigma_-)\right]\Psi + \varphi^\dagger\left(i\partial_0 + \frac{\nabla^2}{4m}\right)\varphi. \tag{4.87}$$

The interacting part is

$$\mathcal{L}_I = g\left(\Psi^\dagger\sigma_+\Psi\varphi + h.c\right) + \mu\Psi^\dagger\sigma_3\Psi - \varphi^\dagger\left(i\partial_0 + \frac{\nabla^2}{4m}\right)\varphi. \tag{4.88}$$

Note that the interacting part generates self energy corrections to the boson propagator which, by virtue of Eq. (4.84), cancel against the kinetic term of boson field. We have also included the chemical potential term in $\mathcal{L}_I$. This is motivated by the fact that near $d = 4$ the system reduces to a non-interacting Bose gas and $\mu \to 0$. We will count μ as a quantity of $O(\epsilon)$.

The Feynman rules are quite simple. The fermion and boson propagators are

$$G(p_0, \boldsymbol{p}) = \frac{i}{p_0^2 - E_p^2 - \phi_0^2}\begin{bmatrix} p_0 + E_p & -\phi_0 \\ -\phi_0 & p_0 - E_p \end{bmatrix}, \tag{4.89}$$

$$D(p_0, \boldsymbol{p}) = \frac{i}{p_0 - E_p/2}, \tag{4.90}$$

and the fermion-boson vertices are $ig\sigma^{\pm}$. Insertions of the chemical potential are $i\mu\sigma_3$. Both g^2 and μ are corrections of order ϵ. In order to verify that the ϵ expansion is well defined we have to check that higher order diagrams do not generate powers of $1/\epsilon$. Studying the superficial degree of divergence of diagrams involving the propagators given in Eq. (4.89) one can show that there is only a finite number of one-loop diagrams that generate $1/\epsilon$ terms.

The leading order diagrams that contribute to the effective potential are shown in Fig. 4.8. The first diagram is the free fermion loop which is $O(1)$. The second diagram is the μ insertion which is $O(1)$ because the loop diagram is divergent in $d = 4$. The two-loop diagram is $O(\epsilon)$ because of the factor of g^2 from the vertices. The free fermion loop diagram is

$$V_0 = i\int\frac{dp_0}{2\pi}\int\frac{d^dp}{(2\pi)^d}\log\left[p_0^2 - E_p^2 - \phi_0^2\right] = -\int\frac{d^dp}{(2\pi)^d}\sqrt{E_p^2 + \phi_0^2}. \tag{4.91}$$

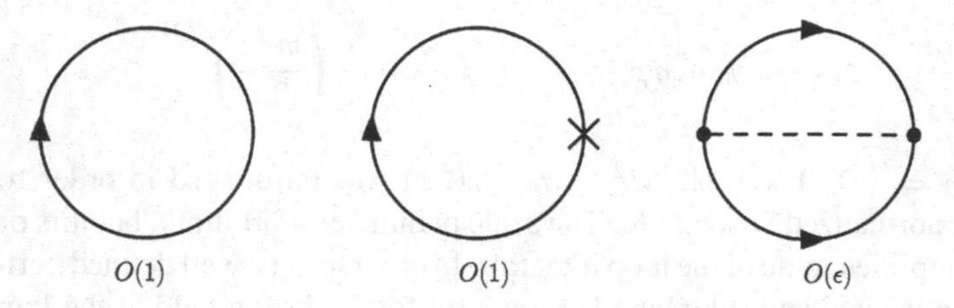

Fig. 4.8 Leading order contributions to the effective potential in the ϵ expansion. *Solid lines* are fermion propagators, *dashed lines* are boson propagators, and the *cross* is an insertion of the chemical potential

The integral can be computed analytically. Expanding to first order in $\epsilon = 4 - d$ we get

$$V_0 = \frac{\phi_0}{3}\left[1 + \frac{7 - 3(\gamma + \log(2))}{6}\,\epsilon\right]\left(\frac{m\phi_0}{2\pi}\right)^{d/2}. \tag{4.92}$$

The μ insertion is given by

$$V_1 = \mu \int \frac{d^d p}{(2\pi)^d}\frac{E_p}{\sqrt{E_p^2 + \phi_0^2}}. \tag{4.93}$$

Again, the integral can be computed analytically. The result is

$$V_1 = -\frac{\mu}{\epsilon}\left[1 + \frac{1 - 2(\gamma - \log(2))}{4}\,\epsilon\right]\left(\frac{m\phi_0}{2\pi}\right)^{d/2} \tag{4.94}$$

Nishida and Son also computed the two-loop contribution shown in Fig. 4.8. The result is

$$V_2 = -C\epsilon\left(\frac{m\phi_0}{2\pi}\right)^{d/2}, \tag{4.95}$$

where $C \simeq 0.14424$. We can now determine the minimum of the effective potential. We find

$$\phi_0 = \frac{2\mu}{\epsilon}\left[1 + (3C - 1 + \log(2))\,\epsilon + O(\epsilon^2)\right]. \tag{4.96}$$

The value of $V = V_0 + V_1 + V_2$ at ϕ_0 determines the pressure and $n = \partial P/\partial\mu$ gives the density. We find

$$n = \frac{1}{\epsilon}\left[1 - \frac{1}{4}\,(2\gamma - 1 - \log(2)) + O(\epsilon^2)\right]\left(\frac{m\phi_0}{2\pi}\right)^{d/2}. \tag{4.97}$$

For comparison, the density of a free Fermi gas in d dimensions is

$$n = \frac{2}{(4\pi)^{d/2}} \frac{k_F^d}{\Gamma\left(1+\frac{d}{2}\right)}. \qquad (4.98)$$

This equation determines the relation between $\epsilon_F \equiv k_F^2/(2m)$ and the density. We get

$$\epsilon_F = \frac{2\pi}{m}\left[\frac{n}{2}\Gamma\left(\frac{d}{2}+1\right)\right]^{2/d}. \qquad (4.99)$$

We determine ϵ_F for the interacting gas by inserting n from Eq. (4.97) into Eq. (4.99). The universal parameter is $\xi = \mu/\epsilon_F$. We find

$$\xi = \frac{1}{2}\epsilon^{3/2} + \frac{1}{16}\epsilon^{5/2}\log(\epsilon) - 0.025\epsilon^{5/2} + \ldots = 0.475 \quad (\epsilon = 1), \qquad (4.100)$$

which agrees quite well with the result of fixed node quantum Monte Carlo calculations. The calculation has been extended to $O(\epsilon^{7/2})$ by Arnold et al. [47]. Unfortunately, the next term is very large and it appears necessary to combine the expansion in $4-\epsilon$ dimensions with a $2+\epsilon$ expansion in order to extract useful results. The ϵ expansion has also been applied to the calculation of the gap [45], the critical temperature [48] and the critical chemical potential imbalance [46, 49].

4.5 QCD and its Symmetries

4.5.1 Introduction

Before we discuss QCD at large baryon density we would like to provide a quick review of QCD and the symmetries of QCD. The elementary degrees of freedom are quark fields $\psi^a_{\alpha,f}$ and gluons A^a_μ. Here, a is color index that transforms in the fundamental representation for fermions and in the adjoint representation for gluons. Also, f labels the quark flavors u, d, s, c, b, t. In practice, we will focus on the three light flavors up, down and strange. The QCD lagrangian is

$$\mathcal{L} = \sum_f^{N_f} \bar\psi_f(i\not{D} - m_f)\psi_f - \frac{1}{4}G^a_{\mu\nu}G^a_{\mu\nu}, \qquad (4.101)$$

where the field strength tensor is defined by

$$G^a_{\mu\nu} = \partial_\mu A^a_\nu - \partial_\nu A^a_\mu + g f^{abc} A^b_\mu A^c_\nu, \qquad (4.102)$$

and the covariant derivative acting on quark fields is

$$i\not{D}\psi = \gamma^{\mu}\left(i\partial_{\mu} + gA_{\mu}^{a}\frac{\lambda^{a}}{2}\right)\psi. \tag{4.103}$$

QCD has a number of remarkable properties. Most remarkably, even though QCD accounts for the rich phenomenology of hadronic and nuclear physics, it is an essentially parameter free theory. To first approximation, the masses of the light quarks u, d, s are too small to be important, while the masses of the heavy quarks c, b, t are too heavy. If we set the masses of the light quarks to zero and take the masses of the heavy quarks to be infinite then the only parameter in the QCD lagrangian is the coupling constant, g. Once quantum corrections are taken into account g becomes a function of the scale at which it is measured. If the scale is large then the coupling is small, but in the infrared the coupling becomes large. This is the famous phenomenon of asymptotic freedom. Since the coupling depends on the scale the dimensionless parameter g is traded for a dimensionful scale parameter Λ_{QCD}. Since Λ_{QCD} is the only dimensionful quantity in QCD with massless fermions it is not really a parameter of QCD, but reflects our choice of units. In standard units, $\Lambda_{QCD} \simeq 200\,\text{MeV} \simeq 1\,\text{fm}^{-1}$.

Another important feature of the QCD lagrangian are its symmetries. First of all, the lagrangian is invariant under local gauge transformations $U(x) \in SU(3)_c$

$$\psi(x) \to U(x)\psi(x), \qquad A_{\mu}(x) \to U(x)A_{\mu}U^{\dagger}(x) + iU(x)\partial_{\mu}U^{\dagger}(x), \tag{4.104}$$

where $A_{\mu} = A_{\mu}^{a}(\lambda^{a}/2)$. In the QCD ground state at zero temperature and density the local color symmetry is confined. This implies that all excitations are singlets under the gauge group.

The dynamics of QCD is completely independent of flavor. This implies that if the masses of the quarks are equal, $m_u = m_d = m_s$, then the theory is invariant under arbitrary flavor rotations of the quark fields

$$\psi_f \to V_{fg}\psi_g, \tag{4.105}$$

where $V \in SU(3)$. This is the well known flavor (isospin) symmetry of the strong interactions. If the quark masses are not just equal, but equal to zero, then the flavor symmetry is enlarged. This can be seen by defining left and right-handed fields

$$\psi_{L,R} = \frac{1}{2}(1 \pm \gamma_5)\psi. \tag{4.106}$$

In terms of L/R fields the fermionic lagrangian is

$$\mathcal{L} = \bar{\psi}_L(i\not{D})\psi_L + \bar{\psi}_R(i\not{D})\psi_R + \bar{\psi}_L M\psi_R + \bar{\psi}_R M\psi_L, \tag{4.107}$$

where $M = \text{diag}(m_u, m_d, m_s)$. We observe that if quarks are massless, $m_u = m_d = m_s = 0$, then there is no coupling between left and right handed fields. As a consequence, the lagrangian is invariant under independent flavor transformations of the left and right handed fields.

$$\psi_{L,f} \to L_{fg}\psi_{L,g}, \qquad \psi_{R,f} \to R_{fg}\psi_{R,g}, \tag{4.108}$$

where $(L, R) \in SU(3)_L \times SU(3)_R$. In the real world, of course, the masses of the up, down and strange quarks are not zero. Nevertheless, since $m_u, m_d \ll m_s < \Lambda_{QCD}$ QCD has an approximate chiral symmetry.

In the QCD ground state at zero temperature and density the flavor symmetry is realized, but the chiral symmetry is spontaneously broken by a quark-anti-quark condensate $\langle \bar{\psi}_L \psi_R + \bar{\psi}_R \psi_L \rangle$. As a result, the observed hadrons can be approximately assigned to representations of the $SU(3)_V$ flavor group, but not to representations of $SU(3)_L \times SU(3)_R$. Nevertheless, chiral symmetry has important implications for the dynamics of QCD at low energy. Goldstone's theorem implies that the breaking of $SU(3)_L \times SU(3)_R \to SU(3)_V$ is associated with the appearance of an octet of (approximately) massless pseudoscalar Goldstone bosons. Chiral symmetry places important restrictions on the interaction of the Goldstone bosons. These constraints are obtained most easily from the low energy effective chiral lagrangian. At leading order we have

$$\mathcal{L} = \frac{f_\pi^2}{4} \text{Tr}\left[\partial_\mu \Sigma \partial^\mu \Sigma^\dagger\right] + \left[B\text{Tr}(M\Sigma^\dagger) + h.c.\right] + \ldots, \tag{4.109}$$

where $\Sigma = \exp(i\phi^a \lambda^a / f_\pi)$ is the chiral field, f_π is the pion decay constant and M is the mass matrix. Expanding Σ in powers of the pion, kaon and eta fields ϕ^a we can derive the leading order chiral perturbation theory results for Goldstone boson scattering and the coupling of Goldstone bosons to external fields. Higher order corrections originate from loops and higher order terms in the effective lagrangian.

Finally, we observe that the QCD lagrangian has two $U(1)$ symmetries,

$$U(1)_B: \qquad \psi_L \to e^{i\phi}\psi_L, \qquad \psi_R \to e^{i\phi}\psi_R \tag{4.110}$$

$$U(1)_A: \qquad \psi_L \to e^{i\alpha}\psi_L, \qquad \psi_R \to e^{-i\alpha}\psi_R. \tag{4.111}$$

The $U(1)_B$ symmetry is exact even if the quarks are not massless. Superficially, it appears that the $U(1)_A$ symmetry is explicitly broken by the quark masses and spontaneously broken by the quark condensate. However, there is no Goldstone boson associated with spontaneous $U(1)_A$ breaking. The reason is that at the quantum level the $U(1)_A$ symmetry is broken by an anomaly. The divergence of the $U(1)_A$ current is given by

$$\partial^\mu j^5_\mu = \frac{N_f g^2}{16\pi^2} G^a_{\mu\nu} \tilde{G}^a_{\mu\nu}, \tag{4.112}$$

where $\tilde{G}^a_{\mu\nu} = \epsilon_{\mu\nu\alpha\beta} G^a_{\alpha\beta}/2$ is the dual field strength tensor.

4.5.2 QCD at Finite Density

In the real world the quark masses are not equal and the only exact global symmetries of QCD are the $U(1)_f$ flavor symmetries associated with the conservation of the number of up, down, and strange quarks. If we take into account the weak interactions then flavor is no longer conserved and the only exact symmetries are the $U(1)_B$ of baryon number and the $U(1)_Q$ of electric charge.

In the following we study hadronic matter at non-zero baryon density. We will mostly focus on systems at non-zero baryon chemical potential but zero electron $U(1)_Q$ chemical potential. We should note that in the context of neutron stars we are interested in situations when the electric charge, but not necessarily the electron chemical potential, is zero. Also, if the system is in equilibrium with respect to strong, but not to weak interactions, then non-zero flavor chemical potentials may come into play.

The partition function of QCD at non-zero baryon chemical potential is given by

$$Z = \sum_i \exp\left(-\frac{E_i - \mu N_i}{T}\right), \tag{4.113}$$

where i labels all quantum states of the system, E_i and N_i are the energy and baryon number of the state i. If the temperature and chemical potential are both zero then only the ground state contributes to the partition function. All other states give contributions that are exponentially small if the volume of the system is taken to infinity. In QCD there is a mass gap for states that carry baryon number. As a consequence there is an onset chemical potential

$$\mu_c = \min_i (E_i/N_i), \tag{4.114}$$

such that the partition function is independent of μ for $\mu < \mu_c$. For $\mu > \mu_c$ the baryon density is non-zero. If the chemical potential is just above the onset chemical potential we can describe QCD, to first approximation, as a dilute gas of non-interacting nucleons. In this approximation $\mu_c = m_N$. Of course, the interaction between nucleons cannot be neglected. Without it, we would not have stable nuclei. As a consequence, nuclear matter is self-bound and the energy per baryon in the ground state is given by

$$\frac{E_N}{N} - m_N \simeq -15\,\text{MeV}. \tag{4.115}$$

The onset transition is a first order transition at which the baryon density jumps from zero to nuclear matter saturation density, $\rho_0 \simeq 0.14\,\text{fm}^{-3}$. The first order transition continues into the finite temperature plane and ends at a critical endpoint at $T = T_c \simeq 10$ MeV, see Fig. 4.9.

Nuclear matter is a complicated many-body system and, unlike the situation at zero density and finite temperature, there is little information from numerical simulations

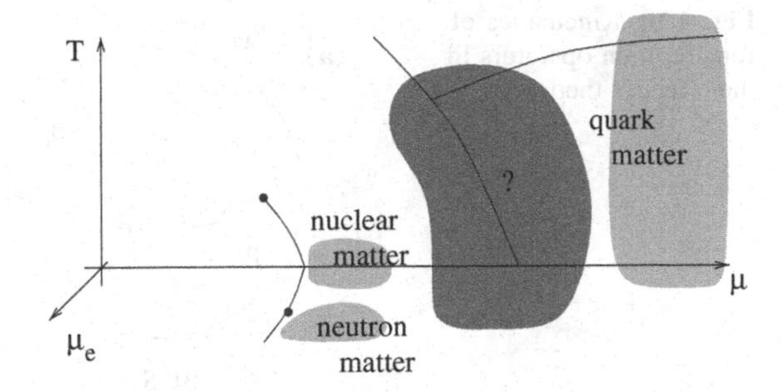

Fig. 4.9 Schematic phase diagram of hadronic matter as a function of the baryon and electron chemical potentials and temperature

on the lattice. This is related to the so-called 'sign problem'. At non-zero chemical potential the euclidean fermion determinant is complex and standard Monte-Carlo techniques based on importance sampling fail. Recently, some progress has been made in simulating QCD for small μ and $T \simeq T_c$ [50–52], but the regime of small temperature remains inaccessible.

In neutron stars there is a non-zero electron chemical potential and matter is neutron rich. Pure neutron matter has positive pressure and is stable at arbitrarily low density. As we emphasized in Sect. 4.4 dilute neutron matter has universal properties that can be explored using atomic systems. As the density increases three and four-body interactions as well as short range forces become more important and effective field theory methods are no longer applicable.

If the density is much larger than nuclear matter saturation density, $\rho \gg \rho_0$, we expect the problem to simplify. In this regime it is natural to use a system of non-interacting quarks as a starting point [53]. The low energy degrees of freedom are quark excitations and holes in the vicinity of the Fermi surface. Since the Fermi momentum is large, asymptotic freedom implies that the interaction between quasi-particles is weak. We shall see that this does not imply that the phase diagram is simple, but it does imply that the phase structure can be studied in a systematic fashion.

4.6 Effective Field Theory Near the Fermi Surface

4.6.1 High Density Effective Theory

The QCD Lagrangian in the presence of a chemical potential is given by

$$\mathcal{L} = \bar{\psi}\left(i\not{D} + \mu\gamma_0 - M\right)\psi - \frac{1}{4}G^a_{\mu\nu}G^a_{\mu\nu}, \tag{4.116}$$

where $D_\mu = \partial_\mu + igA_\mu$ is the covariant derivative, M is the mass matrix and μ is the baryon chemical potential. If the baryon chemical potential is large, $\mu \gg \Lambda_{QCD}$,

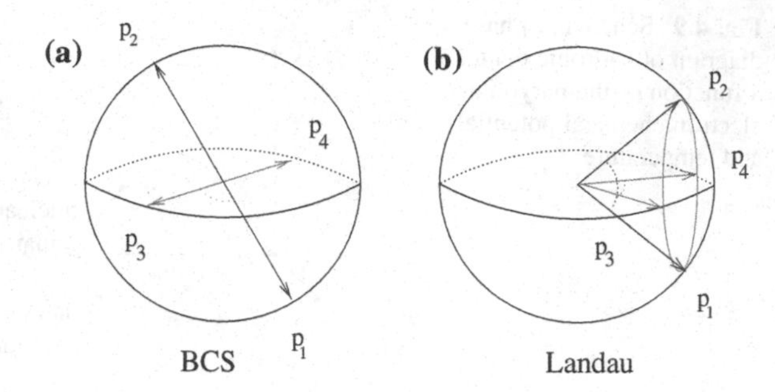

Fig. 4.10 Kinematics of four-fermion operators in the effective theory

then we expect the effective coupling to be small and weak coupling methods to be applicable. We shall see, however, that the weak coupling expansion is not a simple expansion in the number of loops. Effective field theory methods are useful in constructing a systematic weak coupling expansion.

The main observation is that the relevant low energy degrees of freedom are particle and hole excitations in the vicinity of the Fermi surface. We shall describe these excitations in terms of the field $\psi_v(x)$, where v is the Fermi velocity. At tree level, the quark field ψ can be decomposed as $\psi = \psi_{v,+} + \psi_{v,-}$ where $\psi_{v,\pm} = P_{v,\pm}\psi$ with $P_{v,\pm} = \frac{1}{2}(1 \pm \boldsymbol{\alpha} \cdot \hat{v})\psi$. Note that $P_{v,\pm}$ is a projector on states with positive/negative energy. To leading order in $1/\mu$ we can eliminate the field ψ_- using its equation of motion. The lagrangian for the ψ_+ field is given by [54–56]

$$\mathcal{L} = \psi_v^\dagger \left(iv \cdot D - \frac{D_\perp^2}{2\mu} - \frac{g\sigma_{\mu\nu}G_\perp^{\mu\nu}}{4\mu} + \dots \right) \psi_v - \frac{1}{4} G_{\mu\nu}^a G_{\mu\nu}^a + \dots . \quad (4.117)$$

with $v_\mu = (1, \boldsymbol{v})$. Note that v labels patches on the Fermi surface, and that the number of these patches grows as μ^2. The leading order $v \cdot D$ interaction does not connect quarks with different v, but soft gluons can be exchanged between quarks in different patches. In addition to that, there are four, six, etc. fermion operators that contain fermion fields with different velocity labels. These operators are constrained by the condition that the sum of the velocities has to be zero.

In the case of four-fermion operators there are two kinds of interactions that satisfy this constraint, see Fig. 4.10. The first possibility is that both the incoming and outgoing fermion momenta are back-to-back. This corresponds to the BCS interaction

$$\mathcal{L} = \frac{1}{\mu^2} \sum_{v',\Gamma,\Gamma'} V_l^{\Gamma\Gamma'} R_l^{\Gamma\Gamma'}(\boldsymbol{v} \cdot \boldsymbol{v}') \left(\psi_v \Gamma \psi_{-v}\right) \left(\psi_{v'}^\dagger \Gamma' \psi_{-v'}^\dagger\right), \quad (4.118)$$

where $\boldsymbol{v} \cdot \boldsymbol{v}' = \cos\theta$ is the scattering angle, $R_l^{\Gamma\Gamma'}(x)$ is a set of orthogonal polynomials, and Γ, Γ' determine the color, flavor and spin structure. The second possibility is that the final momenta are equal to the initial momenta up to a rotation around the axis defined by the sum of the incoming momenta. The relevant four-fermion operator is

$$\mathcal{L} = \frac{1}{\mu^2} \sum_{v',\Gamma,\Gamma'} F_l^{\Gamma\Gamma'} R_l^{\Gamma\Gamma'}(\boldsymbol{v}\cdot\boldsymbol{v}')\left(\psi_v \Gamma \psi_{v'}\right)\left(\psi_{\tilde{v}}^\dagger \Gamma' \psi_{\tilde{v}'}^\dagger\right). \tag{4.119}$$

In a system with short range interactions only the quantities $F_l(0)$ are known as Fermi liquid parameters.

4.6.2 Hard Loops

The effective field theory expansion is complicated by the fact that the number of patches $N_v \sim \mu^2/\Lambda^2$ grows with the chemical potential. This implies that some higher order contributions that are suppressed by $1/\mu^2$ can be enhanced by powers of N_v. The natural solution to this problem is to sum the leading order diagrams in the large N_v limit [57]. For gluon n-point functions this corresponds to the well known hard dense loop approximation [58–60].

The simplest example is the gluon two point function. At leading order in g and $1/\mu$ we have

$$\Pi_{\mu\nu}^{ab}(p) = 2g^2 N_f \frac{\delta^{ab}}{2} \sum_{\boldsymbol{v}} v_\mu v_\nu \int \frac{d^4k}{(2\pi)^4} \frac{1}{(k_0 - l_k)(k_0 + p_0 - l_{k+p})}, \tag{4.120}$$

where $l_k = \boldsymbol{v}\cdot\boldsymbol{k}$. We note that taking the momentum of the external gluon to zero automatically selects forward scattering. We also observe that the gluon can interact with fermions of any Fermi velocity so that the polarization function involves a sum over all patches. After performing the k_0 integration we get

$$\Pi_{\mu\nu}^{ab}(p) = 2g^2 N_f \frac{\delta^{ab}}{2} \sum_{\boldsymbol{v}} v_\mu v_\nu \int \frac{d^2 l_\perp}{(2\pi)^2} \int \frac{dl_k}{2\pi} \frac{l_p}{p_0 - l_p} \frac{\partial n_k}{\partial l_k}, \tag{4.121}$$

where n_k is the Fermi distribution function. We note that the l_k integration is automatically restricted to small momenta. The integral over the transverse momenta $l_\perp$, on the other hand, diverges quadratically with the cutoff $\Lambda_\perp$. We observe, however, that the sum over patches and the integral over $l_\perp$ can be combined into an integral over the entire Fermi surface

$$\frac{1}{2\pi} \sum_{\boldsymbol{v}} \int \frac{d^2 l_\perp}{(2\pi)^2} = \frac{\mu^2}{2\pi^2} \int \frac{d\Omega}{4\pi}. \tag{4.122}$$

This means that the transverse momentum integral is extended all the way up to μ. Because the energy of the fermions is small but the loop momentum is large the

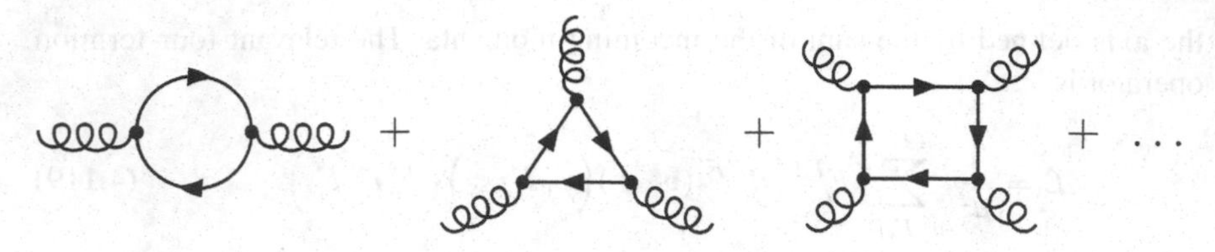

Fig. 4.11 Hard dense loop contribution to gluon n-point functions

integral is referred to as a hard dense loop. We find

$$\Pi^{ab}_{\mu\nu}(p) = 2m^2\delta^{ab}\int\frac{d\Omega}{4\pi}v_\mu v_\nu\left\{1-\frac{p_0}{p_0-l_p}\right\}, \tag{4.123}$$

where we have defined the effective gluon mass $m^2 = N_F g^2\mu^2/(4\pi^2)$. This result has the same structure as the non-relativistic expression given in Eq. (4.33), but the tadpole contribution is missing. As a consequence, Eq. (4.123) is not transverse. In the relativistic theory the tadpole contribution originates from the $D_\perp^2/(2\mu)$ in the effective lagrangian. The tadpole is proportional to the total density and corresponds to a counterterm [54]

$$\mathcal{L} = \frac{1}{2}m^2\int\frac{d\Omega}{4\pi}(A_\perp)^2. \tag{4.124}$$

Putting everything together we find

$$\Pi^{ab}_{\mu\nu}(p) = 2m^2\delta^{ab}\int\frac{d\Omega}{4\pi}\left\{\delta_{\mu 0}\delta_{\nu 0}-\frac{v_\mu v_\nu p_0}{p_0-l_p}\right\}. \tag{4.125}$$

The gluonic three-point function shown in Fig. 4.11 can be computed in the same fashion. We find

$$\Gamma^{abc}_{\mu\nu\alpha}(p,q,r) = igf^{abc}2m^2\int\frac{d\Omega}{4\pi}v_\mu v_\alpha v_\beta\left\{\frac{q_0}{(q\cdot v)(p\cdot v)}-\frac{r_0}{(r\cdot v)(p\cdot v)}\right\}, \tag{4.126}$$

where p, q, r are the incoming gluon momenta ($p+q+r=0$). We note that in the case of the three point function, as well as in all higher n-point functions, there are no tadpole or counterterm contributions. There is a simple generating functional for these loop integrals which is known as the hard dense loop (HDL) effective action [61]

$$\mathcal{L}_{HDL} = -\frac{m^2}{2}\sum_v G^a_{\mu\alpha}\frac{v^\alpha v^\beta}{(v\cdot D)^2}G^a_{\mu\beta}. \tag{4.127}$$

This is a gauge invariant, but non-local, effective lagrangian.

4.6.3 Non-Fermi Liquid Effective Field Theory

In this section we shall study the effective field theory in the regime $\omega < m$ where ω is the excitation energy and m is the effective gluon mass [62]. In the previous section we argued that hard dense loops have to be resummed in order to obtain a consistent low energy expansion. The effective lagrangian is given by

$$\mathcal{L} = \psi_v^\dagger \left(iv \cdot D - \frac{D_\perp^2}{2\mu} \right) \psi_v - \frac{1}{4} G_{\mu\nu}^a G_{\mu\nu}^a + \mathcal{L}_{HDL} + \mathcal{L}_{4f} + \ldots . \tag{4.128}$$

Since electric fields are screened the interaction at low energies is dominated by the exchange of magnetic gluons. The transverse gauge boson propagator is

$$D_{ij}(k) = -\frac{i(\delta_{ij} - \hat{k}_i \hat{k}_j)}{k_0^2 - \boldsymbol{k}^2 + i\frac{\pi}{2} m^2 \frac{k_0}{|\boldsymbol{k}|}}, \tag{4.129}$$

where we have assumed that $|k_0| < |\boldsymbol{k}|$. We observe that the propagator becomes large in the regime $|k_0| \sim |\boldsymbol{k}|^3/m^2$. If the energy is small, $|k_0| \ll m$, then the typical energy is much smaller than the typical momentum,

$$|\boldsymbol{k}| \sim (m^2 |k_0|)^{1/3} \gg |k_0|. \tag{4.130}$$

This implies that the gluon is very far off its energy shell and not a propagating state. We can compute loop diagrams containing quarks and transverse gluons by picking up the pole in the quark propagator, and then integrate over the cut in the gluon propagator using the kinematics dictated by Eq. (4.130). In order for a quark to absorb the large momentum carried by a gluon and stay close to the Fermi surface the gluon momentum has to be transverse to the momentum of the quark. This means that the term $k_\perp^2/(2\mu)$ in the quark propagator is relevant and has to be kept at leading order. Equation (4.130) shows that $k_\perp^2/(2\mu) \gg k_0$ as $k_0 \to 0$. This means that the pole of the quark propagator is governed by the condition $k_\| \sim k_\perp^2/(2\mu)$. We find

$$k_\perp \sim g^{2/3} \mu^{2/3} k_0^{1/3}, \qquad k_\| \sim g^{4/3} \mu^{1/3} k_0^{2/3}. \tag{4.131}$$

In the low energy regime propagators and vertices can be simplified even further. The quark and gluon propagators are

$$S_{\alpha\beta}(p) = \frac{i\delta_{\alpha\beta}}{p_0 - p_\| - \frac{p_\perp^2}{2\mu} + i\epsilon \mathrm{sgn}(p_0)}, \tag{4.132}$$

$$D_{ij}(k) = \frac{i\delta_{ij}}{k_\perp^2 - i\frac{\pi}{2} m^2 \frac{k_0}{k_\perp}}, \tag{4.133}$$

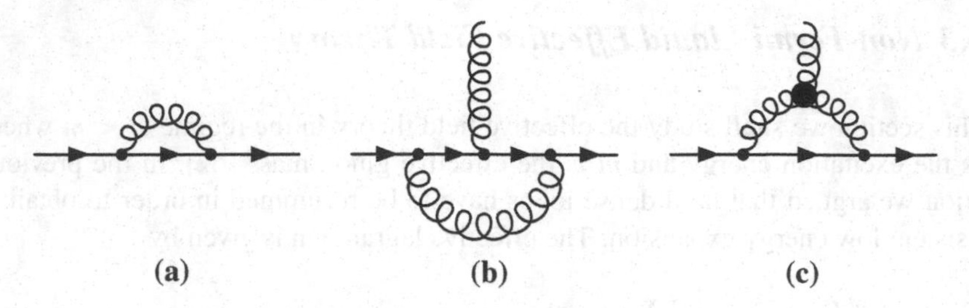

Fig. 4.12 One-loop contributions to the quark self energy and the quark-gluon vertex. The *black blob* in the third diagram denotes the HDL gluon three point function. In the magnetic regime the graphs scale as $\omega \log(\omega)$, $\omega^{1/3}$ and $\omega^{2/3}$, respectively

and the quark gluon vertex is $gv_i(\lambda^a/2)$. Higher order corrections can be found by expanding the quark and gluon propagators as well as the HDL vertices in powers of the small parameter $\epsilon \equiv (k_0/m)$.

The regime in which all momenta, including external ones, satisfy the scaling relation (4.131) is completely perturbative, i.e. graphs with extra loops are always suppressed by extra powers of $\epsilon^{1/3}$. One way to see this is to rescale the fields in the effective lagrangian so that the kinetic terms are scale invariant under the transformation $(x_0, x_{||}, x_\perp) \to (\epsilon^{-1}x_0, \epsilon^{-2/3}x_{||}, \epsilon^{-1/3}x_\perp)$. The scaling behavior of the fields is $\psi \to \epsilon^{5/6}\psi$ and $A_i \to \epsilon^{5/6}A_i$. We find that the scaling dimension of all interaction terms is positive. The quark gluon vertex scales as $\epsilon^{1/6}$, the HDL three gluon vertex scales as $\epsilon^{1/2}$, and the four gluon vertex scales as ϵ. Since higher order diagrams involve at least one pair of quark gluon vertices the expansion involves positive powers of $\epsilon^{1/3}$.

As a simple example we consider the fermion self energy in the limit $p_0 \to 0$. The one-loop diagram is (Fig. 4.12a)

$$\begin{aligned}\Sigma(p) = -ig^2 C_F \int \frac{dk_0}{2\pi} \int \frac{dk_\perp^2}{(2\pi)^2} \frac{k_\perp}{k_\perp^3 + i\eta k_0} \\ \times \int \frac{dk_{||}}{2\pi} \frac{\Theta(p_0 + k_0)}{k_{||} + p_{||} - (k_\perp + p_\perp)^2/(2\mu) + i\epsilon},\end{aligned} \tag{4.134}$$

with $C_F = (N_c^2 - 1)/(2N_c)$ and $\eta = (\pi/2)m^2$. This expression shows a number of interesting features. First we observe that the longitudinal and transverse momentum integrations factorize. The longitudinal momentum integral can be performed by picking up the pole in the quark propagator. The result is independent of the external momenta and only depends on the external energy. The transverse momentum integral is logarithmically divergent. We find [63–67]

$$\Sigma(p) = \frac{g^2}{9\pi^2} p_0 \log\left(\frac{\Lambda}{|p_0|}\right), \tag{4.135}$$

where Λ is a cutoff for the $k_\perp$ integral. We showed that the logarithmic divergence can be absorbed in the parameters of the effective theory [57]. In order to fix the corresponding counterterm we have include electric gluon exchanges For $k_0 \ll m$ the electric gluon propagator is given by

$$D_{00}(k) = -\frac{i}{k^2 + 2m^2}. \tag{4.136}$$

Higher order corrections can be obtained by expanding the full HDL expression in powers of k_0/m. The electric contribution is dominated by large momenta and does not contribute to fractional powers or logarithms of k_0. We get

$$\Sigma(p) = ig^2 C_F \int \frac{dk_0}{2\pi} \int \frac{dk_\perp^2}{(2\pi)^2} \frac{1}{k_\perp^2 + 2m^2} \times \int \frac{dk_{||}}{2\pi} \frac{\Theta(p_0 + k_0)}{k_{||} + p_{||} - (k_\perp + p_\perp)^2/(2\mu) + i\epsilon}. \tag{4.137}$$

This contribution scales as $p_0 \log(\Lambda/m)$. The logarithm of the cutoff cancels the logarithm in Eq. (4.135). We get [67]

$$\Sigma(p_0) = \frac{g^2}{9\pi^2}\left[p_0 \log\left(\frac{4\sqrt{2}m}{\pi|p_0|}\right) + p_0 + i\frac{\pi}{2}|p_0| \right]. \tag{4.138}$$

Finally, there are contributions from the hard regime in which both the energy and the momentum of the gluon are large, $k_0 \sim |\boldsymbol{k}| \geq m$. This regime corresponds to the HDL term in the fermion self energy [60, 68]. The HDL term gives an $O(g^2)$ correction to the low energy parameters v_F and $\delta\mu$.

The logarithmic term in the fermion self energy leads to a breakdown of Fermi liquid theory. The quasi-particle velocity

$$v(p_0) = \frac{1}{1 + \Sigma'(p_0)} \tag{4.139}$$

and the wave function renormalization go to zero logarithmically as the quasi-particle energy goes to zero. One physical consequence of this behavior is an anomalous $T\log(T)$ term in the specific heat [67, 69]. The effective theory can also be used to study perturbative corrections in other quantities. We find, in particular, a QCD version of Migdal's theorem. Migdal showed that vertex corrections to the electron-phonon coupling are suppressed by the ratio of the electron mass to the mass of the positive ions [4]. In the Landau damping regime of QCD loop corrections to the quark-gluon vertex are suppressed by powers of $\epsilon^{1/3}$ (Fig. 4.12b, c).

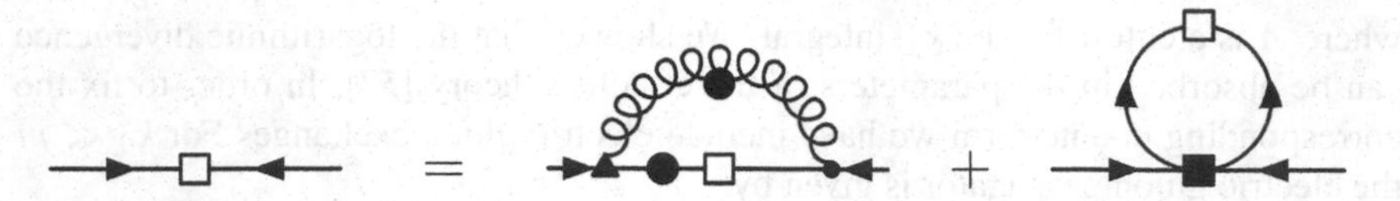

Fig. 4.13 Gap equation for the superfluid gap in a theory with long range interactions

4.6.4 Color Superconductivity

In Sect. 4.3.1 we showed that the particle-particle scattering amplitude in the BCS channel $q(\boldsymbol{p}) + q(-\boldsymbol{p}) \to q(\boldsymbol{p}') + q(-\boldsymbol{p}')$ is special. The total momentum of the pair vanishes and as a consequence loop corrections to the scattering amplitude are enhanced. This implies that all ladder diagrams have to be summed. Crossed ladders, vertex corrections, etc. involve momenta in the regime $k_\perp \gg k_{||} \gg k_0$ and are perturbative.

If the interaction in the particle-particle channel is attractive then the BCS singularity leads to the formation of a pair condensate and to a gap in the fermion spectrum. The gap can be computed by solving a Dyson-Schwinger equation for the anomalous (particle-particle) self energy. In QCD the interaction is attractive in the color anti-triplet channel. The structure of the gap is simplest in the case of two flavors. In that case, there is a unique color anti-symmetric spin zero gap term of the form

$$\langle \psi_i^a C\gamma_5 \psi_j^b \rangle = \phi \epsilon^{3ab} \epsilon_{ij}. \tag{4.140}$$

Here, a, b labels color and i, j flavor. The gap equation is given by (Fig. 4.13)

$$\begin{aligned} \Delta(p_0) = -ig^2 C_A \int \frac{dk_0}{2\pi} \int \frac{dk_\perp^2}{(2\pi)^2} \frac{k_\perp}{k_\perp^3 + \eta(k_0 - p_0)} \\ \times \int \frac{dk_{||}}{2\pi} \frac{\Delta(k_0)}{k_0^2 + k_{||}^2 + \Delta(k_0)^2}, \end{aligned} \tag{4.141}$$

where $C_A = 2/3$ is a color factor. Like the normal self energy, the anomalous self energy $\Delta(p)$ is dominantly a function of energy. We carry out the integrals over $k_\perp$ and $k_{||}$ and analytically continue to imaginary energy $p_4 = ip_0$. The euclidean gap equation is [70–73]

$$\Delta(p_4) = \frac{g^2}{18\pi^2} \int dk_4 \log\left(\frac{\Lambda_{BCS}}{|p_4 - k_4|}\right) \frac{\Delta(k_4)}{\sqrt{k_4^2 + \Delta(k_4)^2}}. \tag{4.142}$$

The scale Λ_{BCS} is sensitive to electric gluon exchange. In the anomalous self energy the logarithmic divergence does not cancel between magnetic and electric gluon exchanges. The reason is that the magnetic contribution is proportional to $\delta_{ij} v_i v_j$

in the normal self energy and $\delta_{ij} v_i(-v_j)$ in the anomalous case. The logarithmic dependence on the cutoff is absorbed by the BCS four-fermion operator. A simple matching calculation gives $\Lambda_{BCS} = 256\pi^4 (2/N_f)^{5/2} g^{-5} \mu$ [68]. The solution to the gap equation was found by Son [70]

$$\Delta(x) = \Delta_0 \sin\left(\frac{g}{3\sqrt{2}\pi} x\right) \tag{4.143}$$

where $x = \log(2\Lambda_{BCS}/(p_4 + \omega_p)$ and $\omega_p^2 = p_4^2 + \Delta_0^2$. The gap on the Fermi surface is

$$\Delta_0 \simeq 2\Lambda_{BCS} \exp\left(-\frac{\pi^2+4}{8}\right) \exp\left(-\frac{3\pi^2}{\sqrt{2}g}\right). \tag{4.144}$$

This result is correct up to $O(g)$ corrections to the pre-exponent. In order to achieve this accuracy the $g^2 p_0 \log(p_0)$ term in the normal self energy, Eq. (4.135), has to be included in the gap equation [68, 74, 75].

The order parameter is slightly more complicated in QCD with $N_f = 3$ massless flavors. The energetically preferred phase is the color-flavor-locked (CFL) phase described by [76]

$$\langle \psi_i^a C\gamma_5 \psi_j^b \rangle = \phi(\delta_i^a \delta_j^b - \delta_j^a \delta_i^b). \tag{4.145}$$

In the CFL phase there are eight fermions with gap Δ_{CFL} and one fermion with gap $2\Delta_{CFL}$. The CFL gap is given by $\Delta_{CFL} = 2^{-1/3}\Delta_0$ [77]. The CFL phase has a number of remarkable properties [76, 78]. Most notably, chiral symmetry is broken in the CFL phase and the low energy spectrum contains a flavor octet of pseudoscalar bosons. We shall study the dynamics of these Goldstone modes in Sect. 4.7.

4.6.5 Mass Terms

Mass terms modify the parameters in the effective lagrangian. These parameters include the Fermi velocity, the effective chemical potential, the screening mass, the BCS terms and the Landau parameters. At tree level the correction to the Fermi velocity and the chemical potential are given by

$$v_F = 1 - \frac{m^2}{2p_F^2}, \qquad \delta\mu = -\frac{m^2}{2p_F}. \tag{4.146}$$

The shift in the Fermi velocity also affects the coupling $g v_F$ of a magnetic gluon to quarks. It is important to note that at leading order in g this the only mass correction to the coupling. This is not entirely obvious, as one can imagine a process in which the quark emits a gluon, makes a transition to a virtual high energy state, and then couples back to a low energy state by a mass insertion. This process would give an $O(m/\mu)$ correction to g, but it vanishes in the forward direction [79].

Quark masses modify quark-quark scattering amplitudes and the corresponding Landau and BCS type four-fermion operators. Consider quark-quark scattering in the forward direction, $v + v' \to v + v'$. At tree level in QCD this process receives contribution from the direct and exchange graph. In the effective theory the direct term is reproduced by the collinear interaction while the exchange terms has to be matched against a four-fermion operator. The spin-color-flavor symmetric part of the exchange amplitude is given by

$$\mathcal{M}(v, v'; v, v') = \frac{C_F}{4N_c N_f} \frac{g^2}{p_F^2} \left\{ 1 - \frac{m^2}{p_F^2} \frac{x}{1-x} \right\} \tag{4.147}$$

where $C_F = (N_c^2 - 1)/(2N_c)$ and $x = \hat{v} \cdot \hat{v}'$ is the scattering angle. We observe that the amplitude is independent of x in the limit $m \to 0$. Mass corrections are singular as $x \to 1$. The means that the Landau coefficients F_l contain logarithms of the cutoff. We note that there is one linear combination of Landau coefficients, $F_0 - F_1/3$, which is cutoff independent.

Equations (4.146, 4.147) are valid for $N_f \geq 1$ degenerate flavors. Spin and color anti-symmetric BCS amplitudes require at least two different flavors. Consider BCS scattering $v + (-v) \to v' + (-v')$ in the helicity flip channel $L + L \to R + R$. The color-anti-triplet amplitude is given by

$$\mathcal{M}(v, -v; v', -v') = \frac{C_A}{4} \frac{g^2}{p_F^2} \frac{m_1 m_2}{p_F^2}. \tag{4.148}$$

where m_1 and m_2 are the masses of the two quarks and $C_A = (N_c + 1)/(2N_c)$. We observe that the scattering amplitude is independent of the scattering angle. This means that at leading order in g and m only the s-wave potential V_0 is non-zero.

In order to match Green functions in the high density effective theory to an effective chiral theory of the CFL phase we need to generalize our results to a complex mass matrix of the form $\mathcal{L} = -\bar{\psi}_L M \psi_R - \bar{\psi}_R M^\dagger \psi_L$, see Fig. 4.14. The $\delta\mu$ term is

$$\mathcal{L} = -\frac{1}{2p_F} \left(\psi_{L+}^\dagger M M^\dagger \psi_{L+} + \psi_{R+}^\dagger M^\dagger M \psi_{R+} \right). \tag{4.149}$$

and the four-fermion operator in the BCS channel is

$$\mathcal{L} = \frac{g^2}{32 p_F^4} \left(\psi_{i,L}^{a\,\dagger} C \psi_{j,L}^{b\,\dagger} \right) \left(\psi_{k,R}^{c} C \psi_{l,R}^{d} \right) [(\boldsymbol{\lambda})^{ac} (\boldsymbol{\lambda})^{bd} (M)_{ik} (M)_{jl}] + \left(L \leftrightarrow R, M \leftrightarrow M^\dagger \right). \tag{4.150}$$

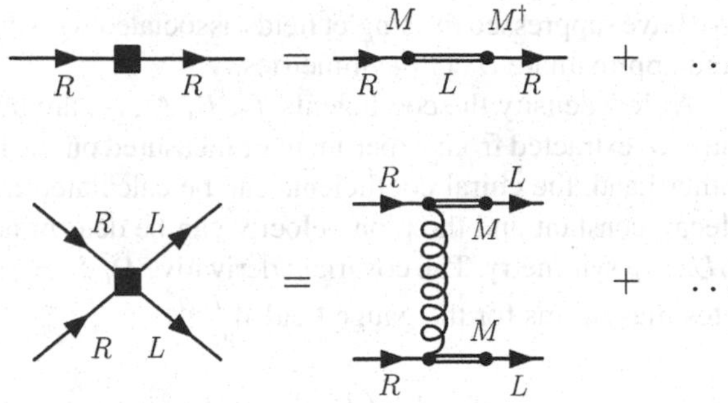

Fig. 4.14 Mass terms in the high density effective theory. The *first* diagram shows a $O(MM^\dagger)$ term that arises from integrating out the ψ_- field in the QCD lagrangian. The *second* diagram shows a $O(M^2)$ four-fermion operator which arises from integrating out ψ_- and hard gluon exchanges

4.7 Chiral Theory of the CFL Phase

4.7.1 Introduction

For energies smaller than the gap the only relevant degrees of freedom are the Goldstone modes associated with spontaneously broken global symmetries. The quantum numbers of the Goldstone modes depend on the symmetries of the order parameter. In the following we shall concentrate on the CFL phase. Goldstone modes determine the specific heat, transport properties, and the response to external fields for temperatures less than T_c. As we shall see, Goldstone modes also determine the phase structure as a function of the quark masses.

4.7.2 Chiral Effective Field Theory

In the CFL phase the pattern of chiral symmetry breaking is identical to the one at $T = \mu = 0$. This implies that the effective lagrangian has the same structure as chiral perturbation theory. The main difference is that Lorentz-invariance is broken and only rotational invariance is a good symmetry. The effective lagrangian for the Goldstone modes is given by [80]

$$\begin{aligned}\mathcal{L}_{\rm eff} = {} & \frac{f_\pi^2}{4}\mathrm{Tr}\big[\nabla_0\Sigma\nabla_0\Sigma^\dagger - v_\pi^2\partial_i\Sigma\partial_i\Sigma^\dagger\big] + \big[B\mathrm{Tr}(M\Sigma^\dagger) + h.c.\big] \\ & + \big[A_1\mathrm{Tr}(M\Sigma^\dagger)\mathrm{Tr}(M\Sigma^\dagger) + A_2\mathrm{Tr}(M\Sigma^\dagger M\Sigma^\dagger) \\ & + A_3\mathrm{Tr}(M\Sigma^\dagger)\mathrm{Tr}(M^\dagger\Sigma) + h.c.\big] + \ldots . \end{aligned} \tag{4.151}$$

Here $\Sigma = \exp(i\phi^a\lambda^a/f_\pi)$ is the chiral field, f_π is the pion decay constant and M is a complex mass matrix. The chiral field and the mass matrix transform as $\Sigma \to L\Sigma R^\dagger$ and $M \to LMR^\dagger$ under chiral transformations $(L, R) \in SU(3)_L \times SU(3)_R$.

We have suppressed the singlet fields associated with the breaking of the exact $U(1)_V$ and approximate $U(1)_A$ symmetries.

At low density the coefficients f_π, B, $A_i, \ldots$ are non-perturbative quantities that have to extracted from experiment or measured on the lattice. At large density, on the other hand, the chiral coefficients can be calculated in perturbative QCD. The pion decay constant and the pion velocity can be determined by gauging the $SU(3)_L \times SU(3)_R$ symmetry. The covariant derivative $D_\mu\Sigma = \partial_\mu\Sigma + iW_\mu^L\Sigma - i\Sigma W_\mu^R$ generates mass terms for the gauge field $W_\mu^{L,R}$,

$$\mathcal{L} = \frac{f_\pi^2}{4}\left(\frac{1}{2}(W_0^L - W_0^R)^2 - \frac{v_\pi^2}{2}(W_i^L - W_i^R)^2\right). \tag{4.152}$$

The electric and magnetic screening masses in the CFL phase can be determined as in Sect. 4.3.4. The main difference is that in the CFL phase there are nine different fermion modes, and that not all of these modes have the same gap. There is also mixing between flavor and color gauge fields. It is easiest to compute the screening for the color gauge fields. The electric screening mass is

$$\begin{aligned}\Pi_{00} = -2i \int \frac{d^4p}{(2\pi)^4} \Bigg\{ & \frac{7}{6}\frac{p_0^2+\epsilon_p^2}{(p_0^2-\epsilon_p^2-\Delta_8^2)(p_0^2-\epsilon_p^2-\Delta_8^2)} \\ & + \frac{1}{3}\frac{p_0^2+\epsilon_p^2}{(p_0^2-\epsilon_p^2-\Delta_8^2)(p_0^2-\epsilon_p^2-\Delta_1^2)} - \frac{1}{3}\frac{\Delta_8^2}{(p_0^2-\epsilon_p^2-\Delta_8^2)(p_0^2-\epsilon_p^2-\Delta_8^2)} \\ & - \frac{1}{3}\frac{\Delta_8\Delta_1}{(p_0^2-\epsilon_p^2-\Delta_8^2)(p_0^2-\epsilon_p^2-4\Delta_1^2)}\Bigg\}. \end{aligned} \tag{4.153}$$

The first terms comes from particle-hole diagrams with two octet quasi-particles while the second term comes from diagrams with one octet and one singlet quasi-particle. There is no coupling of an octet field to two singlet particles. The third and fourth term are the corresponding contributions from particle-particle and hole-hole pairs. In the CFL phase $\Delta_1 = 2\Delta_8 \equiv 2\Delta$. The four integrals in (4.153) give

$$\Pi_{00} = 2\left\{\frac{7}{6} + \frac{1}{3} - \frac{1}{3} - \frac{4\log(2)}{9}\right\}\left(\frac{\mu^2}{4\pi^2}\right) = \frac{21 - 8\log(2)}{18}\left(\frac{\mu^2}{2\pi^2}\right) \tag{4.154}$$

The magnetic mass can be computed in the same fashion. As in Sect. 4.3.4 we have to add the contribution from the tadpole and the structure of the gauge field mass term is $m_D^2(A_0^2 - \mathbf{A}^2/3)$. Mixing between flavor and color gauge fields was studied in [80, 81]. The result is that there is no screening for the vector field $W_L + W_R$ and that the screening mass for the axial field $W_L - W_R$ is equal to the mass of the color gauge field. We conclude that [82]

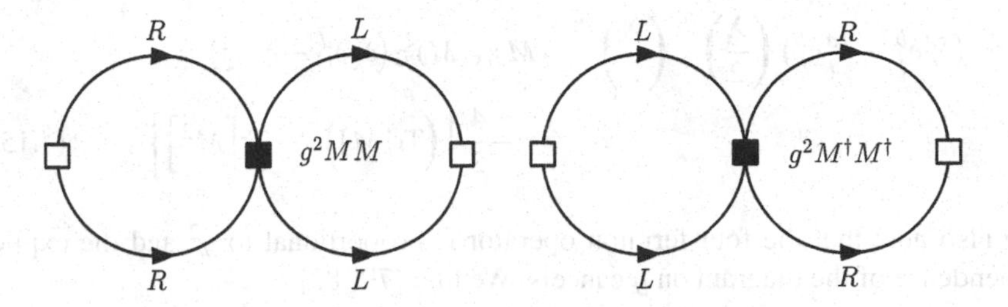

Fig. 4.15 Contribution of the $O(M^2)$ BCS four-fermion operator to the condensation energy in the CFL phase

$$f_\pi^2 = \frac{21 - 8\log(2)}{18}\left(\frac{p_F^2}{2\pi^2}\right), \qquad v_\pi^2 = \frac{1}{3}. \tag{4.155}$$

Mass terms are determined by the operators studied in Sect. 4.6.5. We observe that both Eqs. (4.149) and (4.150) are quadratic in M. This implies that $B = 0$ in perturbative QCD. B receives non-perturbative contributions from instantons, but these effects are small if the density is large, see [83].

We also note that $X_L = MM^\dagger/(2p_F)$ and $X_R = M^\dagger M/(2p_F)$ in Eq. (4.149) act as effective chemical potentials for left and right-handed fermions, respectively. Formally, the effective lagrangian has an $SU(3)_L \times SU(3)_R$ gauge symmetry under which $X_{L,R}$ transform as the temporal components of non-abelian gauge fields. We can implement this approximate gauge symmetry in the CFL chiral theory by promoting time derivatives to covariant derivatives [81],

$$\nabla_0 \Sigma = \partial_0 \Sigma + i\left(\frac{MM^\dagger}{2p_F}\right)\Sigma - i\Sigma\left(\frac{M^\dagger M}{2p_F}\right). \tag{4.156}$$

The BCS four-fermion operator in Eq. (4.150) contributes to to the condensation energy in the CFL phase, see Fig. 4.15. The diagram is proportional to the square of the superfluid density

$$\begin{aligned}\langle \psi^a_{i,L} C \psi^b_{j,L}\rangle &= \left(\delta^a_i\delta^b_j - \delta^a_j\delta^b_i\right)\int \frac{d^4p}{(2\pi)^4}\frac{\Delta(p_0)}{p^2 - \epsilon_p^2 - \Delta^2(p_0)} \\ &= \left(\delta^a_i\delta^b_j - \delta^a_j\delta^b_i\right)\Delta\frac{3\sqrt{2}\pi}{g}\left(\frac{\mu^2}{2\pi^2}\right). \end{aligned}\tag{4.157}$$

We note that the superfluid density is sensitive to energies $p_0 > \Delta$ and the energy dependence of the gap has to be kept. The color-favor factor is

$$(\delta_i^a\delta_j^b - \delta_j^a\delta_i^b)\left(\frac{\lambda}{2}\right)^{ac}\left(\frac{\lambda}{2}\right)^{bd} \quad (M)_{ik}(M)_{jl}(\delta_k^c\delta_l^d - \delta_l^c\delta_k^d)$$
$$= -\frac{4}{3}\left\{\left(\mathrm{Tr}[M]\right)^2 - \mathrm{Tr}\left[M^2\right]\right\}. \quad (4.158)$$

We also note that the four-fermion operator is proportional to g^2 and the explicit dependence of the diagram on g cancels. We find [79, 82]

$$\Delta\mathcal{E} = -\frac{3\Delta^2}{4\pi^2}\left\{\left(\mathrm{Tr}[M]\right)^2 - \mathrm{Tr}\left[M^2\right]\right\} + \left(M \leftrightarrow M^\dagger\right). \quad (4.159)$$

This term can be matched against the A_i terms in the effective lagrangian. The result is [79, 82]

$$A_1 = -A_2 = \frac{3\Delta^2}{4\pi^2}, \qquad A_3 = 0. \quad (4.160)$$

We can now summarize the structure of the chiral expansion in the CFL phase. The effective lagrangian has the form

$$\mathcal{L} \sim f_\pi^2\Delta^2\left(\frac{\partial_0}{\Delta}\right)^k\left(\frac{\vec{\partial}}{\Delta}\right)^l\left(\frac{MM^\dagger}{p_F\Delta}\right)^m\left(\frac{MM}{p_F^2}\right)^n(\Sigma)^o(\Sigma^\dagger)^p. \quad (4.161)$$

Loop graphs in the effective theory are suppressed by powers of $\partial/(4\pi f_\pi)$. Since the pion decay constant scales as $f_\pi \sim p_F$ Goldstone boson loops are suppressed compared to higher order contact terms. We also note that the quark mass expansion is controlled by $m^2/(p_F\Delta)$. This is means that the chiral expansion breaks down if $m^2 \sim p_F\Delta$. This is the same scale at which BCS calculations find a transition from the CFL phase to a less symmetric state.

4.7.3 Kaon Condensation

Using the chiral effective lagrangian we can determine the dependence of the order parameter on the quark masses. We will focus on the physically relevant case $m_s > m_u = m_d$. Because the main expansion parameter is $m_s^2/(p_F\Delta)$ increasing the quark mass is roughly equivalent to lowering the density. The effective potential for the order parameter is

$$V_{\mathrm{eff}} = \frac{f_\pi^2}{4}\mathrm{Tr}\left[2X_L\Sigma X_R\Sigma^\dagger - X_L^2 - X_R^2\right] - A_1\left[\left(\mathrm{Tr}(M\Sigma^\dagger)\right)^2 - \mathrm{Tr}\left((M\Sigma^\dagger)^2\right)\right]. \quad (4.162)$$

The first term contains the effective chemical potential $\mu_s = m_s^2/(2p_F)$ and favors states with a deficit of strange quarks (with strangeness $S = -1$). The second term favors the neutral ground state $\Sigma = 1$. The lightest excitation with positive strangeness is the K^0 meson. We therefore consider the ansatz $\Sigma = \exp(i\alpha\lambda_4)$ which allows the order parameter to rotate in the K^0 direction. The vacuum energy is

$$V(\alpha) = -f_\pi^2\left(\frac{1}{2}\left(\frac{m_s^2 - m^2}{2p_F}\right)^2 \sin(\alpha)^2 + (m_K^0)^2(\cos(\alpha) - 1)\right), \tag{4.163}$$

where $(m_K^0)^2 = (4A_1/f_\pi^2)m(m + m_s)$. Minimizing the vacuum energy we obtain

$$\cos(\alpha) = \begin{cases} 1 & \mu_s < m_K^0 \\ \frac{(m_K^0)^2}{\mu_s^2} & \mu_s > m_K^0 \end{cases} \tag{4.164}$$

The hypercharge density is

$$n_Y = f_\pi^2 \mu_s \left(1 - \frac{(m_K^0)^4}{\mu_s^4}\right). \tag{4.165}$$

This result has the same structure as the charge density of a weakly interacting Bose condensate. Using the perturbative result for A_1 we can get an estimate of the critical strange quark mass. We find

$$m_s(crit) = 3.03 \cdot m_d^{1/3} \Delta^{2/3}, \tag{4.166}$$

from which we obtain $m_s(crit) \simeq 70$ MeV for $\Delta \simeq 50$ MeV. This result suggests that strange quark matter at densities that can be achieved in neutron stars is kaon condensed. We also note that the difference in condensation energy between the CFL phase and the kaon condensed state is not necessarily small. For $\mu_s \to \Delta$ we have $\sin(\alpha) \to 1$ and $V(\alpha) \to f_\pi^2\Delta^2/2$. Since f_π^2 is of order $\mu^2/(2\pi^2)$ this is comparable to the condensation energy in the CFL phase.

The strange quark mass breaks the $SU(3)$ flavor symmetry to $SU(2)_I \times U(1)_Y$. In the kaon condensed phase this symmetry is spontaneously broken to $U(1)_Q$. If isospin is an exact symmetry there are two exactly massless Goldstone modes [84], the K^0 and the K^+. Isospin breaking leads to a small mass for the K^+. The phase structure as a function of the strange quark mass and non-zero lepton chemical potentials was studied by Kaplan and Reddy [85], see Fig. 4.16. We observe that if the lepton chemical potential is non-zero charged kaon and pion condensates are also possible.

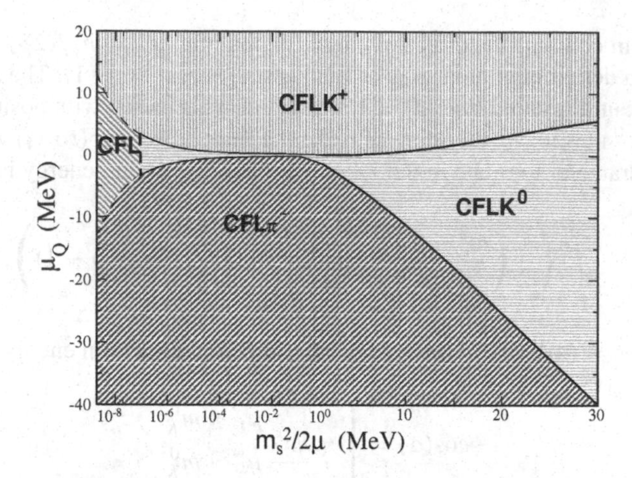

Fig. 4.16 Phase structure of CFL matter as a function of the effective chemical potential $\mu_s = m_s^2/(2p_F)$ and the lepton chemical potential μ_Q, from Kaplan and Reddy [85]. A typical value of μ_s in a neutron star is 10 MeV

4.7.4 Fermions in the CFL Phase

So far we have only studied Goldstone modes in the CFL phase. However, as the strange quark mass is increased it is possible that some of the fermion modes become light or even gapless [86]. In order to study this question we have to include fermions in the effective field theory. The effective lagrangian for fermions in the CFL phase is [87, 88]

$$\mathcal{L} = \mathrm{Tr}\big(N^\dagger i v^\mu D_\mu N\big) - D\mathrm{Tr}\big(N^\dagger v^\mu \gamma_5 \{\mathcal{A}_\mu, N\}\big) - F\mathrm{Tr}\big(N^\dagger v^\mu \gamma_5 [\mathcal{A}_\mu, N]\big) \\ + \frac{\Delta}{2}\Big\{\Big(\mathrm{Tr}\,(N_L N_L) - [\mathrm{Tr}\,(N_L)]^2\Big) - (L \leftrightarrow R) + h.c.\Big\}. \quad (4.167)$$

$N_{L,R}$ are left and right handed baryon fields in the adjoint representation of flavor $SU(3)$. The baryon fields originate from quark-hadron complementarity [78]. We can think of N as describing a quark which is surrounded by a diquark cloud, $N_L \sim q_L\langle q_L q_L\rangle$. The covariant derivative of the nucleon field is given by $D_\mu N = \partial_\mu N + i[\mathcal{V}_\mu, N]$. The vector and axial-vector currents are

$$\mathcal{V}_\mu = -\frac{i}{2}\{\xi\partial_\mu\xi^\dagger + \xi^\dagger\partial_\mu\xi\}, \qquad \mathcal{A}_\mu = -\frac{i}{2}\xi(\nabla_\mu\Sigma^\dagger)\xi, \quad (4.168)$$

where ξ is defined by $\xi^2 = \Sigma$. It follows that ξ transforms as $\xi \to L\xi U(x)^\dagger = U(x)\xi R^\dagger$ with $U(x) \in SU(3)_V$. For pure $SU(3)$ flavor transformations $L = R = V$ we have $U(x) = V$. F and D are low energy constants that determine the baryon axial coupling. In perturbative QCD we find $D = F = 1/2$.

The effective theory given in Eq. (4.167) can be derived from QCD in the weak coupling limit. However, the structure of the theory is completely determined by chiral symmetry, even if the coupling is not weak. In particular, there are no free parameters in the baryon coupling to the vector current. Mass terms are also strongly constrained by chiral symmetry. The effective chemical potentials (X_L, X_R) appear as left and right-handed gauge potentials in the covariant derivative of the nucleon field. We have

$$D_0 N = \partial_0 N + i[\Gamma_0, N], \qquad (4.169)$$
$$\Gamma_0 = -\frac{i}{2}\{\xi\,(\partial_0 + iX_R)\,\xi^\dagger + \xi^\dagger\,(\partial_0 + iX_L)\,\xi\},$$

where $X_L = MM^\dagger/(2p_F)$ and $X_R = M^\dagger M/(2p_F)$ as before. (X_L, X_R) covariant derivatives also appears in the axial vector current given in Eq. (4.168).

We can now study how the fermion spectrum depends on the quark mass. In the CFL state we have $\xi = 1$. For $\mu_s = 0$ the baryon octet has an energy gap Δ and the singlet has gap 2Δ. The correction to this result comes from the term

$$\mathrm{Tr}\big(N^\dagger[\hat{\mu}_s, N]\big) = \frac{\mu_s}{2}\big((\Xi^-)^\dagger(\Xi^-)+(\Xi^0)^\dagger(\Xi^0)-(p)^\dagger(p)-(n)^\dagger(n)\big), \quad (4.170)$$

where $\hat{\mu}_s = \mu_s \mathrm{diag}(0, 0, 1)$. We observe that the excitation energy of the proton and neutron is lowered, $\omega_{p,n} = \Delta - \mu_s$, while the energy of the cascade states Ξ^-, Ξ^0 particles is raised, $\omega_\Xi = \Delta + \mu_s$. All other excitation energies are independent of μ_s. As a consequence we find gapless (p, n) excitations at $\mu_s = \Delta$.

This result is also easily derived in microscopic models [89]. The EFT perspective is nevertheless useful. In microscopic models the shift of the non-strange modes arises from a color chemical potential which is needed in order to neutralize the system. The effective theory is formulated directly in terms of gauge invariant variables and no color chemical potentials are needed. The shift in the non-strange modes is due to the fact that the gauge invariant fermion fields transform according to the adjoint representation of flavor $SU(3)$.

The situation is more complicated when kaon condensation is taken into account. In the kaon condensed phase there is mixing in the $(p, \Sigma^+, \Sigma^-, \Xi^-)$ and $(n, \Sigma^0, \Xi^0, \Lambda^8, \Lambda^0)$ sector. For $m_K^0 \ll \mu_s \ll \Delta$ the spectrum is given by

$$\omega_{p\Sigma^\pm\Xi^-} = \begin{cases} \Delta \pm \frac{3}{4}\mu_s, \\ \Delta \pm \frac{1}{4}\mu_s, \end{cases} \qquad \omega_{n\Sigma^0\Xi^0\Lambda} = \begin{cases} \Delta \pm \frac{1}{2}\mu_s, \\ \Delta, \\ 2\Delta. \end{cases} \qquad (4.171)$$

Numerical results for the eigenvalues are shown in Fig. 4.17. We observe that mixing within the charged and neutral baryon sectors leads to level repulsion. There are two modes that become light in the CFL window $\mu_s \leq 2\Delta$. One mode is a linear

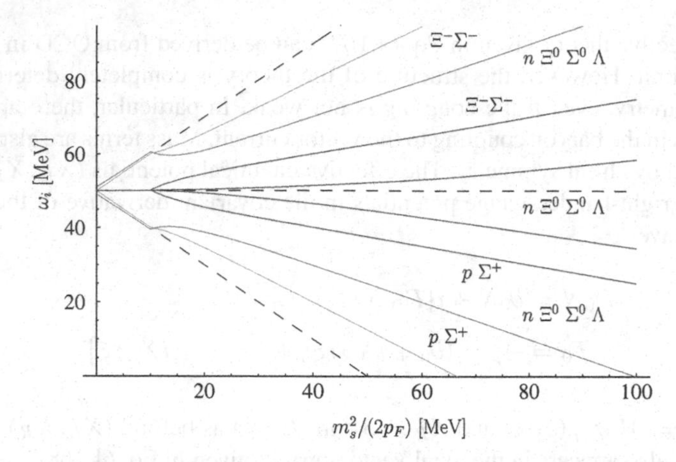

Fig. 4.17 This figure shows the fermion spectrum in the CFL phase. For $m_s = 0$ there are eight fermions with gap Δ and one fermion with gap 2Δ (*not shown*). Without kaon condensation gapless fermion modes appear at $\mu_s = \Delta$ (*dashed lines*). With kaon condensation gapless modes appear at $\mu_s = 4\Delta/3$

combination of the proton and Σ^+ and the other mode is a linear combination of the neutral baryons $(n, \Sigma^0, \Xi^0, \Lambda^8, \Lambda^0)$.

4.7.5 Meson Supercurrent State

Recently, several groups have shown that gapless fermion modes lead to instabilities in the current-current correlation function [90, 91]. Motivated by these results we have examined the stability of the kaon condensed phase against the formation of a non-zero current [92, 93]. Consider a spatially varying $U(1)_Y$ rotation of the maximal kaon condensate

$$U(x)\xi_{K^0}U^\dagger(x) = \begin{pmatrix} 1 & 0 & 0 \\ 0 & 1/\sqrt{2} & ie^{i\phi_K(x)}/\sqrt{2} \\ 0 & ie^{-i\phi_K(x)}/\sqrt{2} & 1/\sqrt{2} \end{pmatrix}. \quad (4.172)$$

This state is characterized by non-zero currents

$$\mathcal{V} = \frac{1}{2}(\nabla\phi_K)\begin{pmatrix} 0 & 0 & 0 \\ 0 & 1 & 0 \\ 0 & 0 & -1 \end{pmatrix}, \quad \mathcal{A} = \frac{1}{2}(\nabla\phi_K)\begin{pmatrix} 0 & 0 & 0 \\ 0 & 0 & -ie^{i\phi_K} \\ 0 & ie^{-i\phi_K} & 0 \end{pmatrix}. \quad (4.173)$$

In the following we compute the vacuum energy as a function of the kaon current $J_K = \nabla\phi_K$. The meson part of the effective lagrangian gives a positive contribution

$$\mathcal{E} = \frac{1}{2} v_\pi^2 f_\pi^2 j_K^2. \tag{4.174}$$

A negative contribution can arise from gapless fermions. In order to determine this contribution we have to calculate the fermion spectrum in the presence of a non-zero current. The relevant part of the effective lagrangian is

$$\begin{aligned}\mathcal{L} &= \mathrm{Tr}\left(N^\dagger i v^\mu D_\mu N\right) + \mathrm{Tr}\left(N^\dagger \gamma_5 \left(\rho_A + \boldsymbol{v} \cdot \boldsymbol{\mathcal{A}}\right) N\right) \\ &\quad + \frac{\Delta}{2} \left\{\mathrm{Tr}\,(NN) - \mathrm{Tr}\,(N)\,\mathrm{Tr}\,(N) + h.c.\right\}, \end{aligned} \tag{4.175}$$

where we have used $D = F = 1/2$. The covariant derivative is $D_0 N = \partial_0 N + i[\rho_V, N]$ and $D_i N = \partial_i N + i\boldsymbol{v} \cdot [\boldsymbol{\mathcal{V}}, N]$ with $\boldsymbol{\mathcal{V}}, \boldsymbol{\mathcal{A}}$ given in Eq. (4.173) and

$$\rho_{V,A} = \frac{1}{2} \left\{ \xi \frac{M^\dagger M}{2p_F} \xi^\dagger \pm \xi^\dagger \frac{M M^\dagger}{2p_F} \xi \right\}. \tag{4.176}$$

The vector potential ρ_V and the vector current $\boldsymbol{\mathcal{V}}$ are diagonal in flavor space while the axial potential ρ_A and the axial current $\boldsymbol{\mathcal{A}}$ lead to mixing. The fermion spectrum is quite complicated. The dispersion relation of the lowest mode is approximately given by

$$\omega_l = \Delta + \frac{(l - l_0)^2}{2\Delta} - \frac{3}{4}\mu_s - \frac{1}{4}\boldsymbol{v} \cdot \boldsymbol{j}_K, \tag{4.177}$$

where $l = \boldsymbol{v} \cdot \boldsymbol{p} - p_F$ and we have expanded ω_l near its minimum $l_0 = (\mu_s + \boldsymbol{v} \cdot \boldsymbol{j}_K)/4$. Equation (4.177) shows that there is a gapless mode if $\mu_s > 4\Delta/3 - j_K/3$. The contribution of the gapless mode to the vacuum energy is

$$\mathcal{E} = \frac{\mu^2}{\pi^2} \int dl \int \frac{d\Omega}{4\pi}\, \omega_l \theta(-\omega_l), \tag{4.178}$$

where $d\Omega$ is an integral over the Fermi surface. The integral in Eq. (4.178) receives contributions from one of the pole caps on the Fermi surface. The result has exactly the same structure as the energy functional of a non-relativistic two-component Fermi liquid with non-zero polarization, see [94]. Introducing dimensionless variables

$$x = \frac{j_K}{a\Delta}, \qquad h = \frac{3\mu_s - 4\Delta}{a\Delta}. \tag{4.179}$$

we can write $\mathcal{E} = c\mathcal{N} f_h(x)$ with

$$f_h(x) = x^2 - \frac{1}{x}\left[(h + x)^{5/2}\Theta(h + x) - (h - x)^{5/2}\Theta(h - x)\right]. \tag{4.180}$$

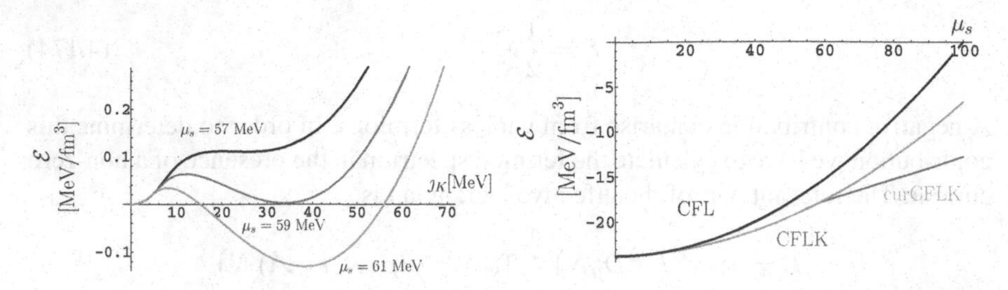

Fig. 4.18 *Left panel* Energy density as a function of the current j_K for several different values of $\mu_s = m_s^2/(2p_F)$ close to the phase transition. *Right panel* Ground state energy density as a function of μ_s. We show the CFL phase, the kaon condensed CFL (KCFL) phase, and the supercurrent state (curKCFL)

We have defined the constants

$$c = \frac{2}{15^4 c_\pi^3 v_\pi^6}, \qquad \mathcal{N} = \frac{\mu^2 \Delta^2}{\pi^2}, \qquad a = \frac{2}{15^2 c_\pi^2 v_\pi^4}, \tag{4.181}$$

where $c_\pi = (21 - 8\log(2))/36$ is the numerical coefficient that appears in the weak coupling result for f_π. According to the analysis in [94] the function $f_h(x)$ develops a non-trivial minimum if $h_1 < h < h_2$ with $h_1 \simeq -0.067$ and $h_2 \simeq 0.502$. In perturbation theory we find $a = 0.43$ and the kaon condensed ground state becomes unstable for $(\Delta - 3\mu_s/4) < 0.007\Delta$.

The energy density as a function of the current and the groundstate energy density as a function of μ_s are shown in Fig. 4.18. In these plots we have included the contribution of a baryon current j_B, as suggested in [93]. In this case we have to minimize the energy with respect to two currents. The solution is of the form $j_B \sim j_K$. The figure shows the dependence on j_K for the optimum value of j_B. We have not properly implemented electric charge neutrality. Since the gapless mode is charged, enforcing electric neutrality will significantly suppress the magnitude of the current. We have also not included the possibility that the neutral mode becomes gapless. This will happen at somewhat larger values of μ_s.

We note that the ground state has no net current. This is clear from the fact that the ground state satisfies $\delta\mathcal{E}/\delta(\nabla\phi_K) = 0$. As a consequence the meson current is canceled by an equal but opposite contribution from gapless fermions. We also expect that the ground state has no chromomagnetic instabilities. The kaon current is equivalent to an external gauge field. By minimizing the thermodynamic potential with respect to j we ensure that the second derivative, and therefore the screening mass, is positive.

4.8 Conclusion: The Many Uses of Effective Field Theory

Strongly correlated quantum many body systems play a role in many different branches of physics, atomic physics, condensed matter physics, nuclear and particle physics. One of the main themes of these lectures is the idea that effective field theories provide a unified description of systems that involve vastly different scales. For example, nuclear physicists studying neutron matter have learned a great deal from studying cold atomic gases (and vice versa). Similarly, progress in understanding non-Fermi liquid behavior in strongly correlated electronic systems has been helpful in understanding dense quark matter in QCD. It is our hope that these lecture notes will play a small part in fostering exchange of ideas between different communities in the future.

Acknowledgments I would like to thank the organizers of the Trento school, Janos Polonyi and Achim Schwenk, for doing such an excellent job in putting the school together, and all the students who attended the school for turning it into a stimulating experience. This work was supported in part by US DOE grant DE-FG02-03ER41260.

References

1. Polchinski, J.: Effective field theory and the Fermi surface. Lectures presented at TASI 92, Boulder, CO, hep-th/9210046
2. Manohar, A.V.: Effective field theories. In: Schladming 1996, Perturbative and nonperturbative aspects of quantum field theory, hep-ph/9606222
3. Kaplan, D.B.: Five lectures on effective field theory. Lectures delivered at the 17th National Nuclear Physics Summer School 2005, nucl-th/0510023
4. Abrikosov, A.A., Gorkov, L.P., Dzyaloshinski, I.E.: Methods of Quantum Field Theory in Statistical Physics. Prentice-Hall, Englewood Cliffs (1963)
5. Fetter, A.L., Walecka, J.D.: Quantum Theory of Many Particle Systems. McGraw Hill, New York (1971)
6. Negele, J.W., Orland, H.: Quantum Many Particle Systems. Perseus Books, Reading (1988)
7. Rajagopal, K., Wilczek, F.: The condensed matter physics of QCD. In: Shifman, M. (ed.) At the Frontier of Particle Physics, Boris Ioffe Festschrift. World Scientific, Singapore (2001) [hep-ph/0011333]
8. Alford, M.G.: Ann. Rev. Nucl. Part. Sci. **51**, 131 (2001) [hep-ph/0102047]
9. Beane, S.R., Bedaque, P.F., Haxton, W.C., Phillips, D.R., Savage, M.J.: From hadrons to nuclei: crossing the border. In: Shifman, M. (ed.) At the Frontier of Particle Physics, Boris Ioffe Festschrift. World Scientific, Singapore (2001) [nucl-th/0008064]
10. Bedaque, P.F., van Kolck, U.: Ann. Rev. Nucl. Part. Sci. **52**, 339 (2002) [nucl-th/0203055]
11. Epelbaum, E.: Prog. Part. Nucl. Phys. **57**, 654 (2006) [nucl-th/0509032]
12. Kaplan, D.B., Savage, M.J., Wise, M.B.: Nucl. Phys. B **534**, 329 (1998) [nucl-th/9802075]
13. Kapusta, J.: Finite Temperature Field Theory. Cambridge University Press, Cambridge (1989)
14. LeBellac, M.: Thermal Field Theory. Cambridge University Press, Cambridge (1996)
15. Lee, T.D., Yang, C.N.: Phys. Rev. **105**, 1119 (1957)
16. Huang, K., Yang, C.N.: Phys. Rev. **105**, 767 (1957)
17. Hammer, H.W., Furnstahl, R.J.: Nucl. Phys. A **678**, 277 (2000) [nucl-th/0004043]
18. Platter, L., Hammer, H.W., Meissner, U.G.: Nucl. Phys. A **714**, 250 (2003) [nucl-th/0208057]
19. Pines, D.: The Theory of Quantum Liquids. Addison-Wesley, Menlo Park (1966)
20. Baym, G., Pethick, C.: Landau Fermi Liquid Theory. Wiley, New York (1991)
21. Landau, L.D., Lifshitz, E.M.: Physical Kinetics (Course of Theoretical Physics, vol. X). Pergamon Press, New York (1981)
22. Shankar, R.: Rev. Mod. Phys. **66**, 129 (1994)
23. Papenbrock, T., Bertsch, G.F.: Phys. Rev. C **59**, 2052 (1999) [nucl-th/9811077]
24. Marini, M., Pistolesi, F., Strinati, G.C.: Eur. Phys. J. B **1**, 151 (1998) [cond-mat/9703160]

25. Gorkov, L.P., Melik-Barkhudarov, T.K.: Sov. Phys. JETP **13**, 1018 (1961)
26. Greiter, M., Wilczek, F., Witten, E.: Mod. Phys. Lett. B **3**, 903 (1989)
27. Son, D.T.: unpublished, hep-ph/0204199
28. Son, D.T., Wingate, M.: Ann. Phys. **321**, 197 (2006) [cond-mat/0509786]
29. Anderson, P.W.: Basic Notions of Condensed Matter Physics. Benjamin/Cummings, Menlo Park (1984)
30. Weinberg, S.: The Quantum Theory of Fields, vol. II. Cambridge University Press, Cambridge (1995)
31. Regal, C.: Ph.D. Thesis, University of Colorado (2005), cond-mat/0601054
32. Beane, S.R., Savage, M.J.: Nucl. Phys. A **717**, 91 (2003) [nucl-th/0208021]
33. Lee, D., Schäfer, T.: Phys. Rev. C **72**, 024006 (2005) [nucl-th/0412002]; Phys. Rev. C **73**, 015201 (2006) [nucl-th/0509017]; Phys. Rev. C **73**, 015202 (2006) [nucl-th/0509018]
34. Chen, J.W., Kaplan, D.B.: Phys. Rev. Lett. **92**, 257002 (2004) [hep-lat/0308016]
35. Beane, S.R., Bedaque, P.F., Parreno, A., Savage, M.J.: Phys. Lett. B **585**, 106 (2004) [hep-lat/0312004]
36. Lee, D.: Phys. Rev. B **73**, 115112 (2006) [cond-mat/0511332]
37. Carlson, J., Morales, J.J., Pandharipande, V.R., Ravenhall, D.G.: Phys. Rev. C **68**, 025802 (2003) [nucl-th/0302041]
38. Bulgac, A., Drut, J.E., Magierski, P.: Phys. Rev. Lett. **96**, 090404 (2006) [cond-mat/0505374]
39. Burovski, E., Prokof'ev, N., Svistunov, B., Troyer, M.: Phys. Rev. Lett. **96**, 160402 (2006) [cond-mat/0602224]
40. Furnstahl, R.J., Hammer, H.W.: Ann. Phys. **302**, 206 (2002) [nucl-th/0208058]
41. Nikolic, P., Sachdev, S.: Phys. Rev. A **75**, 033608 (2007), cond-mat/0609106
42. Steele, J.V.: unpublished, nucl-th/0010066
43. Schäfer, T., Kao, C.W., Cotanch, S.R.: Nucl. Phys. A **762**, 82 (2005) [nucl-th/0504088]
44. Nussinov, Z., Nussinov, S.: unpublished, cond-mat/0410597
45. Nishida, Y., Son, D.T.: Phys. Rev. Lett. **97** (2006) 050403, cond-mat/0604500
46. Nishida, Y., Son, D.T.: Phys. Rev. A **75** (2007) 063617, cond-mat/0607835
47. Arnold, P., Drut, J.E., Son, D.T.: Phys. Rev. A **75** (2007) 043605, cond-mat/0608477
48. Nishida, Y.: preprint, cond-mat/0608321
49. Rupak, G., Schäfer, T., Kryjevski, A.: Phys. Rev. A **75** (2007) 023606 cond-mat/0607834
50. Fodor, Z., Katz, S.D.: JHEP **0203**, 014 (2002) [hep-lat/0106002]
51. de Forcrand, P., Philipsen, O.: Nucl. Phys. B **642**, 290 (2002) [hep-lat/0205016]
52. Allton, C.R., et al.: Phys. Rev. D **66**, 074507 (2002) [hep-lat/0204010]
53. Collins, J.C., Perry, M.J.: Phys. Rev. Lett. **34**, 1353 (1975)
54. Hong, D.K.: Phys. Lett. B **473**, 118 (2000) [hep-ph/981251]
55. Hong, D.K.: Nucl. Phys. B **582**, 451 (2000) [hep-ph/9905523]
56. Nardulli, G.: Riv. Nuovo Cim. **25N3**, 1 (2002) [hep-ph/0202037]
57. Schäfer, T., Schwenzer, K.: Phys. Rev. D **70**, 054007 (2004) [hep-ph/0405053]
58. Braaten, E., Pisarski, R.D.: Nucl. Phys. B **337**, 569 (1990)
59. Blaizot, J.P., Ollitrault, J.Y.: Phys. Rev. D **48**, 1390 (1993) [hep-th/9303070]
60. Manuel, C.: Phys. Rev. D **53**, 5866 (1996) [hep-ph/9512365]
61. Braaten, E., Pisarski, R.D.: Phys. Rev. D **45**, 1827 (1992)
62. Schäfer, T., Schwenzer, K.: Phys. Rev. Lett. **97**, (2006) 092301, hep-ph/0512309
63. Vanderheyden, B., Ollitrault, J.Y.: Phys. Rev. D **56**, 5108 (1997)
64. Manuel, C.: Phys. Rev. D **62**, 076009 (2000)
65. Brown, W.E., Liu, J.T., Ren, H.C.: Phys. Rev. D **62**, 054013 (2000) [hep-ph/0003199]
66. Boyanovsky, D., de Vega, H.J.: Phys. Rev. D **63**, 034016 (2001) [hep-ph/0009172]
67. Ipp, A., Gerhold, A., Rebhan, A.: Phys. Rev. D **69**, 011901 (2004)
68. Schäfer, T.: Nucl. Phys. A **728**, 251 (2003) [hep-ph/0307074]
69. Holstein, T., Norton, A.E., Pincus, P.: Phys. Rev. B **8**, 2649 (1973)
70. Son, D.T.: Phys. Rev. D **59**, 094019 (1999) [hep-ph/9812287]
71. Schäfer, T., Wilczek, F.: Phys. Rev. D **60**, 114033 (1999) [hep-ph/9906512]
72. Pisarski, R.D., Rischke, D.H.: Phys. Rev. D **61**, 074017 (2000) [nucl-th/9910056]

73. Hong, D.K., Miransky, V.A., Shovkovy, I. A., Wijewardhana, L.C.: Phys. Rev. D **61**, 056001 (2000) [hep-ph/9906478]
74. Brown, W.E., Liu, J.T., Ren, H.C.: Phys. Rev. D **61**, 114012 (2000) [hep-ph/9908248]
75. Wang, Q., Rischke, D.H.: Phys. Rev. D **65**, 054005 (2002) [nucl-th/0110016]
76. Alford, M., Rajagopal, K., Wilczek, F.: Nucl. Phys. B **537**, 443 (1999) [hep-ph/9804403]
77. Schäfer, T.: Nucl. Phys. B **575**, 269 (2000) [hep-ph/9909574]
78. Schäfer, T., Wilczek, F.: Phys. Rev. Lett. **82**, 3956 (1999) [hep-ph/9811473]
79. Schäfer, T.: Phys. Rev. D **65**, 074006 (2002) [hep-ph/0109052]
80. Casalbuoni, R., Gatto, D.: Phys. Lett. B **464**, 111 (1999) [hep-ph/9908227]
81. Bedaque, P.F., Schäfer, T.: Nucl. Phys. A **697**, 802 (2002) [hep-ph/0105150]
82. Son, D.T., Stephanov, M.: Phys. Rev. D **61**, 074012 (2000) [hep-ph/9910491], erratum: hep-ph/0004095
83. Schäfer, T.: Phys. Rev. D **65**, 094033 (2002) [hep-ph/0201189]
84. Schäfer, T., Son, D.T., Stephanov, M.A., Toublan, D., Verbaarschot, J.J.: Phys. Lett. B **522**, 67 (2001) [hep-ph/0108210]
85. Kaplan, D.B., Reddy, S.: Phys. Rev. D **65**, 054042 (2002) [hep-ph/0107265]
86. Alford, M., Kouvaris, C., Rajagopal, K.: Phys. Rev. Lett. **92**, 222001 (2004) [hep-ph/0311286]
87. Kryjevski, A., Schäfer, T.: Phys. Lett. B **606**, 52 (2005) [hep-ph/0407329]
88. Kryjevski, A., Yamada, D.: Phys. Rev. D **71**, 014011 (2005) [hep-ph/0407350]
89. Alford, M., Kouvaris, C., Rajagopal, K.: Phys. Rev. D **71**, 054009 (2005) [hep-ph/0406137]
90. Huang, M., Shovkovy, I.A.: Phys. Rev. D **70**, 051501 (2004) [hep-ph/0407049]
91. Casalbuoni, R., Gatto, R., Mannarelli, M., Nardulli, G., Ruggieri, M.: Phys. Lett. B **605**, 362 (2005) [hep-ph/0410401]
92. Schäfer, T.: Phys. Rev. Lett. **96**, 012305 (2006) [hep-ph/0508190]
93. Kryjevski, A.: Phys. Rev. D **77**, (2008) 014018, hep-ph/0508180
94. Son, D.T., Stephanov, M.A.: Phys. Rev. A **74**, 013614 (2006), cond-mat/0507586

Chapter 5
Renormalization Group and Fermi Liquid Theory for Many-Nucleon Systems

Bengt Friman, Kai Hebeler and Achim Schwenk

5.1 Introduction

In these lecture notes we discuss developments using renormalization group (RG) methods for strongly interacting Fermi systems and their application to neutron matter. We rely on material from the review of Shankar [1], the lecture notes by Polchinski [2], and work on the functional RG, discussed in the lectures of Gies [3] and in the recent review by Metzner et al. [4]. The lecture notes are intended to show the strengths and flexibility of the RG for nucleonic matter, and to explain the ideas in more detail.

We start these notes with an introduction to Landau's theory of normal Fermi liquids [5–7], which make the concept of a quasiparticle very clear. Since Landau's work, this concept has been successfully applied to a wide range of many-body systems. In the quasiparticle approximation it is assumed that the relevant part of

---

B. Friman (✉)
Theory Division, GSI Helmholtzzentrum für Schwerionenforschung GmbH,
64291 Darmstadt, Germany
e-mail: b.friman@gsi.de

K. Hebeler · A. Schwenk
TRIUMF, 4004 Wesbrook Mall, Vancouver, BC, V6T 2A3, Canada
e-mail: hebeler.4@osu.edu

K. Hebeler
Department of Physics, The Ohio State University, Columbus, OH 43210, USA

A. Schwenk
ExtreMe Matter Institute EMMI,
GSI Helmholtzzentrum für Schwerionenforschung GmbH,
64291 Darmstadt, Germany
e-mail: schwenk@physik.tu-darmstadt.de

A. Schwenk
Institut für Kernphysik, Technische Universität Darmstadt
64289 Darmstadt, Germany

J. Polonyi and A. Schwenk (eds.), *Renormalization Group and Effective Field Theory Approaches to Many-Body Systems*, Lecture Notes in Physics 852,
DOI: 10.1007/978-3-642-27320-9_5,

the excitation spectrum of the one-body propagator can be incorporated as an effective degree of freedom, a quasiparticle. In Landau's theory of normal Fermi liquids this assumption is well motivated, and the so-called background contributions to the one-body propagator are included in the low-energy couplings of the theory. In microscopic calculations and in applications of the RG to many-body systems, the quasiparticle approximation is physically motivated and widely used due to the great reduction of the calculational effort.

5.2 Fermi Liquid Theory

5.2.1 Basic Ideas

Much of our understanding of strongly interacting Fermi systems at low energies and temperatures goes back to the seminal work of Landau in the late 1950s [5–7]. Landau was able to express macroscopic observables in terms of microscopic properties of the elementary excitations, the so-called quasiparticles, and their residual interactions. In order to illustrate Landau's arguments here, we consider a uniform system of non-relativistic spin-$1/2$ fermions at zero temperature.

Landau assumed that the low-energy, elementary excitations of the interacting system can be described by effective degrees of freedom, the quasiparticles. Due to translational invariance, the states of the uniform system are eigenstates of the momentum operator. The quasiparticles are much like single-particle states in the sense that for each momentum there is a well-defined quasiparticle energy. We stress, however, that a quasiparticle state is not an energy eigenstate, but rather a resonance with a non-zero width. For quasiparticles close to the Fermi surface, the width is small and the corresponding life-time is large; hence the quasiparticle concept is useful for time scales short compared to the quasiparticle life-time. Landau assumed that there is a one-to-one correspondence between the quasiparticles and the single-particle states of a free Fermi gas. For a superfluid system, this one-to-one correspondence does not exist, and Landau's theory must be suitably modified, as discussed by Larkin and Migdal [8] and Leggett [9]. Whether the quasiparticle concept is useful for a particular system can be determined by comparison with experiment or by microscopic calculations based on the underlying theory.

The one-to-one correspondence starts from a free Fermi gas consisting of N particles, where the ground state is given by a filled Fermi sphere in momentum space, see Fig. 5.1. The particle number density n and the ground-state energy E_0 are given by (with $\hbar = c = 1$)

$$n = \frac{1}{V} \sum_{\mathbf{p}\sigma} n^0_{\mathbf{p}\sigma} = \frac{k_\mathrm{F}^3}{3\pi^2} \quad \text{and} \quad E_0 = \sum_{\mathbf{p}\sigma} \frac{\mathbf{p}^2}{2m} n^0_{\mathbf{p}\sigma} = \frac{3}{5} \frac{k_\mathrm{F}^2}{2m} N, \tag{5.1}$$

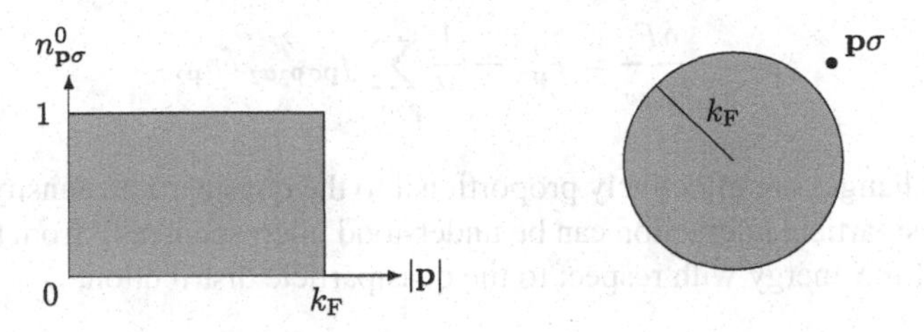

Fig. 5.1 Zero-temperature distribution function of a free Fermi gas in the ground state (*left*) and with one added particle (*right*)

where k_F denotes the Fermi momentum, V the volume, and $n^0_{\mathbf{p}\sigma} = \theta(k_F - |\mathbf{p}|)$ is the Fermi-Dirac distribution function at zero temperature for particles with momentum $\mathbf{p}$, spin projection σ, and mass m. By adding particles or holes, the distribution function is changed by $\delta n_{\mathbf{p}\sigma} = n_{\mathbf{p}\sigma} - n^0_{\mathbf{p}\sigma}$, and the total energy of the system by

$$\delta E = E - E_0 = \sum_{\mathbf{p}\sigma} \frac{\mathbf{p}^2}{2m} \delta n_{\mathbf{p}\sigma}. \tag{5.2}$$

When a particle is added in the state $\mathbf{p}\sigma$, one has $\delta n_{\mathbf{p}\sigma} = 1$ and when a particle is removed (a hole is added) $\delta n_{\mathbf{p}\sigma} = -1$.

In the interacting system the corresponding state is one with a quasiparticle added or removed, and the change in energy is given by

$$\delta E = \sum_{\mathbf{p}\sigma} \varepsilon_{\mathbf{p}\sigma} \, \delta n_{\mathbf{p}\sigma}, \tag{5.3}$$

where $\varepsilon_{\mathbf{p}\sigma} = \delta E / \delta n_{\mathbf{p}\sigma}$ denotes the quasiparticle energy. When two or more quasiparticles are added to the system, an additional term takes into account the interaction between the quasiparticles:

$$\delta E = \sum_{\mathbf{p}\sigma} \varepsilon^0_{\mathbf{p}\sigma} \, \delta n_{\mathbf{p}\sigma} + \frac{1}{2V} \sum_{\mathbf{p}_1\sigma_1, \mathbf{p}_2\sigma_2} f_{\mathbf{p}_1\sigma_1\mathbf{p}_2\sigma_2} \, \delta n_{\mathbf{p}_1\sigma_1} \, \delta n_{\mathbf{p}_2\sigma_2}. \tag{5.4}$$

Here $\varepsilon^0_{\mathbf{p}\sigma}$ is the quasiparticle energy in the ground state. In the next section, we will show that the expansion in δn is general and does not require weak interactions. The small expansion parameter in Fermi liquid theory is the density of quasiparticles, or equivalently the excitation energy, and not the strength of the interaction. This allows a systematic treatment of strongly interacting systems at low temperatures.

The second term in Eq. (5.4), the quasiparticle interaction $f_{\mathbf{p}_1\sigma_1\mathbf{p}_2\sigma_2}$, has no correspondence in the non-interacting Fermi gas. In an excited state with more than one quasiparticle, the quasiparticle energy is modified according to

$$\varepsilon_{\mathbf{p}\sigma} = \frac{\delta E}{\delta n_{\mathbf{p}\sigma}} = \varepsilon^0_{\mathbf{p}\sigma} + \frac{1}{V}\sum_{\mathbf{p}_2\sigma_2} f_{\mathbf{p}\sigma\mathbf{p}_2\sigma_2}\,\delta n_{\mathbf{p}_2\sigma_2}, \tag{5.5}$$

where the changes are effectively proportional to the quasiparticle density.

The quasiparticle interaction can be understood microscopically from the second variation of the energy with respect to the quasiparticle distribution,

$$f_{\mathbf{p}_1\sigma_1\mathbf{p}_2\sigma_2} = V\,\frac{\delta^2 E}{\delta n_{\mathbf{p}_1\sigma_1}\delta n_{\mathbf{p}_2\sigma_2}} = V\,\frac{\delta \varepsilon_{\mathbf{p}_1\sigma_1}}{\delta n_{\mathbf{p}_2\sigma_2}}. \tag{5.6}$$

As discussed in detail in Sect. 5.2.5, this variation diagrammatically corresponds to cutting one of the fermion lines in a given energy diagram and labeling the incoming and outgoing fermion by $\mathbf{p}_1\sigma_1$, followed by a second variation leading to $\mathbf{p}_2\sigma_2$. For the uniform system, the resulting contributions to $f_{\mathbf{p}_1\sigma_1\mathbf{p}_2\sigma_2}$ are quasiparticle reducible in the particle-particle and in the exchange particle-hole (induced interaction) channels, but irreducible in the direct particle-hole (zero sound) channel. The zero-sound reducible diagrams are generated by the particle-hole scattering equation.

In normal Fermi systems, the quasiparticle concept makes sense only for states close to the Fermi surface, where the quasiparticle life-time $\tau_{\mathbf{p}}$ is long. The leading term is quadratic in the momentum difference from the Fermi surface [10], $1/\tau_{\mathbf{p}} \sim (p - k_{\mathrm{F}})^2$, while the dependence of the quasiparticle energy is linear, $\varepsilon_{\mathbf{p}} - \mu \sim (p - k_{\mathrm{F}})$. Thus, the condition

$$|\varepsilon_{\mathbf{p}} - \mu| \gg \frac{1}{\tau_{\mathbf{p}}}, \tag{5.7}$$

which is needed for the quasiparticle to be well defined, is satisfied by states close enough to the Fermi surface. Generally, quasiparticles are useful for time scales $\tau \ll \tau_{\mathbf{p}}$ and thus for high frequencies $|\omega|\tau_{\mathbf{p}} \gg 1$. In particular, states deep in the Fermi sea, which are occupied in the ground-state distribution, do not correspond to well-defined quasiparticles. Accordingly, we refer to the interacting ground state that corresponds to a filled Fermi sea in the non-interacting system as a state with no quasiparticles. In a weakly excited state the quasiparticle distribution $\delta n_{\mathbf{p}\sigma}$ is generally non-zero only for states close to the Fermi surface.

For low-lying excitations, the quasiparticle energy $\varepsilon_{\mathbf{p}\sigma}$ and interaction $f_{\mathbf{p}_1\sigma_1\mathbf{p}_2\sigma_2}$ is needed only for momenta close to the Fermi momentum k_{F}. It is then sufficient to retain the leading term in the expansion of $\varepsilon_{\mathbf{p}\sigma} - \mu$ around the Fermi surface, and to take the magnitude of the quasiparticle momenta in $f_{\mathbf{p}_1\sigma_1\mathbf{p}_2\sigma_2}$ equal to the Fermi momentum. In an isotropic and spin-saturated system ($N_\uparrow = N_\downarrow$), and if the interaction between free particles is invariant under $SU(2)$ spin symmetry (so that there are no non-central contributions, such as $\sim \boldsymbol{\sigma}\cdot\mathbf{p}$ to the energy), we have

$$\varepsilon_{\mathbf{p}\sigma} - \mu = \varepsilon_p - \mu \approx v_{\mathrm{F}}(p - k_{\mathrm{F}}) + \ldots, \tag{5.8}$$

where $v_F = k_F/m^*$ denotes the Fermi velocity and m^* is the effective mass. In addition, the quasiparticle interaction can be decomposed as

$$f_{\mathbf{p}_1\sigma_1\mathbf{p}_2\sigma_2} = f^s_{\mathbf{p}_1\mathbf{p}_2} + f^a_{\mathbf{p}_1\mathbf{p}_2}\,\boldsymbol{\sigma}_1\cdot\boldsymbol{\sigma}_2, \tag{5.9}$$

where

$$f^s_{\mathbf{p}_1\mathbf{p}_2} = \frac{1}{2}\left(f_{\mathbf{p}_1\uparrow\mathbf{p}_2\uparrow} + f_{\mathbf{p}_1\uparrow\mathbf{p}_2\downarrow}\right) \quad \text{and} \quad f^a_{\mathbf{p}_1\mathbf{p}_2} = \frac{1}{2}\left(f_{\mathbf{p}_1\uparrow\mathbf{p}_2\uparrow} - f_{\mathbf{p}_1\uparrow\mathbf{p}_2\downarrow}\right). \tag{5.10}$$

In nuclear physics the notation $f_{\mathbf{p}_1\mathbf{p}_2} = f^s_{\mathbf{p}_1\mathbf{p}_2}$ and $g_{\mathbf{p}_1\mathbf{p}_2} = f^a_{\mathbf{p}_1\mathbf{p}_2}$ is generally used, and the quasiparticle interaction includes additional terms that take into account the isospin dependence and non-central tensor contributions [11–13]. However, for our discussion here, the spin and isospin dependence is not important.

For the uniform system, Eq. (5.6) yields the quasiparticle interaction only for forward scattering (low momentum transfers). In the particle-hole channel, this corresponds to the long-wavelength limit. This restriction, which is consistent with considering low excitation energies, constrains the momenta $\mathbf{p}_1$ and $\mathbf{p}_2$ to be close to the Fermi surface, $|\mathbf{p}_1| = |\mathbf{p}_2| = k_F$. The quasiparticle interaction then depends only on the angle between $\mathbf{p}_1$ and $\mathbf{p}_2$. It is convenient to expand this dependence on Legendre polynomials

$$f^{s/a}_{\mathbf{p}_1\mathbf{p}_2} = f^{s/a}(\cos\theta_{\mathbf{p}_1\mathbf{p}_2}) = \sum_l f^{s/a}_l \, P_l(\cos\theta_{\mathbf{p}_1\mathbf{p}_2}), \tag{5.11}$$

and to define the dimensionless Landau Parameters $F^{s/a}_l$ by

$$F^{s/a}_l = N(0)\, f^{s/a}_l, \tag{5.12}$$

where $N(0) = \frac{1}{V}\sum_{\mathbf{p}\sigma}\delta(\varepsilon_{\mathbf{p}\sigma} - \mu) = m^* k_F/\pi^2$ denotes the quasiparticle density of states at the Fermi surface.

The Landau parameters can be directly related to macroscopic properties of the system. F^s_1 determines the effective mass and the specific heat c_V,

$$\frac{m^*}{m} = 1 + \frac{F^s_1}{3}, \tag{5.13}$$

$$c_V = \frac{m^* k_F}{3}\, k_B^2 T, \tag{5.14}$$

while the compressibility K and incompressibility κ are given by F^s_0,

$$K = -\frac{1}{V}\frac{\partial V}{\partial P} = \frac{1}{n^2}\frac{\partial n}{\partial \mu} = \frac{1}{n^2}\frac{N(0)}{1+F_0^s}, \tag{5.15}$$

$$\kappa = \frac{9}{nK} = -\frac{9V}{n}\frac{\partial P}{\partial V} = 9\frac{\partial P}{\partial n} = \frac{3k_{\rm F}^2}{m^*}(1+F_0^s). \tag{5.16}$$

Moreover, the spin susceptibility χ_m is related to F_0^a,

$$\chi_m = \frac{\partial m}{\partial H} = \beta^2 \frac{N(0)}{1+F_0^a}, \tag{5.17}$$

for spin-1/2 fermions with magnetic moment $\beta = ge/(4m)$ and gyromagnetic ratio g. Finally, a stability analysis of the Fermi surface against small amplitude deformations leads to the Pomeranchuk criteria [14]

$$F_l^{s/a} > -(2l+1). \tag{5.18}$$

For instance $F_0^{s/a} < -1$ implies an instability against spontaneous growth of density/spin fluctuations.

Landau's theory of normal Fermi liquids is an effective low-energy theory in the modern sense [1, 2]. The effective theory incorporates the symmetries of the system and the low-energy couplings can be fixed by experiment or calculated microscopically based on the underlying theory. Fermi Liquid theory has been very successful in describing low-temperature Fermi liquids, in particular liquid ^{3}He. Applications to the normal phase are reviewed, for example, in Baym and Pethick [10] and Pines and Nozières [15], while we refer to Wölfle and Vollhardt [16] for a description of the superfluid phases. The first applications to nuclear systems were pioneered by Migdal [11] and first microscopic calculations for nuclei and nuclear matter by Brown et al. [12]. Recently, advances using RG methods for nuclear forces [17] have lead to the development of a non-perturbative RG approach for nucleonic matter [18], to a first complete study of the spin structure of induced interactions [13], and to new calculations of Fermi liquid parameters [19, 20].

5.2.2 Three-Quasiparticle Interactions

In Sect. 5.2.1, we introduced Fermi liquid theory as an expansion in the density of quasiparticles $\delta n/V$. In applications of Fermi liquid theory to date, even for liquid ^{3}He, which is a very dense and strongly interacting system, this expansion is truncated after the second-order $(\delta n)^2$ term, including only pairwise interactions of quasiparticles [see Eq. (5.4)]. However, for a strongly interacting system, there is a priori no reason that three-body (or higher-body) interactions between quasiparticles are small. In this section, we discuss the convergence of this expansion. Three-quasiparticle interactions arise from iterated two-body forces, leading to three- and higher-body clusters in the linked-cluster expansion, or through many-body forces. While three-body forces play an important role in nuclear physics [21–23], little is known about them in other Fermi liquids. Nevertheless, in strongly interacting systems, the contributions of many-body clusters can in general be significant, leading to potentially important $(\delta n)^3$ terms in the Fermi liquid expansion, also in the absence of three-body forces:

$$\delta E = \sum_1 \varepsilon_1^0 \, \delta n_1 + \frac{1}{2V} \sum_{1,2} f_{1,2}^{(2)} \, \delta n_1 \, \delta n_2 + \frac{1}{6V^2} \sum_{1,2,3} f_{1,2,3}^{(3)} \, \delta n_1 \, \delta n_2 \, \delta n_3. \quad (5.19)$$

Here $f_{1,\ldots,n}^{(n)}$ denotes the n-quasiparticle interaction (the Landau interaction is $f \equiv f^{(2)}$) and we have introduced the short-hand notation $n \equiv \mathbf{p}_n \sigma_n$.

In order to better understand the expansion, Eq. (5.19), around the interacting ground state with N fermions, consider exciting or adding N_q quasiparticles with $N_q \ll N$. The microscopic contributions from many-body clusters or from many-body forces can be grouped into diagrams containing zero, one, two, three, or more quasiparticle lines. The terms with zero quasiparticle lines contribute to the interacting ground state for $\delta n = 0$, whereas the terms with one, two, and three quasiparticle lines contribute to ε_1^0, $f_{1,2}^{(2)}$, and $f_{1,2,3}^{(3)}$, respectively (these also depend on the ground-state density due to the N fermion lines). The terms with more than three quasiparticle lines would contribute to higher-quasiparticle interactions. Because a quasiparticle line replaces a line summed over N fermions when going from ε_1^0 to $f_{1,2}^{(2)}$, and from $f_{1,2}^{(2)}$ to $f_{1,2,3}^{(3)}$, it is intuitively clear that the contributions due to three-quasiparticle interactions are suppressed by N_q/N compared to two-quasiparticle interactions, and that the Fermi liquid expansion is effectively an expansion in N_q/N or n_q/n [15].

Fermi liquid theory applies to normal Fermi systems at low energies and temperatures, or equivalently at low quasiparticle densities. We first consider excitations that conserve the net number of quasiparticles, $\delta N = \sum_{\mathbf{p}\sigma} \delta n_{\mathbf{p}\sigma} = 0$, so that the number of quasiparticles equals the number of quasiholes. This corresponds to the lowest energy excitations of normal Fermi liquids. We denote their energy scale by Δ. Excitations with one valence particle or quasiparticle added start from energies of order the chemical potential μ. In the case of $\delta N = 0$, the contributions of two-quasiparticle interactions are of the same order as the first-order δn term, but three-quasiparticle interactions are suppressed by Δ/μ [24]. This is the reason that Fermi liquid theory with only two-body Landau parameters is so successful in describing even strongly interacting and dense Fermi liquids. This counting is best seen from the variation of the free energy $F = E - \mu N$,

$$\begin{aligned} \delta F &= \delta(E - \mu N) \\ &= \sum_1 (\varepsilon_1^0 - \mu) \, \delta n_1 + \frac{1}{2V} \sum_{1,2} f_{1,2}^{(2)} \, \delta n_1 \, \delta n_2 + \frac{1}{6V^2} \sum_{1,2,3} f_{1,2,3}^{(3)} \, \delta n_1 \, \delta n_2 \, \delta n_3, \end{aligned} \quad (5.20)$$

which for $\delta N = 0$ is equivalent to δE of Eq. (5.19). The quasiparticle distribution is $|\delta n_{\mathbf{p}\sigma}| \sim 1$ within a shell around the Fermi surface $|\varepsilon^0_{\mathbf{p}\sigma} - \mu| \sim \Delta$. The first-order δn term is therefore proportional to Δ times the number of quasiparticles $\sum_{\mathbf{p}\sigma} |\delta n_{\mathbf{p}\sigma}| = N_q \sim N(\Delta/\mu)$, and

$$\sum_1 (\varepsilon^0_1 - \mu)\, \delta n_1 \sim \frac{N\Delta^2}{\mu}. \tag{5.21}$$

Correspondingly, the contribution of two-quasiparticle interactions yields

$$\frac{1}{2V} \sum_{1,2} f^{(2)}_{1,2}\, \delta n_1\, \delta n_2 \sim \frac{1}{V} \langle f^{(2)} \rangle \left(\frac{N\Delta}{\mu} \right)^2 \sim \langle F^{(2)} \rangle \frac{N\Delta^2}{\mu}, \tag{5.22}$$

where $\langle F^{(2)} \rangle = n \langle f^{(2)} \rangle / \mu$ is an average dimensionless coupling on the order of the Landau parameters. Even in the strongly interacting, scale-invariant case $\langle f^{(2)} \rangle \sim 1/k_{\mathrm{F}}$; hence $\langle F^{(2)} \rangle \sim 1$ and the contribution of two-quasiparticle interactions is of the same order as the first-order term. However, the three-quasiparticle contribution is of order

$$\frac{1}{6V^2} \sum_{1,2,3} f^{(3)}_{1,2,3}\, \delta n_1\, \delta n_2\, \delta n_3 \sim \frac{n^2}{\mu} \langle f^{(3)} \rangle \frac{N\Delta^3}{\mu^2} \sim \langle F^{(3)} \rangle \frac{N\Delta^3}{\mu^2}. \tag{5.23}$$

Therefore at low excitation energies this is suppressed by Δ/μ, compared to two-quasiparticle interactions, even if the dimensionless three-quasiparticle interaction $\langle F^{(3)} \rangle = n^2 \langle f^{(3)} \rangle / \mu$ is strong (of order 1). Similarly, higher n-body interactions are suppressed by $(\Delta/\mu)^{n-2}$. Normal Fermi systems at low energies are weakly coupled in this sense. The small parameter is the ratio of the excitation energy per particle to the chemical potential. These considerations hold for all normal Fermi systems where the underlying interparticle interactions are finite range.

The Fermi liquid expansion in Δ/μ is equivalent to an expansion in $N_q/N \sim \Delta/\mu$, the ratio of the number of quasiparticles and quasiholes N_q to the number of particles N in the interacting ground state, or an expansion in the density of excited quasiparticles over the ground-state density, n_q/n. For the case where N_q quasiparticles or valence particles are added to a Fermi-liquid ground state, $\delta N \neq 0$ and the first-order term is

$$\sum_1 \varepsilon^0_1\, \delta n_1 \sim \mu N_q \sim \mu \frac{N\Delta}{\mu} \sim N\Delta, \tag{5.24}$$

while the contribution of two-quasiparticle interactions is suppressed by $N_q/N \sim \Delta/\mu$ and that of three-quasiparticle interactions by $(N_q/N)^2$.

Therefore, either for $\delta N = 0$ or $\delta N \neq 0$, the contributions of three-quasiparticle interactions to normal Fermi systems at low excitation energies are suppressed by the ratio of the quasiparticle density to the ground-state density, or equivalently by the ratio of the excitation energy over the chemical potential. This holds for excitations that conserve the number of particles (excited states of the interacting ground state) as well as for excitations that add or remove particles. This suppression is general and applies to strongly interacting systems even with strong, but finite-range three-body forces. However, this does not imply that the contributions from three-body forces to the interacting ground-state energy (the energy of the core nucleus in the context of shell-model calculations), to quasiparticle energies, or to two-quasiparticle interactions are small. The argument only applies to the effects of residual three-body interactions at low energies.

5.2.3 Microscopic Foundation of Fermi Liquid Theory

A central object in microscopic approaches to many-body systems is the one-body (time-ordered) propagator or Green's function G defined by

$$G(1,2) = -i\, \langle 0|\mathcal{T}\psi(1)\psi^\dagger(2)|0\rangle\,, \tag{5.25}$$

where $|0\rangle$ denotes the ground state of the system, $\mathcal{T}$ is the time-ordering operator, ψ and $\psi^\dagger$ annihilate and create a fermion, respectively, and 1, 2 are short hand for space, time and internal degrees of freedom (such as spin and isospin). For a translationally invariant spin-saturated system that is also invariant under rotations in spin space, the Green's function is diagonal in spin and can be written in momentum space as

$$G(\omega,\mathbf{p})\,\delta_{\sigma_1\sigma_2} = \int d(1-2)\, G(1,2)\, e^{i\omega(t_1-t_2)-i\mathbf{p}\cdot(\mathbf{x}_1-\mathbf{x}_2)} = \frac{\delta_{\sigma_1\sigma_2}}{\omega - \frac{p^2}{2m} - \Sigma(\omega,\mathbf{p})}\,, \tag{5.26}$$

where $\Sigma(\omega,\mathbf{p})$ defines the self-energy. For an introduction to many-body theory and additional details, we refer to the books by Fetter and Walecka [25], Abrikosov et al. [26], Negele and Orland [27], and Altland and Simons [28].

Without interactions the self-energy vanishes and consequently the free Green's function G_0 reads

$$G_0(\omega,\mathbf{p}) = \frac{1}{\omega - \frac{p^2}{2m} + i\delta_{\mathbf{p}}} = \frac{1-n^0_{\mathbf{p}}}{\omega - \frac{p^2}{2m} + i\delta} + \frac{n^0_{\mathbf{p}}}{\omega - \frac{p^2}{2m} - i\delta}\,, \tag{5.27}$$

where $\delta_{\mathbf{p}} = \delta\,\mathrm{sign}(p-k_{\mathrm{F}})$ and δ is a positive infinitesimal. The free Green's function has simple poles, as illustrated in Fig. 5.2, and the imaginary part takes the form

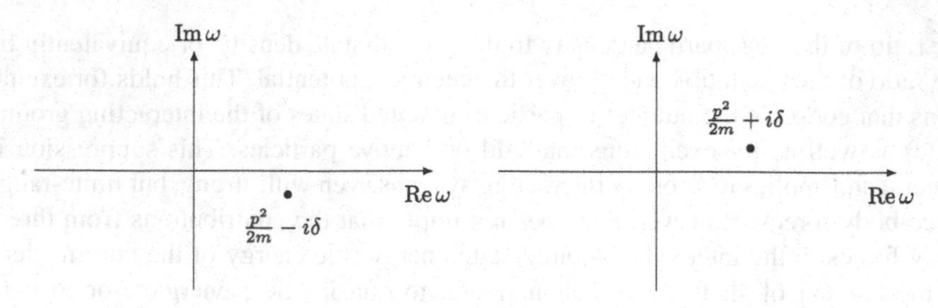

Fig. 5.2 Analytic structure of the free one-body Green's function G_0 in the complex ω plane with simple poles for $p > k_F$ (*left*) and $p < k_F$ (*right*)

$$\operatorname{Im} G_0(\omega, \mathbf{p}) = -\pi\,(1 - 2n^0_{\mathbf{p}})\,\delta\left(\omega - \frac{p^2}{2m}\right). \tag{5.28}$$

The single-particle spectral function $\rho(\omega, \mathbf{p})$ is determined by the imaginary part of the retarded propagator

$$G^R(1,2) = -i\,\theta(t_1 - t_2)\,\langle 0|\{\psi(1), \psi^\dagger(2)\}|0\rangle, \tag{5.29}$$

$$\rho(\omega, \mathbf{p}) = -\frac{1}{\pi}\operatorname{Im} G^R(\omega, \mathbf{p}), \tag{5.30}$$

where $\{\,,\,\}$ denotes the anticommutator. The retarded propagator is analytic in the upper complex ω plane and fulfills Kramers-Kronig relations, which relate the real and imaginary parts. Physically this implies that all modes are propagating forward in time and causality is fulfilled. Therefore, response functions are usually expressed in terms of the retarded propagator.

In a non-interacting system the retarded propagator is given by

$$G_0^R(\omega, \mathbf{p}) = \frac{1}{\omega - \frac{p^2}{2m} + i\delta}, \tag{5.31}$$

$$\operatorname{Im} G_0^R(\omega, \mathbf{p}) = -\pi\,\delta\left(\omega - \frac{p^2}{2m}\right), \tag{5.32}$$

which implies that the free spectral function is a delta function $\rho_0(\omega, \mathbf{p}) = \delta(\omega - \frac{p^2}{2m})$. This simple form follows from the fact that single-particle plane-wave states are eigenstates of the non-interacting Hamiltonian.

In the interacting case the situation is more complicated. Here the quasiparticle energy is given implicitly by the Dyson equation

$$\varepsilon_{\mathbf{p}} = \frac{p^2}{2m} + \Sigma(\varepsilon_{\mathbf{p}}, \mathbf{p}). \tag{5.33}$$

At the chemical potential $\omega = \mu$, the imaginary part of the self-energy vanishes,

$$\mathrm{Im}\, \Sigma(\mu, \mathbf{p}) = 0, \tag{5.34}$$

and the quasiparticle life-time $\tau_{\mathbf{p}} \to \infty$ for $|\mathbf{p}| \to k_{\mathrm{F}}$.[1] For $\omega \neq \mu$, the imaginary part of the self-energy obeys

$$\mathrm{Im}\, \Sigma(\omega, \mathbf{p}) < 0, \quad \text{for } \omega > \mu, \tag{5.35}$$

$$\mathrm{Im}\, \Sigma(\omega, \mathbf{p}) > 0, \quad \text{for } \omega < \mu. \tag{5.36}$$

The retarded self-energy, which enters the retarded Green's function

$$G^R(\omega, \mathbf{p}) = \frac{1}{\omega - \frac{p^2}{2m} - \Sigma^R(\omega, \mathbf{p})}, \tag{5.37}$$

is related to the time-ordered one through

$$\mathrm{Re}\, \Sigma^R(\omega, \mathbf{p}) = \mathrm{Re}\, \Sigma(\omega, \mathbf{p}), \tag{5.38}$$

$$\mathrm{Im}\, \Sigma^R(\omega, \mathbf{p}) = \begin{cases} +\mathrm{Im}\, \Sigma(\omega, \mathbf{p}) < 0, & \text{for } \omega > \mu, \\ -\mathrm{Im}\, \Sigma(\omega, \mathbf{p}) < 0, & \text{for } \omega < \mu. \end{cases} \tag{5.39}$$

Using Eq. (5.30), one finds the general form of the spectral function

$$\rho(\omega, \mathbf{p}) = -\frac{1}{\pi} \frac{\mathrm{Im}\, \Sigma^R(\omega, \mathbf{p})}{\left[\omega - \frac{p^2}{2m} - \mathrm{Re}\, \Sigma^R(\omega, \mathbf{p})\right]^2 + \left[\mathrm{Im}\, \Sigma^R(\omega, \mathbf{p})\right]^2}. \tag{5.40}$$

In the interacting case the single-particle strength is therefore, for a given momentum state, fragmented in energy, and the spectral function provides a measure of the single-particle strength in the eigenstates of the Hamiltonian, with the normalization $\int_{-\infty}^{\infty} d\omega\, \rho(\omega, \mathbf{p}) = 1$.

The spectral or Källén-Lehmann representation of the Green's function,

$$G^R(\omega, \mathbf{p}) = \int_{-\infty}^{\infty} d\omega' \, \frac{\rho(\omega', \mathbf{p})}{\omega - \omega' + i\delta}, \tag{5.41}$$

follows from analyticity and implies that the full propagator is completely determined by the spectral function. Using Eq. (5.41) and the normalization condition, the asymptotic ($|\omega| \to \infty$) behavior of both $G^R(\omega, \mathbf{p})$ and $G(\omega, \mathbf{p}) \sim 1/\omega$ follows. Furthermore, the singularities of the full Green's function (that correspond to eigenvalues of the Hamiltonian) are all located on the real axis and result in a cut along the

[1] At non-zero temperature, the imaginary part of the self-energy never vanishes and the quasiparticle life-time is finite. However, for $T \ll \mu$ and $\omega \approx \mu$, the life-time is large and the quasiparticle concept is useful.

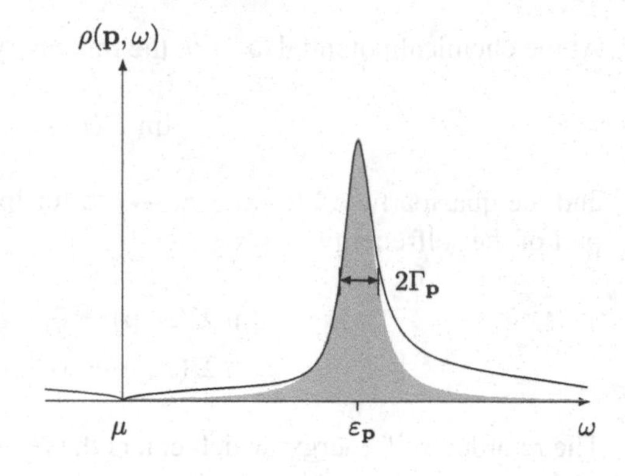

Fig. 5.3 Spectral function $\rho(\omega, \mathbf{p})$ for a given momentum $p \gtrsim k_{\mathrm{F}}$ as a function of frequency ω. The *shaded part* marks the quasiparticle peak, with quasiparticle width $\Gamma_{\mathbf{p}}$ and single-particle strength $z_{\mathbf{p}}$

real axis in the continuum limit. The quasiparticle pole, on the other hand, is located off the real axis, on an unphysical Riemann sheet,[2] with the distance to the real axis given by the quasiparticle width,

$$\begin{aligned} \text{quasiparticle pole } (p \gtrsim k_{\mathrm{F}}): \ \omega_{\mathbf{p}}^{\mathrm{qp}} &= \varepsilon_{\mathbf{p}} - i\Gamma_{\mathbf{p}}, \\ \text{quasihole pole } (p \lesssim k_{\mathrm{F}}): \ \omega_{\mathbf{p}}^{\mathrm{qp}} &= \varepsilon_{\mathbf{p}} + i\Gamma_{\mathbf{p}}. \end{aligned} \tag{5.44}$$

A pole close to the real axis gives rise to a peak in the single-particle strength, as illustrated in Fig. 5.3. Hence, microscopically a quasiparticle or quasihole is identified by a well-defined peak in the spectral function. In other words, the excitation of a quasiparticle corresponds to the coherent excitation of several eigenstates $H|\psi_i\rangle = E_i|\psi_i\rangle$ of the Hamiltonian, with similar energies E_i ($\omega' \approx \varepsilon_{\mathbf{p}}$ in Eq. (5.41)) spread over the quasiparticle width $\Gamma_{\mathbf{p}} = \tau_{\mathbf{p}}^{-1}$. A quasiparticle created at $t = 0$ then propagates in time as

$$|\psi_{\mathrm{qp}}(t)\rangle = \sum_i c_i \, e^{-iE_i t} \, |\psi_i\rangle. \tag{5.45}$$

For short times, $t \ll 1/\Gamma_{\mathbf{p}}$, the eigenstates remain coherent and the quasiparticle is well defined, while for $t \gg 1/\Gamma_{\mathbf{p}}$, the phase coherence is lost and the quasiparticle decays.

[2] This is readily seen by evaluating $G(z = \omega \pm i\delta, \mathbf{p})$ for a spectral function of the quasiparticle form,

$$\rho(\omega, \mathbf{p}) = \frac{1}{\pi} \frac{\Gamma_{\mathbf{p}}/2}{(\omega - \varepsilon_{\mathbf{p}})^2 + \Gamma_{\mathbf{p}}^2/4}, \tag{5.42}$$

using the Källén-Lehmann representation for the Green's function at a complex argument,

$$G(z, \mathbf{p}) = \int_{-\infty}^{\infty} d\omega' \, \frac{\rho(\omega', \mathbf{p})}{z - \omega'}. \tag{5.43}$$

One then finds that $G(\omega \pm i\delta, \mathbf{p}) = (\omega - \varepsilon_{\mathbf{p}} \pm i\Gamma_{\mathbf{p}}/2)^{-1}$. This implies that for ω in the upper half plane the quasiparticle pole is in the lower half plane and vice versa.

Using this definition of a quasiparticle, the full propagator can be formally separated into a quasiparticle part and a smooth background $\phi(\omega, \mathbf{p})$:

$$G(\omega, \mathbf{p}) = \frac{z_{\mathbf{p}}}{\omega - \varepsilon_{\mathbf{p}} + i\Gamma_{\mathbf{p}}} + \phi(\omega, \mathbf{p}), \tag{5.46}$$

where the single-particle strength $z_{\mathbf{p}}$ carried by the quasiparticle is given by

$$z_{\mathbf{p}} = \left[1 - \left.\frac{\partial\Sigma(\omega, \mathbf{p})}{\partial\omega}\right|_{\omega=\varepsilon_{\mathbf{p}}}\right]^{-1} < 1, \tag{5.47}$$

and must be less than unity due to the normalization of the spectral function.

Close to the Fermi surface the width $\Gamma_{\mathbf{p}}$ is small, $\Gamma_{\mathbf{p}} \sim (p-k_{\mathrm{F}})^2$, and consequently quasiparticles are well defined. For processes with a typical time scale $\tau < \tau_{\mathbf{p}} = \Gamma_{\mathbf{p}}^{-1}$, the contribution of the quasiparticle remains coherent, while that of the smooth background is incoherent. Even for very small values of the single-particle strength $z_{\mathbf{p}}$, quasiparticles play a leading role at sufficiently low excitation energies.

5.2.4 Scattering of Quasiparticles

Quasiparticle scattering processes are in general described by the Bethe-Salpeter equation. The quasiparticle scattering amplitude is given by the full four-point function Γ, which includes contributions from scattering in the channels shown in Fig. 5.4. For small $q = (\omega, \mathbf{q})$, the contribution of the direct particle-hole or zero-sound (ZS) channel is singular due to a pinching of the integration contour by the quasiparticle poles of the two intermediate propagators for any external momenta $p_1 = (P+q')/2$ and $p_2 = (P - q')/2$, as discussed below and in detail in Ref. [26]. In contrast, for small q the exchange particle-hole (ZS′) and the particle-particle/hole-hole (BCS) channels are smooth for almost all external momenta.[3]

The Bethe-Salpeter equation that sums all ZS-channel reducible diagrams is shown diagrammatically in Fig. 5.5 and reads

$$\begin{aligned}\Gamma(p_1, p_2; q) &= \widetilde{\Gamma}(p_1, p_2; q) \\ &\quad - i\int \frac{d^4p}{(2\pi)^4}\, \widetilde{\Gamma}(p_1, p; q)\, G(p+q/2)\, G(p-q/2)\, \Gamma(p, p_2; q),\end{aligned} \tag{5.48}$$

where $\widetilde{\Gamma}$ denotes the ZS-channel irreducible four-point function and we suppress the spin of the fermions for simplicity. The singular part of the two intermediate

[3] The BCS singularity for back-to-back scattering, $P = 0$, is discussed in Sect. 5.3.

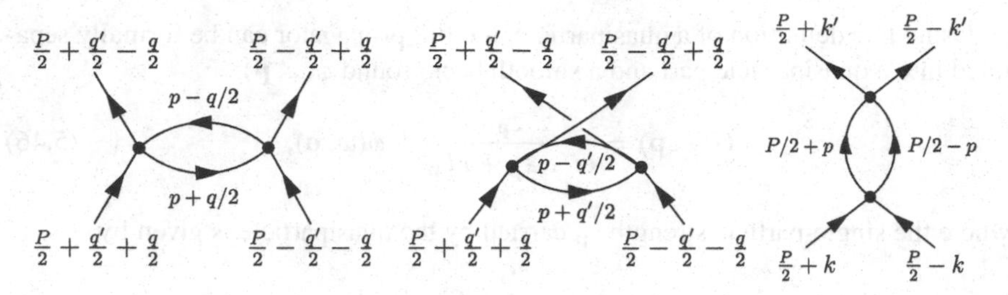

Fig. 5.4 Scattering channels in a many-fermion system. The direct particle-hole or zero-sound (ZS) channel (*left*), the exchange particle-hole (ZS′) channel (*middle*) and the particle-particle/hole-hole (BCS) channel (*right*). The relative incoming and outgoing four-momenta k, k' are related to the momentum transfers q, q' by $k = (q' + q)/2$ and $k' = (q' - q)/2$. The center-of-mass momentum is denoted by P

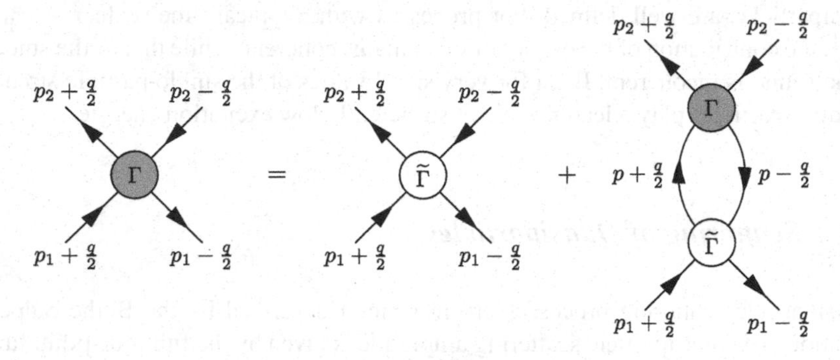

Fig. 5.5 The Bethe-Salpeter equation in the ZS channel, Eq. (5.48)

propagators is obtained by using the quasiparticle representation of the Green's function, Eq. (5.46), and for $q \to 0$ one finds

$$
\begin{aligned}
G(p+q/2)\,G(p-q/2) &= \frac{z_{\mathbf{p}+\mathbf{q}/2}}{\varepsilon+\omega/2-\mu-v_{\mathrm{F}}(|\mathbf{p}+\mathbf{q}|/2-k_{\mathrm{F}})+i\delta_{|\mathbf{p}+\mathbf{q}|/2}} \\
&\times \frac{z_{\mathbf{p}-\mathbf{q}/2}}{\varepsilon-\omega/2-\mu-v_{\mathrm{F}}(|\mathbf{p}-\mathbf{q}|/2-k_{\mathrm{F}})+i\delta_{|\mathbf{p}-\mathbf{q}|/2}} + \phi_2(p), \\
&= 2\pi i\, z_{k_{\mathrm{F}}}^2 \frac{|\mathbf{q}|\cos\theta_{\mathbf{pq}}}{\omega - v_{\mathrm{F}}|\mathbf{q}|\cos\theta_{\mathbf{pq}}}\,\delta(\varepsilon-\mu)\,\delta(|\mathbf{p}|-k_{\mathrm{F}}) + \phi_2(p),
\end{aligned}
\tag{5.49}
$$

where $p = (\varepsilon, \mathbf{p})$. The first term in Eq. (5.49) is the quasiparticle-quasihole part, which constrains the intermediate states to the Fermi surface, and the contribution $\phi_2(p)$ includes at least one power of the smooth background $\phi(p)$. In addition, we have taken $q \to 0$ in all nonsingular terms and neglected the quasiparticle width, which is small close to the Fermi surface.

We observe that the quasiparticle-quasihole part of the particle-hole propagator vanishes in the limit $|\mathbf{q}|/\omega \to 0$. Therefore, we define

$$\Gamma^{\omega}(p_1, p_2) = \lim_{\omega \to 0} \left(\Gamma(p_1, p_2; q) \big|_{|\mathbf{q}|=0} \right), \tag{5.50}$$

$$\Gamma^{q}(p_1, p_2) = \lim_{|\mathbf{q}| \to 0} \left(\Gamma(p_1, p_2; q) \big|_{\omega=0} \right). \tag{5.51}$$

Using the quasiparticle-quasihole representation of the particle-hole propagator, Eq. (5.49), the Bethe-Salpeter equation in the ZS channel takes the form

$$\Gamma(p_1, p_2; q) = \widetilde{\Gamma}(p_1, p_2; q) - i \int \frac{d^4 p}{(2\pi)^4} \widetilde{\Gamma}(p_1, p; q) \, \phi_2(p) \, \Gamma(p, p_2; q)$$
$$+ \frac{z_{k_F}^2 k_F^2}{(2\pi)^3} \int d\Omega_{\mathbf{p}} \, \widetilde{\Gamma}(p_1, p; q) \frac{|\mathbf{q}| \cos \theta_{\mathbf{pq}}}{\omega - v_F |\mathbf{q}| \cos \theta_{\mathbf{pq}}} \Gamma(p, p_2; q). \tag{5.52}$$

In the limit $|\mathbf{q}|/\omega \to 0$, we have for the quasiparticle-quasihole irreducible part of the four-point function Γ^{ω},

$$\Gamma^{\omega}(p_1, p_2) = \widetilde{\Gamma}(p_1, p_2) - i \int \frac{d^4 p}{(2\pi)^4} \widetilde{\Gamma}(p_1, p) \, \phi_2(p) \, \Gamma^{\omega}(p, p_2). \tag{5.53}$$

Using Γ^{ω}, we can then eliminate the ZS-channel irreducible four-point function $\widetilde{\Gamma}$ and the background term ϕ_2 to write the Bethe-Salpeter equation in the form

$$\Gamma(p_1, p_2; q) = \Gamma^{\omega}(p_1, p_2) + \frac{z_{k_F}^2 k_F^2}{(2\pi)^3} \int d\Omega_{\mathbf{p}} \Gamma^{\omega}(p_1, p) \frac{|\mathbf{q}| \cos \theta_{\mathbf{pq}}}{\omega - v_F |\mathbf{q}| \cos \theta_{\mathbf{pq}}} \Gamma(p, p_2; q). \tag{5.54}$$

In the limit $\omega/|\mathbf{q}| \to 0$ one then finds

$$\Gamma^{q}(p_1, p_2) = \Gamma^{\omega}(p_1, p_2) - \frac{z_{k_F}^2 m^* k_F}{(2\pi)^3} \int d\Omega_{\mathbf{p}} \, \Gamma^{\omega}(p_1, p) \, \Gamma^{q}(p, p_2), \tag{5.55}$$

which describes the scattering of quasiparticles. By identifying the quasiparticle interaction with (as justified in Sect. 5.2.5)

$$f_{\mathbf{p}_1 \mathbf{p}_2} = z_{k_F}^2 \, \Gamma^{\omega}(p_1, p_2) \big|_{\omega_1 = \varepsilon_{\mathbf{p}_1}, \, \omega_2 = \varepsilon_{\mathbf{p}_2}}, \tag{5.56}$$

and the quasiparticle scattering amplitude with

$$a_{\mathbf{p}_1 \mathbf{p}_2} = z_{k_F}^2 \, \Gamma^{q}(p_1, p_2) \big|_{\omega_1 = \varepsilon_{\mathbf{p}_1}, \, \omega_2 = \varepsilon_{\mathbf{p}_2}}, \tag{5.57}$$

the Bethe-Salpeter equation for the quasiparticle scattering amplitude reads

$$a_{\mathbf{p}_1 \sigma_1 \mathbf{p}_2 \sigma_2} = f_{\mathbf{p}_1 \sigma_1 \mathbf{p}_2 \sigma_2} - \frac{N(0)}{8\pi} \sum_{\sigma} \int d\Omega_{\mathbf{p}} \, a_{\mathbf{p}_1 \sigma_1 \mathbf{p} \sigma} \, f_{\mathbf{p} \sigma \mathbf{p}_2 \sigma_2}, \tag{5.58}$$

where we have reintroduced spin indices. By expanding the angular dependence of the quasiparticle scattering amplitude on Legendre polynomials,

$$a^{s/a}_{\mathbf{p}_1\mathbf{p}_2} = a^{s/a}(\cos\theta_{\mathbf{p}_1\mathbf{p}_2}) = \sum_l a^{s/a}_l \, P_l(\cos\theta_{\mathbf{p}_1\mathbf{p}_2}), \tag{5.59}$$

and using the corresponding expansion of the quasiparticle interaction, Eq. (5.11), the quasiparticle scattering equation, Eq. (5.58), can be solved analytically for each value of l,

$$A^{s,a}_l = \frac{F^{s,a}_l}{1 + \dfrac{F^{s,a}_l}{2l+1}}, \tag{5.60}$$

with $A^{s,a}_l = N(0)\, a^{s,a}_l$. This analytic solution of the quasiparticle scattering equation, Eq. (5.54), is in general possible only in the limit $q \to 0$.

Quasiparticles are fermionic excitations and therefore obey the Pauli principle. This imposes nontrivial constraints on the Landau parameters, as can be seen by the following argument. The full four-point function must be antisymmetric under exchange of two particles in the initial or final states,

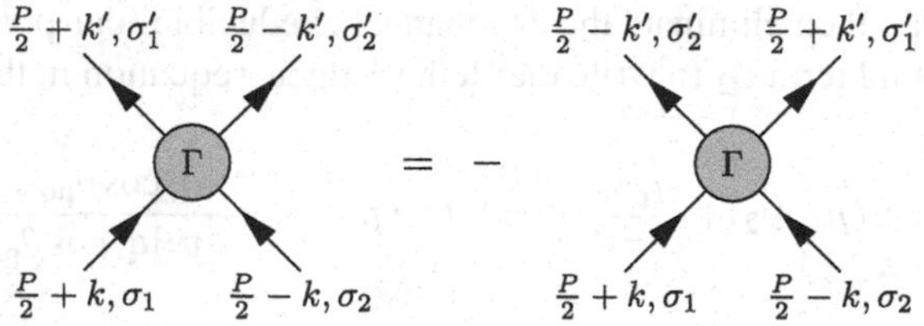

$\frac{P}{2}+k',\sigma_1'$ $\quad \frac{P}{2}-k',\sigma_2'$ $\qquad\qquad \frac{P}{2}-k',\sigma_2'$ $\quad \frac{P}{2}+k',\sigma_1'$

$$\Gamma \quad = \quad - \quad \Gamma \quad , \tag{5.61}$$

$\frac{P}{2}+k,\sigma_1$ $\quad \frac{P}{2}-k,\sigma_2$ $\qquad\qquad \frac{P}{2}+k,\sigma_1$ $\quad \frac{P}{2}-k,\sigma_2$

and therefore the forward scattering amplitude, $q = 0$, of identical particles, $q' = 0$ and $\sigma_1 = \sigma_2$, must vanish. This implies $a_{\mathbf{p}_1\sigma_1\mathbf{p}_1\sigma_1} = 0$, which leads to the Pauli-principle sum rule [7, 29]

$$\sum_l \left(A^s_l + A^a_l\right) = 0. \tag{5.62}$$

The relations given in this section can be generalized to include isospin and tensor forces [12, 13, 30, 31], which play an important role in nuclear systems.

5.2.5 Functional Approach

Functional methods provide a powerful tool for studying many-body systems. We start by discussing the two-particle irreducible (2PI) effective action. This provides a useful framework for an RG approach to many-body systems. For simplicity, we first consider bosonic systems described by a scalar field ϕ and generalize the results later to fermions. Our discussion follows Ref. [32]. We start with the expression for the generating functional W for connected N-point functions

$$W[J,K] = -\ln \int \mathcal{D}\phi \, \exp\Big[iS[\phi] - \int d^4x \, \phi(x)\, J(x) - \frac{1}{2} \int d^4x \, d^4y \, \phi(x)\, K(x,y)\, \phi(y) \Big], \quad (5.63)$$

where $\int \mathcal{D}\phi$ denotes a functional integral over the field ϕ, the action is given by $S[\phi] = \int d^4x \, \mathcal{L}(\phi(x))$, and $J(x)$ and $K(x,y)$ are external sources. In thermodynamic equilibrium, the space-time integral is

$$\int d^4x = \int_0^{-i\beta} dt \int d^3x, \quad (5.64)$$

with inverse temperature $\beta = 1/T$. By taking functional derivatives with respect to J and K, we obtain the expectation value of the field $\overline{\phi} = \langle \phi \rangle$ and the Green's function G respectively,

$$\frac{\delta W[J,K]}{\delta J(x)} = \langle \phi(x) \rangle = \overline{\phi}(x), \quad (5.65)$$

$$\frac{\delta W[J,K]}{\delta K(x,y)} = \frac{1}{2}\Big(\overline{\phi}(x)\overline{\phi}(y) + G(x,y) \Big). \quad (5.66)$$

A double Legendre transform leads to the 2PI effective action $\Gamma[\overline{\phi}, G]$,

$$\Gamma[\overline{\phi}, G] = W[J,K] - \int d^4x \, \overline{\phi}(x)\, J(x) - \frac{1}{2} \int d^4x \, d^4y \Big[\overline{\phi}(x)\overline{\phi}(y) + G(x,y) \Big] K(x,y), \quad (5.67)$$

which is stationary with respect to variations of $\overline{\phi}$ and G for vanishing sources,

$$\frac{\delta \Gamma[\overline{\phi}, G]}{\delta \overline{\phi}(x)} = -J(x) - \int d^4y \, K(x,y)\, \overline{\phi}(y) \overset{J=K=0}{=} 0, \quad (5.68)$$

$$\frac{\delta \Gamma[\overline{\phi}, G]}{\delta G(x,y)} = -\frac{1}{2} K(x,y) \overset{J=K=0}{=} 0. \quad (5.69)$$

An explicit form for Γ in terms of the expectation value $\overline{\phi}$ and the exact Green's function G can be constructed following Refs. [32–34]. This leads to

$$\Gamma[\overline{\phi}, G] = -iS[\overline{\phi}] + \operatorname{Tr} \ln G^{-1} + \operatorname{Tr}\Big[(G_0^{-1} - G^{-1}) G \Big] - \Phi[\overline{\phi}, G]. \quad (5.70)$$

Here the functional Φ is the sum of all 2PI skeleton diagrams (for a discussion of diagrams, see below) and the trace is a short-hand notation for

$$\mathrm{Tr} = \int d^4x \ \mathrm{tr} = \int_0^{-i\beta} dt \int d^3x \ \mathrm{tr}, \tag{5.71}$$

where tr denotes the trace over the internal degrees of freedom, such as spin and isospin. The stationarity of the 2PI effective action in the absence of sources, Eqs. (5.68) and (5.69), leads to the gap and Dyson equations, respectively. With the explicit form for Γ, Eq. (5.70), the Dyson equation is given by

$$\frac{\delta\Gamma[\overline{\phi}, G]}{\delta G} = -G^{-1} + G_0^{-1} + \frac{\delta\Phi[\overline{\phi}, G]}{\delta G} = 0, \tag{5.72}$$

which implies

$$G^{-1} = G_0^{-1} - \Sigma, \tag{5.73}$$

with the self-consistently determined self-energy Σ defined by

$$\Sigma = \frac{\delta\Phi[\overline{\phi}, G]}{\delta G}. \tag{5.74}$$

At the stationary point, the 2PI effective action Γ is proportional to the thermodynamic potential $\Omega = T\Gamma$ (in units with volume $V = 1$) [35]:

$$\begin{aligned}\Omega(\mu, T) = &-\int d^3x \,\mathcal{L}(\overline{\phi}(x)) + \Omega_0(\mu, T) \\ &+T\big[\,\mathrm{Tr}\ln(1 - G_0\Sigma) + \mathrm{Tr}\Sigma G - \Phi(\overline{\phi}, G)\big],\end{aligned} \tag{5.75}$$

where we have introduced the thermodynamic potential of the non-interacting Bose gas,

$$\Omega_0(\mu, T) = T\mathrm{Tr}\ln G_0^{-1} = T\mathrm{tr}\int \frac{d\mathbf{p}}{(2\pi)^3} \ln\Big[1 - \exp\Big[-\beta\Big(\frac{\mathbf{p}^2}{2m} - \mu\Big)\Big]\Big], \tag{5.76}$$

and the third term on the right-hand side of Eq. (5.75) can also be written as $-T\mathrm{Tr}\ln G_0^{-1} + T\mathrm{Tr}\ln G^{-1}$, so that $\Omega_0(\mu, T) + T\mathrm{Tr}\ln(1 - G_0\Sigma) = T\mathrm{Tr}\ln G^{-1}$. Finally, one can verify that the form of the 2PI effective action given by Eq. (5.75), where the self-energy enters explicitly, is stationary with respect to independent variations of G and Σ.

Similarly one has for the thermodynamic potential of a system consisting of fermions (where the expectation value $\langle\psi\rangle$ vanishes in the absence of sources)

$$\Omega(\mu, T) = -T\Big[\mathrm{Tr}\ln G^{-1} + \mathrm{Tr}\Sigma G - \Phi[G]\Big]. \tag{5.77}$$

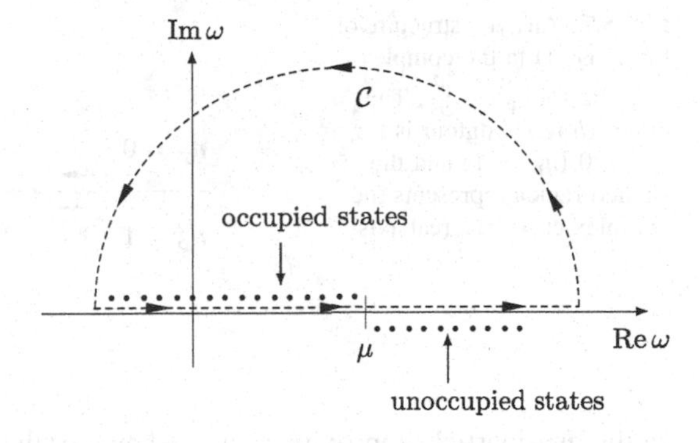

Fig. 5.6 Integration contour in the complex ω plane at $T = 0$

Moreover the energy of a fermionic system can be expressed at $T = 0$ in a form similar to Eq. (5.75),

$$E = E_0 - \mathrm{Tr}\ln(1 - G_0\Sigma) - \mathrm{Tr}\Sigma G + \Phi[G]. \tag{5.78}$$

Here $E_0 = \mathrm{tr}\int \frac{d\mathbf{p}}{(2\pi)^3}\frac{\mathbf{p}^2}{2m} n^0_{\mathbf{p}\sigma}$ is the energy of the non-interacting system, and at $T = 0$ the trace in Eq. (5.78) is given by

$$\mathrm{Tr} = \mathrm{tr}\int \frac{d^4p}{(2\pi)^4 i} = \mathrm{tr}\oint_C \frac{d\omega}{2\pi i}\frac{d\mathbf{p}}{(2\pi)^3}, \tag{5.79}$$

where the integration contour C is shown in Fig. 5.6.

Next, we use the functional integral approach to relate the quasiparticle interaction to the quasiparticle-quasihole irreducible part of the four-point function Γ^ω following Ref. [36]. In the quasiparticle approximation, the full propagator takes the form of Eq. (5.46),

$$G(\omega, \mathbf{p}) = \frac{1}{G_0^{-1}(\omega, \mathbf{p}) - \Sigma(\omega, \mathbf{p})} = \frac{z_{\mathbf{p}}}{\omega - \varepsilon_{\mathbf{p}} + i\delta_{\mathbf{p}}} + \phi(\omega, \mathbf{p}), \tag{5.80}$$

where we have neglected the imaginary part of the self-energy for excitations close to the Fermi surface. As shown in Sect. 5.2.3, the quasiparticle energy $\varepsilon_{\mathbf{p}}$ is given by the self-consistent solution to the Dyson equation, Eq. (5.33),

$$\varepsilon_{\mathbf{p}} = \frac{\mathbf{p}^2}{2m} + \Sigma(\varepsilon_{\mathbf{p}}, \mathbf{p}), \tag{5.81}$$

and the single-particle strength $z_{\mathbf{p}}$ by Eq. (5.47). When a quasiparticle with momentum $\mathbf{p}$ is added to the system, the state is changed from unoccupied to occupied, so that $\delta_{\mathbf{p}} = \delta \to -\delta$, with positive infinitesimal $\delta > 0$. Because the 2PI effective action is stationary with respect to independent variations of G and Σ, we only need to consider changes of E_0 and those induced by variations of G_0 in Eq. (5.78).

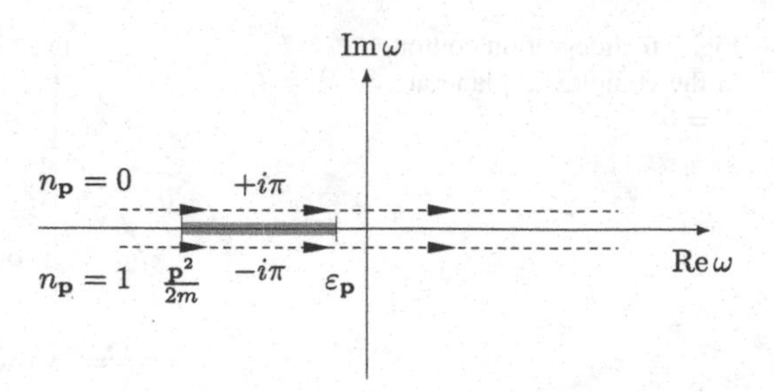

Fig. 5.7 Analytic structure of $\ln(G_0 G^{-1})$ in the complex ω plane for $\varepsilon_\mathbf{p} > \frac{\mathbf{p}^2}{2m}$. The upper (*lower*) contour is for $n_\mathbf{p} = 0$ ($n_\mathbf{p} = 1$) and the shaded region represents the complex cut on the real axis

In the quasiparticle approximation, we have for the argument of the logarithm,

$$1 - G_0(p)\Sigma(p) = G_0(p)\,G^{-1}(p) = \frac{1}{z_\mathbf{p}} \frac{\omega - \varepsilon_\mathbf{p} + i\delta_\mathbf{p}}{\omega - \frac{\mathbf{p}^2}{2m} + i\delta_\mathbf{p}} + \text{smooth parts.} \quad (5.82)$$

Consider the case $\varepsilon_\mathbf{p} > \frac{\mathbf{p}^2}{2m}$. Then the real part of $G_0\,G^{-1}$ is negative for $\frac{\mathbf{p}^2}{2m} < \omega < \varepsilon_\mathbf{p}$, resulting in a cut on the real energy, as shown in Fig. 5.7. When a particle is added to the system, the integration contour changes from above the cut to below, and $\ln(G_0\,G^{-1})$ changes by $-2\pi i$ for $\frac{\mathbf{p}^2}{2m} < \omega < \varepsilon_\mathbf{p}$. As a result, the change in the energy of the system is given by

$$\frac{\delta E}{\delta n_\mathbf{p}} = \frac{\delta E_0}{\delta n_\mathbf{p}} - \frac{\delta}{\delta n_\mathbf{p}} \left(\oint_C \frac{d\omega}{2\pi i} \frac{d\mathbf{p}}{(2\pi)^3} \ln\big(G_0(p)\,G^{-1}(p)\big) \right), \quad (5.83)$$

$$= \frac{\mathbf{p}^2}{2m} + \left(\varepsilon_\mathbf{p} - \frac{\mathbf{p}^2}{2m} \right) = \varepsilon_\mathbf{p}. \quad (5.84)$$

This variation is the quasiparticle energy, as postulated by Landau.

As discussed in Sect. 5.2.1, the quasiparticle interaction $f_{\mathbf{p}_1\mathbf{p}_2}$ is obtained by an additional variation with respect to the occupation number $n_{\mathbf{p}_2}$,

$$f_{\mathbf{p}_1\mathbf{p}_2} = \frac{\delta^2 E}{\delta n_{\mathbf{p}_1}\,\delta n_{\mathbf{p}_2}} = \frac{\delta \varepsilon_{\mathbf{p}_1}}{\delta n_{\mathbf{p}_2}}. \quad (5.85)$$

Using the Dyson equation, Eq. (5.81), the variation of the quasiparticle energy $\varepsilon_{\mathbf{p}_1}$ with respect to the occupation number $n_{\mathbf{p}_2}$ yields

$$\frac{\delta \varepsilon_{\mathbf{p}_1}}{\delta n_{\mathbf{p}_2}} = \left.\frac{\delta \Sigma(\omega_1, \mathbf{p}_1)}{\delta n_{\mathbf{p}_2}}\right|_{\omega_1 = \varepsilon_{\mathbf{p}_1}} + \left.\frac{\partial \Sigma(\omega_1, \mathbf{p}_1)}{\partial \omega_1}\right|_{\omega_1 = \varepsilon_{\mathbf{p}_1}} \frac{\delta \varepsilon_{\mathbf{p}_1}}{\delta n_{\mathbf{p}_2}}. \quad (5.86)$$

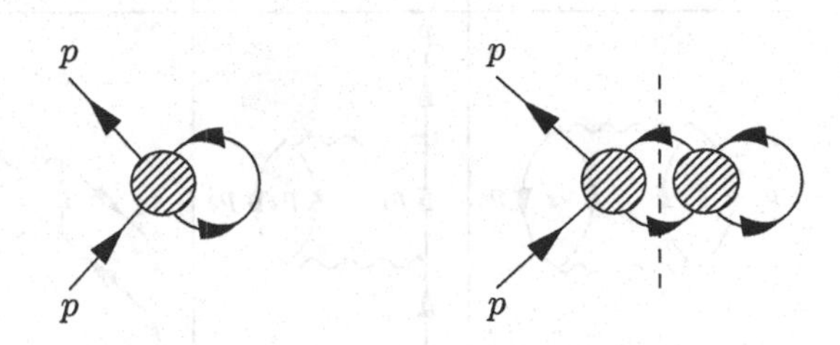

Fig. 5.8 Only 2PI skeleton diagrams contribute to the self-energy Σ, as shown on the *left*. Non-skeleton diagrams, such as the one on the *right*, are not part of Σ. Therefore, the kernel $\widetilde{\Gamma} = \delta\Sigma/\delta G$ (denoted by the *shaded blob*) is particle-hole irreducible in the ZS channel

This can be expressed with the single-particle strength $z_{\mathbf{p}}$ as

$$\frac{\delta\varepsilon_{\mathbf{p}_1}}{\delta n_{\mathbf{p}_2}} = z_{\mathbf{p}_1} \left. \frac{\delta\Sigma(\omega_1, \mathbf{p}_1)}{\delta n_{\mathbf{p}_2}} \right|_{\omega_1 = \varepsilon_{\mathbf{p}_1}} . \tag{5.87}$$

Furthermore, it follows from the definition of Σ through the Φ functional, $\Sigma(p) = \delta\Phi[G]/\delta G(p)$, that the self-energy consists of skeleton diagrams. Therefore, we can write

$$\left. \frac{\delta\Sigma(\omega_1, \mathbf{p}_1)}{\delta n_{\mathbf{p}_2}} \right|_{\omega_1 = \varepsilon_{\mathbf{p}_1}} = \int \frac{d^4 p}{(2\pi)^4 i} \left. \frac{\delta\Sigma(\omega_1, \mathbf{p}_1)}{\delta G(p)} \frac{\delta G(p)}{\delta n_{\mathbf{p}_2}} \right|_{\omega_1 = \varepsilon_{\mathbf{p}_1}} . \tag{5.88}$$

The variation of Σ with respect to G selects one of the internal lines of the diagrams contributing to the self-energy. As a result, the kernel

$$\widetilde{\Gamma}(p_1, p_2) = \frac{\delta\Sigma(p_1)}{\delta G(p_2)}, \tag{5.89}$$

must be particle-hole irreducible in the ZS channel. Otherwise the corresponding diagram in Σ would not have been a 2PI skeleton diagram. This is illustrated in Fig. 5.8.

Using the quasiparticle form of the full propagator, Eq. (5.80), there are two contributions to $\delta G(p_1)/\delta n_{\mathbf{p}_2}$ (for details, see also Ref. [36]). One for $p_1 = p_2$ ($\omega_2 = \varepsilon_{\mathbf{p}_2}$), which results in a shift of the quasiparticle pole across the integration contour, and one for $p_1 \neq p_2$, which corresponds to a variation of the self-energy,

$$\frac{\delta G(p_1)}{\delta n_{\mathbf{p}_2}} = (2\pi)^4 i \, z_{\mathbf{p}_2} \, \delta(\mathbf{p}_1 - \mathbf{p}_2)\delta(\omega_1 - \varepsilon_{\mathbf{p}_2}) + \frac{\delta\Sigma(p_1)}{\delta n_{\mathbf{p}_2}} G^2(p_1). \tag{5.90}$$

The $G^2(p_1)$ part in the second term is equivalent to the non-singular contribution $\phi_2(p_1)$ of Eq. (5.49). By inserting this expression for $\delta G(p_1)/\delta n_{\mathbf{p}_2}$ in Eq. (5.88), one finds the integral equation

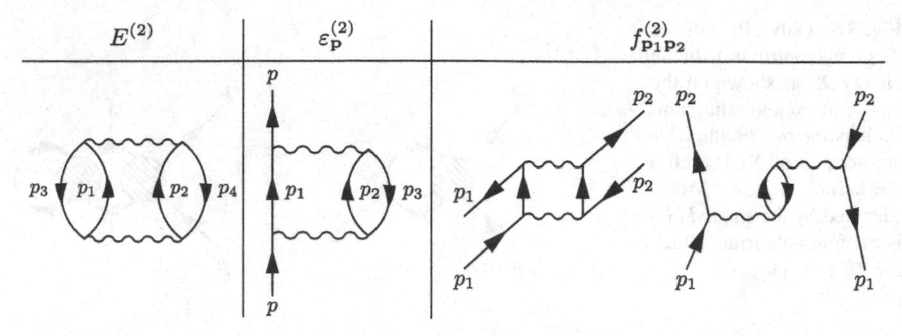

Fig. 5.9 The variation of the second-order energy diagram $E^{(2)}$ (*left*) with respect to the quasiparticle distribution function yields the second-order contribution to the quasiparticle energy $\varepsilon_{\mathbf{p}}^{(2)}$ given by the self-energy diagram (*middle*) and the corresponding two-hole–one-particle diagram. The second variation gives the second-order contributions to the quasiparticle interaction (*right*). These include the particle-particle and particle-hole diagrams shown and a particle-hole diagram that is obtained from the particle-particle one by reversing the arrow on the p_2 line (plus the diagrams obtained from the two-hole–one-particle self-energy contribution)

$$Y(p_1, p_2) = \frac{1}{z_{\mathbf{p}_2}} \frac{\delta \Sigma(\omega_1, \mathbf{p}_1)}{\delta n_{\mathbf{p}_2}}\bigg|_{\omega_1=\varepsilon_{\mathbf{p}_1}} \tag{5.91}$$

$$= \widetilde{\Gamma}(p_1, p_2) + \int \frac{d^4 p}{(2\pi)^4 i}\, \widetilde{\Gamma}(p_1, p)\, G^2(p)\, Y(p, p_2). \tag{5.92}$$

A comparison with Eq. (5.53) leads to the identification

$$Y(p_1, p_2) = \Gamma^{\omega}(p_1, p_2) = \frac{1}{z_{\mathbf{p}_2}} \frac{\delta \Sigma(\omega_1, \mathbf{p}_1)}{\delta n_{\mathbf{p}_2}}\bigg|_{\omega_1=\varepsilon_{\mathbf{p}_1}}. \tag{5.93}$$

Using Eq. (5.87), this implies

$$f_{\mathbf{p}_1\mathbf{p}_2} = \frac{\delta \varepsilon_{\mathbf{p}_1}}{\delta n_{\mathbf{p}_2}} = z_{\mathbf{p}_1} z_{\mathbf{p}_2}\, \Gamma^{\omega}(p_1, p_2) = \quad [\text{vertex } \Gamma^{\omega} \text{ with legs } \sqrt{z_{\mathbf{p}_2}}, \sqrt{z_{\mathbf{p}_2}}, \sqrt{z_{\mathbf{p}_1}}, \sqrt{z_{\mathbf{p}_1}}]\,. \tag{5.94}$$

This provides a microscopic basis for calculating the quasiparticle energy $\varepsilon_{\mathbf{p}} = \frac{\delta E}{\delta n_{\mathbf{p}_1}}$ and the quasiparticle interaction $f_{\mathbf{p}_1\mathbf{p}_2} = \frac{\delta^2 E}{\delta n_{\mathbf{p}_1} \delta n_{\mathbf{p}_2}}$ and a justification for the identification of the quasiparticle interaction as in Eq. (5.56).

The contributions to the quasiparticle interaction can be understood by considering the variation of the second-order energy diagram, as shown in Fig. 5.9. The resulting diagrams in $f_{\mathbf{p}_1\mathbf{p}_2}$ are quasiparticle-quasihole reducible in the BCS and ZS′ channels,

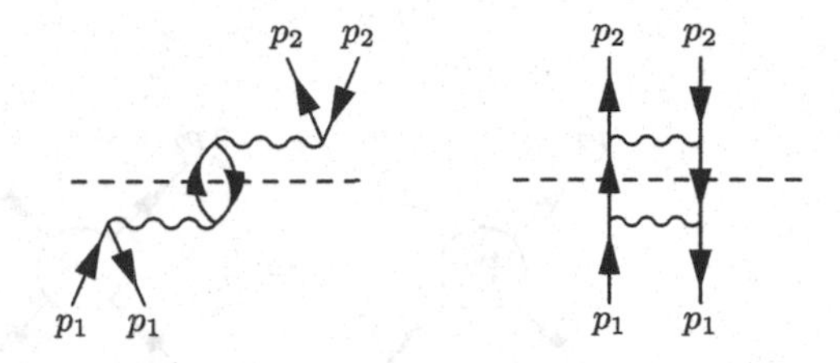

Fig. 5.10 Second-order diagrams that are not generated by variations of the energy diagram

but irreducible in the ZS channel (see Fig. 5.4 for a definition of the BCS, ZS and ZS′ channels). The ZS reducible diagrams, which are shown for the second-order example in Fig. 5.10, are included in the scattering amplitude $a_{\mathbf{p}_1\mathbf{p}_2}$ but not in the quasiparticle interaction. Because the ZS and ZS′ channels are related by exchange [37], the quasiparticle scattering amplitude is antisymmetric, but the quasiparticle interaction is not.

The quasiparticle-quasihole reducible diagrams in the ZS channel are summed by the Bethe-Salpeter equation, Eq. (5.55), which yields the fully reducible four-point function Γ^q, given the quasiparticle-quasihole irreducible one Γ^ω, as shown in Fig. 5.11. The fully reducible four-point function Γ^q corresponds to the quasiparticle scattering amplitude. The four-point function can also be obtained by summing diagrams that are quasiparticle-quasihole reducible in the ZS′ channel. The corresponding Bethe-Salpeter equation is shown in Fig. 5.12, where the irreducible term is the quasiparticle interaction in the ZS′ channel, the exchange of Γ^ω denoted by $\overline{\Gamma^\omega}$.

The kinematics in the integral term on the right-hand side of Fig. 5.12 requires as input the quasiparticle scattering amplitude Γ^q and the quasiparticle interaction Γ^ω at finite q. Therefore, we can generalize Γ^q on the left-hand side to finite q. If we then take the limit $|\mathbf{q}|/\omega \to 0$, all quasiparticle-quasihole reducible terms in the ZS channel vanish. In this limit, Γ^q on the left-hand side of Fig. 5.12 is replaced by Γ^ω, and the first term on the right hand side, $\overline{\Gamma^\omega}$, is reduced to the driving term I, which is quasiparticle-quasihole irreducible in both ZS and ZS′ channels. As a result, we obtain an integral equation for Γ^ω that sums quasiparticle-quasihole reducible diagrams in the ZS′ channel. This is shown diagrammatically in Fig. 5.13, where the second term on the right hand side is the induced interaction of Babu and Brown [37] (see also Ref. [19]). For $p_1 = p_2$, this integral equation has the simple form,

$$\Gamma^\omega(p_1, p_1) = I(p_1, p_1) - P_\sigma \frac{z_{k_\mathrm{F}}^2 m^* k_\mathrm{F}}{(2\pi)^3} \sum_\sigma \int d\Omega_\mathbf{p}\, \Gamma^\omega(p_1, p)\, \Gamma^q(p, p_1). \quad (5.95)$$

The induced interaction accounts for the contributions to the quasiparticle interaction due to the polarization of the medium and is necessary for the antisymmetry of the quasiparticle scattering amplitude Γ^q.

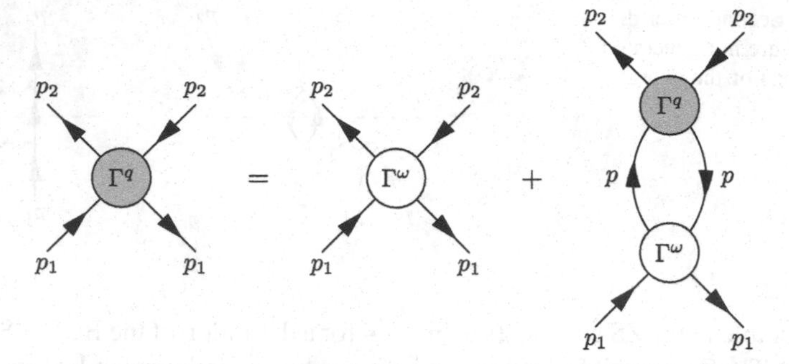

Fig. 5.11 The Bethe-Salpeter equation for the quasiparticle scattering amplitude, Eq. (5.55), which sums diagrams reducible in the ZS channel. The intermediate-state propagators include only the quasiparticle-quasihole part. All other contributions are included in Γ^ω

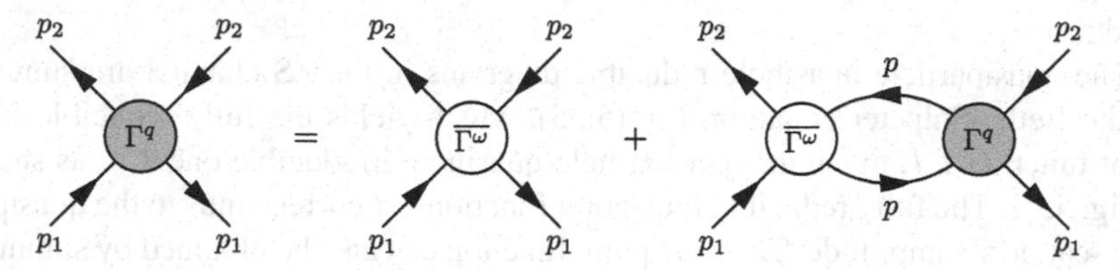

Fig. 5.12 The Bethe-Salpeter equation for the quasiparticle scattering amplitude in the exchange (ZS′) channel. In this channel, the solution requires as input the quasiparticle scattering amplitude Γ^q and the quasiparticle interaction Γ^ω at finite q. As in Fig. 5.11, the intermediate-state propagators include only the quasiparticle-quasihole part

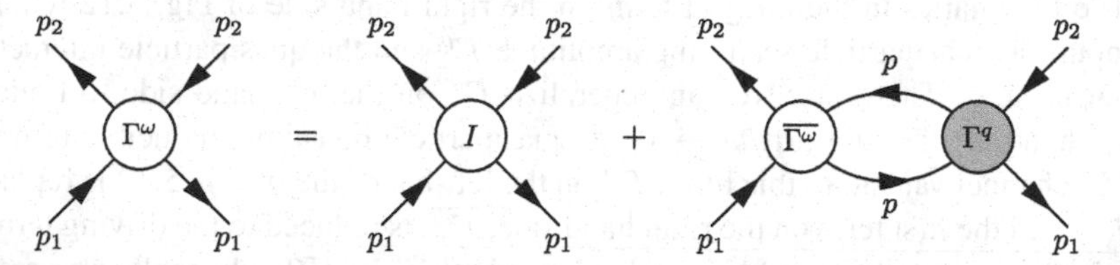

Fig. 5.13 Integral equation for the quasiparticle interaction Γ^ω that sums quasiparticle-quasihole reducible diagrams in the ZS′ channel. The second term on the *right hand side* is the induced interaction of Babu and Brown [37]

5.3 Functional RG Approach to Fermi Liquid Theory

In this section we apply the functional RG [38] to calculate the properties of a Fermi liquid. The basic idea is to renormalize the quasiparticle energy and the quasiparticle interaction, as one sequentially integrates out the excitations of the system. Thereby, one starts with the high-lying states and integrates down to low excitation energy. The functional RG leads to an infinite set of coupled differential equations for the n-point functions, which in practical calculations must be truncated. We return to this question below, when we discuss applications.

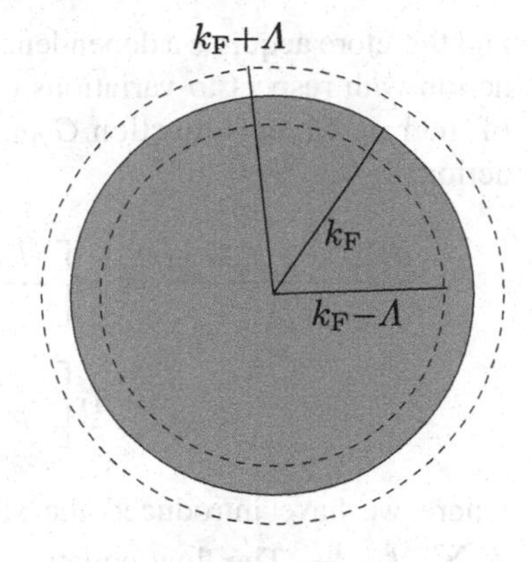

Fig. 5.14 The cutoffs above and below the Fermi surface separate high- and low-lying excitations of the many-fermion system

In zero-temperature Fermi systems, the low-lying states are those near the Fermi surface. Consequently, at some intermediate step of the calculation, the high-lying states, far above and far below the Fermi surface have been integrated out, while the states near the Fermi surface are not yet included. This is schematically illustrated in Fig. 5.14. However, in the first part of this section we keep the discussion more general and allow for finite temperatures. In the second part, we apply the RG approach to a Fermi liquid at zero temperature and also specify the detailed form of the regulator adapted to Fermi systems at zero temperature.

The Green's function

$$G_{\Lambda 0}^{-1}(\omega, \mathbf{p}) = \omega - \frac{\mathbf{p}^2}{2m} - R_\Lambda(\mathbf{p}), \tag{5.96}$$

with regulator $R_\Lambda(\mathbf{p})$ defines the free theory at a finite cutoff scale Λ. The corresponding effective action is given by (see Eq. (5.70) and the corresponding expression for fermions, Eq. (5.77))

$$\Gamma_\Lambda(\psi^*, \psi, G) = -i S_\Lambda(\psi^*, \psi) - \mathrm{Tr}\ln(-G^{-1}) - \mathrm{Tr}[G_{\Lambda 0}^{-1} G - 1] + \Phi(\psi^*, \psi, G), \tag{5.97}$$

with the action

$$S_\Lambda(\psi^*, \psi) = -i\, T \sum_n \int \frac{d^3 p}{(2\pi)^3}\, \psi^*(\omega_n, \mathbf{p}) \left(\omega_n - \frac{\mathbf{p}^2}{2m} - R_\Lambda(\mathbf{p}) \right) \psi(\omega_n, \mathbf{p}), \tag{5.98}$$

and Matsubara frequency $\omega_n = (2n+1) i\pi T$. At the stationary point, the propagator G satisfies the Dyson equation (see Eqs. (5.73) and (5.74))

$$G_\Lambda^{-1} = G_{\Lambda 0}^{-1} - \left.\frac{\delta \Phi(\psi^*, \psi, G)}{\delta G}\right|_{G=G_\Lambda}, \tag{5.99}$$

and therefore acquires a dependence on Λ. Due to the stationarity of the 2PI effective action with respect to variations of ψ, ψ^*, and G, only the explicit Λ dependence of the bare Green's function $G_{\Lambda 0}$ contributes to the flow equation for the effective action:

$$\begin{aligned}\frac{d\Gamma_\Lambda(\psi^*,\psi,G_\Lambda)}{d\Lambda} &= \mathrm{Tr}\left[\frac{\delta\Gamma_\Lambda(\psi^*,\psi,G_\Lambda)}{\delta G_{\Lambda 0}^{-1}(p)}\frac{dG_{\Lambda 0}^{-1}(p)}{d\Lambda}\right] \\ &= \mathrm{Tr}\left[\psi_p^*\,\frac{dR_\Lambda(\mathbf{p})}{d\Lambda}\,\psi_p\right] + \mathrm{Tr}\left[G_\Lambda(p)\,\frac{dR_\Lambda(\mathbf{p})}{d\Lambda}\right], \end{aligned} \tag{5.100}$$

where we have introduced the short-hand notations $\psi_p = \psi(\omega_n,\mathbf{p})$ and $\mathrm{Tr} = T\sum_n\int\frac{d\mathbf{p}}{(2\pi)^3}$. The flow equation follows almost trivially from the stationarity of the 2PI effective action, while within a 1PI scheme,[4] the derivation of the flow equation is somewhat more involved [38, 39]. The flow equation for the two-point function in the 1PI scheme is obtained by varying Eq. (5.100) with respect to ψ and ψ^*. Using

$$\Gamma_{p,p}^{\Lambda(2)} = \frac{\delta^2\Gamma_\Lambda}{\delta\psi_p^*\,\delta\psi_p} = G_\Lambda^{-1}(p), \tag{5.101}$$

and

$$\frac{\delta^2 G_\Lambda(p')}{\delta\psi_p^*\,\delta\psi_p} = -G_\Lambda(p')\,\Gamma_{p,p,p',p'}^{\Lambda(4)}\,G_\Lambda(p') \quad \text{with} \quad \Gamma_{p,p,p',p'}^{\Lambda(4)} = \frac{\delta^2\Gamma_{p,p}^{\Lambda(2)}}{\delta\psi_{p'}^*\,\delta\psi_{p'}}, \tag{5.102}$$

one finds

$$\frac{d\Gamma_{p,p}^{\Lambda(2)}}{d\Lambda} = -\frac{dR_\Lambda(\mathbf{p})}{d\Lambda} - \mathrm{Tr}\left[\Gamma_{p,p,p',p'}^{\Lambda(4)}\,G_\Lambda(p')\,\frac{dR_\Lambda(\mathbf{p}')}{d\Lambda}\,G_\Lambda(p')\right]. \tag{5.103}$$

Next, we briefly discuss the flow equation in the 2PI scheme and make a connection between the two schemes for the two-point function. Here we follow the discussion of Dupuis [40]. The starting point is the observation that the 2PI functional Φ does not flow, when ψ and G are treated as free variables

$$\left.\frac{d\Phi(\psi,G)}{d\Lambda}\right|_{\psi,G} = 0. \tag{5.104}$$

As discussed in the previous section, the functional $\Phi(\psi,G)$ generates the particle-hole irreducible n-point functions through variations with respect to the Green's

[4] The 1PI effective action is obtained by constraining the Green's function in the 2PI effective action to the solution of the Dyson equation, Eq. (5.99).

function (see also Ref. [34])

$$\Phi^{\Lambda(2n)}_{p_1,p_2,\ldots,p_{2n-1},p_{2n}} = \frac{\delta^n \Phi(\psi, G)}{\delta G(p_1, p_2)\cdots \delta G(p_{2n-1}, p_{2n})}. \tag{5.105}$$

After the variation, the Green's functions satisfy the Dyson equation, Eq. (5.99). Consequently, the flow of the 2PI vertices $\Phi^{(n)}$ results only from the Λ dependence of the Green's function:

$$\frac{d}{d\Lambda}\Phi^{\Lambda(2n)}_{p_1,p_2,\ldots,p_{2n-1},p_{2n}} = \mathrm{Tr}\left[\Phi^{\Lambda(2n+2)}_{p_1,p_2,\ldots,p_{2n-1},p_{2n},q,q}\frac{dG_\Lambda(q)}{d\Lambda}\right]. \tag{5.106}$$

Combined with the self-energy $\Sigma(p) = \Phi^{\Lambda(2)}_{p,p}$, so that $\Gamma^{\Lambda(2)}_{p,p} = G^{-1}_{\Lambda 0}(p) - \Phi^{\Lambda(2)}_{p,p}$ and $\frac{d\Gamma^{\Lambda(2)}_{p,p}}{d\Lambda} = -\frac{dR_\Lambda(\mathbf{p})}{d\Lambda} - \frac{d\Phi^{\Lambda(2)}_{p,p}}{d\Lambda}$, one finds

$$\frac{d\Gamma^{\Lambda(2)}_{p,p}}{d\Lambda} = -\frac{d\,R_\Lambda(\mathbf{p})}{d\Lambda} + \mathrm{Tr}\left[\Phi^{\Lambda(4)}_{p,p,p',p'}\,G_\Lambda(p')\,\frac{d\Gamma^{\Lambda(2)}_{p',p'}}{d\Lambda}\,G_\Lambda(p')\right]. \tag{5.107}$$

The flow equations for the two-point function in the 1PI and 2PI schemes, Eqs. (5.103) and (5.107), are equivalent, as can be shown in a straightforward calculation, making use of the Bethe-Salpeter equation for scattering of two particles of vanishing total momentum [41],

$$\Gamma^{\Lambda(4)}_{p_1,p_1,p_2,p_2} = \Phi^{\Lambda(4)}_{p_1,p_1,p_2,p_2} + \mathrm{Tr}\left[\Phi^{\Lambda(4)}_{p_1,p_1,p',p'}\,G_\Lambda(p')\,G_\Lambda(p')\,\Gamma^{\Lambda(4)}_{p',p',p_2,p_2}\right]. \tag{5.108}$$

The relation between the two schemes is illustrated diagrammatically in Fig. 5.15. The particle-hole reducible diagrams can be shifted between the four-point function and the regulator insertion on the fermion line. For more details on the relation between the RG approaches based on 1PI and 2PI functionals the reader is referred to Ref. [40].

We are now in a position to connect with Fermi liquid theory, by deriving a flow equation for the quasiparticle energy closely related to 1PI equation, Eq. (5.103). To this end, we first define the energy functional following Eq. (5.78)

$$E_\Lambda(G) = E_{\Lambda 0} - \mathrm{Tr}\ln(G_{\Lambda 0}\,G^{-1}) - \mathrm{Tr}\big[(G^{-1}_{\Lambda 0} - G^{-1})\,G\big] + \Phi(G), \tag{5.109}$$

which at the stationary point equals the (zero-temperature) ground-state energy at the cutoff scale Λ. The trace Tr is defined as in Eq. (5.79), $E_{\Lambda 0}$ is the energy of the non-interacting system at the scale Λ,

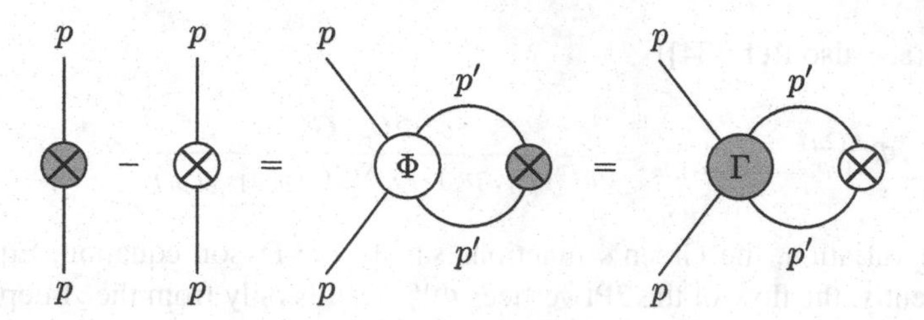

Fig. 5.15 Graphical representation of the 1PI and 2PI flow equations for the two-point function, Eqs. (5.103) and (5.107). The *internal lines* represent full propagators G_Λ. The *filled crossed circles* represent $\frac{d\Gamma^{\Lambda(2)}}{d\Lambda}$ and the unfilled ones $\frac{dG_{\Lambda 0}^{-1}}{d\Lambda} = -\frac{dR_\Lambda}{d\Lambda}$

$$E_{\Lambda 0} = \mathrm{tr} \int \frac{d^3p}{(2\pi)^3} \left[\frac{\mathbf{p}^2}{2m} + R_\Lambda(\mathbf{p}) \right] n^0_{\mathbf{p},\Lambda}, \tag{5.110}$$

with $n^0_{\mathbf{p},\Lambda} = \theta(k_{\mathrm{F}} - \Lambda - p)$, and the free Green's function $G_{\Lambda 0}$ is given by

$$G_{\Lambda 0}(\omega, \mathbf{p}) = \frac{1 - n^0_{\mathbf{p},\Lambda}}{\omega - \frac{\mathbf{p}^2}{2m} - R_\Lambda(\mathbf{p}) + i\delta} + \frac{n^0_{\mathbf{p},\Lambda}}{\omega - \frac{\mathbf{p}^2}{2m} - R_\Lambda(\mathbf{p}) - i\delta}. \tag{5.111}$$

Furthermore, in the quasiparticle approximation the one-particle Green's function of the interacting system is of the form

$$G_\Lambda(\omega, \mathbf{p}) = z_{\mathbf{p}} \left[\frac{1 - n^0_{\mathbf{p},\Lambda}}{\omega - \tilde{\varepsilon}_{\mathbf{p}} + i\delta} + \frac{n^0_{\mathbf{p},\Lambda}}{\omega - \tilde{\varepsilon}_{\mathbf{p}} - i\delta} \right] + \phi(\mathbf{p}, \omega), \tag{5.112}$$

where the quasiparticle energy $\tilde{\varepsilon}_{\mathbf{p}}$ is given by

$$\tilde{\varepsilon}_{\mathbf{p}} = \frac{\mathbf{p}^2}{2m} + \Sigma_\Lambda(\tilde{\varepsilon}_{\mathbf{p}}, \mathbf{p}) + R_\Lambda(\mathbf{p}). \tag{5.113}$$

The flow equation that follows from the energy functional, Eq. (5.109), reads

$$\frac{dE_\Lambda}{d\Lambda} = \frac{dE_{\Lambda 0}}{d\Lambda} - \mathrm{Tr}\left[(G_{\Lambda 0} - G_\Lambda) \frac{dR_\Lambda}{d\Lambda} \right] = \mathrm{Tr}\left[G_\Lambda \frac{dR_\Lambda}{d\Lambda} \right]. \tag{5.114}$$

By varying Eq. (5.114) with respect to the quasiparticle occupation number, we obtain a flow equation for the quasiparticle energy

$$\frac{d\tilde{\varepsilon}_{\mathbf{p}}}{d\Lambda} = \frac{\delta}{\delta n_{\mathbf{p}}}\frac{dE_\Lambda}{d\Lambda} = \frac{\delta}{\delta n_{\mathbf{p}}}\left(\mathrm{tr}\int \frac{d^4p'}{(2\pi)^4 i}\, G_\Lambda(p')\,\frac{dR_\Lambda(\mathbf{p}')}{d\Lambda}\right)$$
$$= z_{\mathbf{p}}\,\frac{dR_\Lambda(\mathbf{p})}{d\Lambda} + \mathrm{tr}\int \frac{d^4p'}{(2\pi)^4 i}\,\frac{1}{z_{\mathbf{p}'}}\, f_{\mathbf{p}\mathbf{p}'}\, G_\Lambda^2(p')\,\frac{dR_\Lambda(\mathbf{p}')}{d\Lambda}, \quad (5.115)$$

where we have used Eqs. (5.90), (5.93) and (5.94). Combined with Eq. (5.113), the right-hand side of the flow equation for the quasiparticle energy, Eq. (5.115), can be written as

$$\frac{d\tilde{\varepsilon}_{\mathbf{p}}}{d\Lambda} = z_{\mathbf{p}}\left(\frac{\partial \Sigma_\Lambda(\tilde{\varepsilon}_{\mathbf{p}},\mathbf{p})}{\partial\Lambda} + \frac{dR_\Lambda(\mathbf{p})}{d\Lambda}\right) \equiv \frac{d\varepsilon_{\mathbf{p}}}{d\Lambda} + z_{\mathbf{p}}\,\frac{dR_\Lambda(\mathbf{p})}{d\Lambda}, \quad (5.116)$$

where we have identified $d\varepsilon_{\mathbf{p}}/d\Lambda$ with $z_{\mathbf{p}}\,\partial\Sigma_\Lambda(\tilde{\varepsilon}_{\mathbf{p}},\mathbf{p})/\partial\Lambda$. In the limit $\Lambda \to 0$, $\varepsilon_{\mathbf{p}}$ approaches the solution of the Dyson equation, Eq. (5.81).

For our purpose, we use following regulator (adapted from Ref. [42])

$$R_\Lambda(\mathbf{p}) = \frac{\mathbf{p}^2}{2m}\left[\frac{1}{\Theta_\epsilon(|\mathbf{p}| - (k_{\mathrm{F}}+\Lambda)) + \Theta_\epsilon(k_{\mathrm{F}} - \Lambda - |\mathbf{p}|)} - 1\right], \quad (5.117)$$

where $\lim_{\epsilon\to 0}\Theta_\epsilon(x) \to \Theta(x)$ at the end of the calculation. The regulator suppresses low-lying single-particle modes with momenta in the range $k_F - \Lambda < |\mathbf{p}| < k_F + \Lambda$ (see Fig. 5.14). In the limit of a sharp cutoff, $\epsilon \to 0$, one finds [42]

$$G_\Lambda^2(p)\,\frac{dR_\Lambda(\mathbf{p})}{d\Lambda} = -\frac{\delta(|\mathbf{p}| - (k_{\mathrm{F}}+\Lambda)) + \delta(|\mathbf{p}| - (k_{\mathrm{F}}-\Lambda))}{\omega - \frac{\mathbf{p}^2}{2m} - \Sigma_\Lambda(\omega,\mathbf{p})}. \quad (5.118)$$

Keeping only the quasiparticle contribution in the second term of Eq. (5.115) and canceling the trivial renormalization due to the explicit regulator term in $\tilde{\varepsilon}_{\mathbf{p}}$, we find[5]

$$\frac{d\varepsilon_{\mathbf{p}}}{d\Lambda} = -\mathrm{tr}\int \frac{d^3p'}{(2\pi)^3}\, f_{\mathbf{p}\mathbf{p}'}\, n^0_{\mathbf{p}',\Lambda}\Big[\delta(|\mathbf{p}'| - (k_{\mathrm{F}}+\Lambda)) + \delta(|\mathbf{p}'| - (k_{\mathrm{F}}-\Lambda))\Big]. \quad (5.119)$$

This flow equation for the quasiparticle energy, Eq. (5.119), is illustrated diagrammatically in Fig. 5.16.

The four-point vertex that enters the flow equation for the self-energy is the quasiparticle-quasihole irreducible quasiparticle interaction $f_{\mathbf{p}\mathbf{p}'}$. This result can be understood by recognizing that the quasiparticle-quasihole reducible contributions of the full four-point vertex $\Gamma^{\Lambda(4)}_{p,p,p',p'}$ in Eq. (5.103) do not contribute for the kinematics relevant to the self-energy, for $|\mathbf{q}|/\omega = 0$. We can therefore replace the full four-point vertex in Eq. (5.103) by Γ^ω, which for quasiparticle kinematics is proportional to $f_{\mathbf{p}\mathbf{p}'}$.

[5] The resulting flow equation, Eq. (5.119), is consistent with Eq. (5.5), if we make the natural identification $\delta n_{\mathbf{p}} = -n^0_{\mathbf{p}}\,\delta(|\mathbf{p}| - (k_F - \Lambda))\,d\Lambda$.

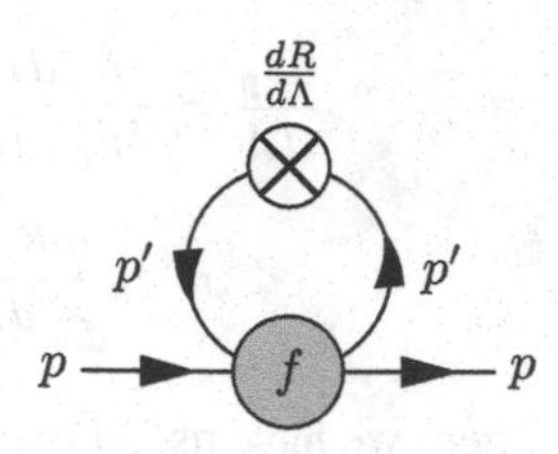

Fig. 5.16 Flow equation for the self-energy in the quasiparticle approximation

The flow equation for the four-point function,

$$\Gamma^{(4)}_{p'_1,p'_2,p_1,p_2} = \frac{\delta^4 \Gamma}{\delta\psi^*_{p'_2}\,\delta\psi^*_{p'_1}\,\delta\psi_{p_2}\,\delta\,\psi_{p_1}}, \tag{5.120}$$

is obtained by functionally differentiating Eq. (5.100) twice with respect to ψ^* and twice with respect to ψ.[6] For details we refer the reader to Ref. [43]. The resulting flow equation is illustrated diagrammatically in Fig. 5.17. There are two types of contributions to the flow equation: those involving two four-point functions (where all three channels, the particle-hole ZS and ZS′ channels, as well as the particle-particle/hole-hole BCS channel contribute) and one obtained by closing two legs of the six-point function.

The particle-hole channels take into account contributions to the quasiparticle scattering amplitude and interaction due to long-range density and spin-density excitations, whereas the BCS channel builds up contributions from the coupling to high-momentum states and due to pairing correlations. For the application to neutron matter, we start the many-body calculation from low-momentum interactions $V_{\text{low}\,k}$ [17], for which the particle-particle channel is perturbative at nuclear densities, except for low-lying pairing correlations [22, 44]. This is in contrast to hard potentials, where the coupling to high momenta renders all channels non-perturbative.

We therefore solve the flow equations for the four-point function shown in Fig. 5.17 including only the particle-hole contributions (the first and second diagrams). After the RG flow, we calculate the low-lying pairing correlations by solving the quasiparticle BCS gap equation. In our first study [18], we neglected the contribution from the six-point function to the flow equation (the last term in Fig. 5.17) and approximated the internal Green's functions by the quasiparticle part (the first term on the right-hand-side of Eq. (5.112)). The resulting flow equations for the quasiparticle scattering amplitude

$$a(\mathbf{q},\mathbf{q}';\Lambda) = z_{k_F}^2\, \Gamma^{(4)}_{p-\frac{q}{2},p'+\frac{q}{2},p+\frac{q}{2},p'-\frac{q}{2}}\Big|_{\omega=\omega'=\varepsilon_F,\, q_0=0}, \tag{5.121}$$

and the quasiparticle interaction $f(\mathbf{q},\mathbf{q}';\Lambda)$ are given by [18]:

[6] For vanishing external sources, all vertices with an odd number of external fermion lines vanish.

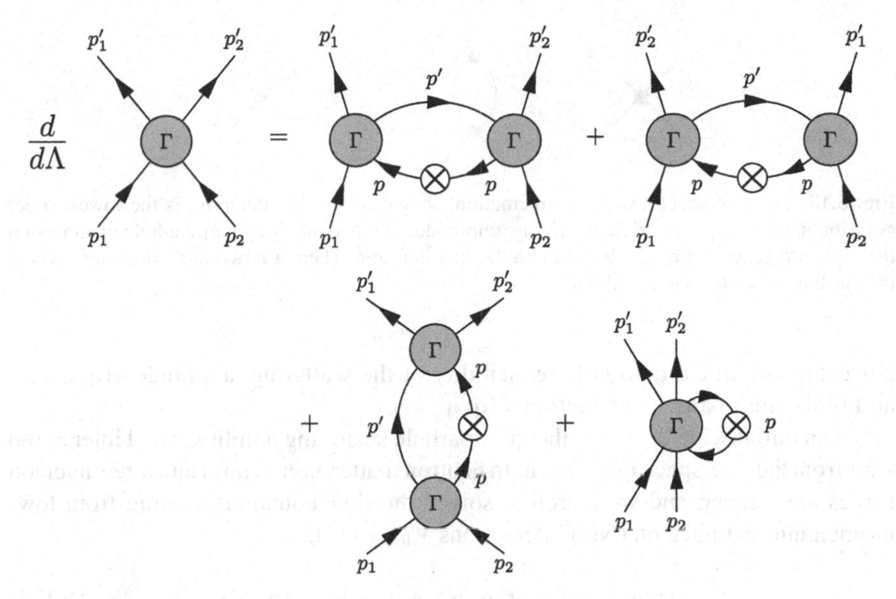

Fig. 5.17 Flow equation for the four-point function $\Gamma^{(4)}$. For each of the first three diagrams on the *right-hand side*, there is an additional diagram, where the *unfilled crossed circle* is inserted on the other internal line

$$\begin{aligned}\frac{d}{d\Lambda}a(\mathbf{q},\mathbf{q}';\Lambda) = z_{k_F}^2\frac{d}{d\Lambda}\Bigg[g\int_{\text{fast},\Lambda}\frac{d^3\mathbf{p}''}{(2\pi)^3}\frac{n_{\mathbf{p}''+\mathbf{q}/2}-n_{\mathbf{p}''-\mathbf{q}/2}}{\varepsilon_{\mathbf{p}''+\mathbf{q}/2}-\varepsilon_{\mathbf{p}''-\mathbf{q}/2}}\Bigg]\\ \times a\Big(\mathbf{q},\frac{\mathbf{p}+\mathbf{p}'}{2}+\frac{\mathbf{q}'}{2}-\mathbf{p}'';\Lambda\Big)a\Big(\mathbf{q},\mathbf{p}''-\frac{\mathbf{p}+\mathbf{p}'}{2}+\frac{\mathbf{q}'}{2};\Lambda\Big)\\ +\frac{d}{d\Lambda}f(\mathbf{q},\mathbf{q}';\Lambda),\end{aligned} \tag{5.122}$$

$$\begin{aligned}\frac{d}{d\Lambda}f(\mathbf{q},\mathbf{q}';\Lambda) = -z_{k_F}^2\frac{d}{d\Lambda}\Bigg[g\int_{\text{fast},\Lambda}\frac{d^3\mathbf{p}''}{(2\pi)^3}\frac{n_{\mathbf{p}''+\mathbf{q}'/2}-n_{\mathbf{p}''-\mathbf{q}'/2}}{\varepsilon_{\mathbf{p}''+\mathbf{q}'/2}-\varepsilon_{\mathbf{p}''-\mathbf{q}'/2}}\Bigg]\\ \times a\Big(\mathbf{q}',\frac{\mathbf{p}+\mathbf{p}'}{2}+\frac{\mathbf{q}}{2}-\mathbf{p}'';\Lambda\Big)a\Big(\mathbf{q}',\mathbf{p}''-\frac{\mathbf{p}+\mathbf{p}'}{2}+\frac{\mathbf{q}}{2};\Lambda\Big).\end{aligned} \tag{5.123}$$

Here the spin labels and the spin trace in the flow equation have been suppressed. In a spin-saturated system the spin dependence of f and a is of the form of Eq. (5.9), when non-central forces are neglected. We note that for $\mathbf{q} = \mathbf{q}' = 0$ and identical spins the contributions from the ZS and ZS′ channels to the scattering amplitude a cancel, as required by the Pauli principle. Thus, the resulting quasiparticle interaction and the corresponding Fermi liquid parameters satisfy the Pauli-principle sum rules, see Eq. (5.62). Moreover, the flow equations, Eqs. (5.122) and (5.123), yield the

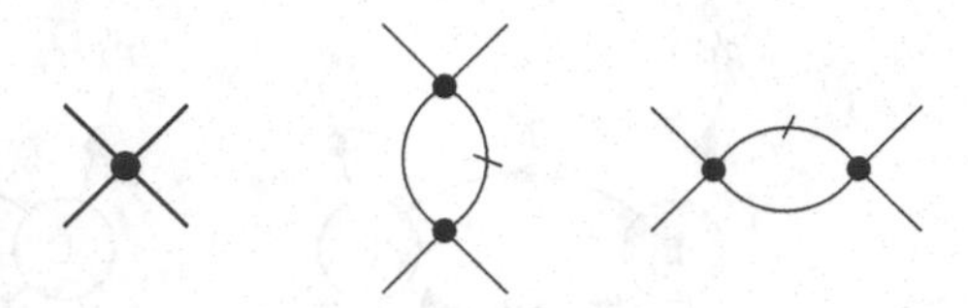

Fig. 5.18 The free-space two-body interaction, shown by the *left* diagram, is the lowest order contribution to the quasiparticle scattering amplitude. The two one-loop particle-hole diagrams on the *right* are generated by the RG flow in the first iteration. There are two more diagrams, where the slash is on the other internal line

correct quasiparticle-quasihole reducibility of the scattering amplitude $a(\mathbf{q}, \mathbf{q}'; \Lambda)$ and of the quasiparticle interaction $f(\mathbf{q}, \mathbf{q}'; \Lambda)$.

At an initial scale $\Lambda = \Lambda_0$, the quasiparticle scattering amplitude and interaction start from the free-space interaction. In neutron matter, non-central and three-nucleon forces are weaker, and we therefore solved the flow equations starting from low-momentum two-nucleon (NN) interactions $V_{\text{low}\,k}$ [17],

$$a(\mathbf{q}, \mathbf{q}'; \Lambda_0) = f(\mathbf{q}, \mathbf{q}'; \Lambda_0) = V_{\text{low}\,k}(\mathbf{q}, \mathbf{q}'), \tag{5.124}$$

including only scalar and spin-spin interactions, which dominate at densities below nuclear saturation density.

It is instructive to study the RG method diagrammatically, to understand how many-body correlations are generated by the flow equations. We discuss the set of diagrams, which is generated by the flow equation for the scattering amplitude, Eq. (5.122). We denote the antisymmetrized free-space two-body interaction Γ_{vac}, the initial condition for the flow equation, by a dot (see Fig. 5.18). Starting from Γ_{vac}, we integrate out the first shell $\delta\Lambda$ of high-lying particle-hole excitations to obtain the effective scattering amplitude at the lower scale $\Lambda_1 = \Lambda_0 - \delta\Lambda$. The RG flow includes contributions from both particle-hole channels, which leads to the four diagrams, two of which are shown in Fig. 5.18. The lines marked by a slash are restricted by the regulator to momenta in the shell $\Lambda_1 \leqslant k < \Lambda_0$.

The four-point vertex used in each iteration of the flow equation is the one obtained in the previous iteration. Thus, in the second iteration, one finds the one-loop diagrams shown in Fig. 5.18, but now with the momentum of the marked line in the second shell $\Lambda_2 = \Lambda_1 - \delta\Lambda \leqslant k < \Lambda_1$. In addition, the RG flow generates the two-loop diagrams shown in Fig. 5.19. Here, lines with momenta in the first shell are marked by one slash and those with momenta in the second shell by two slashes. For every diagram shown, there are three more diagrams obtained by moving the slash or the two slashes from a particle (hole) line to the corresponding hole (particle) line. The first four diagrams illustrate the coupling between the two particle-hole channels generated by the RG equations.

We observe that the four diagrams of third order in the vacuum interaction shown in Fig. 5.20 are missing in this scheme. These diagrams are obtained when the last term in Fig. 5.17 (the contribution of the six-point function to the flow of the four-point

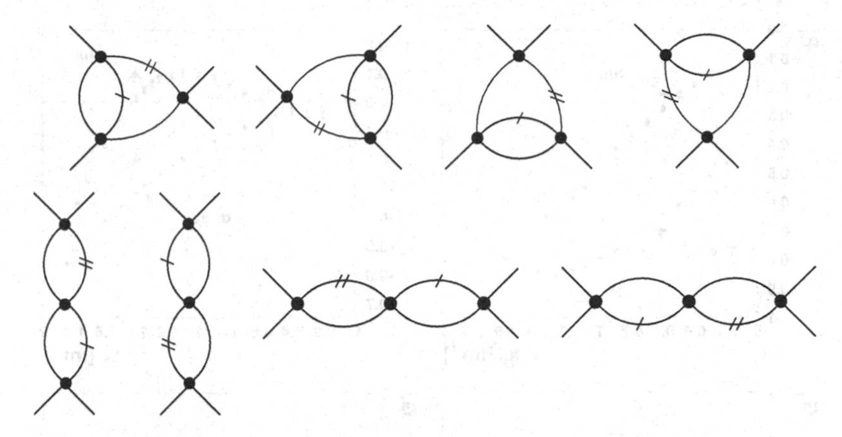

Fig. 5.19 Two-loop diagrams that contribute to the scattering amplitude after two iterations of the flow equations

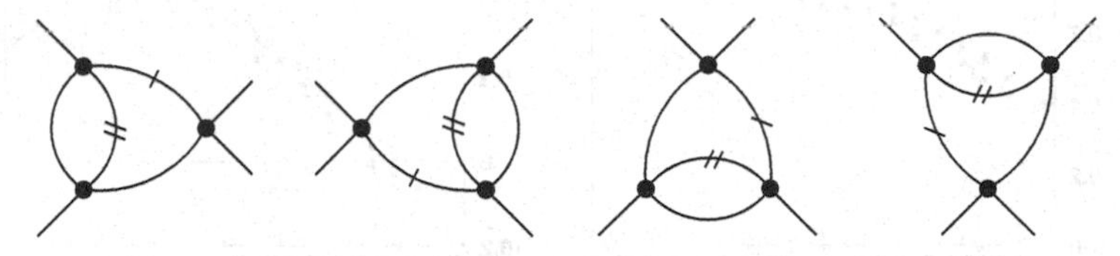

Fig. 5.20 The two-loop planar diagrams that are not generated by the RG flow after two iterations

function) is included. This shows that the flow equations reproduce the full one-loop particle-hole phase space exactly, augmented by a large antisymmetric subset of the particle-hole parquet diagrams. In a truncation scheme where the BCS channel and the six-point function in Fig. 5.17 is included, the RG sums all planar diagrams using the RG.

5.3.1 Fermi Liquid Parameters and Scattering Amplitude

In Fig. 5.21, we show the resulting $l = 0$ and $l = 1$ Fermi liquid parameters as a function of the Fermi momentum $k_{\rm F}$. In this section we use the notation common in nuclear physics,

$$f_{\mathbf{p}\sigma\mathbf{p}'\sigma'} = f_{\mathbf{pp}'} + g_{\mathbf{pp}'}\,\boldsymbol{\sigma}\cdot\boldsymbol{\sigma}', \tag{5.125}$$

$$a_{\mathbf{p}\sigma\mathbf{p}'\sigma'} = a_{\mathbf{pp}'} + b_{\mathbf{pp}'}\,\boldsymbol{\sigma}\cdot\boldsymbol{\sigma}', \tag{5.126}$$

with $F_l = N_0\, f_l$ and $G_l = N_0\, g_l$. The cutoff scale of the starting $V_{\text{low}\,k}$ is taken as $\Lambda = \sqrt{2}\,k_{\rm F}$. This choice has the advantage that scattering to high-lying states in the particle-particle channel is uniformly accounted for at different densities.

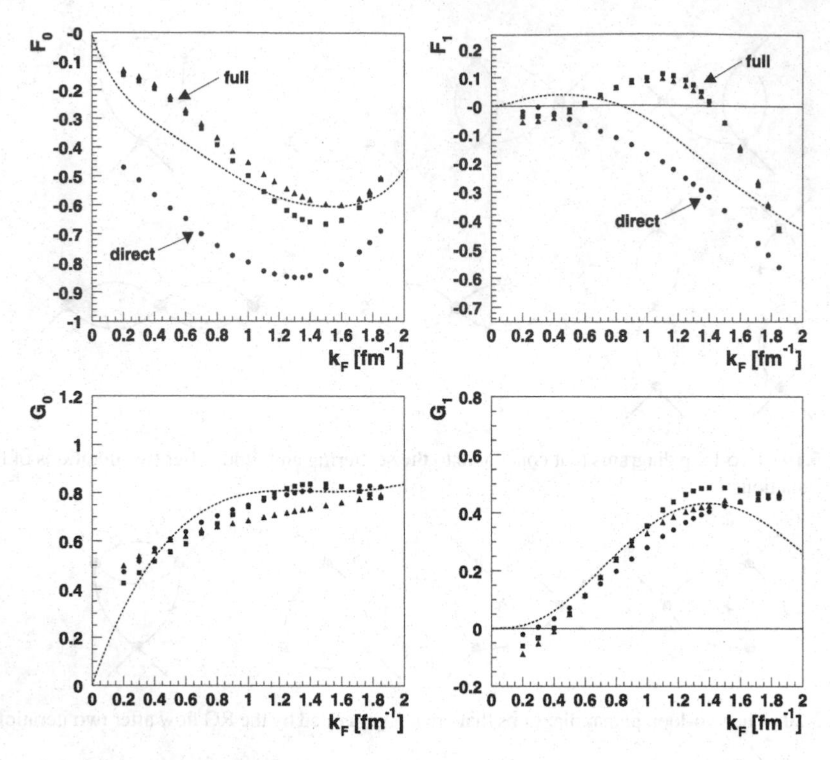

Fig. 5.21 The $l = 0$ and $l = 1$ Fermi liquid parameters versus the Fermi momentum k_F. The *dots* denote the direct contribution only ($z_{k_\mathrm{F}} = 1$, but including the effective mass in the density of states), whereas the *squares* (static z_{k_F} factor) and the *triangles* (adaptive z_{k_F} factor) are calculated from the full RG solution. The results of Wambach et al. [46] are given for comparison as dashed lines

The low-momentum interaction $V_{\mathrm{low}\,k}$ then drives the flow in the ZS$'$ channel for the quasiparticle interaction, and thus our results include effects of the induced interaction. A further advantage of the RG approach is that the calculations can be performed without truncating the expansion of the quasiparticle interaction, Eq. (5.11), at some l.

The flow equations were solved with two different assumptions for the z_{k_F} factor. In one case, we use a static, density-independent mean value of $z^2_{k_F} = 0.9$ which remains unchanged under the RG. In the other case, we compute the z_{k_F} factor dynamically, by assuming that the change of the effective mass from the initial one, based on the momentum dependence of $V_{\mathrm{low}\,k}$, is due to the z_{k_F} factor alone [18]. Generally, we find a very good agreement between our results and the ones obtained using the polarization potential model by Ainsworth et al. [45, 46]. There are minor differences in the value of the effective mass, which is treated self-consistently in the RG approach. We note that, in the density range $0.6\,\mathrm{fm}^{-1} < k_\mathrm{F} < 1.4\,\mathrm{fm}^{-1}$, we find that the effective mass at the Fermi surface exceeds unity. The quasiparticle interaction was also calculated taking into account induced interactions [47–49].

In these papers, the value for the $l = 0$ spin-dependent parameter G_0 is in good agreement with ours, while there are differences for the spin-independent F_0. We stress the important role of the large G_0 for the induced interaction. This Landau parameter causes the strong spin-density correlations, which in turn enhance the Landau parameter F_0 and consequently the incompressibility of neutron matter.

The results can be qualitatively understood by inspecting the explicit spin dependence of the RG equation for the quasiparticle interaction, Eqs. (5.122) and (5.123) for $q = 0$,

$$\Lambda \frac{d}{d\Lambda} a(q = 0, \mathbf{q}'; \Lambda) = -\Theta(q' - 2\Lambda) \left(\frac{1}{2} \beta_{\mathrm{ZS}'}[a, \mathbf{q}', \Lambda] + \frac{3}{2} \beta_{\mathrm{ZS}'}[b, \mathbf{q}', \Lambda] \right), \tag{5.127}$$

$$\Lambda \frac{d}{d\Lambda} b(q = 0, \mathbf{q}'; \Lambda) = -\Theta(q' - 2\Lambda) \left(\frac{1}{2} \beta_{\mathrm{ZS}'}[a, \mathbf{q}', \Lambda] - \frac{1}{2} \beta_{\mathrm{ZS}'}[b, \mathbf{q}', \Lambda] \right), \tag{5.128}$$

where we have introduced the β functions $\beta_{\mathrm{ZS}'}[a, \mathbf{q}'; \Lambda]$ and $\beta_{\mathrm{ZS}'}[b, \mathbf{q}'; \Lambda]$ for the contribution from density and spin-density fluctuations in the ZS$'$ channel to the RG flow. In this qualitative argument we neglect Fermi liquid parameters with $l \geqslant 1$. The flow equations, Eqs. (5.122) and (5.123), show that the β functions are quadratic in the four-point functions a and b, while Eq. (5.124) implies that the initial values for a and b are given by the lowest order contribution to the quasiparticle interaction. At a typical Fermi momentum $k_{\mathrm{F}} = 1.0\,\mathrm{fm}^{-1}$, we observe that the initial F_0 and G_0 are similar in absolute value, $|F_0| \approx |G_0| \approx 0.8$. Consequently, there is a cancellation between the contributions due to the spin-independent and spin-dependent parameters in Eq. (5.128), while in Eq. (5.127) both contributions are repulsive. Thus, one expects a relatively small effect of the RG flow on G_0 and a substantial renormalization of F_0, in agreement with our results.

The RG approach enables us to compute the scattering amplitude for general scattering processes on the Fermi surface, without making further assumptions for the dependence on the particle-hole momentum transfers q and q'. The scattering amplitude at finite momentum transfer is of great interest for calculating transport processes and superfluidity, as discussed in the following section. The flow equations treat the dependence on the momenta q and q' on an equal footing and maintain the symmetries of the scattering amplitude. For scattering on the Fermi surface, $\mathbf{q}$, $\mathbf{q}'$ and $\mathbf{P}$ are orthogonal, and they are restricted to $q^2 + q'^2 + P^2 \leqslant 4\,k_{\mathrm{F}}^2$. Therefore, in Ref. [18] we approximated $a(\mathbf{q}, \mathbf{q}'; \Lambda) = a(q^2, q'^2; \Lambda)$ for the solution of the flow equations to extrapolate off the Fermi surface. On the $V_{\mathrm{low}\,k}$ level, we checked that the $\mathbf{q} \cdot \mathbf{q}'$ dependence is small for neutrons.

5.3.2 *Superfluidity in Neutron Matter*

Superfluidity plays a key role in strongly-interacting many-body systems. Pairing in infinite matter impacts the cooling of isolated neutron stars [50] and of the neutron star crust [51], and is used to develop non-empirical energy-density functionals [52]. In this section, we discuss superfluidity in neutron matter, with particular attention to induced interactions using the RG approach.

Figure 5.22 shows the superfluid pairing gaps in neutron matter, obtained by solving the BCS gap equation with a free spectrum. In this approximation, the $^1S_0$ superfluid gap $\Delta(k)$ is determined from the gap equation,

$$\Delta(k) = -\frac{1}{\pi} \int dp\, p^2 \, \frac{V(k,p)\,\Delta(p)}{\sqrt{\xi^2(p) + \Delta^2(p)}}, \tag{5.129}$$

where $V(k,p)$ is the free-space NN interaction in the $^1S_0$ channel, $\xi(p) \equiv p^2/(2m) - \mu$, and for a free spectrum the chemical potential is given by $\mu = k_F^2/(2m)$. At low densities (in the crust of neutron stars), neutrons form a $^1S_0$ superfluid. At higher densities, the S-wave interaction is repulsive and neutrons pair in the $^3P_2$ channel (with a small coupling to $^3F_2$ due to the tensor force). Figure 5.22 demonstrates that the $^1S_0$ BCS gap is practically independent of nuclear interactions, and therefore strongly constrained by NN phase shifts [53]. This includes a very weak cutoff dependence for low-momentum interactions $V_{\text{low}\,k}$ with sharp or sufficiently narrow smooth regulators with $\Lambda > 1.6\,\text{fm}^{-1}$. The inclusion of N^{2}LO three-nucleon (3N) forces leads to a reduction of the $^1S_0$ BCS gap for Fermi momenta $k_F > 0.6\,\text{fm}^{-1}$, where the gap is decreasing [22]. Two-nucleon interactions are well constrained by scattering data for relative momenta $k \lesssim 2\,\text{fm}^{-1}$ [17]. The model dependencies at higher momenta show up prominently in Fig. 5.22 in the $^3P_2$–$^3F_2$ gaps for Fermi momenta $k_F > 2\,\text{fm}^{-1}$ [54].

Understanding many-body effects beyond the BCS level constitutes an important open problem. For recent progress and a survey of results, see for instance Ref. [55]. At low densities, induced interactions due to particle-hole screening and vertex corrections are significant even in the perturbative $k_F a$ limit [56] and lead to a reduction of the S-wave gap by a factor $(4e)^{-1/3} \approx 0.45$,

$$\frac{\Delta}{\varepsilon_F} = \frac{8}{e^2} \exp\left\{ \left(\quad + \quad + \quad + \cdots \right)^{-1} \right\}$$

$$= (4e)^{-1/3}\, \frac{8}{e^2} \exp\left\{ \frac{\pi}{2 k_F a} + \mathcal{O}(k_F a) \right\}. \tag{5.130}$$

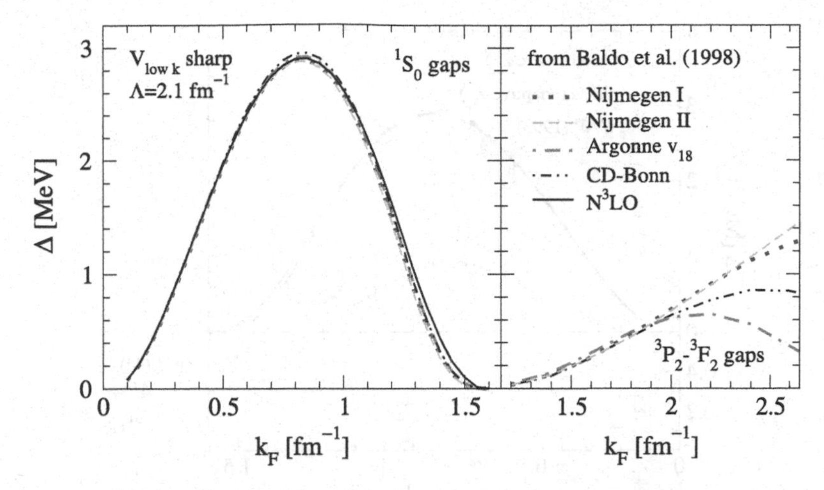

Fig. 5.22 The $^1S_0$ (*left panel*) and $^3P_2-^3F_2$ (*right panel*) superfluid pairing gaps Δ at the Fermi surface as a function of Fermi momentum k_F in neutron matter. The gaps are obtained from charge-dependent NN interactions at the BCS level. For details see Refs. [53, 54]

This reduction is due to spin fluctuations, which are repulsive for spin-singlet pairing and overwhelm attractive density fluctuations.[7]

Here, we discuss the particle-hole RG (phRG) approach to this problem using the BCS-channel-irreducible quasiparticle scattering amplitude as pairing interaction [18]. The quasiparticle scattering amplitude is obtained, as discussed in the previous section, by solving the flow equations in the particle-hole channels, Eqs. (5.122) and (5.123), starting from $V_{\text{low}\,k}$. This builds up many-body correlations from successive momentum shells, on top of an effective interaction with particle/hole polarization effects from all previous shells, and thereby efficiently includes induced interactions to low-lying states in the vicinity of the Fermi surface beyond a perturbative calculation.

The results for the $^1S_0$ gap are shown in Fig. 5.23, where induced interactions lead to a factor 3–4 reduction to a maximal gap $\Delta \approx 0.8$ MeV [18]. Similar values to those of Wambach et al. [46] are found. In addition, for the lower densities, the phRG is consistent with the dilute result[8] $\Delta/\Delta_0 = (4e)^{-1/3}$, and at the larger densities the dotted band indicates the uncertainty due to an approximate self-energy treatment.

Non-central spin-orbit and tensor interactions are crucial for $^3P_2-^3F_2$ superfluidity. Without a spin-orbit interaction, neutrons would form a $^3P_0$ superfluid instead. The first perturbative calculation of non-central induced interactions shows that $^3P_2$ gaps below 10 keV are possible (while second-order contributions to the pairing

[7] In finite systems, the spin and density response differs. In nuclei with cores, the low-lying response is due to surface vibrations. Consequently, induced interactions may be attractive, because the spin response is weaker [57].

[8] For $k_F \approx 0.4\,\text{fm}^{-1}$, neutron matter is close to the universal regime, but theoretically simpler due to an appreciable effective range $k_F r_e \approx 1$ [58].

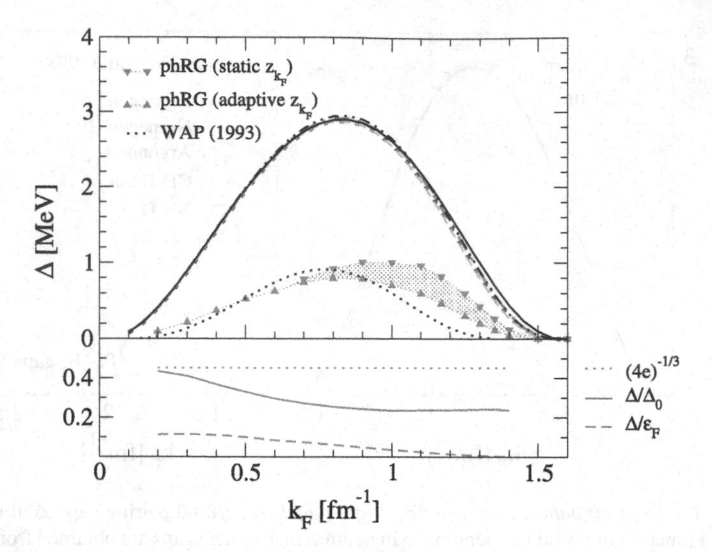

Fig. 5.23 *Top panel* Comparison of the $^1S_0$ BCS gap to the results including polarization effects through the particle-hole RG (phRG), for details see Ref. [18], and to the results of Wambach et al. [46]. *Lower panel* Comparison of the full superfluid gap Δ to the BCS gap Δ_0 and to the Fermi energy ε_F

interaction are not substantial $|V_{ind}/V_{low\,k}| < 0.5$) [13]. This arises from a repulsive induced spin-orbit interaction due to the mixing with the stronger spin-spin interaction. As a result, neutron P-wave superfluidity (in the interior of neutron stars) may be reduced considerably below earlier estimates. This implies that low-mass neutron stars cool slowly [50]. Smaller values for the $^3P_2$ gap compared to Fig. 5.22 are also required for consistency with observations in a minimal cooling scenario [59].

5.4 Outlook

In these lecture notes we discussed RG approaches to Fermi liquids, based on the ideas of Shankar [1] and Polchinski [2], and formulated the flow equation in the framework of a functional RG [38]. The appeal of RG methods is that they allow a systematic study of non-perturbative correlations at different length scales. These lecture notes show that the RG approach developed for neutron matter in Ref. [18] is equivalent to a functional RG. We reviewed results for the quasiparticle interaction and the superfluid pairing gaps in neutron matter, starting from low-momentum interactions and focusing on the effects of induced interactions due to long-range particle-hole fluctuations.

There are many important problems for RG methods applied to many-nucleon systems. These include the role of non-central interactions and of many-body forces. For example, we recently showed that novel non-central interactions are generated by polarization of the nuclear medium [13]. A non-perturbative RG treatment of these effects, which have important consequences for neutron stars, is of great interest. In recent years an alternative RG scheme, the similarity RG (SRG) [60, 61] has emerged as a powerful tool for exploring the scale dependence of nuclear interactions and the role of many-body forces in nuclear systems. Moreover, the SRG can be applied directly to solve the many-body problem [62]. The SRG approach is formulated as a continuous matrix transformation acting on the Hamiltonian. An important open question is to understand the relation of the SRG to functional and field theory methods.

Acknowledgments This work was supported in part by NSERC and by the Alliance Program of the Helmholtz Association (HA216/EMMI).

References

1. Shankar, R.: Rev. Mod. Phys. **66**, 129 (1994)
2. Polchinski, J.: In: Harvey, J., Polchinski, J. (eds.): Proceedings of the 1992 Theoretical Advanced Studies Institute in Elementary Particle Physics. World Scientific, Singapore (1993). hep-th/9210046
3. Gies, H.: Lectures in this Volume. hep-ph/0611146
4. Metzner, W., Salmhofer, M., Honerkamp, C., Meden, V., Schönhammer, K.: arXiv:1105.5289
5. Landau, L.D.: Sov. Phys. JETP **3**, 920 (1957)
6. Landau, L.D.: Sov. Phys. JETP **5**, 101 (1957)
7. Landau, L.D.: Sov. Phys. JETP **8**, 70 (1959)
8. Larkin, A., Migdal, A.B.: Sov. Phys. JETP **17**, 1146 (1963)
9. Leggett, A.J.: Phys. Rev. A **140**, 1869 (1965); ibid. **147**, 119 (1966)
10. Baym, G., Pethick, C.J.: Landau Fermi Liquid Theory: Concepts and Applications. Wiley, New York (1991)
11. Migdal, A.B.: Theory of Finite Fermi Systems and Applications to Atomic Nuclei. Interscience, New York (1967)
12. Bäckman, S.-O., Brown, G.E., Niskanen, J.: Phys. Rept. **124**, 1 (1985)
13. Schwenk, A., Friman, B.: Phys. Rev. Lett. **92**, 082501 (2004)
14. Pomeranchuk, I.Y.: Sov. Phys. JETP **8**, 361 (1959)
15. Pines, D., Nozières, P.: The Theory of Quantum Liquids, vol. 1, Advanced Book Classics. Westview Press, Boulder (1999)
16. Vollhardt, D., Wölfle, P.: The Superfluid Phases of Helium 3. Taylor and Francis, London (1990)
17. Bogner, S.K., Kuo, T.T.S., Schwenk, A.: Phys. Rept. **386**, 1 (2003); Bogner, S.K., Furnstahl, R.J., Schwenk, A.: Prog. Part. Nucl. Phys. **65**, 94 (2010)
18. Schwenk, A., Friman, B., Brown, G.E.: Nucl. Phys. A **713**, 191 (2003)
19. Schwenk, A., Brown, G.E., Friman, B.: Nucl. Phys. A **703**, 745 (2002)
20. Kaiser, N.: Nucl. Phys. A **768**, 99 (2006); Holt, J.W., Kaiser, N., Weise, W.: Nucl. Phys. A **870–871**, 1 (2011)
21. Epelbaum, E., Hammer, H.-W., Meißner, U.-G.: Rev. Mod. Phys. **81**, 1773 (2009)
22. Bogner, S.K., Schwenk, A., Furnstahl, R.J., Nogga, A.: Nucl. Phys. A **763**, 59 (2005); Hebeler, K., Schwenk, A.: Phys. Rev. C **82**, 014314 (2010); Hebeler, K., Bogner, S.K., Furnstahl, R.J., Nogga, A., Schwenk, A.: Phys. Rev. C **83**, 031301(R) (2011)
23. Otsuka, T., Suzuki, T., Holt, J.D., Schwenk, A., Akaishi, Y.: Phys. Rev. Lett. **105**, 032501 (2010); Holt, J.D., Otsuka, T., Schwenk, A., Suzuki, T.: arXiv:1009.5984

24. Pethick, C.J.: In: Mahantappa, K.T., Brittin, W.E. (eds.) Lectures in Theoretical Physics, vol. XI-B, Boulder, Colorado, 1968, p. 187. Gordon and Breach, New York (1969)
25. Fetter, A.L., Walecka, J.D.: Quantum Theory of Many-Particle Systems. Dover, New York (2003)
26. Abrikosov, A.A., Gor'kov, L.P., Dzyaloshinski, I.E.: Methods of Quantum Field Theory in Statistical Physics. Dover, New York (1963)
27. Negele, J.W., Orland, H.: Quantum Many-Particle Systems: Advanced Book Classics. Westview Press, Boulder (1998)
28. Altland, A., Simons, B.: Condensed Matter Field Theory. Cambridge University Press, Cambridge (2007)
29. Abrikosov, A.A., Khalatnikov, I.M.: Rept. Prog. Phys. **22**, 329 (1959)
30. Bäckman, S.-O., Sjöberg, O., Jackson, A.D.: Nucl. Phys. A **321**, 10 (1979)
31. Friman, B.L., Dhar, A.K.: Phys. Lett. B **85**, 1 (1979)
32. Cornwall, J.M., Jackiw, R., Tomboulis, E.: Phys. Rev. D **10**, 2428 (1974)
33. Luttinger, J.M., Ward, J.C.: Phys. Rev. **118**, 1417 (1960)
34. Baym, G.: Phys. Rev. **127**, 1391 (1962)
35. Wetterich, C.: Phys. Rev. B **75**, 085102 (2007)
36. Nozières, P.: Theory of Interacting Fermi Systems. Westview Press, Boulder (1997)
37. Babu, S., Brown, G.E.: Ann. Phys. **78**, 1 (1973)
38. Wetterich, C.: Phys. Lett. B **301**, 90 (1993)
39. Berges, J., Tetradis, N., Wetterich, C.: Phys. Rept. **363**, 223 (2002)
40. Dupuis, N.: Eur. Phys. J. B **48**, 319 (2005)
41. Hebeler, K.: Ph.D. Thesis, Technische Universität Darmstadt (2007)
42. Morris, T.M.: Nucl. Phys. B **458**, 477 (1996)
43. Ellwanger, U.: Z. Phys. C **62**, 503 (1994)
44. Bogner, S.K., Furnstahl, R.J., Ramanan, S., Schwenk, A.: Nucl. Phys. A **773**, 203 (2006); Ramanan, S., Bogner, S.K., Furnstahl, R.J.: Nucl. Phys. A **797**, 81 (2007)
45. Ainsworth, T.L., Wambach, J., Pines, D.: Phys. Lett. B **222**, 173 (1989)
46. Wambach, J., Ainsworth, T.L., Pines, D.: Nucl. Phys. A **555**, 128 (1993)
47. Bäckman, S.-O., Källman, C.-G., Sjöberg, O.: Phys. Lett. B **43**, 263 (1973)
48. Jackson, A.D., Krotschek, E., Meltzer, D.E., Smith, R.A.: Nucl. Phys. A **386**, 125 (1982)
49. Cao, L.G., Lombardo, U., Schuck, P.: Phys. Rev. C **74**, 064301 (2006)
50. Yakovlev, D.G., Pethick, C.J.: Annu. Rev. Astron. Astrophys. **42**, 169 (2004); Blaschke, D., Grigorian, H., Voskresensky, D.N.: Astron. Astrophys. **424**, 979 (2004); Page, D., Lattimer, J.M., Prakash, M., Steiner, A.W.: Astrophys. J. Suppl. **155**, 623 (2004)
51. Cackett, E.M., Wijnands, R., Linares, M., Miller, J.M., Homan, J., Lewin, W.: Mon. Not. R. Astron. Soc. **372**, 479 (2006); Cackett, E.M., Wijnands, R., Miller, J.M., Brown, E.F., Degenaar, N.: Astrophys. J. **687**, L87 (2008); Brown, E.F., Cumming, A.: Astrophys. J. **698**, 1020 (2009)
52. Lesinski, T., Duguet, T., Bennaceur, K., Meyer, J.: Eur. Phys. J. A **40**, 121 (2009); Hebeler, K., Duguet, T., Lesinski, T., Schwenk, A.: Phys. Rev. C **80**, 044321 (2009); Lesinski, T., Hebeler, K., Duguet, T., Schwenk, A.: J. Phys. G **39**, 015108 (2012)
53. Hebeler, K., Schwenk, A., Friman, B.: Phys. Lett. B **648**, 176 (2007)
54. Baldo, M., Elgaroy, O., Engvik, L., Hjorth-Jensen, M., Schulze, H.-J.: Phys. Rev. C **58**, 1921 (1998)
55. Gezerlis, A., Carlson, J.: Phys. Rev. C **81**, 025803 (2010)
56. Gorkov, L.P., Melik-Barkhudarov, T.K.: Sov. Phys. JETP **13**, 1018 (1961); Heiselberg, H., Pethick, C.J., Smith, H., Viverit, L.: Phys. Rev. Lett. **85**, 2418 (2000)
57. Barranco, F., Broglia, R.A., Colo, G., Gori, G., Vigezzi, E., Bortignon, P.F.: Eur. Phys. J. A **21**, 57 (2004); Pastore, A., Barranco, F., Broglia, R.A., Vigezzi, E.: Phys. Rev. C **78**, 024315 (2008)
58. Schwenk, A., Pethick, C.J.: Phys. Rev. Lett. **95**, 160401 (2005)
59. Page, D., Prakash, M., Lattimer, J.M., Steiner, A.W.: Phys. Rev. Lett. **106**, 081101 (2011); Shternin, P.S., Yakovlev, D.G., Heinke, C.O., Ho, W.C.G., Patnaude, D.J.: Mon. Not. R. Astron. Soc. **412**, L108 (2011)

60. Glazek, S.D., Wilson, K.G.: Phys. Rev. D **48**, 5863 (1993); Wegner, F.: Ann. Phys. (Leipzig) **3**, 77 (1994)
61. Bogner, S.K., Furnstahl, R.J., Perry, R.J.: Phys. Rev. C **75**, 061001(R) (2007)
62. Tsukiyama, K., Bogner, S.K., Schwenk, A.: Phys. Rev. Lett. **106**, 222502 (2011)

Chapter 6
Introduction to the Functional RG and Applications to Gauge Theories

Holger Gies

Prologue

This lecture course is intended to fill the gap between graduate courses on quantum field theory and specialized reviews or forefront-research articles on functional renormalization group approaches to quantum field theory and gauge theories.

These lecture notes are meant for advanced students who want to get acquainted with modern renormalization group (RG) methods as well as functional approaches to quantum gauge theories. In the first lecture, the functional renormalization group is introduced with a focus on the flow equation for the effective average action. The second lecture is devoted to a discussion of flow equations and symmetries in general, and flow equations and gauge symmetries in particular. The third lecture deals with the flow equation in the background formalism which is particularly convenient for analytical computations of truncated flows. The fourth lecture concentrates on the transition from microscopic to macroscopic degrees of freedom; even though this is discussed here in the language and the context of QCD, the developed formalism is much more general and will be useful also for other systems. Sections which have an asterisk $*$ in the section title contain more advanced material and may be skipped during a first reading.

This is not a review. I apologize for many omissions of further interesting and important aspects of this field (and their corresponding references). General reviews and more complete reference lists can be found in [1–7, 50]. A guide to more specialized literature is given in the Further-Reading subsections at the end of some sections.

H. Gies (✉)
Theoretisch-Physikalisches Institut, Friedrich-Schiller-Universität Jena,
Max-Wien-Platz 1, 07743 Jena, Germany
e-mail: Holger.Gies@uni-jena.de

H. Gies
Institute for Theoretical Physics, Heidelberg University,
Philosophenweg 16, 69120 Heidelberg, Germany

J. Polonyi and A. Schwenk (eds.), *Renormalization Group and Effective Field Theory Approaches to Many-Body Systems*, Lecture Notes in Physics 852,
DOI: 10.1007/978-3-642-27320-9_6,

6.1 Introduction

Quantum and statistical field theory are two sides of the same medal, representing the fundament on which modern physics is built. Both branches have been molded by the concept of the renormalization group. The renormalization group deals with the physics of scales. A central theme is the understanding of the macroscopic physics at long distances or low momenta in terms of the fundamental microscopic interactions. Bridging this gap from micro to macro scales requires a thorough understanding of quantum or statistical fluctuations on all the scales in between.

Field theories with gauge symmetry are of central importance, in particular, since all elementary particle-physics interactions are described by gauge theories. Moreover, nonabelian gauge theories—and most prominently quantum chromodynamics (QCD)—are in many respects paradigmatic, since they exhibit numerous features that are encountered in various field theoretical systems both in particle physics as well as condensed-matter systems. During the transition from micro to macro scales, these theories turn from weak to strong coupling, the relevant degrees of freedom are changed, the realization of the fundamental symmetries is different on the various scales, and the phase diagram is expected to have a rich structure, being formed by different and competing collective phenomena.

A profound understanding of gauge theories thus requires not just one but a whole toolbox of field theoretical methods. In addition to analytical perturbative methods for weak coupling and numerical lattice gauge theory for arbitrary couplings, *functional methods* begin to bridge the gap, since they are not restricted to weak coupling and can still largely be treated analytically. Functional methods aim at the computation of generating functionals of correlation functions, such as the that governs the dynamics of the macroscopic expectation values of the fields. These generating functionals contain all relevant physical information about a theory, once the microscopic fluctuations have been integrated out.

The functional RG combines this functional approach with the RG idea of treating the fluctuations not all at once but successively from scale to scale [8, 9]. Instead of studying correlation functions after having averaged over all fluctuations, only the *change* of the correlation functions as induced by an infinitesimal momentum shell of fluctuations is considered. From a structural viewpoint, this allows to transform the functional-integral structure of standard field theory formulations into a functional differential structure [10–13]. This goes along not only with a better analytical and numerical accessibility and stability, but also with a great flexibility of devising approximations adapted to a specific physical system. In addition, structural investigations of field theories from first principles such as proofs of renormalizability can more elegantly and efficiently be performed with this strategy [12, 14–16].

The central tool of the functional RG is given by a flow equation. This flow equation describes the evolution of correlation functions or their generating functional under the influence of fluctuations. It connects a well-defined initial quantity, e.g., the microscopic correlation functions in a perturbative domain, in an exact manner with the desired full correlation functions after having integrated out the fluctuations. Hence, solving the flow equation corresponds to solving the full theory.

The complexity of quantum gauge theories and QCD are a serious challenge for all field theoretical methods. The construction of flow equations for gauge theories has to take special care of the gauge symmetry, i.e., the invariance of the theory under *local* transformations of the fields in coordinate space [17–21]. The beauty of gauge symmetry is turned into the beast of complex dynamical equations and nontrivial symmetry constraints both of which have to be satisfied by the flow of the correlation functions. Nevertheless, the success of the functional RG in many branches of physics as described in other lectures of this volume makes it a promising tool also for gauge theories. The recent rapid development of functional methods and their application to gauge theories and QCD also in the strong-coupling domain confirm this expectation. In combination with and partly complementary to other field theoretical methods, the functional RG has the potential to shed light on some of the still hardly accessible parameter regions of quantum gauge theories.

6.2 Functional RG Approach to Quantum Field Theory

6.2.1 Basics of QFT

In quantum field theory (QFT), all physical information is stored in correlation functions. For instance, consider a collider experiment with two incident beams and $(n-2)$ scattering products. All information about this process can be obtained from the *n-point function*, a correlator of n quantum fields. In QFT, we obtain this correlator by definition from the product of n field operators at different spacetime points $\varphi(x_n)$ averaged over all possible field configurations (quantum fluctuations).

In Euclidean QFT, the field configurations are weighted with an exponential of the action $S[\varphi]$,

$$\langle \varphi(x_1)\ldots\varphi(x_n)\rangle := \mathcal{N}\int \mathcal{D}\varphi\, \varphi(x_1)\ldots\varphi(x_n)\, \mathrm{e}^{-S[\varphi]}, \tag{6.1}$$

where we fix the normalization $\mathcal{N}$ by demanding that $\langle 1\rangle = 1$. We assume that Minkowski-valued correlators can be defined from the Euclidean ones by analytic continuation. We also assume that a proper regularized definition of the measure can be given (for instance, using a spacetime lattice discretization), which we formally write as $\int \mathcal{D}\varphi \to \int_\Lambda \mathcal{D}\varphi$; here, Λ denotes an ultraviolet (UV) cutoff. This regularized measure should also preserve the symmetries of the theory: for a symmetry transformation U which acts on the fields, $\varphi \to \varphi^U$, and leaves the action invariant, $S[\varphi] \to S[\varphi^U] \equiv S[\varphi]$, the invariance of the measure implies

$$\int_\Lambda \mathcal{D}\varphi \to \int_\Lambda \mathcal{D}\varphi^U \equiv \int_\Lambda \mathcal{D}\varphi. \tag{6.2}$$

For simplicity, let φ denote a real scalar field; the following discussion also holds for other fields such as fermions with minor modifications. All n-point correlators are summarized by the generating functional $Z[J]$,

$$Z[J] \equiv \mathrm{e}^{W[J]} = \int \mathcal{D}\varphi\, \mathrm{e}^{-S[\varphi]+\int J\varphi}, \tag{6.3}$$

with source term $\int J\varphi = \int \mathrm{d}^D x\, J(x)\varphi(x)$. All n-point functions are obtained by functional differentiation:

$$\langle \varphi(x_1)\ldots\varphi(x_n)\rangle = \frac{1}{Z[0]}\left(\frac{\delta^n Z[J]}{\delta J(x_1)\ldots\delta J(x_n)}\right)_{J=0}. \tag{6.4}$$

Once the generating functional is computed, the theory is solved.

In Eq. (6.3), we have also introduced the generating functional of *connected correlators*,[1] $W[J] = \ln Z[J]$, which, loosely speaking, is a more efficient way to store the physical information. An even more efficient information storage is obtained by a Legendre transform of $W[J]$: the *effective action* Γ:

$$\Gamma[\phi] = \sup_J \left(\int J\phi - W[J]\right). \tag{6.5}$$

For any given ϕ, a special $J \equiv J_{\text{sup}} = J[\phi]$ is singled out for which $\int J\phi - W[J]$ approaches its supremum. Note that this definition of Γ automatically guarantees that Γ is convex. At $J = J_{\text{sup}}$, we get

$$\begin{aligned} 0 &\stackrel{!}{=} \frac{\delta}{\delta J(x)}\left(\int J\phi - W[J]\right) \\ \Rightarrow \quad \phi &= \frac{\delta W[J]}{\delta J} = \frac{1}{Z[J]}\frac{\delta Z[J]}{\delta J} = \langle\varphi\rangle_J. \end{aligned} \tag{6.6}$$

This implies that ϕ corresponds to the expectation value of φ in the presence of the source J. The meaning of Γ becomes clear by studying its derivative at $J = J_{\text{sup}}$

$$\frac{\delta\Gamma[\phi]}{\delta\phi(x)} = -\int_y \frac{\delta W[J]}{\delta J(y)}\frac{\delta J(y)}{\delta\phi(x)} + \int_y \frac{\delta J(y)}{\delta\phi(x)}\phi(y) + J(x) \stackrel{(6.6)}{=} J(x). \tag{6.7}$$

This is the *quantum equation of motion* by which the effective action $\Gamma[\phi]$ governs the dynamics of the field expectation value, taking the effects of all quantum fluctuations into account.

From the definition of the generating functional, we can straightforwardly derive an equation for the effective action:

---

[1] In this short introduction, we use but make no attempt at fully explaining the standard QFT nomenclature; for the latter, we refer the reader to any standard QFT textbook, such as [22, 23].

$$e^{-\Gamma[\phi]} = \int_{\Lambda} \mathcal{D}\varphi \exp\left(-S[\phi+\varphi] + \int \frac{\delta\Gamma[\phi]}{\delta\phi}\varphi\right). \tag{6.8}$$

Here, we have performed a shift of the integration variable, $\varphi \to \varphi + \phi$. We observe that the effective action is determined by a nonlinear first-order functional differential equation, the structure of which is itself a result of a functional integral. An exact determination of $\Gamma[\phi]$ and thus an exact solution has so far been found only for rare, special cases.

As a first example of a functional technique, a solution of Eq. (6.8) can be attempted by a *vertex expansion* of $\Gamma[\phi]$,

$$\Gamma[\phi] = \sum_{n=0}^{\infty} \frac{1}{n!} \int d^D x_1 \dots d^D x_n \, \Gamma^{(n)}(x_1, \dots, x_n)\, \phi(x_1) \dots \phi(x_n), \tag{6.9}$$

where the expansion coefficients $\Gamma^{(n)}$ correspond to the *one-particle irreducible* (*1PI*) *proper vertices*. Inserting Eq. (6.9) into Eq. (6.8) and comparing the coefficients of the field monomials results in an infinite tower of coupled integro-differential equations for the $\Gamma^{(n)}$: the Dyson-Schwinger equations. This functional method of constructing approximate solutions to the theory via truncated Dyson-Schwinger equations, i.e., via a finite truncation of the series truncation of the series Eq. (6.9) has its own merits and advantages; their application to gauge theories is well developed; see, e.g., [24–27]. Here, we proceed by amending the RG idea to functional techniques in QFT.

6.2.2 RG Flow Equation

A versatile approach to the computation of Γ is based on RG concepts [13]. Whereas a computation via Eq. (6.8) or via Dyson-Schwinger equations corresponds to integrating-out all fluctuations at once, we can implement Wilson's idea of integrating out modes momentum shell by momentum shell.

In terms of Γ, we are looking for an interpolating actionf Γ_k, which is also called *effective average action*, with a momentum-shell parameter k, such that Γ_k for $k \to \Lambda$ corresponds to the bare action to be quantized; the full quantum action Γ should be approached for $k \to 0$,

$$\Gamma_{k\to\Lambda} \simeq S_{\text{bare}}, \quad \Gamma_{k\to 0} = \Gamma. \tag{6.10}$$

This can indeed be constructed from the generating functional. For this, let us define the IR regulated functional

$$
\begin{aligned}
e^{W_k[J]} \equiv Z_k[J] &:= \exp\left(-\Delta S_k\left[\frac{\delta}{\delta J}\right]\right) Z[J] \\
&= \int_\Lambda \mathcal{D}\varphi\, e^{-S[\varphi]-\Delta S_k[\varphi]+\int J\varphi},
\end{aligned} \tag{6.11}
$$

where

$$
\Delta S_k[\varphi] = \frac{1}{2}\int \frac{d^D q}{(2\pi)^D}\, \varphi(-q) R_k(q) \varphi(q) \tag{6.12}
$$

is a regulator term which is quadratic in φ and can be viewed as a momentum-dependent mass term. The regulator function $R_k(q)$ should satisfy

$$
\lim_{q^2/k^2 \to 0} R_k(q) > 0, \tag{6.13}
$$

which implements an IR regularization. For instance, if $R_k \sim k^2$ for $q^2 \ll k^2$, the regulator screens the IR modes in a mass-like fashion, $m^2 \sim k^2$. Furthermore,

$$
\lim_{k^2/q^2 \to 0} R_k(q) = 0, \tag{6.14}
$$

which implies that the regulator vanishes for $k \to 0$. As an immediate consequence, we automatically recover the standard generating functional as well as the full effective action in this limit: $Z_{k\to 0}[J] = Z[J]$ and $\Gamma_{k\to 0} = \Gamma$. The third condition is

$$
\lim_{k^2 \to \Lambda \to \infty} R_k(q) \to \infty, \tag{6.15}
$$

which induces that the functional integral is dominated by the stationary point of the action in this limit. This justifies the use of a saddle-point approximation which filters out the classical field configuration and the bare action, $\Gamma_{k\to\Lambda} \to S + \text{const}$. A sketch of a typical regulator that satisfies these three requirements is shown in Fig. 6.1. Incidentally, the regulator is frequently written as

$$
R_k(p^2) = p^2 r(p^2/k^2), \tag{6.16}
$$

where $r(y)$ is a dimensionless regulator shape function with a dimensionless momentum argument. The requirements (6.13)–(6.15) translate in a obvious manner into corresponding requirements for $r(y)$.

Since we already know that the interpolating functional Γ_k exhibits the correct limits, let us now study the intermediate trajectory. We start with the generating functional $W_k[J]$, using the abbreviations

$$
t = \ln\frac{k}{\Lambda}, \qquad \partial_t = k\frac{d}{dk}. \tag{6.17}
$$

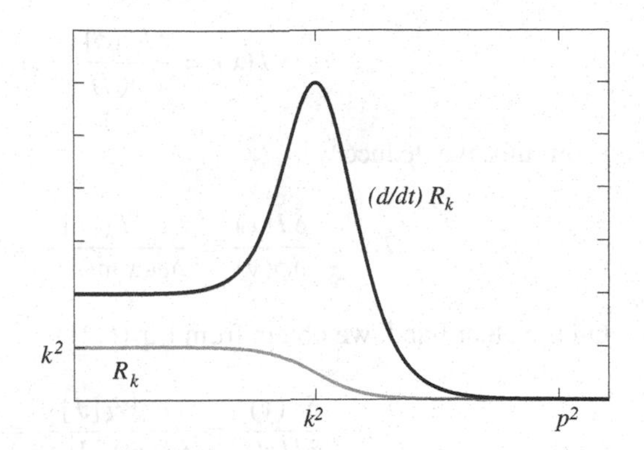

Fig. 6.1 Sketch of a regulator function $R_k(p^2)$ (*lower curve*) and its derivative $\partial_t R_k(p^2)$ (*upper curve*). Whereas the regulator provides for an IR regularization for all modes with $p^2 \lesssim k^2$, its derivative implements the Wilsonian idea of integrating out fluctuations within a momentum shell near $p^2 \simeq k^2$

Keeping the source J fixed, i.e., k independent, we obtain

$$\begin{aligned}\partial_t W_k[J] &= -\frac{1}{2}\int \mathcal{D}\varphi\, \varphi(-q)\, \partial_t R_k(q)\, \varphi(q) \mathrm{e}^{-S-\Delta S+\int J\varphi} \\ &= -\frac{1}{2}\int \frac{\mathrm{d}^D q}{(2\pi)^D}\, \partial_t R_k(q)\, G_k(-q,q) + \partial_t \Delta S_k[\phi].\end{aligned} \tag{6.18}$$

Here, we have defined the *connected* propagator

$$G_k(p,q) = \left(\frac{\delta^2 W_k}{\delta J \delta J}\right)(p,q) = \langle \varphi(p)\varphi(q)\rangle - \langle \varphi(p)\rangle\langle\varphi(q)\rangle. \tag{6.19}$$

(Note that we frequently change from coordinate to momentum space or vice versa by Fourier transformation for reasons of convenience.) Now, we are in a position to define the interpolating effective action Γ_k by a slightly modified Legendre transform,[2]

$$\Gamma_k[\phi] = \sup_J \left(\int J\phi - W_k[J]\right) - \Delta S_k[\phi]. \tag{6.20}$$

Since we later want to study Γ_k as a functional of a k-independent field ϕ, it is clear from Eq. (6.20) that the source $J \equiv J_{\text{sup}} = J[\phi]$ for which the supremum is approached is necessarily k dependent. As before, we get at $J = J_{\text{sup}}$:

$$\phi(x) = \langle\varphi(x)\rangle_J = \frac{\delta W_k[J]}{\delta J(x)}. \tag{6.21}$$

The quantum equation of motion receives a regulator modification,

[2] Now, only the "sup" part of Γ_k is convex. For finite k, any non-convexity of Γ_k must be of the form of the last regulator term of Eq. (6.20).

$$J(x) = \frac{\delta \Gamma_k[\phi]}{\delta\phi(x)} + \big(R_k\phi\big)(x). \tag{6.22}$$

From this, we deduce[3]:

$$\frac{\delta J(x)}{\delta\phi(y)} = \frac{\delta^2 \Gamma_k[\phi]}{\delta\phi(x)\delta\phi(y)} + R_k(x, y). \tag{6.23}$$

On the other hand, we obtain from Eq. (6.21):

$$\frac{\delta\phi(y)}{\delta J(x')} = \frac{\delta^2 W_k[J]}{\delta J(x')\delta J(y)} \equiv G_k(y, x'). \tag{6.24}$$

This implies the important identity

$$\begin{aligned}\delta(x - x') = \frac{\delta J(x)}{\delta J(x')} &= \int \mathrm{d}^D y \, \frac{\delta J(x)}{\delta\phi(y)} \frac{\delta\phi(y)}{\delta J(x')} \\ &= \int \mathrm{d}^D y \, (\Gamma_k^{(2)}[\phi] + R_k)(x, y) G_k(y, x'),\end{aligned} \tag{6.25}$$

or, in operator notation,

$$\mathbb{1} = (\Gamma_k^{(2)} + R_k)\, G_k. \tag{6.26}$$

Here, we have introduced the short-hand notation

$$\Gamma_k^{(n)}[\phi] = \frac{\delta^n \Gamma_k[\phi]}{\delta\phi \ldots \delta\phi}. \tag{6.27}$$

Collecting all ingredients, we can finally derive the flow equation for Γ_k for fixed ϕ and at $J = J_{\text{sup}}$ [13]:

$$\begin{aligned}\partial_t \Gamma_k[\phi] &= -\partial_t W_k[J]|_\phi + \int (\partial_t J)\phi - \partial_t \Delta S_k[\phi] = -\partial_t W_k[J]|_J - \partial_t \Delta S_k[\phi] \\ &\overset{(6.18)}{=} \frac{1}{2} \int \frac{\mathrm{d}^D q}{(2\pi)^D} \, \partial_t R_k(q)\, G_k(-q, q) \\ &\overset{(6.26)}{=} \frac{1}{2} \,\mathrm{Tr} \left[\partial_t R_k \left(\Gamma_k^{(2)}[\phi] + R_k \right)^{-1} \right].\end{aligned} \tag{6.28}$$

This flow equation forms the starting point of all our further investigations. Hence, let us carefully discuss a few of its properties, as they are apparent already at this stage:

---

[3] In case of fermionic Grassmann-valued fields, the following ϕ derivative should act on Eq. (6.22) from the right.

- The flow equation is a functional differential equation for Γ_k. In contrast to Eq. (6.8), no functional integral has to be performed to reveal the full structure of the equation.
- We have *derived* the flow equation from the standard starting point of QFT: the generating functional. But a different—if not inverse—perspective is also legitimate. We may *define* QFT based on the flow equation. For given suitable initial conditions, for instance, by defining the bare action at a high UV cutoff scale $k = \Lambda$, the flow equation defines a trajectory to the full quantum theory described by the full effective action Γ. In the case of additional symmetries, the QFT-defining flow equation may be supplemented by symmetry constraints to the effective action.
- The purpose of the regulator is actually twofold: by construction, the occurrence of R_k in the denominator of Eq. (6.28) guarantees the IR regularization by construction. In addition to this and thanks to the conditions (6.13) and (6.14), the derivative $\partial_t R_k$ occurring in the numerator of Eq. (6.28) ensures also UV regularization, since its predominant support lies on a smeared momentum shell near $p^2 \sim k^2$. A typical shape of the regulator and its derivative is depicted in Fig. 6.1. The peaked structure of $\partial_t R_k$ implements nothing but the Wilsonian idea of integrating over momentum shells and implies that the flow is localized in momentum space.
- The solution to the flow equation (6.28) corresponds to an RG trajectory in *theory space*. The latter is a space of all action functionals spanned by all possible invariant operators of the field. The two ends of the trajectory are given by the initial condition $\Gamma_{k\to\Lambda} = S_{\text{bare}}$, and the full effective action $\Gamma_{k\to 0} = \Gamma$.
- Apart from the conditions (6.13)–(6.15), the regulator can be chosen arbitrarily. Of course, the precise form of the trajectory depends on the regulator R_k. The variation of the trajectory with respect to R_k reflects the RG scheme dependence of a non-universal quantity; see Fig. 6.2. Nevertheless, the final point on the trajectory is independent of R_k as is guaranteed by Eqs. (6.13)–(6.15).
- The flow equation has a one-loop structure, but is nevertheless an exact equation, as is signaled by the occurrence of the exact propagator in the loop; see Fig. 6.3. The one-loop structure is a direct consequence of ΔS_k being quadratic in the field operator φ which is coupled to the source [28].
- Perturbation theory can immediately be re-derived from the flow equation. For instance, imposing the loop expansion on Γ_k, $\Gamma_k = S + \hbar\Gamma_k^{1-\text{loop}} + \mathcal{O}(\hbar^2)$, it becomes obvious that, to one-loop order, $\Gamma_k^{(2)}$ can be replaced by $S^{(2)}$ on the right-hand side of Eq. (6.28). From this, we infer:

$$\partial_t \Gamma_k^{1-\text{loop}} = \frac{1}{2}\,\text{Tr}\left[\partial_t R_k \left(S^{(2)} + R_k\right)^{-1}\right] = \frac{1}{2}\,\partial_t\,\text{Tr}\,\ln(S^{(2)} + R_k)$$

$$\Rightarrow \Gamma^{1-\text{loop}} = S + \frac{1}{2}\,\text{Tr}\,\ln S^{(2)} + \text{const.}$$

The last formula corresponds to the standard one-loop effective action, as it should.

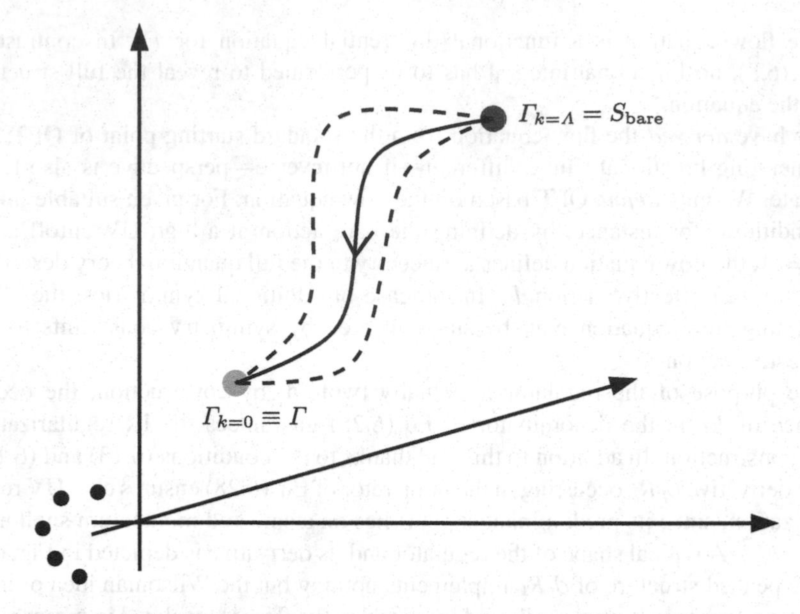

Fig. 6.2 Sketch of the RG flow in theory space. Each axis labels a different operator which spans the effective action, e.g., ϕ^2, $(\partial\phi)^2$, etc. Once the initial conditions in terms of the bare action $\Gamma_{k=\Lambda} = S_{\text{bare}}$ are given, the solution to the flow equation (6.28) is a trajectory (*solid line*) in this space of action functionals, ending at the full quantum effective action $\Gamma \equiv \Gamma_{k=0}$. A change of the regulator R_k can modify the trajectory (*dashed lines*), but the end point Γ stays always the same

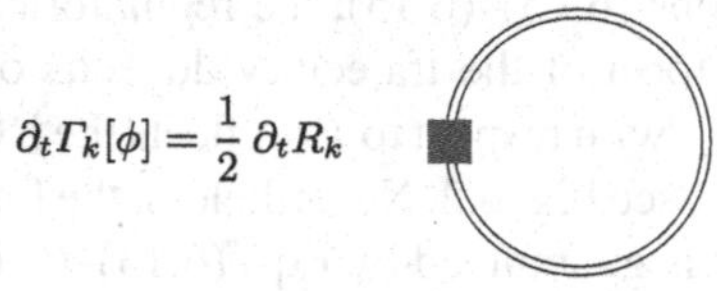

$$\partial_t \Gamma_k[\phi] = \frac{1}{2}\,\partial_t R_k$$

Fig. 6.3 Diagrammatic representation of the flow equation (6.28): the flow of Γ_k is given by a one-loop form, involving the full propagator $G_k = (\Gamma_k^{(2)} + R_k)^{-1}$ (*double line*) and an operator insertion in the form of $\partial_t R_k$ (*filled box*)

In practice, the versatility of the flow equation is its most important strength: beyond perturbation theory, various systematic approximation schemes exist which can be summarized under the *method of truncations*.

A first example for such an approximation scheme has already been given above in Eq. (6.9), the vertex expansion, which now reads

$$\Gamma_k[\phi] = \sum_{n=0}^{\infty} \frac{1}{n!} \int \mathrm{d}^D x_1 \ldots \mathrm{d}^D x_n\, \Gamma_k^{(n)}(x_1, \ldots, x_n)\, \phi(x_1) \ldots \phi(x_n). \qquad (6.29)$$

Upon inserting this expansion into the flow equation (6.28), we obtain flow equations for the vertex functions $\Gamma_k^{(n)}$ which interpolate between the bare and the fully dressed vertices. These equations are similar but not identical to Dyson-Schwinger equations, as will be discussed in more detail below.

As a second example, let us introduce the *operator expansion* which constructs the effective action from operators of increasing mass dimension. Focusing in particular on derivative operators, we arrive at the gradient expansion. For instance, for a theory with one real scalar field, we obtain

$$\Gamma_k = \int \mathrm{d}^D x \left[V_k(\phi) + \frac{1}{2} Z_k(\phi)\, (\partial_\mu \phi)^2 + \mathcal{O}(\partial^4) \right], \tag{6.30}$$

where, for instance, $V_k(\phi)$ corresponds to the effective potential.

Further examples of this type can easily be constructed by combining these two expansions in various ways. Formally, expansions of the effective action should be *systematic*; this implies that a classification scheme exists which classifies all possible building blocks of the effective action and relates them to a definite order in the expansion. Truncations based on such expansions should also be *consistent* in the sense that, once the maximal order of the truncated expansion is chosen, all terms up to this order are kept in the flow equation.

Systematics and consistency are, of course, only a necessary condition for the construction of a reliable truncation—they are not necessarily sufficient. As a word of caution, let us stress that these two conditions do not guarantee a rapid convergence of the truncated effective action towards the true result. As for any expansion, convergence properties have to be checked separately, in order to estimate or even control the truncation errors.

Only one general recipe for the construction of a truncation exists: let your truncation scheme be guided by physics by making sure that the truncation includes the most relevant degrees of freedom of a given problem.

6.2.3 Euclidean Anharmonic Oscillator

Let me illustrate the capabilities of the flow equation by a simple example: a $0+1$ dimensional real scalar field theory, or, in other words, the Euclidean quantum mechanical anharmonic oscillator. This system has first been studied by RG techniques in [29]. The bare action that we want to quantize is

$$S = \int d\tau \left(\frac{1}{2}\dot{x}^2 + \frac{1}{2}\omega^2 x^2 + \frac{\lambda}{24} x^4 \right), \tag{6.31}$$

with $\omega^2, \lambda > 0$.[4] We will mainly be interested in the determination of the ground state energy which we expect to be predominantly influenced by the effective potential. Hence, we consider the truncation

$$\Gamma_k[x] = \int d\tau \left(\frac{1}{2}\dot{x}^2 + V_k(x) \right). \tag{6.32}$$

For a concrete computation, we have to choose a regulator which conforms to the conditions (6.13)–(6.15). The following choice is not only simple and convenient, it is also an *optimal* choice for the present problem, since it improves the stability properties of our flow equation [32],

$$R_k(p) = (k^2 - p^2)\theta(k^2 - p^2), \tag{6.33}$$

which implies that $\partial_t R_k = 2k^2\,\theta(k^2 - p^2)$. On the right-hand side of the flow equation (6.28), we need $\Gamma_k^{(2)} = (-\partial_\tau^2 + V_k''(x))\delta(\tau - \tau')$. In order to project the flow equation onto the flow of the effective potential, it suffices to consider the special case $x = \text{const.}$, for which the right-hand side can immediately be Fourier transformed, e.g., $-\partial_\tau^2 \to p_\tau^2$. We obtain the flow of the effective potential:

$$\begin{aligned} \partial_t V_k(x) &= \frac{1}{2}\int_{-\infty}^{\infty} \frac{dp_\tau}{2\pi} \frac{2k^2\theta(k^2 - p_\tau^2)}{k^2 + V_k''(x)} \\ \Rightarrow \frac{\mathrm{d}}{\mathrm{d}k} V_k(x) &= \frac{1}{\pi}\frac{k^2}{k^2 + V_k''(x)}. \end{aligned} \tag{6.34}$$

This is a partial differential equation for the effective potential. Aiming at the ground state energy, it suffices to study a polynomial expansion of the potential,

$$V_k(x) = \frac{1}{2}\omega_k^2 x^2 + \frac{1}{24}\lambda_k x^4 + \cdots + \tilde{E}_k, \tag{6.35}$$

where the effects of the fluctuations are now encoded in a scale-dependent frequency ω_k and coupling λ_k; their initial conditions at $k = \Lambda$ are given by the parameters ω and λ in the bare action (6.31). Unfortunately, $\tilde{E}_k$ is not identical to the desired ground state energy $E_{0,k}$, but differs by further x-independent contributions. This becomes already clear by looking at the UV limit $k \to \Lambda$, where the regulator term becomes $\sim \frac{1}{2}\Lambda^2 x^2$, contributing a harmonic-oscillator-like ground state energy $\sim \frac{1}{2}\Lambda$ to $\tilde{E}$. In order to extract the true ground state energy from the flow of $\tilde{E}_k$,

$$\frac{\mathrm{d}}{\mathrm{d}k}\tilde{E}_k = \frac{1}{\pi}\frac{k^2}{k^2 + \omega_k^2}, \tag{6.36}$$

[4] Also the double-well potential with $\omega^2 < 0$ can be studied with RG techniques, see [30, 31].

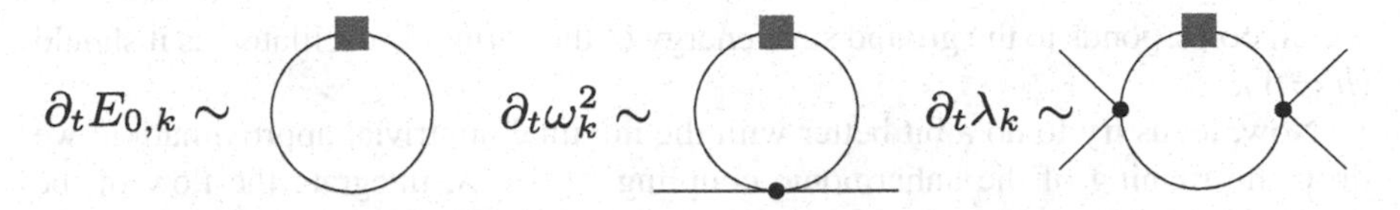

Fig. 6.4 Diagrammatic representation of Eqs. (6.37)–(6.39). The diagrams look similar to one-loop perturbative diagrams with all internal propagators and vertices being fully dressed quantities. One internal line always carries the regulator insertion $\partial_t R_k$ (*filled box*). (One further diagram for $\partial_t \lambda_k$ involving a 6-point vertex is dropped, as in Eq.(6.39))

we can perform a controlled subtraction to avoid the build-up of the unphysical contributions: in the limit $\lambda = \omega = 0$, the ground state energy has to stay zero, $E_{0,k} = 0$. This fixes the subtraction term for Eq. (6.36), which then reads:

$$\frac{\mathrm{d}}{\mathrm{d}k} E_{0,k} = \frac{1}{\pi}\left(\frac{k^2}{k^2+\omega_k^2} - 1\right), \tag{6.37}$$

Expanding Eq. (6.34) to higher orders, yields

$$\frac{\mathrm{d}}{\mathrm{d}k}\omega_k^2 = -\frac{2}{\pi}\frac{k^2}{(k^2+\omega_k^2)^2}\frac{\lambda_k}{2}, \tag{6.38}$$

$$\frac{\mathrm{d}}{\mathrm{d}k}\lambda_k = \frac{24}{\pi}\frac{k^2}{(k^2+\omega_k^2)^3}\left(\frac{\lambda_k}{2}\right)^2 + \ldots, \tag{6.39}$$

where the ellipsis in the last equation denotes contributions from higher-order terms $\sim x^6$, which we neglect here for simplicity. We have boiled the flow equation down to a coupled set of first-order ordinary differential equations, which can be viewed as the RG β functions of the generalized couplings $E_{0,k}, \omega_k, \lambda_k$. These equations can diagrammatically be displayed as in Fig. 6.4. The diagrams look very similar to one-loop perturbative diagrams, but there are important differences: all internal lines and vertices correspond to full propagators and full vertices (in our simple truncation here, the vertex represents a full running λ_k and the propagators contain the running ω_k^2). Furthermore, one propagator in each loop carries a regulator insertion, implying the replacement $G \to G_k \partial_t R_k G_k$ in comparison with perturbative diagrams.

It is instructive, to solve these three equations (6.37)–(6.39) with various approximations. We begin with dropping the anharmonic coupling $\lambda = 0$, implying that $\omega_k = \omega_{k=\Lambda} \equiv \omega$; integrating the remaining flow of the ground state energy yields

$$E_0 \equiv E_{0,k=0} = \int_0^\infty \mathrm{d}k \frac{\mathrm{d}}{\mathrm{d}k} E_{0,k} \overset{(6.37)}{=} \frac{1}{2}\omega, \tag{6.40}$$

which corresponds to the ground state energy of the harmonic oscillator, as it should ($\hbar = 1$).

Now, let us try to do a bit better with the minimal nontrivial approximation: we drop the running of the anharmonic coupling $\lambda_k \to \lambda$, integrate the flow of the frequency (6.38), insert the resulting ω_k into Eq. (6.37) and integrate the energy. Expanding the result perturbatively in λ, we find[5]

$$E_0 = \frac{1}{2}\omega + \frac{3}{4}\omega\left(\frac{\lambda}{24\omega^3}\right) - \frac{3(8\pi^2+29)}{16\pi^2}\omega\left(\frac{\lambda}{24\omega^3}\right)^2 + \dots, \tag{6.41}$$

which can immediately be compared with direct 2nd-order perturbation theory [33],

$$E_0^{\mathrm{PT}} = \frac{1}{2}\omega + \frac{3}{4}\omega\left(\frac{\lambda}{24\omega^3}\right) - \frac{21}{8}\omega\left(\frac{\lambda}{24\omega^3}\right)^2 + \dots. \tag{6.42}$$

Our first-order "one-loop" result agrees exactly with perturbation theory, whereas the second-order "two-loop" coefficient has the right sign and order of magnitude but comes out too small with a $\sim$20% error. However, it should be kept in mind that we have obtained this two-loop estimate from a cheap calculation which involved only a one-loop integral with an RG-improved propagator.

Of course, we do not have to stop with perturbation theory. We can look at the full integrated result based on Eqs. (6.37) and (6.38) for any value of λ. For instance, let us boldly study the strong-coupling limit, where the asymptotics is known to be of the form,

$$E_0 = \left(\frac{\lambda}{24}\right)^{1/3}\left[\alpha_0 + \mathcal{O}\left(\lambda^{-2/3}\right)\right]. \tag{6.43}$$

The constant α_0 has been determined in [34] to a high precision by means of large-order variational perturbation theory: $\alpha_0 = 0.66798\dots$.

With the simplest nontrivial approximation based on Eqs. (6.37) and (6.38), we obtain $\alpha_0^{(6.37),(6.38)} = 0.6920\dots$, which differs from the full solution by merely 4%. Now, solving all three equations (6.37)–(6.39) simultaneously, we find $\alpha_0^{(6.37)-(6.39)} = 0.6620\dots$, which corresponds to a 1% error.

A plot of the ground state energy is depicted in Fig. 6.5 (left panel) for $\omega = 2$ and $\lambda = 0\dots 100$. Whereas the perturbative estimate (dotted line) is a good approximation for small λ, it becomes useless for larger coupling. Already the simplest approximation based on Eqs. (6.37) and (6.38) (short-dashed/blue line) is a reasonable estimate of the exact result (solid/black line), obtained from the exact integration of the Schrödinger equation as quoted in [35]. The flow equation estimate based on Eqs. (6.37)–(6.39) (long-dashed/red line) is hardly distinguishable from the exact result. For better visibility, the relative errors of these estimates are shown in Fig. 6.5 (right panel). Over the whole range of λ, the error of our estimate based on Eqs. (6.37)–(6.39) (long-dashed/red line) does not exceed 0.3%.

[5] I am grateful to A. Wipf for analytically determining the 2nd-order coefficient.

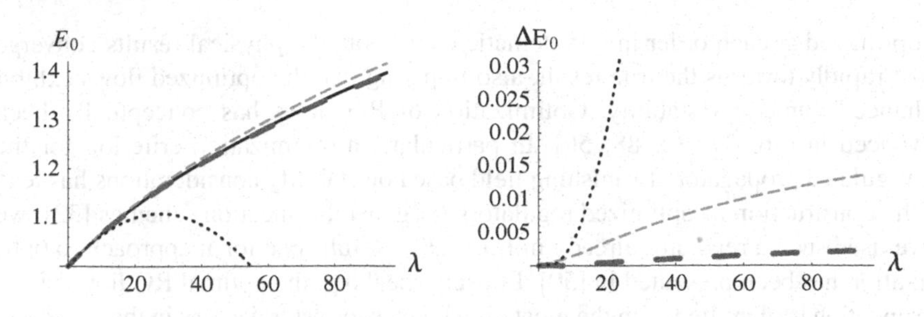

Fig. 6.5 *Left panel* Ground state energy of the anharmonic oscillator for $\omega = 2$ versus the anharmonic coupling λ: exact result (*solid/black line*), 2-loop perturbation theory (*dotted/black line*), flow-equation estimate based on Eqs. (6.37) and (6.38) (*short-dashed/blue line*) or on Eqs. (6.37)–(6.39) (*long-dashed/red line*). *Right panel* Relative error of the different estimates

The quality of the result is remarkable in view of the extremely simple approximation of the full flow. In particular, the truncation to a low-order polynomial potential does not seem to be justified at large coupling. In fact, there is no reason, why the dropped higher-order terms, e.g., $\sim x^6$, should be small compared to the terms kept.

The lesson to be learned is the following: it does not really matter whether the terms dropped are small compared to the terms kept. It only matters whether their influence on the terms that belong to the truncation is small or large.

6.2.4 Further Reading: Regulator Dependence and Optimization*

The reliability of the solution of a truncated RG flow is an important question that needs to be addressed in detail for each application. In absence of an obvious small expansion parameter, the convergence of any systematic and consistent expansion of the effective action can be checked by studying the quantitative influence of higher-order terms. Whereas such computations can become rather extensive, an immediate check can be performed by regulator studies. For the exact flow, physical observables evaluated at $k = 0$ are, of course, regulator independent by construction. However, truncations generically induce spurious regulator dependencies, the amount of which provides a measure for the importance of higher-order terms outside a given truncation. Resulting regulator dependencies of physical observables can thus be used for a quantitative error estimate in a rather direct manner.

Moreover, the freedom to choose the regulator function within the mild set of conditions provided by Eqs. (6.13)–(6.15) can also actively be used for an optimization of the flow. A truncated flow is optimized if the results for physical observables lie as close as possible to the true results; the influence of operators outside a given truncation scheme on the physical results is then minimized. If a truncated flow is optimized at each order in a systematic expansion, the physical results converge most rapidly towards the true result, also implying that the optimized flows exhibit enhanced numerical stability. Optimization of RG flows has conceptually been advanced in [36, 37, 32, 88, 50]. In particular, an optimization criterion for the IR regulated propagator at vanishing field based on stability considerations has lead to the construction of optimized regulators for general truncation schemes [32]; we have used such a regulator already in Eq. (6.33). A full functional approach to optimization has been presented in [50]. Loosely speaking, the optimal RG flow within a truncation is identified with the most direct, i.e., shortest trajectory in theory space.

Since optimization helps improving quantitative predictions, enhances numerical stability and can at the same time be used to reduce technical effort, it is of high practical relevance. Optimization is therefore recommended for all modern applications of the functional RG.

6.3 Functional RG for Gauge Theories

6.3.1 RG Flow Equations and Symmetries

Before we embark on the complex machinery of quantum field theories with non-abelian gauge symmetries, let us study the interplay between flow equations and symmetries from a more general perspective with emphasis on the structural aspects.

Consider a QFT which is invariant under a continuous symmetry transformation which can be realized linearly on the fields; let $\mathcal{G}$ be the generator of an infinitesimal version of this transformation, i.e., $\mathcal{G}\phi$ is linear in ϕ. For example, a global O(N) symmetry in a QFT for N-component scalar fields is generated by

$$\mathcal{G}^a = -f^{abc} \int \mathrm{d}^D x \, \phi^b(x) \frac{\delta}{\delta \phi^c(x)}. \tag{6.44}$$

Incidentally, a local symmetry would be generated by the analogue of Eq. (6.44) without the spacetime integral. Together with the invariance of the measure under this symmetry, cf. Eq. (6.2), the invariance of the QFT can be stated by

$$0 = \frac{1}{Z} \int \mathcal{D}\varphi \, \mathcal{G} \, \mathrm{e}^{-S + \int J\varphi}. \tag{6.45}$$

In other words, a transformation of the action can be undone by a transformation of the measure.

What does this symmetry of the QFT imply for the effective action? First, we observe that Eq. (6.45) yields

$$0 = \frac{1}{Z} \int \mathcal{D}\varphi \left(-(\mathcal{G}S) + \int J(\mathcal{G}\varphi) \right) e^{-S+\int J\varphi}$$
$$= -\langle \mathcal{G}S \rangle_J + e^{-W[J]} \int J(\mathcal{G}\varphi)|_{\varphi = \frac{\delta}{\delta J}} e^{W[J]}. \qquad (6.46)$$

Performing the Legendre transform, we find at $J = J_{\text{sup}} \equiv J[\phi]$:

$$0 = -\langle \mathcal{G}S \rangle_{J[\phi]} + \int \frac{\delta \Gamma}{\delta \phi} \mathcal{G}\phi. \qquad (6.47)$$

The last term is nothing but $\mathcal{G}\Gamma$, and we obtain the important identity (Ward identity)

$$\mathcal{G}\Gamma[\phi] = \langle \mathcal{G}S \rangle_{J[\phi]}. \qquad (6.48)$$

It demonstrates that the effective action is invariant under a symmetry if the bare action as well as the measure are invariant.

On this level, the statement sounds rather trivial, but it can readily be generalized to the effective average action by keeping track of the regulator term,

$$\mathcal{G}\Gamma_k[\phi] = \langle \mathcal{G}(S + \Delta S_k) \rangle_{J[\phi]} - \mathcal{G}\Delta S_k[\phi]. \qquad (6.49)$$

Here, we learn that the whole RG trajectory Γ_k is invariant under the symmetry if $\mathcal{G}S = 0$ and $\mathcal{G}\Delta S_k = 0$. This is the case if the regulator preserves the symmetry. For instance for the globally O(N)-symmetric theory, a regulator of the form

$$\Delta S_k = \frac{1}{2} \int \frac{d^D p}{(2\pi)^D} \varphi^a(-p) \delta^{ab} R_k(p) \varphi^b(p) \qquad (6.50)$$

is such an invariant regulator.

Note the following crucial point: since the regulator vanishes, $\Delta S_k \to 0$, for $k \to 0$, Eq. (6.49) appears to tell us that $\mathcal{G}\Gamma_{k=0}[\phi] = \langle \mathcal{G}S \rangle_{J[\phi]}$ always holds, implying that the symmetry is always restored at $k = 0$ even for a non-symmetric regulator. This is indeed true, but requires that the initial conditions at $k = \Lambda$ have to be carefully chosen in a highly non-symmetric manner, since

$$\mathcal{G}\Gamma_\Lambda[\phi] = \langle \mathcal{G}S \rangle + \langle \mathcal{G}\Delta S_{k=\Lambda} \rangle - \mathcal{G}\Delta S_{k=\Lambda}. \qquad (6.51)$$

Here, the last two terms do generally not cancel each other. Therefore, even for non-symmetric regulators, the Ward identity tells us how to choose initial conditions with non-symmetric UV counterterms, such that the latter are exactly eaten up by non-symmetric flow contributions, see Fig. 6.6.

In general, it is advisable to use a symmetric regulator, since the space of symmetric actions is smaller, implying that fewer couplings have to be studied in a truncation. However, if a symmetric regulator is not available, the flow equation

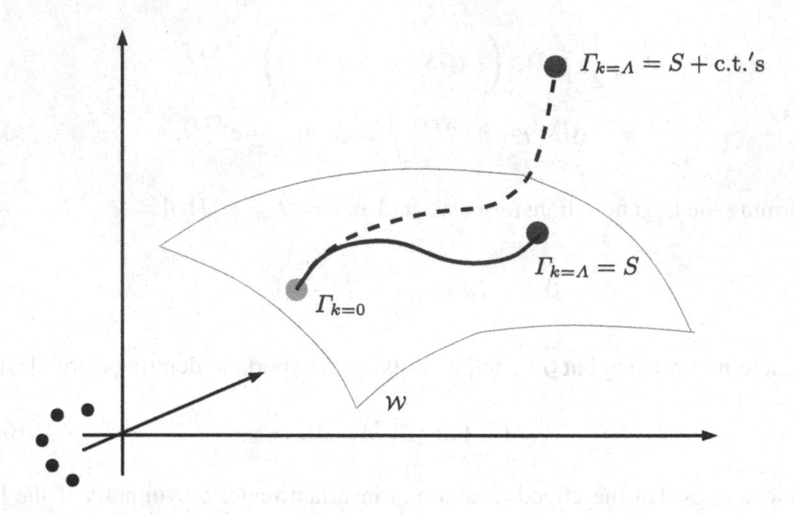

Fig. 6.6 Sketch of the RG flow in theory space with symmetries: the symmetry relation cuts out a hypersurface $\mathcal{W}$ where the action is invariant under the given symmetry. A symmetric regulator keeps the invariance explicitly, such that the RG trajectory always stays inside $\mathcal{W}$ (*solid line*). A non-symmetric regulator can still be used if non-symmetric counterterms (c.t.'s) are chosen such that they eat up the non-symmetric flow contributions (*dashed line*) by virtue of the Ward identity Eq. (6.51). Also in this case, the resulting effective action is invariant, $\Gamma_{k=0} \in \mathcal{W}$

together with the Ward identity can still be used. In fact, in gauge theories there is no simple formalism with a symmetric regulator.

6.3.2 Basics of Gauge Theories

Let us review a few basic elements of quantum gauge theories for reasons of completeness and in order to introduce our notation.

Consider the classical Yang-Mills action,

$$S_{\text{YM}} = \int d^D x \, \frac{1}{4} F^a_{\mu\nu} F^a_{\mu\nu}, \tag{6.52}$$

with the field strength

$$F^a_{\mu\nu} = \partial_\mu A^a_\nu - \partial_\nu A^a_\mu + g f^{abc} A^b_\mu A^c_\nu, \tag{6.53}$$

where the gauge field A^a_μ carries an internal-symmetry index (color), and f^{abc} are the structure constants of a compact non-abelian Lie group. The hermitean generators T^a of this group form a Lie algebra and satisfy

$$[T^a, T^b] = \mathrm{i} f^{abc} T^c, \tag{6.54}$$

e.g., $a, b, c, = 1, 2, \ldots, N_c^2 - 1$ for the group $\mathrm{SU}(N_c)$. (In the fundamental representation, the T^a are hermitean $N_c \times N_c$ matrices which can be normalized to $\mathrm{tr}[T^a, T^b] = \frac{1}{2}\delta^{ab}$.)

The naive attempt to define the corresponding quantum gauge theory by

$$Z[J] \stackrel{?}{=} \int \mathcal{D}A\, \mathrm{e}^{-S_{\mathrm{YM}}[A] + \int J_\mu^a A_\mu^a} \tag{6.55}$$

fails and generically leads to ill-defined quantities plagued by infinities. The reason is that the measure $\mathcal{D}A_\mu^a$ contains a huge redundancy, since many gauge-field configurations A_μ^a are physically equivalent. Namely, the action is invariant under the local symmetry

$$A_\mu^a \to A_\mu^a - \partial_\mu \omega^a + g f^{abc} \omega^b A_\mu^c \equiv A_\mu^a + \delta A_\mu^a, \tag{6.56}$$

where $\omega^a(x)$ is considered to be infinitesimal and differentiable, but otherwise arbitrary. The set of all possible transformations forms the corresponding gauge group. The generator of this symmetry is

$$\mathcal{G}_A^a(x) = D_\mu^{ab} \frac{\delta}{\delta A_\mu^b}, \tag{6.57}$$

where $D_\mu^{ab} = \partial_\mu \delta^{ab} + g f^{abc} A_\mu^c$ denotes the covariant derivative in adjoint representation. It is a simple exercise to show that $\int \mathrm{d}^D x\, \omega^b \mathcal{G}_A^b A_\mu^a = \delta A_\mu^a$. The full symmetry transformation for non-infinitesimal $\omega^a(x)$ can be written as ($A_\mu = A_\mu^a T^a$)

$$A_\mu \to A_\mu^\omega = U A_\mu U^{-1} - \frac{\mathrm{i}}{g} (\partial_\mu U) U^{-1}, \tag{6.58}$$

with $U = U[\omega] = \mathrm{e}^{-\mathrm{i} g \omega^a T^a}$ being an element of the Lie group. Field configurations which are connected by Eq. (6.58) are *gauge equivalent* and form the *gauge orbit*:

$$[A_\mu^{\mathrm{orbit}}] = \{A_\mu^\omega \,|\, A_\mu = A_\mu^{\mathrm{ref}},\ \ U[\omega] \in \mathrm{SU}(N_c)\}. \tag{6.59}$$

Here, A_μ^{ref} is a reference gauge field which is representative for the orbit.

In order to define the quantum theory, we would like to dispose of a measure which picks one representative gauge-field configuration out of each orbit. This is intended by choosing a gauge-fixing condition,

$$\mathsf{F}^a[A] = 0. \tag{6.60}$$

For instance, $\mathsf{F}^a = \partial_\mu A_\mu^a$ is an example for a Lorentz covariant gauge-fixing condition. Ideally, Eq. (6.60) should be satisfied by only one A_μ^a of each orbit.

(As discussed below, this is actually impossible for standard smooth gauge-fixing conditions, owing to topological obstructions [38]).

Gauge fixing can be implemented in the generating functional by means of the Faddeev-Popov trick which is usually derived from

$$1 = \int \mathcal{D}\mathsf{F}^a\, \delta[\mathsf{F}^a] = \int \mathrm{d}\mu(\omega)\, \delta[\mathsf{F}^a(A^\omega)]\, \det\left(\frac{\delta \mathsf{F}^a[A^\omega]}{\delta \omega^b}\right), \tag{6.61}$$

where $\mathrm{d}\mu$ denotes the invariant Haar measure for an integration over the gauge group manifold (at each spacetime point). This rule is reminiscent to the corresponding rule for variable substitution in an ordinary integral, $1 = \int \mathrm{d}f\, \delta(f) = \int \mathrm{d}x\, \delta(f(x)) \left|\frac{\mathrm{d}f}{\mathrm{d}x}\right|$, for $f(x)$ having only one zero. But as already this simple comparison shows, Eq. (6.61) is accompanied by the tacit assumption that $\mathsf{F}^a = 0$ picks only one representative and that the Faddeev-Popov determinant $\det(\delta\mathsf{F}/\delta\omega) > 0$. Both assumptions are generally not true; and both these properties, namely, that several gauge copies on the same gauge orbit all satisfy a given standard gauge-fixing condition and that the Faddeev-Popov determinant is not positive definite, are characteristic of the famous Gribov problem [39]. A direct solution to this problem in the functional integral formalism is by no means simple; nevertheless, let us assume here that a solution exists which renders the gauge-fixed functional integral well defined such that we can proceed with deriving the flow equation. We will return to this problem later in the discussion of the flow equation.

As discussed in standard textbooks, it is now possible to show that the Faddeev-Popov determinant is gauge invariant, $\Delta_{\mathrm{FP}}[A^\omega] \equiv \det \frac{\delta \mathsf{F}^a[A^\omega]}{\delta \omega^b} = \Delta_{\mathrm{FP}}[A]$. As a consequence, $\delta[\mathsf{F}^a[A]]\Delta_{\mathrm{FP}}[A]$ can be inserted into the functional integral, such that the redundancy introduced by gauge symmetry is removed at least for perturbative amplitudes; this renders the perturbative amplitudes well defined and S matrix elements are, in fact, independent of the gauge-fixing condition. As a result, the Euclidean gauge-fixed generating functional, replacing the naive attempt Eq. (6.55), becomes

$$Z[J] = \mathrm{e}^{W[J]} = \int \mathcal{D}A\, \Delta_{\mathrm{FP}}[A]\, \delta[\mathsf{F}^a[A]]\, \mathrm{e}^{-S_{\mathrm{YM}} + \int JA}. \tag{6.62}$$

The additional terms can be brought into the exponent:

$$\delta[\mathsf{F}^a[A]] \to \mathrm{e}^{-\frac{1}{2\alpha}\int \mathrm{d}^D x \mathsf{F}^a \mathsf{F}^a}\Big|_{\alpha\to 0} \equiv \mathrm{e}^{-S_{\mathrm{gf}}[A]}, \tag{6.63}$$

where we have used a Gaußian representation of the δ functional. The exponentiation of the Faddeev-Popov determinant can be done with Grassmann-valued anticommuting real *ghost* fields c, $\bar{c}$, yielding

$$\Delta_{\mathrm{FP}}[A] = \int \mathcal{D}\bar{c}\mathcal{D}c\, \mathrm{e}^{-\int \mathrm{d}^D x\, \bar{c}^a \frac{\delta F^a}{\delta \omega^b} c^b} \equiv \int \mathcal{D}\bar{c}\mathcal{D}c\, \mathrm{e}^{-S_{\mathrm{gh}}}. \tag{6.64}$$

The ghost fields transform homogeneously,

$$c^a \to c^a + g f^{abc} \omega^b c^c, \quad \bar{c}^a \to \bar{c}^a + g f^{abc} \omega^b \bar{c}^c, \tag{6.65}$$

as induced by a corresponding generator,

$$\mathcal{G}^a_{\mathrm{gh}} = -g f^{abc} \left(c^c \frac{\delta}{\delta c^b} + \bar{c}^c \frac{\delta}{\delta \bar{c}^b} \right). \tag{6.66}$$

In perturbative S matrix elements, the ghosts can be shown to cancel unphysical redundant gauge degrees of freedom.

Now, we generalize the generating functional by coupling sources also to the ghosts, in order to treat them on the same footing as the gauge field,

$$Z[J, \eta, \bar{\eta}] = \mathrm{e}^{W[J,\eta,\bar{\eta}]} = \int \mathcal{D}A \mathcal{D}c \mathcal{D}\bar{c}\, \mathrm{e}^{-S_{\mathrm{YM}} - S_{\mathrm{gh}} - S_{\mathrm{gf}} + \int JA + \int \bar{\eta} c - \int \bar{c} \eta}. \tag{6.67}$$

The construction of the effective action Γ now proceeds in a standard fashion,

$$\Gamma[A, \bar{c}, c] = \sup_{J,\eta,\bar{\eta}} \left(\int JA + \int \bar{\eta} c - \int \bar{c} \eta - W[J, \eta, \bar{\eta}] \right). \tag{6.68}$$

Since Γ is the result of a gauge-fixed construction, it is not manifestly gauge invariant. Gauge invariance is now encoded in a constraint given by the Ward identity Eq. (6.48) applied to the present case with the generator

$$\mathcal{G}^a(x) = \mathcal{G}^a_{\mathrm{A}}(x) + \mathcal{G}^a_{\mathrm{gh}}(x) = D^{ab}_\mu \frac{\delta}{\delta A^b_\mu} - g f^{abc} \left(c^c \frac{\delta}{\delta c^b} + \bar{c}^c \frac{\delta}{\delta \bar{c}^b} \right). \tag{6.69}$$

Since $\mathcal{G}^a S_{\mathrm{YM}} = 0$, the Ward identity boils down to

$$\mathcal{W} := \mathcal{G}^a \Gamma[A, \bar{c}, c] - \langle \mathcal{G}^a (S_{\mathrm{gf}} + S_{\mathrm{gh}}) \rangle = 0, \tag{6.70}$$

which in the context of nonabelian gauge theories is also commonly referred to as Ward-Takahashi identity (WTI). For instance in the Landau gauge,

$$\mathsf{F}^a[A] = \partial_\mu A^a_\mu, \quad \frac{\delta \mathsf{F}^a[A^\omega]}{\delta \omega^b} = -\partial_\mu D^{ab}_\mu[A], \tag{6.71}$$

together with the gauge parameter $\alpha \to 0$, cf. Eq. (6.63), we can work out this identity more explicitly by computing $\mathcal{G}^a S_{\mathrm{gf,gh}}$, e.g., in momentum space. This is done in Subsect. 6.3.4. As an example for how the Ward-Takahashi identity constrains the effective action, let us consider a gluon mass term, $\Gamma_{\mathrm{mass}} = \frac{1}{2} \int m_{\mathrm{A}}^2 A^a_\mu A^a_\mu$. It can be shown order by order in perturbation theory that the Ward-Takahashi identity enforces this gluon mass to vanish, $m_{\mathrm{A}}^2 = 0$; for more details, see Subsect. 6.3.4.

Hence, the gluon is protected against acquiring a mass by perturbative quantum fluctuations because of gauge invariance.

6.3.3 RG Flow Equation for Gauge Theories

From the gauge-fixed generating functional, the RG flow equation for Γ_k can straightforwardly be derived along the lines of Subsect. 6.2.2. Using a regulator term,

$$\begin{aligned}\Delta S_k &= \frac{1}{2}\int \frac{d^D p}{(2\pi)^D} A^a_\mu(-p)\,(R_{k,A})^{ab}_{\mu\nu}(p)\,A^b_\nu(p)\\ &= +\int \frac{d^D p}{(2\pi)^D} \bar{c}^a(p)\,(R_{k,\mathrm{gh}})^{ab}(p)\,c^b(p),\end{aligned} \tag{6.72}$$

we obtain the flow equation

$$\begin{aligned}\partial_t \Gamma_k[A,\bar{c},c] &= \frac{1}{2}\,\mathrm{Tr}\,\partial_t R_{k,A}[(\Gamma_k^{(2)}+R_k)^{-1}]_A - \mathrm{Tr}\partial_t R_{k,\mathrm{gh}}[(\Gamma_k^{(2)}+R_k)^{-1}]_{\mathrm{gh}}\\ &\equiv \frac{1}{2}\,\mathrm{STr}\,\partial_t R_k(\Gamma_k^{(2)}+R_k)^{-1}.\end{aligned} \tag{6.73}$$

The minus sign in front of the ghost term arises because of the anti-commuting nature of these Grassmann-valued fields; the super-trace in the second line of Eq. (6.73) takes this sign into account. In our notation in Eq. (6.73), $\Gamma_k^{(2)}$ is also matrix-valued in field space, i.e., with respect to $(A,\bar{c},c)$; therefore, $[(\Gamma^{(2)}+R_k)^{-1}]_A$ denotes the gluon component of the full inverse of $\Gamma^{(2)}+R_k$ (and not just the inverse of $\delta^2\Gamma_k/\delta A\delta A + R_{k,A}$).

Is this a gauge-invariant flow? Manifest gauge invariance is certainly lost, because the regulator is not gauge invariant; e.g., at small p, the regulator—here being similar to a mass term—is forbidden by gauge invariance, as discussed above.[6] But manifest gauge invariance is anyway lost, owing to the gauge-fixing procedure. Gauge symmetry is encoded in the Ward-Takahashi identity. From this viewpoint, the regulator is merely another source of explicit gauge-symmetry breaking, giving rise to further terms in the Ward-Takahashi identity, the form of which we can directly read off from Eq. (6.49):

$$\mathcal{W}_k := \mathcal{G}\Gamma_k + \mathcal{G}\Delta S_k - \langle \mathcal{G}(S_{\mathrm{gf}} + S_{\mathrm{gh}} + \Delta S_k)\rangle = 0. \tag{6.74}$$

[6] One may wonder whether a gauge-invariant flow can be set up with a gauge-invariant regularization procedure. In fact, this is an active line of research, and various promising formalisms have been developed so far [40–42]. However, the price to be paid for the resulting simple gauge constraints comes in the form of nontrivial Nielsen identities, non-localities or extensive algebraic constructions. For practical application, we thus consider the standard formulation described here as the most efficient approach so far.

Owing to the additional regulator terms, this equation is called *modified Ward-Takahashi identity* (mWTI). With ΔS_k being quadratic in the field variables, these regulator-dependent terms have a one-loop structure, since $\langle \mathcal{G}\Delta S_k\rangle - \mathcal{G}\Delta S_k$ corresponds to an integral over the connected 2-point function with a regulator insertion $\sim R_k$. Since the standard Ward-Takahashi identity already involves loop terms (cf. Subsect. 6.3.4), the solution to Eq. (6.74) is no more difficult to find than that of the standard WTI.

As before, we observe that

$$\lim_{k\to 0} \mathcal{W}_k \equiv \mathcal{W}, \tag{6.75}$$

such that a solution to the mWTI $\mathcal{W}_k = 0$ satisfies the standard WTI, $\mathcal{W} = 0$, if the regulator is removed at $k = 0$. Such a solution is thus gauge invariant. Loosely speaking, the mWTI $\mathcal{W}_k = 0$ defines a modified gauge invariance that reduces to the physical gauge invariance for $k \to 0$. For a discussion of these modified symmetry constraints from different perspectives, see e.g., [18–21, 43–46, 50], or the review [3].

One further important observation is that the flow of the mWTI satisfies

$$\partial_t \mathcal{W}_k = -\frac{1}{2}\, G_k^{AB}\, \partial_t R_k^{AC}\, G_k^{CD}\, \frac{\delta}{\delta \Phi^B}\frac{\delta}{\delta \Phi^D}\, \mathcal{W}_k. \tag{6.76}$$

Here, we have used the collective field variable $\Phi = (A, \bar{c}, c)$. The collective indices $A, B, C, \ldots$ label these components, and denote all discrete indices (color, Lorentz, etc.) as well as momenta; e.g., the flow equation reads in this notation: $\partial_t \Gamma_k = \frac{1}{2}\partial_t R_k^{AB} G_k^{BA}$. The derivation of Eq. (6.76) from Eq. (6.74) is indeed straightforward and a worthwhile exercise.

Let us draw an important conclusion based on Eq. (6.76): if we manage to find an effective action Γ_k which solves the mWTI $\mathcal{W}_k = 0$ at some scale k, then also the flow of the mWTI vanishes, $\partial_t \mathcal{W}_k = 0$. In other words, the mWTI is a fixed point under the RG flow. Now, if this Γ_k is connected with $\Gamma_{k'}$ at another scale k' by the flow equation, also $\Gamma_{k'}$ satisfies the mWTI at this new scale, $\mathcal{W}_{k'} = 0$. Gauge invariance at some scale therefore implies gauge invariance at all other scales, if the corresponding Γ_k's solve the flow equation. The whole concept of the mWTI is sketched and summarized in Fig. 6.7.

Unfortunately, the picture is not as rosy as it seems for a simple practical reason: in the general case, we will not be able to solve the flow equation exactly. Hence, the identity (6.76) and thus $\partial_t \mathcal{W}_k = 0$ will be violated on the same level of accuracy. This problem is severe if $\partial_t \mathcal{W}_k = 0$ is violated by RG-relevant operators, see Fig. 6.8; the latter are forbidden in the perturbative gauge-invariant theory.

In the perturbative domain where naive power-counting holds, an RG relevant operator is potentially given by the gluon mass term, $\frac{1}{2}\int m_A^2 A_\mu^a A_\mu^a$. Gauge invariance in the form of the standard WTI enforces $m_A^2 = 0$ as a consequence of $\mathcal{W} = 0$, as mentioned above. By contrast, such a bosonic mass term in a system without gauge symmetry would receive large contributions from fluctuations; perturbative diagrams

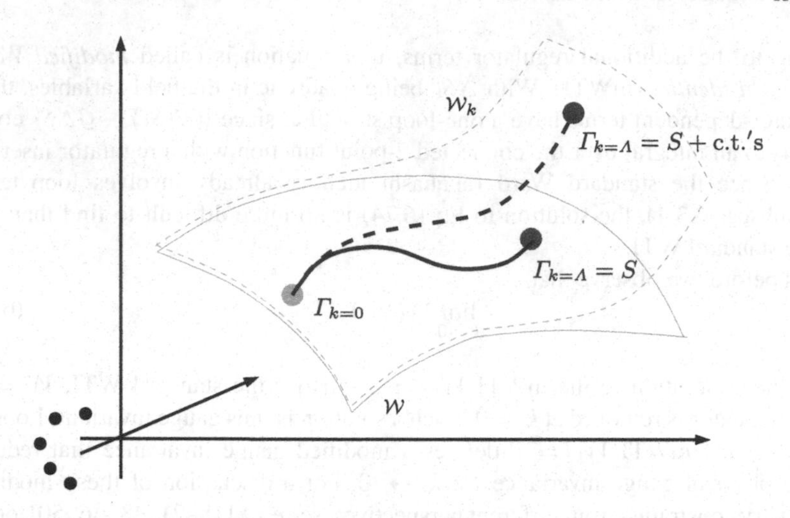

Fig. 6.7 Sketch of an RG flow with gauge symmetry in theory space: the standard Ward identity $\mathcal{W}$ cuts out a hypersurface of gauge-invariant action functionals. A gauge-invariant trajectory (*solid line*) would lie completely within this hypersurface. Instead, the presence of the regulator leads to the mWTI $\mathcal{W}_k$, cutting out a different hypersurface, which approaches $\mathcal{W}$ in the limit $k \to 0$. A solution to the flow equation stays within the $\mathcal{W}_k$ hypersurface (*dashed line*), implying gauge invariance of the final full action $\Gamma \equiv \Gamma_{k=0}$

are typically quadratically divergent in such systems. In a naively truncated RG flow of a gauge system, we can therefore expect the gluon mass to become large if the gauge symmetry is not respected, $m_A^2 \sim g^2 \Lambda^2$. Now, the mWTI assists to control the situation: since gauge symmetry is not manifestly present in our RG flow, we cannot expect the gluon mass to vanish at all values of k. As discussed in more detail in Subsect. 6.3.4, the gluon mass becomes of order $m_A^2 \sim g^2 k^2$, as can be determined from $\mathcal{W}_k = 0$. As long as perturbative power-counting holds, this implies that $m_A^2 \to 0$ for $k \to 0$, and the gauge constraint becomes satisfied in the limit when the regulator is removed.

This consideration demonstrates that the mWTI can turn a potentially dangerous relevant operator, which may appear in some truncation, into an irrelevant harmless operator which dies out in the limit of vanishing regulator. From another viewpoint, the mWTI tells us precisely the right amount of gauge-symmetry breaking that we have to put at the UV scale Λ in the form of counterterms, such that this explicit breaking is ultimately eaten up by the fluctuation-induced breaking terms from the regulator, ending up with a perfectly gauge-invariant effective action.

This simple gluon-mass example teaches an important lesson: for the construction of a truncated gauge-invariant flow, the flow equation and the mWTI should be solved simultaneously within the truncation. As a standard recipe [47, 48], the flow equation can first be used to determine the flow of all *independent* operators (e.g., transverse propagators and transverse vertex projections) which are not constrained by $\mathcal{W}_k = 0$. Then use the mWTI to compute the remaining dependent operators (e.g., gluon mass, longitudinal terms). In this manner, the gauge constraint is explicitly solved on the

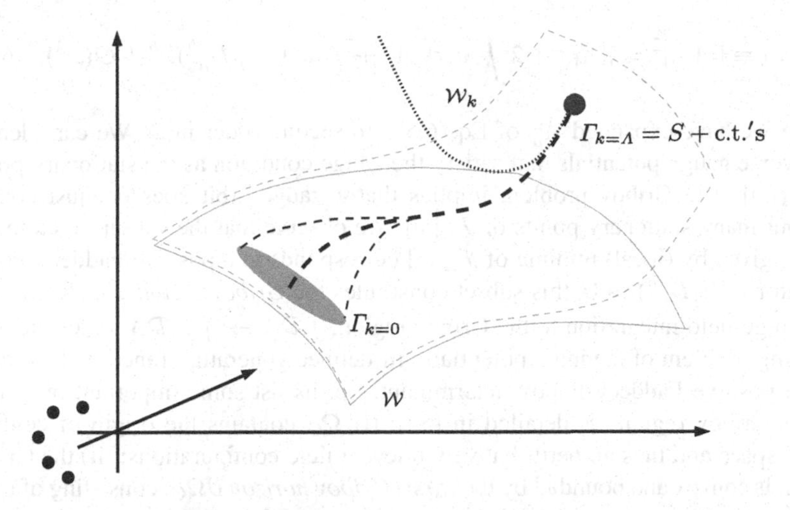

Fig. 6.8 Sketch of an RG flow with gauge symmetry in theory space: a truncation of the effective action introduces an error, implying a possible range of estimates for $\Gamma_{k=0}$ as depicted by the extended ellipse. The error can also have a component orthogonal to the $\mathcal{W}$ or $\mathcal{W}_k$ hypersurface, if the Ward identities are solved on the same level of accuracy as the truncated flow. A violation of the mWTI by RG relevant operators is particularly dangerous, since it can quickly drive the system away from the physical solution (*dotted line*) and thus must be avoided

truncation and gauge-invariance of the truncation is guaranteed.[7] It turns out that the use of the mWTI (instead of the flow equation itself) for the comsputation of a dependent operator generically corresponds to a resummation of a larger class of diagrams [49].

For the remainder of this subsection, let us return to the Gribov problem discussed below Eq. (6.61). The fact that standard gauge fixings do not uniquely pick exactly one representative of each gauge orbit and that the Faddeev-Popov determinant hence is not positive makes the nonperturbative definition of the functional integral problematic. Any nonperturbative method which is related to the functional integral such as the flow equation therefore appears to face the same problem. As an example, let us concentrate on the Landau gauge and consider the set of all gauge-field configurations that satisfy the gauge-fixing condition $\mathsf{F}^a = \partial_\mu A^a_\mu = 0$; the Faddeev-Popov operator then is $\delta \mathsf{F}^a[A^\omega]/\delta\omega^b = -\partial_\mu D^{ab}_\mu[A]$. A perturbative expansion around $g \to 0$ goes along with the Faddeev-Popov operator at the origin of configuration space, $-\partial_\mu D^{ab}_\mu[A] \to -\partial^2$; since this Laplacian is a positive operator, the Gribov problem does not play a role in perturbation theory to any finite order.

Moving away from the origin, it is useful to consider the following gauge-fixing functional, corresponding to the L_2 norm of the gauge potential along the gauge orbit,

[7] An alternative option could be to use only the flow equation together with a regulator that does automatically suppress artificial relevant operators. In fact, this is conceivable in the framework of optimization [50].

$$\mathcal{F}_A[\omega] \equiv ||A^\omega||^2 = ||A||^2 + 2\int_x \omega^a \partial_\mu A^a_\mu + \int_x \omega^a(-\partial_\mu D^{ab}_\mu)\omega^b + \mathcal{O}(\omega^3), \quad (6.77)$$

where we have expanded A^ω_μ of Eq. (6.58) to second order in ω. We can identify transverse gauge potentials that satisfy the gauge condition as the stationary points of $\mathcal{F}_A[\omega]$. The Gribov problem implies that a gauge orbit does not just contain one but many stationary points of $\mathcal{F}_A[\omega]$. We observe that the subset of stationary points given by (local) minima of $\mathcal{F}_A[\omega]$ corresponds to a positive Faddeev-Popov operator $(-\partial_\mu D^{ab}_\mu) > 0$; this subset constitutes the *Gribov region* Ω_G. Restricting the gauge-field integration to the Gribov region, $\int \mathcal{D}A \to \int_{\Omega_G} \mathcal{D}A$, cures the most pressing problem of having a potentially ill-defined generating functional owing to a non-positive Faddeev-Popov determinant. Let us list some important properties of the Gribov region, as detailed in [51]: (i) Ω_G contains the origin of configuration space and thus all perturbatively relevant field configurations; (ii) the Gribov region is convex and bounded by the *(first) Gribov horizon* $\partial\Omega_G$, consisting of those field configurations for which the lowest eigenvalue of the Faddeev-Popov operator vanishes; hence $\Delta_{FP} = 0$ on $\partial\Omega_G$.

Does this restriction of the gauge-field integration to the Gribov region modify the flow equation? In order to answer this question, let us go back to the functional-integral equation for the effective action (without IR regulator) in Eq. (6.8),

$$e^{-\Gamma[\phi]} = \int \mathcal{D}\varphi \exp\left(-S[\phi+\varphi] + \int \frac{\delta\Gamma[\phi]}{\delta\phi}\varphi\right), \quad (6.78)$$

with the supplementary condition that $\langle\varphi\rangle = 0$, owing to the shifted integration variable, cf. Eq. (6.8). Differentiating both sides with respect to ϕ yields

$$\frac{\delta\Gamma[\phi]}{\delta\phi} = \left\langle \frac{\delta S[\varphi+\phi]}{\delta\phi} \right\rangle_{J[\phi]}, \quad (6.79)$$

which is a compact representation of the Dyson-Schwinger equations. Note that the same equation can be obtained from the following identity:

$$0 = \int \mathcal{D}\varphi \frac{\delta}{\delta\varphi} e^{-S[\phi+\varphi]+\int \frac{\delta\Gamma[\phi]}{\delta\phi}\varphi}, \quad (6.80)$$

which holds, since the integrand is a total derivative. No boundary terms appear here, because the action typically goes to infinity, $S \to \infty$, for an unconstrained field $\varphi \to \infty$.

Now, the crucial point for a quantum gauge theory with Faddeev-Popov gauge fixing is that the identity corresponding to Eq. (6.80) holds also for the constrained integration domain Ω_G. No boundary term arises, simply because the Faddeev-Popov operator and thus the integrand vanishes on the boundary $\partial\Omega_G$. We conclude that the Dyson-Schwinger equations are not modified by the restriction to the Gribov region. Finally, the same argument can be transfered to the flow-equation formalism by noting that the effective average action has to satisfy an identity similar to Eq. (6.78) including the regulator,

$$
\begin{aligned}
&e^{-\Gamma_k[\phi]-\Delta S_k[\phi]} \\
&= \int \mathcal{D}\varphi \exp\left(-S[\phi+\varphi]-\Delta S_k[\phi+\varphi]+\int \frac{\delta(\Gamma_k[\phi]+\Delta S[\phi])}{\delta\phi}\varphi\right),
\end{aligned}
\tag{6.81}
$$

with corresponding IR regulated Dyson-Schwinger equations,

$$
\frac{\delta(\Gamma_k[\phi]+\Delta S_k[\phi])}{\delta\phi} = \left\langle \frac{\delta(S[\varphi+\phi]+\Delta S_k[\varphi+\phi])}{\delta\phi} \right\rangle_{J[\phi]} . \tag{6.82}
$$

The latter can again be obtained from a functional integral over a total derivative similar to Eq. (6.80), indicating that a restriction to the Gribov region does not modify Eq. (6.82). The final step of the argument consists in noting that the scale derivative ∂_t of Eq. (6.82) yields the flow equation (once differentiated with respect to ϕ).

To summarize, solving quantum gauge theories by the construction of correlation functions by means of functional methods (Dyson-Schwinger equations or flow equations) precisely corresponds to an approach with a build-in restriction to the Gribov region as an attempt to solve the Gribov problem. From a flow-equation perspective, the argument can even be turned around: taking the viewpoint that the quantum gauge theory is defined by the flow equation, we can initiate the flow in the perturbative deep UV where the Faddeev-Popov determinant is guaranteed to be positive. Solving the flow, a resulting stable trajectory necessarily stays within the Gribov region.

Let us close this section with the remark that the picture developed so far is not yet complete. The restriction to the Gribov region only removes the problem of the non-positive Faddeev-Popov determinant. It does not guarantee that we have integrated over the configuration space by picking only one representative of each gauge orbit. In fact, even within the Gribov region, there are still Gribov copies. Therefore, the integration domain in gauge configuration space has to be restricted even further by picking the global minimum of $\mathcal{F}_A[\omega]$ in Eq. (6.77). The resulting space is known as the *fundamental modular region* Λ. In practice, the explicit construction of Λ is difficult; for instance, finding the global minimum of the gauge-fixing functional $\mathcal{F}_A[\omega]$ on the lattice, corresponds to an extremely involved spin-glass problem. However, it has been argued in [51] within a stochastic-quantization approach that the problem of Gribov copies within the Gribov region does not affect the correlation functions and their computation. This stresses even further the potential of functional methods for nonperturbative problems in gauge theories.

6.3.4 Ward-Takahashi Identity*

The following subsection is devoted to a detailed discussion of the gauge constraints in the form of the Ward-Takahashi identity (WTI) and its modified counterpart in the presence of the regulator (mWTI). Let us start with the standard WTI which we derived already in a compact notation in Eq. (6.70),

$$\mathcal{W} := \mathcal{G}^a \Gamma[A, \bar{c}, c] - \langle \mathcal{G}^a (S_{\mathrm{gf}} + S_{\mathrm{gh}}) \rangle = 0, \tag{6.83}$$

which represents the gauge-symmetry encoding constraint that the effective action Γ has to satisfy in a gauge-fixed formulation. In order to work out the single terms more explicitly, it is useful to go to momentum space; we use the Fourier conventions

$$A_\mu^a(x) = \int_q \mathrm{e}^{\mathrm{i}qx} A_\mu^a(q), \quad c^a(x) = \int_q \mathrm{e}^{\mathrm{i}qx} c^a(q), \quad \bar{c}^a(x) = \int_q \mathrm{e}^{-\mathrm{i}qx} \bar{c}^a(q), \tag{6.84}$$

where $\int_q \equiv \int \frac{\mathrm{d}^D q}{(2\pi)^D}$. This implies for the functional derivatives, for instance,

$$\frac{\delta}{\delta A_\mu^a(x)} = \int_q \mathrm{e}^{-\mathrm{i}qx} \frac{\delta}{\delta A_\mu^a(q)}, \quad \text{etc.} \tag{6.85}$$

As a result, the generator of gauge transformations $\mathcal{G}^a$ reads in momentum space

$$\begin{aligned} \mathcal{G}^a(p) &= \mathrm{i} p_\mu \frac{\delta}{\delta A_\mu^a(-p)} \\ &-g f^{abc} \int_q \left[A_\mu^c(q) \frac{\delta}{\delta A_\mu^b(q-p)} + c^c(q) \frac{\delta}{\delta c^b(q-p)} + \bar{c}^c(q) \frac{\delta}{\delta \bar{c}^b(q-p)} \right]. \end{aligned} \tag{6.86}$$

This allows us to compute the building blocks of the last two terms in Eq. (6.83) in momentum space; we obtain the gauge transforms

$$\mathcal{G}^a(p) S_{\mathrm{gf}} = \frac{\mathrm{i}}{\alpha} p^2 p_\mu A_\mu^a(p) - \frac{1}{\alpha} g f^{abc} \int_q A_\mu^c(q)(p-q)_\mu (p-q)_\nu A_\nu^b(p-q), \tag{6.87}$$

$$\begin{aligned} \mathcal{G}^a(p) S_{\mathrm{gh}} &= -g f^{abc} \int_q \bar{c}^c(q) p \cdot (q+p) c^b(q+p) \\ &\quad -\mathrm{i} g^2 f^{feb} f^{abc} \int_{q_1, q_2} p_\mu \bar{c}^c(q_1) A_\mu^e(p+q_1+q_2) c^f(q_2). \end{aligned} \tag{6.88}$$

Upon insertion into Eq. (6.83), we arrive at a explicit representation of the WTI in terms of full correlation functions in the presence of a source J which is field dependent by virtue of the Legendre transform, $J = J_{\mathrm{sup}} = J[\phi]$,

$$\begin{aligned}
&\mathcal{G}^a(p)\left[\Gamma_k - \int\left(\frac{1}{2\alpha}(\partial_\mu A^a_\mu)^2 + (\partial_\mu \bar{c}^a) D^{ab}_\mu(A) c^b\right)\right] \\
&= -\frac{1}{\alpha} g f^{abc} \int_q (p+q)_\mu (p+q)_\nu \langle A^c_\mu(-q) A^b_\nu(p+q)\rangle_{\text{con}} \\
&\quad - g f^{abc} \int_{q_1,q_2} p_\mu \big(\delta(p+q_1-q_2) q_{2\mu} \delta^{bf} - \mathrm{i} g f^{bef} A^e_\mu(p+q_1-q_2)\big) \\
&\quad \times \langle \bar{c}^c(q_1) c^f(q_2)\rangle_{\text{con}} \\
&\quad - \mathrm{i} g^2 f^{abc} f^{feb} \int_{q_1,q_2} p_\mu \langle \bar{c}^c(q_1) A^e_\mu(p+q_1-q_2) c^f(q_2)\rangle_{\text{con}},
\end{aligned} \tag{6.89}$$

where $\langle \dots \rangle_{\text{con}}$ denotes only the connected part of the correlation functions, e.g., $\langle \varphi\varphi\rangle_{\text{con}} = \langle \varphi\varphi\rangle - \langle \varphi\rangle\langle \varphi\rangle$. All terms on the right-hand side are loop terms, the last term is even a two-loop term. It should also be stressed that the WTI is expressed here in terms of unrenormalized fields and couplings.

As an example, let us see how the WTI imposes constraints on operators in the effective action. For this, we discuss a gluon mass term in the Landau gauge, $\alpha \to 0$. The gluon-mass operator reads,

$$\Gamma_{\text{mass}} = \frac{1}{2} \int m_A^2\, A^a_\mu A^a_\mu. \tag{6.90}$$

Its gauge transform yields

$$\mathcal{G}^a(p) \Gamma_{\text{mass}} = m_A^2\, \mathrm{i} p_\mu A^a_\mu(p). \tag{6.91}$$

This implies that we have to project the remaining terms of the WTI only onto the operator $\sim p_\mu A^a_\mu(p)$ in order to study the gauge constraint on the gluon mass. Let us do so for the first term on the right-hand side of Eq. (6.89) by way of example; shifting the momentum q by $q \to q - p$, this term reads

$$\begin{aligned}
&-\frac{1}{\alpha} g f^{abc} \int_q q_\mu q_\nu \langle A^c_\mu(p-q) A^b_\nu(q)\rangle_{\text{con}}|_{\alpha\to 0} \\
&= -\frac{1}{\alpha} \int_q P_{\mathrm{L},\mu\nu}(q)\, q^2\, G^{cb}_{\mu\nu}(p-q, q|A, c, \bar{c})|_{\alpha\to 0},
\end{aligned} \tag{6.92}$$

where we have introduced the longitudinal projector $P_{\mathrm{L},\mu\nu} = q_\mu q_\nu / q^2$ as well as the full gluon propagator in the background of all fields $G^{cb}_{\mu\nu}(p-q, q|A, c, \bar{c})$. In view of Eq. (6.91), this propagator is needed only to linear order in the gauge field,

$$\begin{aligned}
G_{\mu\nu}(p-q, q|A, c, \bar{c}) = G_{\mu\nu}(p-q, q) \\
&+ G_{\mu\kappa}(p-q, q-p) V_{3A,\kappa\lambda\rho}(p-q, -p, q) \\
&\times G_{\rho\nu}(-q, q) A_\lambda(p) + \mathcal{O}(A^2, \bar{c}c),
\end{aligned} \tag{6.93}$$

where V_{3A} denotes the full three-gluon vertex, and all momenta are counted as in-flowing. Inserting the order linear in A of Eq. (6.93) into Eq. (6.92), we observe that both gluon propagators are contracted with the longitudinal projector,

$$\begin{aligned}
&-\frac{1}{\alpha} g f^{abc} \int_q q_\mu q_\nu \langle A^c_\mu(p-q) A^b_\nu(q)\rangle_{\text{con}}|_{\alpha\to 0} \\
&= -\frac{1}{\alpha}\int_q q^2 G_{\mathrm{L},\mu\kappa} V_{3A,\kappa\lambda\rho} G_{\mathrm{L},\rho\mu} A_\lambda|_{\alpha\to 0}.
\end{aligned} \tag{6.94}$$

Now, the Landau gauge strictly enforces the gauge fields to be transverse. Any longitudinal modes have to decouple in the Landau-gauge limit; in particular, we have $G_{\mathrm{L}} \sim \alpha \to 0$. As a consequence, the whole expression (6.94) goes to zero linearly with α, at least order by order in perturbation theory. We conclude that this first term on the right-hand side of the WTI (6.89) does not support a nonvanishing value for the gluon mass. Let us mention without proof that the same property can be shown also for all other terms on the right-hand side of Eq. (6.89). Therefore, the WTI enforces the gluon mass term to vanish to any order in perturbation theory, $\Gamma_{\text{mass}} = 0$, implying $m_A^2 = 0$. The WTI protects the zero mass of the gluon against perturbative quantum contributions, because of gauge invariance.

Let us now turn to the modifications of the gauge constraint in the presence of a regulator. We already derived the regulator-modified WTI (mWTI) in Eq. (6.74),

$$\mathcal{W}_k := \mathcal{G}\Gamma_k + \mathcal{G}\Delta S_k - \langle \mathcal{G}(S_{\text{gf}} + S_{\text{gh}} + \Delta S_k)\rangle = 0. \tag{6.95}$$

For an explicit representation, we need the gauge transforms of the regulator terms,

$$\mathcal{G}^a(p)\Delta S_{k,A} = i p_\mu (R_{k,A})^{ab}_{\mu\nu}(p) A^b_\nu(p) - g f^{abc}\int_q A^c_\mu(q)(R_{k,A})^{bd}_{\mu\nu}(p-q) A^d_\nu(p-q), \tag{6.96}$$

$$\mathcal{G}^a(p)\Delta S_{k,\text{gh}} = -g f^{abc}\int_q \bar{c}^c(q)[R_{k,\text{gh}}(q+p) - R_{k,\text{gh}}(q)]c^b(q+p). \tag{6.97}$$

Using our previous result for the standard WTI (6.89), the mWTI can now be displayed in the more explicit form,

$$\begin{aligned}
&\mathcal{G}^a(p)\left[\Gamma_k - \int \left(\frac{1}{2\alpha}(\partial_\mu A^a_\mu)^2 + (\partial_\mu \bar{c}^a) D^{ab}_\mu(A) c^b\right)\right] \\
&= -\frac{1}{\alpha} g f^{abc}\int_q (p+q)_\mu (p+q)_\nu \langle A^c_\mu(-q) A^b_\nu(p+q)\rangle_{\text{con}} \\
&\quad - g f^{abc}\int_{q_1,q_2} p_\mu\big(\delta(p+q_1-q_2) q_{2\mu}\delta^{bf} - i g f^{bef} A^e_\mu(p+q_1-q_2)\big) \\
&\quad \times \langle \bar{c}^c(q_1) c^f(q_2)\rangle_{\text{con}}
\end{aligned}$$

$$-ig^2 f^{abc} f^{feb} \int_{q_1,q_2} p_\mu \langle \bar{c}^c(q_1) A^e_\mu(p+q_1-q_2) c^f(q_2)\rangle_{\text{con}}$$
$$-\frac{1}{2} g f^{abc} \int_q [(R_{k,A})_{\mu\nu}(p+q) - (R_{k,A})_{\mu\nu}(q)] \langle A^c_\mu(-q) A^b_\nu(p+q)\rangle_{\text{con}}$$
$$-g f^{abc} \int_q [R_{k,\text{gh}}(p+q) - R_{k,\text{gh}}(q)] \langle \bar{c}^c(q) c^b(q+p)\rangle_{\text{con}} \Big|_{\alpha\to 0}. \quad (6.98)$$

The last two terms denote the modification of the mWTI in comparison to the standard WTI. These two terms are one-loop terms with a structure similar to the flow equation itself. Both terms vanish in the limit $k \to 0$ and the standard WTI is recovered, as it should. Again, we stress that the mWTI is expressed in terms of unrenormalized fields and couplings.

As an example, it would be straightforward to work out the precise contribution of the regulator terms to the gluon mass which does not vanish in contrast to the WTI result [48, 52]. However, here it suffices to estimate the order of magnitude of this contribution. The structure $R_k(p+q) - R_k(q)$, together with the fact that we need to project only on the terms linear in A and p_μ (cf. Eq. (6.91)), implies that the q integral is peaked around $q^2 \simeq k^2$. The dimensionful scales on the right-hand side for the gluon mass are thus set by k^2, resulting in $m_A^2 \sim g^2 k^2$ with a proportionality coefficient that depends on the regulator. This is a very unusual bosonic-mass running which guarantees that the gluon mass is not a relevant operator in the flow, but vanishes with $k \to 0$.

Let us close this subsection by briefly discussing the connection of the present formalism with the BRST formalism. The latter involves one further conceptual step, emphasizing BRST invariance as a residual invariance of the gauge-fixed functional integral. The corresponding symmetry constraints on the effective action, the *Slavnov-Taylor identities*, have the advantage in the standard formulation that they are bilinear in derivatives of the effective action. This allows for an algebraic resolution of the gauge constraints in contrast to the loop computations necessary for the WTI, cf. Eq. (6.89). If the regulator term is included, modified Slavnov-Taylor identities can still be derived [18, 20, 21, 43], but the result no longer has a bilinear structure. We conclude that the BRST formulation has no real advantage in the case of the RG flow equation, such that the present formalism can fully be recommended also for practical applications.

*6.3.5 Further Reading: Landau-Gauge IR Propagators**

The following paragraphs give a short introduction to recent lines of research, and may serve as a guide to the literature.

The functional RG techniques for gauge theories developed above can now be used for computing the effective action in a vertex expansion, cf. Eq. (6.29). In momentum space, the expansion reads,

$$\Gamma_k[\phi] = \sum_n \frac{1}{n} \int_{p_1,\dots,p_n} \delta(p_1 + \dots + p_n)\, \Gamma_k^{(n)}(p_1,\dots,p_n)\, \phi(p_1)\dots\phi(p_n), \tag{6.99}$$

where $\int_p = \int d^D p/(2\pi)^D$, $\delta(p) = (2\pi)^D \delta^{(D)}(p)$, and $\phi = (A^a_\mu, \bar{c}^a, c^a)$. Inserting Eq. (6.99) into the flow equation (6.73), we obtain an infinite set of coupled first-oder differential equations for the proper vertices $\Gamma_k^{(n)}$. Truncating the expansion at order $n_{\max}$ leaves all equations for the vertices $\Gamma_k^{(n \le n_{\max}-2)}$ unaffected. In order to close this tower of equations, the vertices of order $n_{\max}$ and $n_{\max} - 1$ can either be derived from their truncated equations or taken as bare or constructed by further considerations; see, e.g., [53]. This defines a consistent approximation scheme that can in principle be iterated to arbitrarily high orders in $n_{\max}$.

Let us consider here the lowest nontrivial order,

$$\Gamma_k = \frac{1}{2}\int_q A^a_\mu(-q)\Big[Z_A(q^2) q^2\, P_{\mathrm{T}\mu\nu} + m_k^2 \delta_{\mu\nu}\Big] A^a_\nu(q) + \int_q \bar{c}^a(q)\, Z_{\mathrm{gh}}(q^2) q^2\, c^a(q) + \dots, \tag{6.100}$$

where P_T is the transverse projector, and the ellipsis denotes higher-order vertices and longitudinal gluonic terms. In the following, we will confine ourselves to the Landau gauge $\alpha = 0$ where longitudinal modes decouple completely; moreover, the Landau gauge is known to be a fixed-point of the RG flow [44, 47]. The nontrivial ingredients consist in the fully momentum-dependent wave function renormalizations $Z_A(p^2)$ and $Z_{\mathrm{gh}}(p^2)$, as well as a gluon mass term which has been discussed in detail above. The flow equations for the inverse of the transverse gluon and ghost propagators ($G_k = (\Gamma_k^{(2)} + R_k)^{-1}$),

$$\Gamma^{(2)}_{k,A\mathrm{T}}(p^2) = Z_A(p^2) p^2 + m_k^2, \quad \Gamma^{(2)}_{k,\mathrm{gh}}(p^2) = Z_{\mathrm{gh}}(p^2) p^2, \tag{6.101}$$

are shown in Fig. 6.9, except for diagrams involving quartic vertices. The flow-equation diagrams are reminiscent to those of Dyson-Schwinger equations. But there are two differences: Dyson-Schwinger equations also involve two-loop diagrams, whereas the flow equation imposes its one-loop structure also on the propagator and vertex equations. Second, all internal propagators and vertices are fully dressed quantities in the flow equation, whereas Dyson-Schwinger equations always involve one bare vertex. Both vertex expansions are slightly different infinite-tower expansions of the same generating functional, with the flow equations being amended by the differential RG structure.

A truncation at this order requires information about the triple and quartic vertices. In a minimalistic approach, they may be taken as bare (possibly accompanied by k-dependent renormalization constants). In general, this nonperturbative approximation is expected to be reliable at weak coupling. For instance, the perturbative

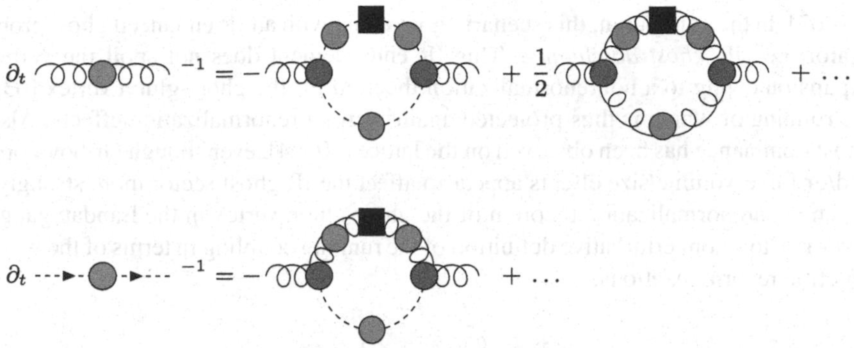

Fig. 6.9 Flow equations for gluon and ghost propagators in a vertex expansion. All internal lines and vertices denote fully dressed quantities, indicated by *filled circles*. To each diagram, there is another one with identical topology but with the regulator insertion occurring at the opposite internal line. The ellipses denote diagrams involving quartic vertices which are not displayed

result is rediscovered at high scales, $k = \Lambda_{\mathrm{p}}$; for momenta p^2 larger than this perturbative scale Λ_{p}^2, the wave function renormalizations yield,

$$Z_A(p^2) \simeq Z_{\Lambda_{\mathrm{p}},A}\left(1 - \eta_A \frac{11 N_{\mathrm{c}} \alpha_{\Lambda_{\mathrm{p}}}}{12\pi} \ln \frac{p^2}{\Lambda_{\mathrm{p}}^2}\right),$$

$$Z_{\mathrm{gh}}(p^2) \simeq Z_{\Lambda_{\mathrm{p}},\mathrm{gh}}\left(1 - \eta_{\mathrm{gh}} \frac{11 N_{\mathrm{c}} \alpha_{\Lambda_{\mathrm{p}}}}{12\pi} \ln \frac{p^2}{\Lambda_{\mathrm{p}}^2}\right), \tag{6.102}$$

where $\eta_A = -13/22$ and $\eta_{\mathrm{gh}} = -9/44$ denote the anomalous dimensions for gluons and ghosts, respectively. $Z_{\Lambda_{\mathrm{p}},A}$, $Z_{\Lambda_{\mathrm{p}},\mathrm{gh}}$ are the normalizations of the fields, and $\alpha_{\Lambda_{\mathrm{p}}}$ is the value of the coupling constant at Λ_{p}.

At first glance, there is no reason why the higher-order vertex structures which are dropped in the minimalistic truncation should not become dominant at strong coupling. However, as we learned from the anharmonic oscillator example in Subsect. 6.2.3, the quality of a low-order truncation does not depend on how large higher-order terms may get but whether they exert a strong influence on the low-order equations. Moreover, mechanisms may exist that systematically suppress higher-order contributions, such as an IR suppression of certain propagators; since higher-order vertex equations involve more propagators, such a propagator suppression would control large classes of diagrams.

In recent years, evidence has been provided that such a suppression is indeed operative in the gluon sector in the Landau gauge: low-order vertex expansions reveal an IR suppressed gluon propagator which renders the contributions from higher gluonic vertices subdominant. This solution has been pioneered in [54] using truncated Dyson-Schwinger equations, see [24, 27] for reviews; IR gluon suppression in the Landau gauge has meanwhile been confirmed by many lattice simulations

[55–62]. In the continuum, this scenario goes along with an IR enhanced ghost propagator, i.e., IR *ghost dominance*. This IR enhancement does not spoil the vertex expansion owing to a nonrenormalization theorem for the ghost-gluon vertex [63], the running of which is thus protected against strong renormalization effects. Also ghost dominance has been observed on the lattice [60, 64], even though Gribov-copy and/or finite-volume/size effects appear to affect the IR ghost sector more strongly.

The nonrenormalization theorem of the ghost-gluon vertex in the Landau gauge gives rise to a nonperturbative definition of the running coupling in terms of the wave function renormalizations,

$$\alpha(p^2) = \frac{g^2}{4\pi} \frac{1}{Z_A(p^2)\, Z_{\mathrm{gh}}^2(p^2)}. \tag{6.103}$$

In the IR, the above-described scenario of ghost dominance and gluon suppression is quantitatively observed in terms of a power-law behavior of the wave function renormalizations,[8]

$$Z_A(p^2) \sim (p^2)^{-2\kappa}, \quad Z_{\mathrm{gh}}(p^2) \sim (p^2)^{\kappa}, \tag{6.104}$$

where κ denotes a positive IR exponent. In the functional RG framework, this solution can be shown to be a fixed point of the flow equations for the propagators, cf. Fig. 6.9, in the momentum regime $k^2 \ll p^2 \ll \Lambda_{\mathrm{QCD}}^2$ [65]. The interrelation of the ghost and gluon propagators owing to the simultaneous occurrence of the exponent κ arises in all functional approaches from self-consistency arguments; as a direct consequence, the running coupling (6.103) approaches a fixed point in the IR, $\alpha(p^2 \ll \Lambda_{\mathrm{QCD}}^2) \to \alpha_*$. For instance, a truncation with a bare ghost-gluon vertex results in $\kappa \simeq 0.595$; possibly induced momentum dependencies of the vertex can lead to slightly lower values $0.5 \lesssim \kappa \lesssim 0.595$ [66]. Regulator dependencies arising in an RG calculation also lie in this range [67]. The IR solutions (6.104) are IR attractive fixed-point solutions for a wide class of initial conditions and momentum-dependencies of the gluon vertices, once the gluon propagator has developed a mass-like structure at intermediate momenta at a few times Λ_{QCD} [52]. Using suitable vertex *ansätze*, full solutions connecting the perturbative UV branch Eq. (6.102) and the IR power-laws (6.104) have been found with Dyson-Schwinger equations, see [27, 68].

Most importantly, the IR power-law behavior featuring gluon suppression and ghost dominance agrees with criteria which are expected to be satisfied in two different scenarios of confinement: the Kugo-Ojima [69] and the Gribov-Zwanziger [39, 70, 71] confinement scenario. In particular, a strongly IR divergent ghost propagator represents a signature of confinement in these scenarios, which describe the absence of color charged asymptotic states and (indirectly) a linear rise of the potential between a static quark-antiquark pair. The study of correlation functions in

---

[8] In the Dyson-Schwinger literature, the gluon and ghost propagator behavior is often characterized by *dressing functions* Z_{DSE}, G_{DSE} which are related to the wave function renormalizations by $Z_A(p^2) = Z_{\mathrm{DSE}}^{-1}(p^2)$ and $Z_{\mathrm{gh}}(p^2) = G_{\mathrm{DSE}}^{-1}(p^2)$ for $k \to 0$.

connection with these scenarios of low-energy gauge theories clearly demonstrates the potential of functional methods to access even the strongly-coupled gauge sector by analytical means.

6.4 Background-Field Flows

Imagine for a second that we knew nothing about computing loops and constructing amplitudes in some sort of expansion which involves a perturbative or even a fully dressed propagator. If we knew only the degrees of freedom of our gauge system and the symmetries we would be trying to write down an effective action in terms of all possible gauge-invariant gluon operators, such as $F^a_{\mu\nu}F^a_{\mu\nu}$, $F^a_{\mu\nu}(D_\kappa D_\kappa)^{ab}F^b_{\mu\nu}$ etc. and determine the coefficients, e.g., in a manner similar to chiral perturbation theory. The result would be gauge invariant by construction, and we would never worry about Ward identities and how gauge invariance can be encoded in a nontrivial manner in a gauge-fixed formulation.

6.4.1 Background-Field Formalism

The background-field formalism aims precisely at the construction of such an effective action, nevertheless by computing loops and integrating out fluctuations in a special gauge-fixed manner. Here is a rough sketch of the idea:

(1) Introduce an auxiliary, non-dynamical field $\bar{A}^a_\mu$ (background field) with its own auxiliary symmetry transformation $\bar{\mathcal{G}}$.
(2) Construct a gauge-fixed QFT which has broken invariance under the standard gauge transformation $\mathcal{G}$ but a manifest invariance under $(\mathcal{G}+\bar{\mathcal{G}})$.
(3) Let the full Γ inherit the symmetry properties in the end by setting $A=\bar{A}$ after the gauge-fixed calculation.

Let us start with (1): we introduce $\bar{A}^a_\mu$ and corresponding covariant derivatives $\bar{D}^{ab}_\mu = \partial_\mu\delta^{ab} + gf^{acb}\bar{A}^c_\mu$, and the generator of symmetry transformations

$$\bar{\mathcal{G}}^a(x) = \bar{D}^{ab}_\mu \frac{\delta}{\delta \bar{A}^b_\mu}, \tag{6.105}$$

which we call the background transformation. Note that the ghosts are not affected by $\bar{\mathcal{G}}$. Together with Eq. (6.69), it is obvious that $(A-\bar{A})$ now transforms homogeneously under $\mathcal{G}+\bar{\mathcal{G}}$,

$$\int d^D y\, \omega^b(y)(\mathcal{G}+\bar{\mathcal{G}})^b(y)(A-\bar{A})^a_\mu(x) = gf^{abc}\omega^b(x)(A-\bar{A})^c_\mu(x). \tag{6.106}$$

As step (2), we choose a gauge fixing F^a which fixes the $\mathcal{G}$ symmetry but is invariant under $\mathcal{G}+\bar{\mathcal{G}}$:

$$\begin{aligned}\mathsf{F}^a &= \bar{D}^{ab}_\mu (A^b_\mu - \bar{A}^b_\mu) \\ &\Rightarrow (\mathcal{G}+\bar{\mathcal{G}})S_{\text{gf}} = \frac{1}{2\alpha}\,(\mathcal{G}+\bar{\mathcal{G}}) \int \mathrm{d}^D x\, [\bar{D}(A-\bar{A})]^2 = 0.\end{aligned} \tag{6.107}$$

In fact, the gauge-fixing term is invariant under the combined transformation. With the Faddeev-Popov operator

$$\frac{\delta \mathsf{F}^a}{\delta \omega^b} = -\bar{D}^{ac}_\mu D^{cb}_\mu, \tag{6.108}$$

it is also straightforward to show that

$$(\mathcal{G}+\bar{\mathcal{G}})S_{\text{gh}} = -(\mathcal{G}+\bar{\mathcal{G}}) \int \mathrm{d}^D x\, \bar{c}^a \bar{D}^{ac}_\mu D^{cb}_\mu c^b = 0. \tag{6.109}$$

Finally, let us merely sketch step (3): the price to be paid so far is that Γ now depends on A and $\bar{A}$. But at the end of the calculation, we can identify $A = \bar{A}$, such that

$$0 = (\mathcal{G}+\bar{\mathcal{G}})\Gamma[A,\bar{A}]\big|_{A=\bar{A}} = \mathcal{G}\Gamma[A,A], \tag{6.110}$$

where the first equality holds by construction and the second arises from setting $A = \bar{A}$. Now, it is possible to prove that the background effective action with $A = \bar{A}$ reduces precisely to the standard effective action, $\Gamma[A,A] \equiv \Gamma[A]$ [23, 72, 73]. Therefore, Eq. (6.110) verifies the desired gauge-invariant construction of $\Gamma[A]$. This $\Gamma[A]$ thus only consists of gauge-invariant building blocks. Of course, there is still a nontrivial constraint which becomes visible if we go away from the limit $A = \bar{A}$; namely, $\mathcal{G}\Gamma[A,\bar{A}]$ has to satisfy the standard WTI [45].

6.4.2 Background-Field Flow Equation

The desired properties of the effective action expressed by Eq. (6.110) can be maintained in the construction of the corresponding RG flow equation, if also the regulator satisfies

$$(\mathcal{G}+\bar{\mathcal{G}})\Delta S_k = 0. \tag{6.111}$$

This holds, e.g., for the choice

$$\Delta S_k = \frac{1}{2}\int (A-\bar{A}) R_{k,A}(\bar{\Delta}_A)(A-\bar{A}) + \int \bar{c} R_{k,\text{gh}}(\bar{\Delta}_{\text{gh}}) c, \tag{6.112}$$

where $\bar{\Delta}_{A,\mathrm{gh}}$ are operators that can depend on $\bar{A}$ and transform homogeneously. For the gluon sector, a suitable choice can, for instance, be given by

$$(\bar{\Delta}_A)^{ac}_{\mu\nu} = \{-\bar{D}^{ab}_\kappa \bar{D}^{bc}_\kappa \delta_{\mu\nu}, -\bar{D}^{ab}_\kappa \bar{D}^{bc}_\kappa \delta_{\mu\nu} + 2ig(\bar{F}^b_{\mu\nu} T^b)^{ac}, \dots\}, \tag{6.113}$$

where the first form corresponds to the background-covariant Laplacian, and the second also contains the spin-one coupling to the background field. For the ghost sector, the Laplacian is also an option, $\bar{\Delta}_{\mathrm{gh}} = -\bar{D}^{ab}_\kappa \bar{D}^{bc}_\kappa$. The resulting flow equation in the background-field gauge reads [74]

$$\partial_t \Gamma_k[A, \bar{c}, c, \bar{A}] = \frac{1}{2} \mathrm{STr} \left\{ \partial_t R_k(\bar{\Delta}) [\Gamma^{(2)}_k[A, \bar{A}] + R_k(\bar{\Delta})]^{-1} \right\}. \tag{6.114}$$

Here, it is a temptation to set $A = \bar{A}$ in the flow equation; however, the above construction tells us that this should be done only at the end of the calculation at $k = 0$.

Let us parameterize (suppressing ghosts for a moment) the effective action as [74]

$$\Gamma_k[A, \bar{A}] = \Gamma^{\mathrm{inv}}_k[A] + \Gamma^{\mathrm{gauge}}_k[A, \bar{A}], \tag{6.115}$$

where Γ^{inv}_k is a gauge-invariant functional, and $\Gamma^{\mathrm{gauge}}_k$ denotes the gauge-non-invariant remainder that satisfies $\Gamma^{\mathrm{gauge}}_k[A, \bar{A} = A] = 0$, cf. Eq. (6.110). Considering the second functional derivative with respect to A,

$$\Gamma^{(2)}_k[A, \bar{A}] = \Gamma^{\mathrm{inv}(2)}_k[A] + \Gamma^{\mathrm{gauge}(2)}_k[A, \bar{A}], \tag{6.116}$$

we observe that $\Gamma^{\mathrm{inv}(2)}_k[A]$ must be singular. This is because of the zero modes associated with gauge invariance: a variation with respect to A which points tangentially to the gauge orbit has to leave Γ^{inv}_k invariant, corresponding to a flat direction. For the flow equation (6.114) to be well defined, the contribution of $\Gamma^{\mathrm{gauge}(2)}_k$ to the denominator in Eq. (6.114) has to lift these flat directions. This demonstrates that $\Gamma^{\mathrm{gauge}}_k$ must not be dropped from a truncation, even though we may ultimately be interested only in Γ^{inv}_k.

On the other hand, the mWTI does not impose any constraint on Γ^{inv}_k,

$$0 = \mathcal{W}_k[\Gamma_k] \equiv \mathcal{W}_k[\Gamma^{\mathrm{gauge}}_k], \tag{6.117}$$

which implies that we have the full freedom to choose any gauge-invariant functional as an ansatz for Γ^{inv}_k, and its solution will solely be determined by the flow equation.

To summarize, the background formalism facilitates the construction of a gauge-invariant RG flow in which the manifestly gauge-invariant parts of the effective action Γ^{inv}_k can be separated from the gauge-dependent parts $\Gamma^{\mathrm{gauge}}_k$. In practice, we can construct our ansatz for Γ^{inv}_k by picking gauge-invariant building blocks, in a similar manner as we would do for other effective field theories. In addition, we have to construct (a truncation for) Γ_k^{gauge} with the aid of the mWTI (6.117), in order to lift the gauge zero modes in the flow equation.

6.4.3 Running Coupling

The background formalism also provides for a convenient nonperturbative definition of the running coupling. As a general remark, let us stress that there is no unique definition of the running coupling in the nonperturbative domain. Universality of the running coupling holds only near fixed points; e.g., only the one-loop coefficient of the perturbative β function is definition and scheme independent (in a mass-independent regularization scheme, also the two-loop coefficient is universal). Hence, any result for the running coupling in the nonperturbative domain has to be understood strictly in the context of its definition.

The definition within the background formalism follows, for instance, from the RG invariance of the background-covariant derivative, which is a gauge-covariant building block of Γ_k^{inv},

$$\bar{D}_\mu^{ab}[\bar{A}] = \partial_\mu \delta^{ab} + \bar{g} f^{abc} \bar{A}_\mu^c. \tag{6.118}$$

Here, we have used the notation $\bar{g}$ for the bare coupling. Obviously, the first term $\partial_\mu \delta^{ab}$ is RG invariant. Hence, also the product of $\bar{g}$ and $\bar{A}$ must be RG invariant [72],

$$\partial_t (\bar{g} \bar{A}_\mu^c) = 0, \quad \Rightarrow \quad \bar{g} \bar{A}_\mu^c = g \bar{A}_{\text{R},\mu}^c, \tag{6.119}$$

where g now denotes the renormalized coupling and $\bar{A}_\text{R}$ the renormalized background field. Renormalization of the background field is described by a wave function renormalization factor, $\bar{A}_\text{R} = Z_k^{1/2} \bar{A}$. Consequently, also the running of the coupling is tied to the same wave function renormalization,

$$g^2 = Z_k^{-1} \bar{g}^2. \tag{6.120}$$

We obtain for the β function of the running coupling

$$\beta_{g^2} \equiv \partial_t g^2 = \eta g^2, \quad \eta = -\partial_t \ln Z_k, \tag{6.121}$$

where η denotes the anomalous dimension of the background field. The wave function renormalization of the background field can be read off from the kinetic term of the background gauge potential,

$$\Gamma_k^{\text{inv}}[A] = \int \frac{Z_k}{4} F_{\mu\nu}^a F_{\mu\nu}^a + \ldots, \tag{6.122}$$

where the dots represent further terms in the truncation. The running coupling is thus linked to the lowest-order term of an operator expansion of the effective action.

According to its definition, the running coupling can be viewed as the response coefficient to excitations about the background field. If the background field, as a natural choice, is associated with the vacuum state the running coupling measures the coupling between the vacuum and excitations, for instance, in the form of (effective) quark and gluon fluctuations.

6.4.4 Truncated Background Flows

Let us study the background flow in a simple approximation, following [75]. In particular, we will be satisfied by a minimal truncation for Γ_k^{gauge}. For this, we expand Γ^{gauge} to leading order in $A - \bar{A}$,

$$\Gamma_k^{\text{gauge}}[A, \bar{A}] = \int (A - \bar{A})_\mu^a \, M_{\mu\nu}^{ab} \, (A - \bar{A})_\nu^b + \mathcal{O}((A - \bar{A})^3). \tag{6.123}$$

Then, we fix $M_{\mu\nu}^{ab}$ by the tree-level order of the mWTI:

$$\Gamma_k^{\text{gauge}}[A, \bar{A}] = \frac{1}{2\alpha} \int (A - \bar{A})_\mu^a \, (-\bar{D}_\mu^{ac} \bar{D}_\nu^{cb}) \, (A - \bar{A})_\nu^b + \dots, \tag{6.124}$$

which is just the classical gauge-fixing term. This approximation has an important consequence: to this order,

$$\Gamma_k^{(2)} = \Gamma_k^{\text{inv}(2)} + \Gamma_k^{\text{gauge}(2)} \tag{6.125}$$

is independent of $(A - \bar{A})$, and we can set $A = \bar{A}$ under the flow for all k. The form of Γ_k^{gauge} in Eq. (6.124) is just enough to lift the gauge zero modes. This gives us an approximate flow for Γ_k^{inv},

$$\begin{aligned} \partial_t \Gamma_k^{\text{inv}}[A] = \frac{1}{2} \operatorname{Tr} & \left\{ \partial_t R_{k,A}(\Delta_A) \left[\Gamma_k^{\text{inv}(2)} + \frac{1}{\alpha}(-DD) + R_k(\Delta) \right]_A^{-1} \right\} \\ & - \operatorname{Tr} \left\{ \partial_t R_{k,\text{gh}}(\Delta_{\text{gh}}) \left[\Gamma_k^{\text{inv}(2)} + R_k(\Delta) \right]_{\text{gh}}^{-1} \right\}, \end{aligned} \tag{6.126}$$

where we have dropped the bars on the right-hand side, since $A = \bar{A}$. So far, we have suppressed the ghost fields $c, \bar{c}$. If the dependence of Γ_k^{inv} on the ghosts is such that ghosts are contracted with homogeneously transforming color tensors, the mWTI is satisfied to the same level of accuracy as by the ghost-independent part. The classical ghost action in background-field gauge reads $S_{\text{gh}} = -\int \mathrm{d}^D x \, \bar{c}^a \bar{D}^{ab} D^{bc} c^c$; hence, the lowest-order approximation for our truncation is given by the classical term at $A = \bar{A}$,

$$\Gamma^{\text{inv}}_{k,\text{gh}}[A, \bar{c}, c] = -\int \mathrm{d}^D x\, \bar{c}^a D^{ab} D^{bc} c^c. \tag{6.127}$$

Let us now concentrate on the running coupling in the background gauge. As discussed above, this can be read off from the kinetic terms of the gauge field which we will therefore choose as the only nontrivial part of our simplest nontrivial truncation,

$$\Gamma^{\text{inv}}_{k,\text{glue}}[A] = \int \frac{Z_k}{4}\, F^a_{\mu\nu} F^a_{\mu\nu}, \tag{6.128}$$

where the running of the wave function renormalization remains to be calculated. This is done by projecting the right-hand side of the flow equation onto this kinetic operator only; any other operator generated by the flow, e.g., containing ghost fields, is dropped. Hence, we can set $c, \bar{c} = 0$ in $\Gamma^{(2)}$ (of course, *after* functional differentiation), implying that $\Gamma^{(2)}$ becomes block-diagonal with respect to gluon and ghost sectors. In the gluon sector, we obtain

$$\delta^2 \Gamma^{\text{inv}}_k|_{\text{glue}} = Z_k \int \mathrm{d}^D x\, \delta A^a_\mu \left[\mathcal{D}^{ab}_{\mathrm{T},\mu\nu} + D^{ac}_\mu D^{cb}_\nu - \frac{1}{\alpha} D^{ac}_\mu D^{cb}_\nu \right] \delta A^b_\nu, \tag{6.129}$$

where we have introduced the notation

$$\mathcal{D}^{ab}_{\mathrm{T},\mu\nu} = -D^{ac}_\kappa D^{cb}_\kappa \delta_{\mu\nu} + 2\bar{g} f^{abc} F^c_{\mu\nu}, \tag{6.130}$$

for the covariant spin-one Laplacian. Using the Feynman gauge $\alpha = 1$ here for simplicity, the gluon sector thus reduces to

$$(\Gamma^{\text{inv}(2)}_k)^{ab}_{\mu\nu}|_{\text{glue}} = Z_k \mathcal{D}^{ab}_{\mathrm{T},\mu\nu}. \tag{6.131}$$

From Eq. (6.127), we can immediately read off the form of $\Gamma^{(2)}$ in the ghost sector,[9]

$$\delta^2 \Gamma^{\text{inv}}_k|_{\text{gh}} = -\int \mathrm{d}^D x\, \delta\bar{c}^a D^{ac} D^{cb} \delta c^b, \quad \Rightarrow \quad (\Gamma^{\text{inv}(2)}_k)^{ab}|_{\text{gh}} = -D^{ac}_\kappa D^{cb}_\kappa. \tag{6.132}$$

Upon insertion into Eq. (6.126), the flow equation boils down to

$$\partial_t \Gamma^{\text{inv}}_k[A] = \frac{1}{2} \operatorname{Tr} \left[\partial_t R_{k,A} (Z_k \mathcal{D}_{\mathrm{T}} + R_{k,A})^{-1} \right] - \operatorname{Tr} \left[\partial_t R_{k,\text{gh}} (-D^2 + R_{k,\text{gh}})^{-1} \right]. \tag{6.133}$$

---

[9] Be aware of footnote 3 on p. xxx.

A convenient choice for the regulator is given by (cf. Eq. (6.16)

$$R_{k,A} = Z_k \mathcal{D}_\mathrm{T} r(\mathcal{D}_\mathrm{T}/k^2), \quad R_{k,\mathrm{gh}} = (-D^2) r(-D^2/k^2), \quad \lim_{y\to 0} r(y) \to \frac{1}{y}. \tag{6.134}$$

The insertion of the wave function renormalization Z_k into the regulator is useful, since it maintains the invariance of the flow equation under RG rescalings of the fields (the ghost wave function renormalization has been set to $Z_{k,\mathrm{gh}} = 1$ in our truncation). The choice $r(y) \to \frac{1}{y}$ implies that the IR modes with $p^2 \lesssim k^2$ are regulated by acquiring a mass term $\sim k^2$; the identification of the one-loop running is more straightforward with this choice. Then, the flow equation reads

$$\begin{aligned}\partial_t \Gamma_k^{\mathrm{inv}}[A] &= \frac{1}{2}\,\mathrm{Tr}\left[\frac{\partial_t (Z_k r(\mathcal{D}_\mathrm{T}/k^2))}{Z_k(1+r(\mathcal{D}_\mathrm{T}/k^2))}\right] - \mathrm{Tr}\left[\frac{\partial_t r(-D^2/k^2)}{1+r(-D^2/k^2)}\right] \\ &=: \mathrm{Tr}\,\mathcal{H}_{Z_k}(\mathcal{D}_\mathrm{T}/k^2) - 2\,\mathrm{Tr}\,\mathcal{H}(-D^2/k^2).\end{aligned} \tag{6.135}$$

The obvious definition of the $\mathcal{H}$ functions in the last line expresses the fact that both terms on the right-hand side are traces over a function of a single operator. It is useful to formally introduce the Laplace transform of the $\mathcal{H}$ functions,

$$\mathcal{H}(y) = \int_0^\infty ds\, \widetilde{\mathcal{H}}(s)\, \mathrm{e}^{-ys}, \tag{6.136}$$

such that the flow equation can be written as

$$\partial_t \Gamma_k^{\mathrm{inv}}[A] = \int_0^\infty ds\, \widetilde{\mathcal{H}}_{Z_k}(s)\, \mathrm{Tr}\, \mathrm{e}^{-s(\mathcal{D}_\mathrm{T}/k^2)} - 2\int_0^\infty ds\, \widetilde{\mathcal{H}}(s)\, \mathrm{Tr}\, \mathrm{e}^{-s(-D^2/k^2)}. \tag{6.137}$$

As a result, we have brought the flow equation into *propertime* form. Note that this was possible because of the particular operator structure resulting from our truncation: we were able to choose the regulators such that they depend on the operators which appear as entries of the block-diagonal $\Gamma^{(2)}$. Such a propertime form of the flow can, for instance, always be established to leading order in a derivative expansion of the RG flow [76]. In fact, a wide class of truncations can be mapped onto a propertime form [28, 77] which is computationally advantageous; moreover, propertime flows have extensively and successfully been used in the literature [78–83]. In the present case, we are finally dealing with traces over operator exponentials, so-called *heat kernels*, which can conveniently be dealt with by standard methods. The heat kernels of the covariant Laplacians occurring above have frequently been studied in the literature; see, e.g., [77]; here we merely need the term quadratic in the field strength,

$$\operatorname{Tr} e^{-s(\mathcal{D}_T/k^2)}|_{F^2} = \frac{N_c(24-D)}{3} \frac{\bar{g}^2}{(4\pi)^{D/2}} \left(\frac{s}{k^2}\right)^{2-(D/2)} \int d^D x \frac{1}{4} F^a_{\mu\nu} F^a_{\mu\nu}, \tag{6.138}$$

$$\operatorname{Tr} e^{-s(-D^2/k^2)}|_{F^2} = -\frac{N_c}{3} \frac{\bar{g}^2}{(4\pi)^{D/2}} \left(\frac{s}{k^2}\right)^{2-(D/2)} \int d^D x \frac{1}{4} F^a_{\mu\nu} F^a_{\mu\nu}, \tag{6.139}$$

Let us from now on concentrate on four dimensional spacetime, $D = 4$. From Eq. (6.137), we can extract the flow of the wave function renormalization

$$\begin{aligned} \partial_t Z_k &= \frac{20N_c}{3} \frac{\bar{g}^2}{(4\pi)^2} \int_0^\infty ds\, \widetilde{\mathcal{H}}_{Z_k}(s) + \frac{2N_c}{3} \frac{\bar{g}^2}{(4\pi)^2} \int_0^\infty ds\, \widetilde{\mathcal{H}}(s) \\ &= \frac{20N_c}{3} \frac{\bar{g}^2}{(4\pi)^2} \mathcal{H}_{Z_k}(0) + \frac{2N_c}{3} \frac{\bar{g}^2}{(4\pi)^2} \mathcal{H}(0), \end{aligned} \tag{6.140}$$

where we have used Eq. (6.136) for $y = 0$. Together with $\partial_t(Z_k r(\mathcal{D}_T/k^2)) = -Z_k[2\mathcal{D}_T/k^2 r'(\mathcal{D}_T/k^2) + \eta r(\mathcal{D}_T/k^2)]$, and the anomalous dimension $\eta = -\partial_t \ln Z_k$ as defined in Eq. (6.121), we observe that the $\mathcal{H}$ functions for zero argument uniquely yield

$$\mathcal{H}_{Z_k}(0) = 1 - \frac{\eta}{2}, \quad \mathcal{H}(0) = 1, \tag{6.141}$$

independently of the precise form of the regulator shape function $r(y)$ introduced in Eq. (6.134). This is a direct manifestation of the regularization-scheme independence of the one-loop β function coefficient, as will become clear soon.

Introducing the renormalized coupling $g^2 = Z_k^{-1} \bar{g}^2$, cf. Eq. (6.120), we obtain

$$-\eta = \frac{22N_c}{3} \frac{g^2}{(4\pi)^2} - \frac{10N_c}{3} \frac{g^2}{(4\pi)^2} \eta. \tag{6.142}$$

This brings us to our final result for the Yang-Mills β function

$$\beta_{g^2} \equiv \partial_t g^2 = \eta g^2 = -\frac{22N_c}{3} \frac{g^4}{(4\pi)^2} \left(1 - \frac{10N_c}{3} \frac{g^2}{(4\pi)^2}\right)^{-1} \tag{6.143}$$

$$= -\frac{22N_c}{3} \frac{g^4}{(4\pi)^2} - \frac{220N_c^2}{9} \frac{g^6}{(4\pi)^4} - \dots \tag{6.144}$$

It is instructive to compare our result to the full perturbative two-loop β function:

$$\beta_{g^2}^{\text{2-loop}} = -\frac{22N_c}{3} \frac{g^4}{(4\pi)^2} - \frac{204N_c^2}{9} \frac{g^6}{(4\pi)^4} - \dots \tag{6.145}$$

Obviously, we have rediscovered the one-loop result exactly, as we should. Moreover, the two-loop coefficient comes out remarkably well with an error of only 8%. This is astonishing in two ways: first, we have dropped a number of operators that contribute to the two-loop coefficient coupling, e.g., $(F^a_{\mu\nu}F^a_{\mu\nu})^2$ or $(F^a_{\mu\nu}\widetilde{F}^a_{\mu\nu})^2$, and also the ghost wave function renormalization, which hence appear to be less relevant. Second, whereas Eq. (6.145) is a universal result obtained within a mass-independent regularization scheme such as $\overline{\text{MS}}$, our result arises from a mass-dependent regularization scheme with a mass scale k; for such a scheme, the two-loop coefficient generally is not universal. However, in the present truncation, our β function indeed is universal to all orders computed above, owing to the fact that the dependence of the regulator shape function drops out in Eq. (6.141). We conclude that this simple truncation contains already much relevant information about the universal part of this two-loop coefficient. Larger truncations indeed contribute further terms which are partly non-universal, as expected. An exact two-loop calculation based on the functional RG can be found in [84].

In view of the quality of this simple truncation, it is tempting to speculate whether the result also contains reliable nonperturbative information. However, our result for the β function in Eq. (6.143) develops a pole at $g^2 = 3(4\pi)^2/(10N_c)$, clearly signaling the breakdown of the truncation. This could already have been expected from the physics content of the truncation: independently of the coupling strength, the resulting quantum equations of motion are identical to classical Yang-Mills theory. This equations allow for plain-wave solutions, describing freely propagating gluons. The truncation misses the important IR phenomenon of confinement and thus cannot be reliable in the deeply nonperturbative domain. For a stable flow from the UV down to the IR, a larger truncation is required that in addition has the potential to describe the relevant degrees of freedom for confinement.

*6.4.5 Further Reading: IR Running Coupling**

The nonperturbative estimate of the β function for the running coupling derived above has a remarkable property. All higher-loop order corrections can be summed into a geometric series, resulting in the structure of Eq. (6.143). The origin of this structure lies in the fact that we have included the wave function renormalization Z_k in the regulator, cf. Eq. (6.134). The apparent formal reason was that the RG invariance of the flow equations is thus maintained; but in addition, we thereby obtain a result which contains a resummation of a larger class of perturbative diagrams.

From a more physical viewpoint, the wave function renormalization describes the deformation of the perturbative spectrum of fluctuations, $S^{(2)} \sim p^2$, as it is induced by quantum fluctuations; at lower scales, we find the spectrum $\Gamma_k^{(2)} \sim Z_k p^2$. Therefore, the inclusion of Z_k in the regulator leads to a better adjustment of the regulator to the deformation of the spectrum, i.e., to the *spectral flow* of $\Gamma_k^{(2)}$.

If Z_k had not been included in the regulator, we would have found only the one-loop β function in the truncation (6.128) without any higher-loop orders. The latter would have been encoded in the flow of higher-order operators outside this simple truncation. For instance, the inclusion of the higher-order operator $(F^a_{\mu\nu}F^a_{\mu\nu})^2$ would have resulted in a β-function estimate of the form $\beta_{g^2} = \eta g^2$, with

$$\eta = -b_0 \frac{g^2}{(4\pi)^2} - b_1 \frac{g^2}{(4\pi)^2} w_2, \tag{6.146}$$

with some coefficient b_1, and b_0 being the correct one-loop result, and w_2 denoting the generalized coupling of this higher-order operator. In this truncation, all nonperturbative information is contained in the flow of w_2, which in turn can reliably be computed only by including even higher-order operators. A good estimate therefore probably requires a very large truncation. Even if the precise infrared values of the higher couplings $w_2, w_3, \ldots$ may not be very important, their flow exerts a strong influence on the running coupling in this approximation.

Together with the inclusion of Z_k in the regulator, our estimate for the β function would be of the form $\beta_{g^2} = \eta g^2$ with

$$\eta = -\frac{b_0 \frac{g^2}{(4\pi)^2} + b_1 \frac{g^2}{(4\pi)^2} w_2}{1 + d_1 \frac{g^2}{(4\pi)^2} + d_2 \frac{g^2}{(4\pi)^2} w_2}, \tag{6.147}$$

with a further coefficient d_2, and $d_1 < 0$ can be read off from Eq. (6.143). Particularly this d_1 makes an important contribution to the two-loop β function coefficient, as discussed above. Contrary to Eq. (6.146), this equation contains information to all orders in g^2, even for the strict truncation $w_2 = 0$, solely due to the spectral adjustment of the regulator.

Of course, for more complicated truncations, the deformation of the spectrum of $\Gamma_k^{(2)}$ becomes much more involved. A natural generalization would thus be the inclusion of the full $\Gamma_k^{(2)}$ in the regulator. However, any dependence of R_k on the fluctuation would invalidate the derivation of the flow equation in Eq. (6.18), especially spoiling the one-loop structure. But within the background formalism, we can at least include $\bar{\Gamma}_k^{(2)}$ where we have set all field dependence of Γ equal to the background field $\bar{A}$,

$$R_k(\bar{\Gamma}_k^{(2)}) = \bar{\Gamma}_k^{(2)}\, r(\bar{\Gamma}_k^{(2)}/(Z_k k^2), \tag{6.148}$$

such that the regulator is fully adjusted to the spectral flow of the fluctuation operator evaluated at the background field. The resulting flow equation then contains also $\bar{\Gamma}_k^{(2)}$ derivatives; for instance, at $A = \bar{A}$ where $\Gamma_k^{(2)} = \bar{\Gamma}_k^{(2)}$, we find

$$\partial_t \Gamma_k = \frac{1}{2}\,\mathrm{STr}\left[(2-\eta)\frac{-y\, r'(y)}{1+r(y)} + \frac{\partial_t \Gamma_k^{(2)}}{\Gamma_k^{(2)}} \frac{r(y) + y r'(y)}{1+r(y)}\right]_{y=\frac{\Gamma_k^{(2)}}{Z_k k^2}}. \tag{6.149}$$

Despite the additional terms, this form of the flow equation (together with the approximation of setting $A = \bar{A}$ already for finite k) is technically advantageous, since it

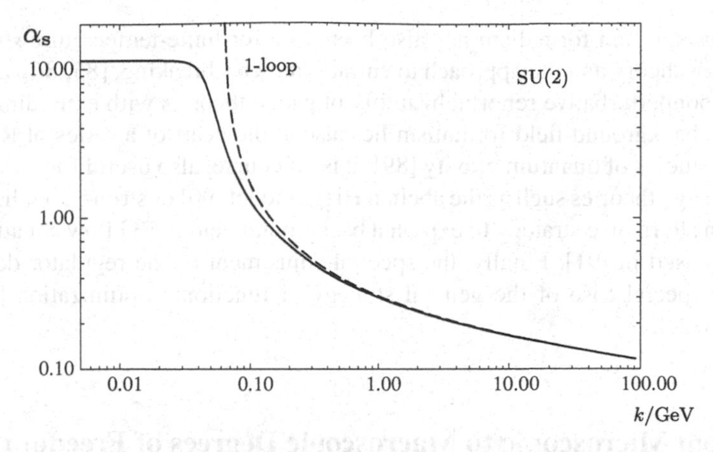

Fig. 6.10 Running coupling α_s versus momentum scale k in GeV for gauge group SU(2), using the initial value $\alpha_s(M_Z) \simeq 0.117$ for illustration. The *solid line* represents the result of an infinite-order resummation of Eq. (6.150) as taken from [77] in comparison with one-loop perturbation theory (*dashed line*)

can be written in generalized propertime form by means of a Laplace transformation, see the discussion in the preceding section.

The additional $\partial_t \Gamma_k^{(2)}$ terms are the generalization of the η term in Eq. (6.141) and thus can support a further resummation of a large class of diagrams. In [77], this has been used together with an operator-expansion truncation involving arbitrary powers of the field strength invariant $(F^a_{\mu\nu} F^a_{\mu\nu})^n$, $n = 1, 2, \ldots$, to get a nonperturbative estimate of the β function. The anomalous dimension has the structure,

$$\eta = -\frac{b_0 \frac{g^2}{(4\pi)^2} + b_1 \frac{g^4}{(4\pi)^4} + b_2 \frac{g^6}{(4\pi)^6} + \ldots}{1 + d_1 \frac{g^2}{(4\pi)^2} + d_2 \frac{g^4}{(4\pi)^4} + d_3 \frac{g^6}{(4\pi)^6} + \ldots}, \tag{6.150}$$

the form of which can approximately be resummed in a closed-form integral expression, see [77]. The resulting β function exhibits a second zero at $g^2 \to g_*^2 > 0$, corresponding to an IR fixed point, see Fig. 6.10. Hence the flow-equation results for the IR running coupling in background gauge based on an operator expansion show strong similarities to those in the Landau gauge based on a vertex expansion mentioned in Subsect. 6.3.5. Within low-order perturbation theory, the universality of the running coupling is a well-known property. Independently of the different definitions of the coupling, the one-loop (and in mass-independent schemes also the two-loop) β-function coefficient is always the same. Here, we also observe a qualitative agreement between the Landau gauge and the background gauge in the nonperturbative IR in the form of an attractive fixed point. This points to a deeper connection between the two gauges which deserves further study and may be traced back to certain non-renormalization properties in the two gauges.

The background formalism has also been used for finite-temperature studies of Yang-Mills theory and the approach to chiral symmetry breaking, [85, 86], and for a study of nonperturbative renormalizability of gauge theories with extra dimensions [87]. The background-field formalism lies also at the heart of a series of RG flow-equation studies of quantum gravity [89]; it is, of course, also useful for the study of abelian gauge theories such as the abelian Higgs model [90] or strong-coupling QED [102]. An alternative strategy to exploit a background field in RG flow equations has been proposed in [91]. Finally, the spectral adjustment of the regulator described here is a special case of the general strategy of functional optimization [50], cf. Subsect. 6.2.4.

6.5 From Microscopic to Macroscopic Degrees of Freedom

A typical feature of strongly interacting field theories is given by the fact that macroscopic degrees of freedom can be very different from microscopic degrees of freedom. For instance in QCD, quarks and gluons represent the microscopic degrees of freedom, whereas macroscopic degrees of freedom are mesons and baryons. The latter are bound states of quarks and gluons. Prominent representatives are the light pseudo-scalar mesons (pions, kaons, ...) which carry bi-fermionic quantum numbers, $\phi \sim \bar{\psi}\psi$. This type of fermionic pairing occurs in many systems, also in condensed-matter physics with strongly correlated electrons. Generically, a strong fermionic (self-)interaction is required for this pairing. In QCD, this is, of course, induced via the interactions with gluons. For an efficient description of the physics, it is advisable to take this transition from microscopic to macroscopic degrees of freedom into account [92–97].

6.5.1 Partial Bosonization

An explicit example for a fermion-to-boson transition is provided by the Hubbard-Stratonovich transformation, or *partial bosonization*. Let us discuss this transformation with the aid of a specific system: the Nambu–Jona-Lasinio (NJL) model [98]. We consider a version with one Dirac fermion, defined by the action

$$S_{\mathrm{F}} = \int \mathrm{d}^4x \left\{ \bar{\psi} \mathrm{i} \partial\!\!\!/ \psi + \frac{1}{2} \lambda [(\bar{\psi}\psi)^2 - (\bar{\psi}\gamma_5\psi)^2] \right\}. \tag{6.151}$$

This model has a $\mathrm{U}(1)\times\mathrm{U}(1)$ symmetry which here plays a similar role as chiral symmetry in QCD; in particular, it protects the fermions against acquiring a mass due to fluctuations.

Partial bosonization is obtained with the aid of the following mixed fermionic-bosonic theory,

$$S_{\mathrm{FB}} = \int \mathrm{d}^4x \left\{ \bar{\psi} \mathrm{i}\partial\!\!\!/ \psi + m^2 \phi^* \phi + h[\bar{\psi} P_{\mathrm{L}} \phi \psi - \bar{\psi} P_{\mathrm{R}} \phi^* \psi] \right\}, \tag{6.152}$$

with $P_{\mathrm{L,R}} = (1/2)(1 \pm \gamma_5)$ being the projectors onto left- and right-handed components of the Dirac fermion. In fact, the model (6.152) is equivalent to that of (6.151) also on the quantum level if

$$m^2 = \frac{h^2}{2\lambda}. \tag{6.153}$$

For a proof, it suffices to realize that

$$\begin{aligned}
\int \mathcal{D}\phi \, \mathrm{e}^{-S_{\mathrm{FB}}} &= \mathrm{e}^{-\int \bar{\psi} \mathrm{i}\partial\!\!\!/ \psi} \int \mathcal{D}\phi \, \mathrm{e}^{-\int \left(\phi^* + \frac{h}{m^2} \bar{\psi} P_{\mathrm{L}} \psi\right) m^2 \left(\phi - \frac{h}{m^2} \bar{\psi} P_{\mathrm{R}} \psi\right)} \\
&\quad \times \mathrm{e}^{-\int \frac{h^2}{m^2} (\bar{\psi} P_{\mathrm{L}} \psi)(\bar{\psi} P_{\mathrm{R}} \psi)} \\
&= \mathcal{N} \, \mathrm{e}^{-\int \left(\bar{\psi} \mathrm{i}\partial\!\!\!/ \psi + \frac{1}{4} \frac{h^2}{m^2} [(\bar{\psi}\psi)^2 - (\bar{\psi}\gamma_5\psi)^2]\right)} \overset{(6.153)}{\equiv} \mathcal{N} \, \mathrm{e}^{-S_{\mathrm{F}}},
\end{aligned} \tag{6.154}$$

where we have used that $(\bar{\psi} P_{\mathrm{L}} \psi)(\bar{\psi} P_{\mathrm{R}} \psi) = (\bar{\psi}\psi)^2 - (\bar{\psi}\gamma_5\psi)^2$, and $\mathcal{N}$ abbreviates the Gaußian integral over ϕ which is a pure number and can be absorbed into the normalization of the remaining fermionic integral $\int \mathcal{D}\bar{\psi}\mathcal{D}\psi \, \mathrm{e}^{-S_{\mathrm{F}}}$. Also on the classical level, the equations of motion display the fermionic pairing, i.e., bosonization,

$$\phi = \frac{h}{m^2} \bar{\psi} P_{\mathrm{R}} \psi, \quad \phi^* = -\frac{h}{m^2} \bar{\psi} P_{\mathrm{L}} \psi. \tag{6.155}$$

Equation (6.152) is the starting point for mean-field theory. The fermionic integral is Gaußian now,

$$\int \mathcal{D}\phi \mathcal{D}\bar{\psi} \mathcal{D}\psi \, \mathrm{e}^{-S_{\mathrm{FB}}} = \int \mathcal{D}\phi \, \mathrm{e}^{-S_{\mathrm{B}}}, \tag{6.156}$$

resulting in the purely bosonic action

$$S_{\mathrm{B}} = \int \mathrm{d}^4x \left\{ m^2 \phi^* \phi - \ln \det[\mathrm{i}\partial\!\!\!/ + h(P_{\mathrm{L}}\phi - P_{\mathrm{R}}\phi^*)] \right\}. \tag{6.157}$$

Mean-field theory now neglects bosonic fluctuations and assumes that the bosonic ground state corresponds to that of the classical bosonic action. Of course, the ln det is still a complicated nonlinear and nonlocal expression; nevertheless, assuming that the ground state is homogeneous in space and time, $\phi = \text{const.}$, the determinant can be computed by standard means [22]. For our purposes, it suffices to know that for

$$\lambda > \frac{8\pi^2}{\Lambda^2} \equiv \lambda_{\mathrm{cr}}, \tag{6.158}$$

with Λ being the UV cutoff, the resulting effective potential $V_{\mathrm{B}}(\phi^*\phi)$ has a nonzero minimum, implying a nonzero vacuum expectation value $\langle \phi \rangle \neq 0$. In the fermionic language, this vacuum expectation value corresponds to a bi-fermionic condensate $\langle\phi\rangle \sim \langle\bar{\psi}\psi\rangle$ (a chiral condensate in the QCD context). The expectation value generates fermion mass terms $\sim m_{\mathrm{f}}\bar{\psi}\gamma_5\psi$ with[10]

$$m_{\mathrm{f}} \sim h\langle\phi\rangle, \tag{6.159}$$

and the U(1)×U(1) symmetry is spontaneously broken to U(1) (fermion number conservation); this implies the existence of one Goldstone boson corresponding to excitations of the phase of the ϕ field. In the QCD context, this scenario corresponds to the spontaneous break-down of chiral symmetry with the pseudo-scalar mesons as Goldstone bosons. For $\lambda < \lambda_{\mathrm{cr}}$, the vacuum expectation value is zero, $\langle\phi\rangle = 0$ and the system remains in the symmetric phase.

These models of NJL type with broken symmetry show many similarities with QCD phenomenology, but there are important caveats from a microscopic viewpoint. First, there is no microscopic four-quark (or higher) self-interaction beyond criticality in QCD, $\lambda|_\Lambda \to 0$. In other words, $\lambda[(\bar{\psi}\psi)^2 - (\bar{\psi}\gamma_5\psi)^2]$ and other four-fermion operators are RG irrelevant; the Euclidean microscopic action for vanishing current quark masses is given by

$$S_{\mathrm{QCD}} = \int \mathrm{d}^D x \left(\frac{1}{4} F^a_{\mu\nu} F^a_{\mu\nu} + \mathrm{i}\bar{\psi} D\!\!\!\!/\,\psi \right), \tag{6.160}$$

where $D_\mu = \partial_\mu + \mathrm{i}\bar{g}\tau^c A^c_\mu$ denotes the gauge covariant derivative in the fundamental representation, which is generated by the τ^a with $\mathrm{tr}[\tau^a\tau^b] = (1/2)\delta^{ab}$. Of course, four-quark operators in the effective action are generated by gluon exchange from fluctuations described by box diagrams; see Fig. 6.11. The resulting β function for the four-quark coupling reads to lowest order

$$\partial_t \lambda \equiv \beta_\lambda = -c_\lambda \frac{1}{k^2} g^4, \tag{6.161}$$

where c_λ is a coefficient which depends on the algebraic structure of the theory and the details of the IR regularization. For definiteness, let us consider QCD with an SU(3) gauge sector but only one massless quark flavor $N_{\mathrm{f}} = 1$, the classical action of which has the same "chiral" symmetry properties as the NJL model used above. For this system, the coefficient c_λ obtains $c_\lambda = 5/(12\pi^2) > 0$ for the regulator (6.33) and a Fierz decomposition as chosen in [100]. Obviously, the four-quark self-interaction λ is asymptotically free as it should be for a QCD-like theory. Of course, for increasing gauge coupling g towards the IR, a naive extrapolation of Eq. (6.161) predicts that the fermionic self-interaction can become critical $\lambda > \lambda_{\mathrm{cr}}$ for some IR scale k_{cr}. If this holds also in the full theory, the quark self-interaction $\lambda[(\bar{\psi}\psi)^2 - (\bar{\psi}\gamma_5\psi)^2]$

[10] The occurrence of γ_5 in the fermion mass term arises from our fermion conventions [99]; these are related to more standard conventions by a discrete chiral rotation.

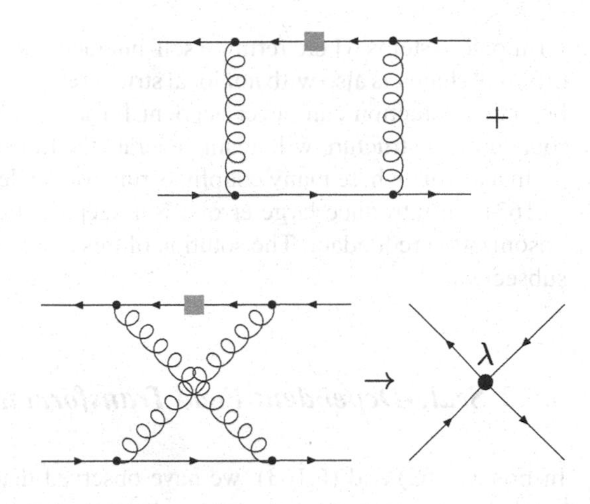

Fig. 6.11 Box diagrams with fundamental QCD interactions generate effective four-fermion self-interactions interactions λ. (Only one diagram per topology is shown; further diagrams with the regulator insertion (*filled box*) attached to other internal lines, of course, also exist.) The resulting flow-equation contribution to the running of λ is given in Eq. (6.161)

becomes strongly RG relevant at scales below k_{cr} and we expect the system to end up in the symmetry-broken phase [101].

This scenario appears to match our expectations for QCD. But in order to arrive at a quantitative description we do not only face the problem of computational control in the nonperturbative domain; moreover, we have to deal with the conceptual problem of how to switch from the description in terms of quarks and gluons to that involving boson fields as well. In view of the Hubbard-Stratonovich transformation, we may be tempted to apply partial bosonization at some scale $k_{\mathrm{B}} < k_{\mathrm{cr}}$. However, it turns out that this leads to a strong spurious dependence on the precise choice of k_{B} in generic truncations. This has to do with the following observation. Let us naively partially bosonize at $k_{\mathrm{B}} < k_{\mathrm{cr}}$. Diagrammatically,

$$k_{\mathrm{B}}: \quad \lambda \quad \Longrightarrow \quad h \cdots\cdots h$$

$$S_{\mathrm{F}} \Longrightarrow S_{\mathrm{FB}}, \tag{6.162}$$

where $\lambda = 0$ in S_{FB} after bosonization. Now, let us perform another RG step and integrate out another momentum shell Δk. Owing to the box diagrams, new quark self-interactions are generated again in this RG step,

$$k - \Delta k: \quad \partial_t \lambda = \qquad \sim g^4 \neq 0. \tag{6.163}$$

In other words, the bosonizing field ϕ which did a perfect job at k_{B} is no longer perfect at $k - \Delta k$; it does no longer bosonize all quark self-interactions which it was introduced for. Incidentally, this problem does not only occur if gauge interactions are present, as Eq. (6.163) seems to suggest. The same problem arises, e.g., in purely fermionic systems where fermion self-interactions are generated by the flow in many different channels also with nonlocal structure; since partial bosonization with a local bosonic interaction can never account for all four-fermion vertices, the remaining four-fermion structure will again generate the full structure in the RG flow.

In a regime where many couplings run fast, neglecting the newly generated terms (6.163) can introduce large errors. But keeping these terms seems to make partial bosonization redundant. The solution of this dilemma is the subject of the following subsection.

6.5.2 Scale-Dependent Field Transformations

In Eqs. (6.162) and (6.163), we have observed that different bosonizing fields are needed to compensate the quark self-interactions at different scales. This points already to a solution of the problem [93]: we promote the bosonizing field ϕ to be scale dependent, $\phi \to \phi_k$, the flow of which can be written as

$$\partial_t \phi_k = C_k[\phi, \psi, \bar{\psi}, \dots]. \tag{6.164}$$

Here, C_k is an a priori arbitrary functional of possibly all fields in the system. For the present problem, the idea is to choose C_k such that the resulting effective action $\Gamma_k[\phi_k]$ does not possess fermionic self-interactions. For more general cases, the functional C_k can be chosen such that the effective action $\Gamma_k[\phi_k]$ becomes simple, since its simplicity is a strong criterion for the proper choice of the relevant degrees of freedom.

We can formulate this idea in a differential fashion: we are looking for a functional C_k which yields a flow of $\Gamma_k[\phi_k]$ taken at fixed ϕ_k, (suppressing further field dependencies on $\bar{\psi}, \psi, \dots$),

$$\partial_t \Gamma_k[\phi_k]|_{\phi_k} = \partial_t \Gamma_k[\phi_k] - \int \frac{\delta \Gamma_k[\phi_k]}{\delta \phi_k} \partial_t \phi_k, \tag{6.165}$$

such that the flow of the fermion self-interaction vanishes for all k, $\partial_t \lambda|_{\phi_k} = 0$. Therefore, if $\lambda = 0$ holds at one scale k, λ stays zero at all scales, and ϕ_k becomes the "perfect" boson an all scales. The right-hand side of Eq. (6.165) can be read as follows: the first term denotes the full RG flow given in terms of a flow equation, whereas the second term with $\partial_t \phi_k = C_k$ characterizes how ϕ_k has to be modified scale by scale in order to partially bosonize fermion self-interactions on all scales. This fixes the functional form of C_k.

Now, we need a flow equation for the effective action $\Gamma_k[\phi_k]$ with scale-dependent field variables. This can indeed be formulated in various ways. Here, we follow a general and flexible exact construction given in [50]. Consider the modified generating functional

$$Z_k[J] = \mathrm{e}^{W_k[J]} = \int \mathcal{D}\varphi \, \mathrm{e}^{-S[\varphi]-\frac{1}{2}\int \varphi_k R_k \varphi_k + \int J\varphi_k}, \tag{6.166}$$

where we have coupled a scale-dependent φ_k to the source and the regulator. This combination guarantees that the resulting flow equation has a one-loop structure [28]. The scale dependence of φ is given by

$$\partial_t \varphi_k = \tilde{\mathcal{C}}_k[\varphi], \tag{6.167}$$

similar to Eq. (6.164) with the difference that Eq. (6.167) is formulated under the functional integral, whereas Eq. (6.164) holds for the fields conjugate to the source J,

$$\phi_k = \langle \phi_k \rangle \equiv \frac{\delta W_k[J]}{\delta J}. \tag{6.168}$$

Hence, $\mathcal{C}_k$ and $\tilde{\mathcal{C}}_k$ generally are not identical.

The derivation of the flow of $W_k[J]$ is straightforward:

$$\begin{aligned}\partial_t W_k[J] &= \frac{1}{Z_k[J]} \int \mathcal{D}\varphi \left(J\partial_t \varphi_k - \frac{1}{2}\int \varphi_k \partial_t R_k \varphi_k - \int \varphi_k R_k \partial_t \varphi_k \right) \\ &\quad \times \mathrm{e}^{-S[\varphi]-\Delta S_k[\varphi_k]+\int J\varphi_k} \\ &= J\langle \partial_t \varphi_k\rangle - \frac{1}{2}\mathrm{Tr}\partial_t R_k G_k - \int \frac{\delta}{\delta J} R_k \langle \partial_t \varphi_k \rangle \\ &\quad - \int \phi_k R_k \langle \partial_t \varphi_k \rangle - \frac{1}{2}\int \phi_k \partial_t R_k \phi_k. \end{aligned} \tag{6.169}$$

Here, we have defined the propagator similar to Eq. (6.19)

$$G_k(p,q) = \left(\frac{\delta^2 W_k}{\delta J \delta J}\right)(p,q) = \langle \varphi_k(p)\varphi_k(q)\rangle - \phi_k(p)\phi_k(q). \tag{6.170}$$

We have also used the relation $\langle \varphi_k \partial_t \varphi_k \rangle = (\frac{\delta}{\delta J} + \phi_k)\langle \partial_t \varphi_k \rangle$. As usual, we define the effective action by means of a Legendre transformation, this time involving the scale-dependent field variables (cf. Eq. (6.20)),

$$\Gamma_k[\phi_k] = \sup_J \left(\int J\phi_k - W_k[J] \right) - \frac{1}{2}\int \phi_k R_k \phi_k. \tag{6.171}$$

The resulting flow of this effective action is

$$\partial_t \Gamma_k[\phi_k] = \frac{1}{2}\mathrm{Tr}\,\partial_t R_k G_k + \int \left(G_k \frac{\delta}{\delta \phi_k} \right) R_k \langle \partial_t \varphi_k \rangle + \int \frac{\delta \Gamma_k}{\delta \phi_k} (\partial_t \phi_k - \langle \partial_t \varphi_k \rangle), \tag{6.172}$$

where, in the course of the Legendre transformation, all J dependence turns into a ϕ_k dependence by virtue of $J = J_{\text{sup}} = J[\phi_k]$; this also implies $\frac{\delta}{\delta J} = G_k \frac{\delta}{\delta \phi_k}$.

For a given scale-dependent field transformation $\partial_t \varphi_k = \tilde{C}_k[\varphi]$, we can successively work out $\langle \partial_t \varphi_k \rangle$, $\phi_k = \langle \varphi_k \rangle$ and $\partial_t \phi_k = \partial_t \langle \varphi_k \rangle$; in general, the latter is not identical to $\langle \partial_t \varphi_k \rangle$. Following this strategy, we would only then be able to compute the flow of $\Gamma_k[\phi_k]$. Also C_k would be a derived quantity, fixed implicitly by the choice of $\tilde{C}_k$.

By contrast, we can supplement the flow equation (6.172) with a bootstrap argument: since all we want to choose in the end is $\partial_t \phi_k = C_k[\phi]$, the precise form of $\partial_t \varphi = \tilde{C}_k$ need not be known; in fact, $\tilde{C}_k$ does not occur directly in Eq. (6.172), but only in expectation values. Therefore, we simply assume that a suitable $\tilde{C}_k$ exists for a desired C_k such that

$$\langle \partial_t \varphi_k \rangle \stackrel{!}{=} \partial_t \phi_k. \tag{6.173}$$

Of course, this is a highly implicit construction, and in view of the complicated structure of the mapping $\tilde{C}_k \to C_k$, the existence of a suitable $\tilde{C}_k$ for an arbitrary C_k is generally not guaranteed or, at least, difficult to prove. Nevertheless, since the resulting flow equation will, in practice, be used together with a truncation, it is reasonable to assume that Eq. (6.174) can at least be satisfied to the order of the truncation. As a consequence of Eq. (6.174), the flow equation simplifies,

$$\begin{aligned} \partial_t \Gamma_k[\phi_k]|_{\phi_k} &\stackrel{(6.165)}{=} \partial_t \Gamma_k[\phi_k] - \int \frac{\delta \Gamma_k[\phi_k]}{\delta \phi_k} \partial_t \phi_k \\ &= \frac{1}{2} \operatorname{Tr} \partial_t R_k G_k + \int \left(G_k \frac{\delta}{\delta \phi_k} \right) R_k \partial_t \phi_k - \int \frac{\delta \Gamma_k[\phi_k]}{\delta \phi_k} \partial_t \phi_k, \end{aligned} \tag{6.174}$$

which is the desired flow equation for scale-dependent field variables. Apart from the standard first term $\sim \operatorname{Tr} \partial_t R_k G_k$ and the third term which carries the explicit scale dependence of ϕ_k, we encounter the second term which takes care of fluctuation contributions to the renormalization flow of the operator insertion $\partial_t \varphi_k$ in the functional integral. Actually, this second term will generally be subdominant for not too large coupling: first, it is of higher order in the coupling, and second, R_k insertions lead to weaker numerical coefficients than $\partial_t R_k$ insertions for standard regulators. We expect that this term does not induce strong modifications for couplings up to $\mathcal{O}(1)$.

6.5.3 Scale-Dependent Field Transformations for QCD: Rebosonization

Let us now turn back to our original problem of QCD-like systems, where quark self-interactions are generated by gluon exchange. In order to arrive at a mesonic description in the infrared, we now want to apply scale-dependent field transformations that translate the quark self-interactions into the bosonic sector an all scales—a process that may be termed partial *rebosonization*.

In order to illustrate the formalism, let us study one-flavor QCD in a simple truncation. Apart from the standard kinetic terms for the quark and gluons, supplemented by wave function renormalization factors, we include a point-like four-quark self-interaction in the scalar–pseudo-scalar sector[11]

$$\Gamma_{\mathrm{F},k} = \int \mathrm{d}^4x \left(\frac{Z_k}{4} F^a_{\mu\nu} F^a_{\mu\nu} + \mathrm{i} Z_\psi \bar{\psi} D\!\!\!/ \psi + \frac{1}{2}\lambda[(\bar{\psi}\psi)^2 - (\bar{\psi}\gamma_5\psi)^2] \right). \quad (6.175)$$

The initial condition of the four-quark operator in the UV is obviously given by $\lambda|_{k\to\Lambda} \to 0$. As a first step towards rebosonization, we include a complex mesonic scalar field in the truncation on equal footing,

$$\Gamma_k = \Gamma_{\mathrm{F},k} + \int \mathrm{d}^4x \left(Z_\phi \partial_\mu \phi^* \partial_\mu \phi + V(\phi^*\phi) + h[\bar{\psi} P_{\mathrm{L}} \phi \psi - \bar{\psi} P_{\mathrm{R}} \phi^* \psi] \right), \quad (6.176)$$

with a scalar potential $V(\phi^*\phi) = m^2\phi^*\phi + \mathcal{O}((\phi^*\phi)^2)$. The initial conditions for the scalar field at $k \to \Lambda$ need to be chosen such that the scalar has no observable effect on the QCD sector whatsoever. This is easily done by demanding that the Yukawa interaction with the quarks vanishes $h|_{k\to\Lambda} \to 0$. We also choose a large scalar mass $m^2|_{k\to\Lambda} = \mathcal{O}(\Lambda^2)$, ensuring a fast decoupling of the scalar; finally, we set $Z_\phi|_{k\to\Lambda} \to 0$ which makes the scalar non-dynamical at the UV scale. Solving the flow with these initial conditions, the scalars rapidly decouple and only the standard QCD flow remains, revealing the purely formal character of this first step towards rebosonization.

As a second step, we now use the freedom to perform scale-dependent field transformations, as suggested in Eq. (6.164). We promote the field ϕ to be scale dependent, and choose the functional C_k characterizing this scale dependence to be of the form

$$\partial_t \phi_k = \bar{\psi} P_{\mathrm{R}} \psi\, \partial_t \alpha_k, \quad \partial_t \phi_k^* = -\bar{\psi} P_{\mathrm{L}} \psi\, \partial_t \alpha_k, \quad (6.177)$$

with some function α_k to be determined below. At this stage, let us study the consequences of the last term in Eq. (6.174) on the resulting flow; note that we are now dealing with a complex field, such that this term goes over into, $\int \frac{\delta\Gamma_k}{\delta\phi_k}\partial_t\phi_k \to \int \frac{\delta\Gamma_k}{\delta\phi_k}\partial_t\phi_k + \int \frac{\delta\Gamma_k}{\delta\phi_k^*}\partial_t\phi_k^*$. From the Yukawa interaction, we get from this term, together with Eq. (6.177),

[11] Of course, in order to avoid any ambiguity with respect to possible Fierz rearrangements of the four-fermion interactions in the point-like limit, all possible linearly-independent four-fermion interactions, in principle, have to be included in the truncation. For simplicity, we confine ourselves here just to the scalar–pseudo-scalar channel, where chiral condensation is expected to occur. For the four-fermion interactions that will be generated by the flow, we use the Fierz decomposition as proposed in [100].

$$h[\bar{\psi}P_{\mathrm{L}}\phi\psi - \bar{\psi}P_{\mathrm{R}}\phi^*\psi] \to \frac{1}{2}h\partial_t\alpha_k[(\bar{\psi}\psi)^2 - (\bar{\psi}\gamma_5\psi)^2], \tag{6.178}$$

which is a contribution to the flow of the quark self-interaction. Together with contributions from the first term of Eq. (6.174), and neglecting the second term of Eq. (6.174) here and in the following as discussed above, we obtain the flow of the four-quark coupling at fixed ϕ_k,

$$\partial_t\lambda|_{\phi_k} = -c_\lambda\,\frac{1}{k^2}\,g^4 - h\,\partial_t\alpha_k, \tag{6.179}$$

where we used the result given in Eq. (6.161). Choosing the transformation function

$$\partial_t\alpha_k = -c_\lambda\,\frac{1}{k^2}\,\frac{g^4}{h}, \tag{6.180}$$

we obtain

$$\partial_t\lambda|_{\phi_k} = 0, \tag{6.181}$$

which, together with the initial condition $\lambda|_{k\to\Lambda} \to 0$ implies that $\lambda = 0$ holds for all scales k. The scale-dependent transformation Eq. (6.177) thus has removed the point-like four-quark interaction in the scalar–pseudo-scalar sector completely by partial rebosonization. The information about this interaction is transformed into the scalar sector; for instance, the scalar mass term is also subject to the last term of Eq. (6.174):

$$m^2\phi^*\phi \to -m^2\partial_t\alpha_k[\bar{\psi}P_{\mathrm{L}}\phi\psi - \bar{\psi}P_{\mathrm{R}}\phi^*\psi]. \tag{6.182}$$

This yields a contribution to the flow of the Yukawa coupling. Together with the contribution from the first term of Eq. (6.174), i.e., the standard flow term, we obtain

$$\begin{aligned}\partial_t h|_{\phi_k} &= -\frac{1}{2}\,c_h\,g^2 h + m^2\,\partial_t\alpha_k\\ &\overset{(6.180)}{=} -\frac{1}{2}\,c_h\,g^2 h - c_\lambda\,\frac{m^2}{k^2}\,\frac{g^4}{h},\end{aligned} \tag{6.183}$$

where the coefficient c_h is a result of the diagram shown in Fig. 6.12a. For the linear regulator and in the Landau gauge, the result is $c_h = 1/\pi^2$ for SU(3). The second term of Eq. (6.183) accounts for the fact that the fermions couple to a scale-dependent boson. The flow of the scalar mass term is not transformed by the choice of Eq. (6.177); only the standard flow-equation term contributes,

$$\partial_t m^2 = c_m\,k^2\,h^2, \tag{6.184}$$

where the coefficient c_m yields for the regulator (6.33) $c_m = N_c/(8\pi^2)$, resulting from the diagram in Fig. 6.12b. The physical properties of the resulting boson field can best be illustrated with the convenient dimensionless composite coupling

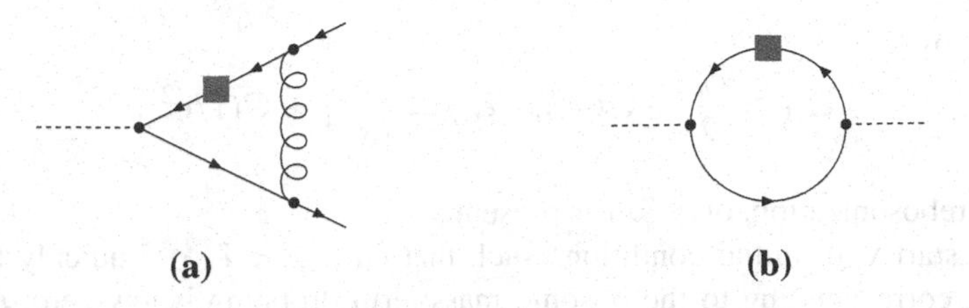

Fig. 6.12 Diagrams contributing to the RG flow of the scalar sector: **a** flow of the Yukawa coupling, see Eq. (6.183); **b** flow of the scalar mass, see Eq. (6.184). Only one diagram per topology is shown; further diagrams exhibit the regulator insertion (*filled box*) attached to other internal lines

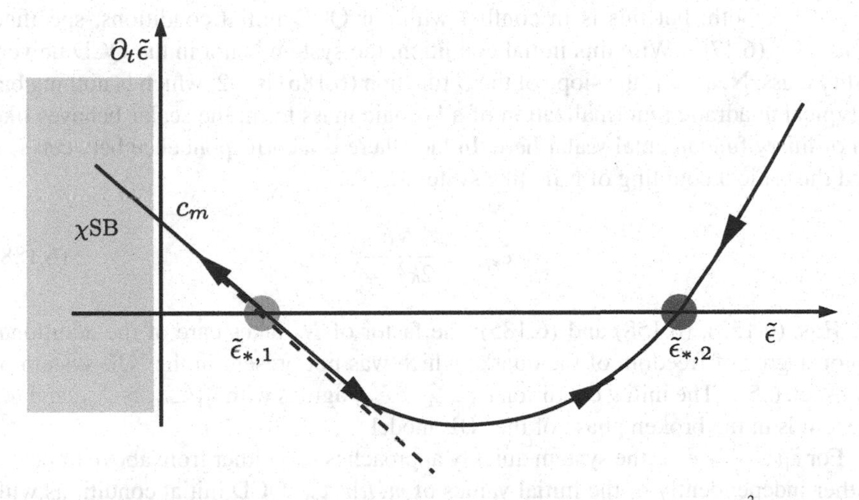

Fig. 6.13 Schematic plot of the β function (6.186) for the composite coupling $\tilde{\epsilon} = \frac{m^2}{k^2h^2}$ with arrows pointing along the flow towards the IR. The fixed point $\tilde{\epsilon}_{*,1}$ is IR repulsive; in its vicinity, the scalar field behaves as a fundamental scalar (*dashed line*). If the flow is initiated with $\tilde{\epsilon}|_{k=\Lambda} < \tilde{\epsilon}_{*,1}$, $\tilde{\epsilon}$ drops quickly below zero and the system runs into the regime with chiral symmetry breaking (χSB). For $\tilde{\epsilon}|_{k=\Lambda} > \tilde{\epsilon}_{*,1}$, the system rapidly approaches the bound-state IR fixed point $\tilde{\epsilon}_{*,2}$, where the scalar exhibits bound-state behavior. QCD initial conditions correspond to $\tilde{\epsilon}|_{k\to\Lambda} \to \infty$

$$\tilde{\epsilon} := \frac{m^2}{k^2h^2}, \tag{6.185}$$

and its β function

$$\partial_t\tilde{\epsilon} = -2\tilde{\epsilon} + c_m + c_h g^2\tilde{\epsilon} + 2c_\lambda g^4\tilde{\epsilon}^2, \tag{6.186}$$

resulting from Eqs. (6.161), (6.183) and (6.184). The last term comes directly from rebosonization. Since all $c_i > 0$, this β function looks like a parabola; see Fig. 6.13, solid line. Without rebosonization, this β function would have corresponded to a straight line, see Fig. 6.13, dashed line.

The β function Eq. (6.186) exhibits two fixed points: $\tilde{\epsilon}_{*,1}$ is IR repulsive and $\tilde{\epsilon}_{*,2}$ is IR attractive. In a small-gauge-coupling expansion, the positions of the fixed points are given by

$$\tilde{\epsilon}_{*,1} \simeq \frac{c_m}{2} + \mathcal{O}(g^2), \quad \tilde{\epsilon}_{*,2} \simeq \frac{1}{c_\lambda g^4} + \mathcal{O}(1/g^2). \tag{6.187}$$

Without rebosonization, only $\tilde{\epsilon}_{*,1}$ is present.

If we start with initial conditions such that $\tilde{\epsilon}|_{k\to\Lambda} < \tilde{\epsilon}_{*,1}$, $\tilde{\epsilon}$ quickly becomes negative, corresponding to the bosonic mass term dropping below zero, $m^2 < 0$. This indicates that the potential develops a nonzero minimum, giving rise to chiral symmetry breaking and quark mass generation. However, we obtain this initial condition $\tilde{\epsilon}_{k\to\Lambda} < \tilde{\epsilon}_{*,1}$ only if either the scalar mass is small or the Yukawa coupling is large or both; but this is in conflict with our QCD initial conditions, specified below Eq. (6.176). With this initial condition, the system is not in the QCD universality class. Near $\tilde{\epsilon}_{*,1}$, the slope of the β function (6.186) is -2, which is nothing but a typical quadratic renormalization of a bosonic mass term; the scalar behaves like an ordinary fundamental scalar here. In fact, there is a correspondence between $\tilde{\epsilon}_{*,1}$ and the critical coupling of NJL-like systems,

$$\tilde{\epsilon}_{*,1} \simeq \frac{Nc}{2k^2\lambda_{\text{cr}}}, \tag{6.188}$$

cf. Eqs. (6.153), (6.158) and (6.185); the factor of N_c takes care of the additional color degree of freedom of the quarks which was not present in the NJL system of Subsect. 6.5.1. The initial condition $\tilde{\epsilon}|_{k\to\Lambda} < \tilde{\epsilon}_{*,1}$ agrees with $\lambda|_{k\to\Lambda} > \lambda_{\text{cr}}$, and the system is in the broken phase of the NJL model.

For $\tilde{\epsilon}|_{k\to\Lambda} > \tilde{\epsilon}_{*,1}$, the system quickly approaches $\tilde{\epsilon}_{*,2}$ either from above or below rather independently of the initial values of $m, h|_{k\to\Lambda}$. QCD initial conditions with large initial m and small initial h correspond to $\tilde{\epsilon}|_{k\to\Lambda} \to \infty$. But also for much smaller initial $\tilde{\epsilon}$, the system rapidly flows to $\tilde{\epsilon}_{*,2}$, and the memory of the precise initial values gets lost. There, the system is solely determined by the gauge coupling g^2 which governs the fixed-point position. This is precisely how it should be in QCD.

Near $\tilde{\epsilon}_{*,2}$, the boson is not really a fully developed degree of freedom. The flow does not at all remind us of the flow of a fundamental scalar, but points to the composite nature of the scalar. This justifies to call $\tilde{\epsilon}_{*,2}$ the bound-state fixed point; for instance, in weakly coupled systems such as QED, the boson at the bound-state fixed point describes a positronium-like bound state.

It is a particular strength of the RG approach with scale-dependent field transformations that one and the same field can describe bound state formation on the one hand and condensate formation as well as meson excitations on the other hand; whether the field behaves as a bound state or as a fundamental scalar is solely governed by the dynamics and the coupling strength of the system.

So far, our analysis of the system was essentially based on weak-coupling arguments, revealing that QCD at initial stages of the flow approaches the bound-state fixed point. But, we still have to answer a crucial question: how does QCD leave the bound-state fixed point and ultimately approach the chiral-symmetry broken regime? The answer is again given by the gauge coupling which controls the whole flow. For

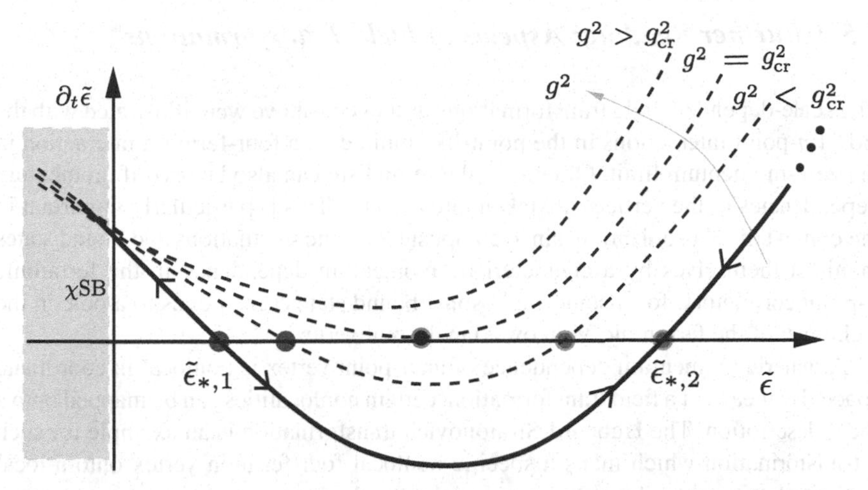

Fig. 6.14 Schematic plot of the β function (6.186) for the composite coupling $\tilde{\epsilon} = \frac{m^2}{k^2h^2}$ with arrows pointing along the flow towards the IR. At weak gauge coupling, QCD-like systems first flow to the bound-state fixed point $\tilde{\epsilon}_{*,2}$ where they remain over a wide range of scales. For increasing gauge coupling g^2, the β function is lifted (*dashed lines*). At the critical coupling $g^2 = g^2_{\rm cr}$, the fixed points are destabilized and the system rapidly runs into the chiral symmetry broken regime (χSB)

increasing gauge coupling, the parabola characterizing the β function (6.186) for $\tilde{\epsilon}$ is lifted, as depicted in Fig. 6.14. At a critical coupling value, $g^2 = g^2_{\rm cr}$, the fixed points $\tilde{\epsilon}_{*,1}$ and $\tilde{\epsilon}_{*,2}$ annihilate each other and the system runs towards the chiral-symmetry broken regime. This transition is unambiguously triggered by gluonic interactions. For instance in the present case of one-flavor QCD with gauge group SU(3), the critical coupling is given by $\alpha_{\rm cr} \equiv \frac{g^2_{\rm cr}}{4\pi} \simeq 0.74$ for the regulator Eq. (6.33). Since this coupling value is not a universal quantity, one should not overemphasize its meaning; however, it is interesting to observe that this coupling strength is in the nonperturbative domain, as expected, but not very deeply. In particular, since loop expansions go along with the expansion parameter α/π, the critical coupling and thus the approach to chiral symmetry breaking appears still to be in reach of weak-coupling methods (not to be confused with perturbation theory).

We conclude that the continuous scale-dependent translation allows for a controllable transition between microscopic to macroscopic degrees of freedom and between different dynamical regimes of a system. From a quantitative viewpoint, no spurious dependence on a bosonization scale, i.e., a scale at which degrees of freedom are discretely changed, is introduced, because field transformations are continuously performed on all scales. This helps maintaining the predictive power of truncated RG flows. As a result, macroscopic parameters can quantitatively be related to microscopic input.

6.5.4 Further Reading: Aspects of Field Transformations*

The scale-dependent field transformations introduced above were illustrated with the aid of n-point interactions in the point-like limit, e.g., a four-fermion interaction in the zero-momentum limit. Of course, the formalism can also be used if momentum dependencies of the vertices are taken into account. This is particularly important in the context of rebosonization, since composite bosonic fluctuations and bound states manifest themselves by a characteristic momentum dependence in the fermionic n-point correlators; for instance, a bosonic bound state corresponds to a pole in the s channel of the fermionic Minkowskian 4-point vertex.

A generic momentum dependence of an n-point vertex is nonlocal in coordinate space. By means of a field transformation, certain nonlocalities can be mapped onto a local description. The Hubbard-Stratonovich transformation is an example for such a transformation which maps a specific nonlocal four-fermion vertex onto a local bosonic theory with a local Yukawa coupling to the fermion.

With the following scale-dependent field transformation, this nonlocal-to-local mapping can be performed on all scales for a four-fermion vertex with s channel momentum dependencies, $\lambda = \lambda(s = q^2)$,

$$\partial_t \phi_k(q) = (\bar{\psi} P_{\mathrm{R}} \psi)(q)\, \partial_t \alpha_k(q), \quad \partial_t \phi_k^*(q) = -(\bar{\psi} P_{\mathrm{L}} \psi)(-q)\, \partial_t \alpha_k(q). \tag{6.189}$$

This results in a flow of the four-point interaction for the transformed fields which reads

$$\partial_t \lambda|_{\phi_k} = \partial_t \lambda - h\, \partial_t \alpha_k(q). \tag{6.190}$$

Obviously, the q dependence of α_k can be chosen such the fluctuation-induced s-channel q dependence of λ is eaten up for all scales and all values of q. Via the momentum-dependent analog of Eq. (6.183), this induces a momentum dependence of the Yukawa interaction $hb \to h(q)$; this can be taken care of by a further generalization of the field transformation,

$$\begin{aligned} \partial_t \phi_k(q) &= (\bar{\psi} P_{\mathrm{R}} \psi)(q)\, \partial_t \alpha_k(q) - \beta_k(q) \phi_k(q), \\ \partial_t \phi_k^*(q) &= -(\bar{\psi} P_{\mathrm{L}} \psi)(-q)\, \partial_t \alpha_k(q) - \beta_k(q) \phi_k^*(q). \end{aligned} \tag{6.191}$$

The resulting flows for the Yukawa coupling and the inverse scalar propagator for the transformed fields are then given by

$$\begin{aligned} \partial_t h(q)|_{\phi_k} &= \partial_t h(q) + \frac{Z_\phi q^2 + m^2}{h}\, \partial_t \lambda(q^2) + h\, \partial_t \beta_k(q), \\ (\partial_t Z_\phi(q) q^2 + \partial_t m^2)|_{\phi_k} &= \partial_t m^2 + 2 \partial_t \beta_k(q)\, (Z_\phi q^2 + m^2). \end{aligned} \tag{6.192}$$

For a given momentum dependence of λ, the first equation can be used to determine the transformation function $\beta_k(q)$.[12] The second equation then fixes the running of the mass and of the wave function renormalization of the transformed field. In this manner, specific nonlocal momentum structures can be transformed from the fermionic interactions into a local boson sector.

This strategy has been used for the abelian gauged NJL model [93], QED [102] and QCD-like systems in [100]. At weak coupling, one typically finds that the bound-state fixed point discussed above of the composite coupling $\tilde{\epsilon}$ has similar counterparts in many other couplings, such as the scalar dimensionless mass m^2/k^2 and the Yukawa coupling h. This is a manifestation of their RG irrelevance at weak coupling, with the dynamics of the scalar sector being fully controlled by the fermions and their gauge interactions. If the gauge coupling becomes large, the bound-state fixed points in all these couplings is destabilized and interactions involving composite bosons can become relevant or even dominant close to a transition into a broken-symmetry regime. Since this switching from irrelevant to relevant is a smooth process under the flow being controlled by the dynamics itself, the predictive power of the computation is maintained.

In QCD, the resulting effective action at IR scales naturally exhibits a sector which is similar to a chiral quark-meson model [80, 103, 104] but with all parameters fixed by the outcome of the RG flow with field transformations. Also the gluonic sector can still be dynamically active and contribute further to the running of the chiral sector.

A further application of the scale-dependent field transformation can be found in [105] where it is shown that the Fierz ambiguities mentioned in the preceding section can be overcome by treating all possible interaction channels and the corresponding scale-dependent bosonic composites on equal footing.

Acknowledgments It is a great pleasure to thank A. Schwenk and J. Polonyi for organizing the ECT∗ school and for creating such a stimulating atmosphere. I am particularly grateful to the students for their active participation, critical questions and detailed discussions which have left their traces in these lecture notes. I would like to thank J. Braun, C.S. Fischer, J. Jaeckel, J.M. Pawlowski, and C. Wetterich for pleasant and fruitful collaborations on some of the topics presented here, and for numerous intense discussions, some essence of which has condensed into these lecture notes. Critical remarks on the manuscript by J. Braun, M. Ghasemkhani, J.M. Pawlowski, and A. Wipf are gratefully acknowledged. This work was supported by the DFG Gi 328/1-3 (Emmy-Noether program).

References

1. Fisher, M.E.: Rev. Mod. Phys. **70**, 653 (1998)
2. Morris, T.R.: Prog. Theor. Phys. Suppl. **131**, 395 (1998)

---

[12] The momentum-independent part can, for instance, be fixed such that $\partial_t Z_\phi(q=k)=0$, ensuring that the approximation of a momentum-independent Z_ϕ is self-consistent.

3. Litim, D.F., Pawlowski, J.M.: In: Krasnitz, A., et al. (eds.) The Exact Renormalization Group, p. 168. World Scientific, Singapore (1999)
4. Aoki, K.: Int. J. Mod. Phys. B **14**, 1249 (2000)
5. Bagnuls, C., Bervillier, C.: Phys. Rept. **348**, 91 (2001)
6. Berges, J., Tetradis, N., Wetterich, C.: Phys. Rept. **363**, 223 (2002)
7. Polonyi, J.: Central Eur. J. Phys. **1**, 1 (2004)
8. Wilson, K.G.: Phys. Rev. B **4**, 3174 (1971); ibid., 3184 (1971)
9. Wilson, K.G., Kogut, J.B.: Phys. Rept. **12**, 75 (1974)
10. Wegner, F.J., Houghton, A.: Phys. Rev. A **8**, 401 (1973)
11. Nicoll, J.F., Chang, T.S.: Phys. Lett. A **62**, 287 (1977)
12. Polchinski, J.: Nucl. Phys. B **231**, 269 (1984)
13. Wetterich, C.: Phys. Lett. B **301**, 90 (1993)
14. Warr, B.J.: Ann. Phys. **183**, 1, 59 (1988)
15. Hurd, T.R.: Commun. Math. Phys. **124**, 153 (1989)
16. Keller, G., Kopper, C., Salmhofer, M.: Helv. Phys. Acta **65**, 32 (1992)
17. Reuter, M., Wetterich, C.: Nucl. Phys. B **391**, 147 (1993)
18. Reuter, M., Wetterich, C.: Nucl. Phys. B **417**, 181 (1994)
19. Bonini, M., D'Attanasio, M., Marchesini, G.: Nucl. Phys. B **418**, 81 (1994); Nucl. Phys. B **421**, 429 (1994)
20. Ellwanger, U.: Phys. Lett. B **335**, 364 (1994)
21. D'Attanasio, M., Morris, T.R.: Phys. Lett. B **378**, 213 (1996)
22. Peskin, M.E., Schröder, D.V.: An Introduction to Quantum Field Theory. Addison-Wesley, Reading (1995)
23. Pokorski, S.: Gauge Field Theories. Cambridge University Press, Cambridge (1987)
24. Alkofer, R., von Smekal, L.: Phys. Rept. **353**, 281 (2001)
25. Roberts, C.D., Schmidt, S.M.: Prog. Part. Nucl. Phys. **45**, S1 (2000)
26. Maris, P., Roberts, C.D.: Int. J. Mod. Phys. E **12**, 297 (2003)
27. Fischer, C.S.: J. Phys. G **32**, R253 (2006)
28. Litim, D.F., Pawlowski, J.M.: Phys. Rev. D **66**, 025030 (2002)
29. Horikoshi, A., Aoki, K.I., Taniguchi, M.A., Terao, H.: arXiv:hep-th/9812050
30. Kapoyannis, A.S., Tetradis, N.: Phys. Lett. A **276**, 225 (2000)
31. Zappala, D.: Phys. Lett. A **290**, 35 (2001)
32. Litim, D.F.: Phys. Lett. B **486**, 92 (2000); Phys. Rev. D **64**, 105007 (2001)
33. Bender, C.M., Wu, T.T.: Phys. Rev. **184**, 1231 (1969)
34. Janke, W., Kleinert, H.: arXiv:quant-ph/9502019
35. Galindo, A., Pascual, P.: Quantum Mechanics, vol. 2. Springer, Berlin (1990)
36. Ball, R.D., Haagensen, P.E., Latorre, J.I., Moreno, E.: Phys. Lett. B **347**, 80 (1995) [arXiv:hep-th/9411122]
37. Liao, S.B., Polonyi, J., Strickland, M.: Nucl. Phys. B **567**, 493 (2000) [arXiv:hep-th/9905206]
38. Singer, I.M.: Commun. Math. Phys. **60**, 7 (1978)
39. Gribov, V.N.: Nucl. Phys. B **139**, 1 (1978)
40. Morris, T.R.: JHEP **0012**, 012 (2000); Arnone, S., Morris, T.R., Rosten, O.J.: arXiv:hep-th/0507154; Rosten, O.J.: arXiv:hep-th/0602229
41. Branchina, V., Meissner, K.A., Veneziano, G.: Phys. Lett. B **574**, 319 (2003)
42. Pawlowski, J.M.: arXiv:hep-th/0310018
43. Bonini, M., D'Attanasio, M., Marchesini, G.: Nucl. Phys. B **437**, 163 (1995)
44. Litim, D.F., Pawlowski, J.M.: Phys. Lett. B **435**, 181 (1998)
45. Freire, F., Litim, D.F., Pawlowski, J.M.: Phys. Lett. B **495**, 256 (2000)
46. Igarashi, Y., Itoh, K., So, H.: Prog. Theor. Phys. **106**, 149 (2001)
47. Ellwanger, U., Hirsch, M., Weber, A.: Z. Phys. C **69**, 687 (1996)
48. Ellwanger, U., Hirsch, M., Weber, A.: Eur. Phys. J. C **1**, 563 (1998)
49. Gies, H., Jaeckel, J., Wetterich, C.: Phys. Rev. D **69**, 105008 (2004)
50. Pawlowski, J.M.: arXiv:hep-th/0512261
51. Zwanziger, D.: Phys. Rev. D **69**, 016002 (2004)

52. Fischer, C.S., Gies, H.: JHEP **0410**, 048 (2004)
53. Blaizot, J.P., Mendez-Galain, R., Wschebor, N.: arXiv:hep-th/0603163; arXiv:hep-th/0512317
54. von Smekal, L., Alkofer, R., Hauck, A.: Phys. Rev. Lett. **79**, 3591 (1997)
55. Cucchieri, A.: Nucl. Phys. B **508**, 353 (1997)
56. Leinweber, D.B., Skullerud, J.I., Williams, A.G., Parrinello, C. [UKQCD Collaboration]: Phys. Rev. D **60**, 094507 (1999) [Erratum-ibid. D **61**, 079901 (2000)]
57. Alexandrou, C., de Forcrand, P., Follana, E.: Phys. Rev. D **63**, 094504 (2001)
58. Langfeld, K., Reinhardt, H., Gattnar, J.: Nucl. Phys. B **621**, 131 (2002)
59. Furui, S., Nakajima, H.: Phys. Rev. D **70**, 094504 (2004)
60. Sternbeck, A., Ilgenfritz, E.M., Mueller-Preussker, M., Schiller, A.: Phys. Rev. D **72**, 014507 (2005)
61. Silva, P.J., Oliveira, O.: hep-lat/0511043
62. Boucaud, P., et al.: hep-ph/0507104
63. Taylor, J.C.: Nucl. Phys. B **33**, 436 (1971)
64. Gattnar, J., Langfeld, K., Reinhardt, H.: Phys. Rev. Lett. **93**, 061601 (2004)
65. Pawlowski, J.M., Litim, D.F., Nedelko, S., von Smekal, L.: Phys. Rev. Lett. **93**, 152002 (2004)
66. Lerche, C., von Smekal, L.: Phys. Rev. D **65**, 125006 (2002)
67. Pawlowski, J.M., Litim, D.F., Nedelko, S., von Smekal, L.: AIP Conf. Proc. **756**, 278 (2005)
68. Fischer, C.S., Alkofer, R.: Phys. Lett. B **536**, 177 (2002)
69. Kugo, T., Ojima, I.: Prog. Theor. Phys. Suppl. **66**, 1 (1979)
70. Zwanziger, D.: Nucl. Phys. B **364**, 127 (1991)
71. Zwanziger, D.: Nucl. Phys. B **399**, 477 (1993)
72. Abbott, L.F.: Nucl. Phys. B **185**, 189 (1981)
73. Dittrich, W., Reuter, M.: Lect. Notes Phys. **244**, 1 (1986)
74. Reuter, M., Wetterich, C.: Phys. Rev. D **56**, 7893 (1997)
75. Reuter, M.: arXiv:hep-th/9602012
76. Litim, D.F., Pawlowski, J.M.: Phys. Lett. B **516**, 197 (2001); Phys. Rev. D **65**, 081701 (2002)
77. Gies, H.: Phys. Rev. D **66**, 025006 (2002)
78. Liao, S.B.: Phys. Rev. D **53**, 2020 (1996); Phys. Rev. D **56**, 5008 (1997)
79. Floreanini, R., Percacci, R.: Phys. Lett. B **356**, 205 (1995)
80. Schaefer, B.J., Pirner, H.J.: Nucl. Phys. A **660**, 439 (1999)
81. Papp, G., Schaefer, B.J., Pirner, H.J., Wambach, J.: Phys. Rev. D **61**, 096002 (2000)
82. Bonanno, A., Zappala, D.: Phys. Lett. B **504**, 181 (2001)
83. Zappala, D.: Phys. Rev. D **66**, 105020 (2002)
84. Pawlowski, J.M.: Int. J. Mod. Phys. A **16**, 2105 (2001)
85. Braun, J., Gies, H.: JHEP **0606**, 024 (2006); arXiv:hep-ph/0512085
86. Braun, J.: Contribution to This Volume (2006)
87. Gies, H.: Phys. Rev. D **68**, 085015 (2003)
88. Canet, L., Delamotte, B., Mouhanna, D., Vidal, J.: Phys. Rev. D **67**, 065004 (2003); Phys. Rev. B **68**, 064421 (2003)
89. Reuter, M.: Phys. Rev. D **57**, 971 (1998); Lauscher, O., Reuter, M.: Phys. Rev. D **65**, 025013 (2002)
90. Litim, D., Wetterich, C., Tetradis, N.: Mod. Phys. Lett. A **12**, 2287 (1997)
91. Litim, D.F., Pawlowski, J.M.: JHEP **0209**, 049 (2002)
92. Polonyi, J., Sailer, K.: Phys. Rev. D **63**, 105006 (2001)
93. Gies, H., Wetterich, C.: Phys. Rev. D **65**, 065001 (2002); Acta Phys. Slov. **52**, 215 (2002)
94. Schwenk, A., Polonyi, J.: nucl-th/0403011
95. Schutz, F., Bartosch, L., Kopietz, P.: Phys. Rev. B **72**, 035107 (2005)
96. Salmhofer, M., Honerkamp, C., Metzner, W., Lauscher, O.: Prog. Theor. Phys. **112**, 943 (2004)
97. Harada, K., Inoue, K., Kubo, H.: nucl-th/0511020
98. Nambu, Y., Jona-Lasinio, G.: Phys. Rev. **122**, 345 (1961); ibid. **124**, 246 (1961); Vaks, V.G., Larkin, A.I.: Zh. Eksp. Teor. Fiz. **40**(1) (1961) [Sov. Phys. JETP **13**, 192 (in English)]
99. Wetterich, C.: Z. Phys. C **48**, 693 (1990)
100. Gies, H., Wetterich, C.: Phys. Rev. D **69**, 025001 (2004)

101. Aoki, K.I., Morikawa, K.I., Sumi, J.I., Terao, H., Tomoyose, M.: Prog. Theor. Phys. **97**, 479 (1997)
102. Gies, H., Jaeckel, J.: Phys. Rev. Lett. **93**, 110405 (2004)
103. Jungnickel, D.U., Wetterich, C.: Phys. Rev. D **53**, 5142 (1996)
104. Schaefer, B.J., Wambach, J.: Nucl. Phys. A **757**, 479 (2005)
105. Jaeckel, J., Wetterich, C.: Phys. Rev. D **68**, 025020 (2003); Jäckel, J.: hep-ph/0309090